AF598431

DEGRADABLE POLYMERS

2nd EDITION

Degradable Polymers

Principles and Applications

2nd Edition

Edited by

GERALD SCOTT

Professor Emeritus in Chemistry and Polymer Science,
Aston University, U.K.

KLUWER ACADEMIC PUBLISHERS
DORDRECHT / BOSTON / LONDON

A C.I.P. Catalogue record for this book is available from the Library of Congress.

ISBN 1-4020-0790-6

Published by Kluwer Academic Publishers,
P.O. Box 17, 3300 AA Dordrecht, The Netherlands.

Sold and distributed in North, Central and South America
by Kluwer Academic Publishers,
101 Philip Drive, Norwell, MA 02061, U.S.A.

In all other countries, sold and distributed
by Kluwer Academic Publishers,
P.O. Box 322, 3300 AH Dordrecht, The Netherlands.

Printed on acid-free paper

Printed in the Netherlands.

CONTENTS

1

WHY DEGRADABLE POLYMERS?

GERALD SCOTT
Aston University
Birmingham B4 7ET, UK

1 Polymers in modern society

Polymers have gained a unique position in modern materials technology for a number of quite different reasons. The development of the inflatable rubber tyre in modern transport would not have been possible without the use of natural, and later synthetic, rubbers as the energy absorbing components. 'Plastics', have largely replaced traditional materials used in packaging because of their better physical properties, notably strength and toughness, lightness and barrier properties. Their ability to protect perishable commodities against spoilage at minimal cost has led to a revolution in the distribution of foodstuffs to the extent that they are now indispensable in modern retailing [1].

Plastics are also energy-efficient compared with traditional materials [1]. It takes twice the weight of paper to effectively protect goods than in the case of polyethylene and if all the plastics currently used in packaging were to be replaced by paper, the effect on the environment would be catastrophic in terms of forest depletion, increased energy utilisation and damage to the environment [2,3]. This ecologically important characteristic of the polyolefins will be discussed in more detail in Section 4.

2 The management of polymer wastes

The volume of plastics, synthetic fibres and rubber that appear as wastes presents disposal authorities with an increasingly serious problem. At one time it was relatively inexpensive to dispose of domestic and industrial wastes in holes in the ground on the peripheries of towns and cities. The reduction in the number of such sites, coupled with the increasing bulk of the wastes means that the cost of transporting packaging wastes to available landfill sites has increased unacceptably. There is also an increasing recognition that society should treat waste as a resource to be re-utilised by 'recycling' to useful products rather than by burying them. Consequently the disposal of packaging waste has become the responsibility of the producer of the packaging. Landfill taxes have

G. Scott (ed.), Degradable Polymers, 2nd Edition, 1-15.

been introduced in the developed countries to recover materials that would have originally gone to landfill from the waste stream for recycling [4].

It is now accepted, for reasons that will become apparent in the following discussion, that the term 'recycling' must be broadened to include not only reprocessing or **mechanical recycling**, but also other methods of conserving the intrinsic value of the materials, including **energy recycling** and **biological recycling**. The alternative ways of constructively utilising waste will be discussed in more detail in Chapter 14. It now seems likely that before the end of the second decade of this century, domestic and industrial wastes will be reused by a combination of these methods and that only wastes with little or no potential value will be disposed of in sanitary landfill [4-6].

2.1 MECHANICAL RECYCLING

During the past decade it has been recognised that the initial enthusiasm shown for mechanical recycling by environmental enthusiasts [7-9] was somewhat misplaced [4]. It was assumed that plastics could be recycled to the original products in the same way as metals and glass. Some industrial products such as automotive components (e.g. battery cases and bumpers) and some packaging (e.g. crates and shrink-wrap film) can be readily recovered from the waste stream [10] and, because they are generally well protected from environmental degradation by antioxidants and stabilisers in their first life, they may be recycled in a 'closed loop' [4]. However, plastics packaging and other disposable products such as plastics plates and cutlery from domestic wastes and retail outlets comprises over 60% of the post-user plastics wastes generated and it is technically much more difficult to reprocess this waste to useful products [10,11]. There are two associated reasons for this. The first is that they are normally highly contaminated by non-polymer components, notably fats, oils and transition metal ions that reduce the quality of the recycled products. The high surface area to mass ratio of packaging makes it difficult and generally quite uneconomic to segregate and cleanse them after collection. 'Kerb-side' collection of individual polymer components of packaging already segregated by the householder offers a potential solution to this practical problem, particularly for the more expensive packaging materials such as the polyesters (PET).

The second reason is not quite so obvious. Approximately one third of the fossil fuel energy that goes into the initial manufacture of plastics products is used in transforming the polymer from pellets to the final product [6,12]. Even more energy is used in recycling, since the recovered polymer has to be first cleansed and shredded or ground. The energy (and cost) of collection, segregation and cleansing, mechanical recycling of plastics packaging waste from domestic sources may be up to twice that used in the fabrication of the original package [13,14]. Since the energy used in manufacture is at present almost entirely derived from fossil fuel resources, mechanical recycling of single-component plastics packaging from domestic sources may actually waste fossil resources rather than save them.

Mechanical recycling of mixed plastics wastes is an even more contentious issue due to the poor mechanical properties and durability of blends of mixed plastics [10,11]. The first can be overcome to some extent by using expensive solid phase dispersants ('compatibilizers') but most recyclers of mixed plastics cannot afford to use this technology and instead rely on thick sections of the recycled product to provide acceptable mechanical performance (e.g. as wood substitutes for park benches, docks, road signs, etc). However, it has been shown in a critical study of the Duales

Deutschland System (DSD) [13] that to compete with conventional materials such as wood and concrete, plastics recycled from domestic waste would have to last 3.3 times as long as the materials they replace. From the known properties of recycled mixed plastics it is most unlikely that this could be achieved [10].

2.2 ENERGY RECYCLING

Polyolefins differ from metals and glass in that when incinerated they produce energy equivalent to the oil from which they were originally manufactured [12]. This then in principle gives them a second life as a source of energy. Unfortunately incineration is looked upon with considerable suspicion in most developed societies since there is evidence that dioxins and other toxic products may be present in the atmosphere downwind from incinerators [4], often situated in or near conurbations. This has been attributed particularly to chlorine-containing polymers that are also difficult to incinerate as a result of the formation of highly corrosive hydrogen chloride in the plant. Polyolefins and other hydrocarbon polymers are rather different since they produce only carbon dioxide and water on complete incineration. There is therefore considerable potential for utilising waste polyolefins in energy generation, particularly in cement or steel furnaces [13].

Alternatives to incineration with energy recovery are to pyrolyse or hydrogenate waste plastics to give liquid fuels or new polymer feed-stocks [4]. The advantage of this approach is that the wastes can be processed under controlled industrial conditions to give portable liquids for use elsewhere. The disadvantages are that substantial amounts of thermal energy have to be used to obtain useful chemicals.

2.3. BIOCYCLING

It will be evident from the above discussion that neither mechanical recycling nor energy recycling provides a complete solution to the problems of plastics wastes in modern society. In particular, plastics litter is normally found in locations that make it prohibitive to collect wastes for mechanical or energy recycling. In this situation, the concept of reabsorbing plastics into the biological cycle is an attractive ecological alternative to locking away polymer wastes in expensive landfill. Technologies for achieving this with the polyolefins were pioneered in universities some time before mechanical or energy recycling were considered to be viable alternatives to landfill disposal.

3 Biodegradable polymers

During the 1960s percipient environmentalists became aware that the increase in volume of synthetic polymers, particularly in the form of one-trip packaging, presented a potential threat to the environment. This was particularly evident in the appearance of persistent plastics packaging litter in the streets, in the countryside and in the seas. Not only was this aesthetically undesirable but it presented a potential threat to animals and birds both on land and in the sea. The first practical response to this threat came from academe rather than industry. Several university groups, as part of their research programmes into polymer stabilisation were aware of the basic scientific reasons for the environmental instability of polymers and were able to develop technologies to induce polymer degradation in existing commercial polymers in a controlled way. Because of the well-understood mechanisms of polymer degradation and stabilisation, the hydrocarbon polymers were selected as the basis for polymers with enhanced but controlled degradation. However, this solution was not well received by industry since considerable research had been invested in developing antioxidants and light stabilisers to produce polymers as environmentally stable as natural products such as metals, glass and wood. It was argued that the way forward was to recover the original materials by recycling and that the development of degradable polymers would threaten the recycling strategy.

3.1 BIODEGRADABLE POLYMERS BASED ON POLYOLEFINS

Table 1 shows some commercial degradable plastics based on polyolefins originally developed in the in the 1970s. In some cases they have been successfully used in agriculture and to a more limited extent in packaging applications for more than 20 years. Regular polyolefins are not biodegradable in an acceptable time, since they are protected by antioxidants and stabilisers incorporated during processing to provide durability during use. However, polyolefins can be made sensitive to heat or light in the presence of oxygen after use to give bioassimilable products [3,16-20]. In the **oxo-biodegradation** process bioassimilation is controlled by abiotic (and occasionally biotic) peroxidation and is mechanistically distinguished from **hydro-biodegradation** in which bio-assimilation is preceded by hydrolysis (e.g. in polysaccharides, polyesters, etc.) [18,21].

Table 1. Early commercial degradable polyolefins

Photolytic polymers

Ethylene-carbon monoxide copolymers: E-CO, Ethylene-vinyl ketone copolymers:
Ecolyte™ (J.E.Guillet) [3,15,16]

Oxo-biodegradable polymers

Antioxidant controlled, transition metal-catalysed photo- and thermooxidisable polymers:
Plastor™ (G.Scott-D.Gilead) [17-20]

PE-Starch blends

PE blended with starch (and in later developments with prooxidants):
Coloroll, St.Lawrence Starch (G.J.L.Griffin) [22]

It is instructive to explore in more detail the reason why polyolefins were initially selected for development as degradable polymers rather than natural products such as cellulose, which was already available commercially in derivatized form as cellulose acetate. The latter was known to be slowly biodegradable but suffered from a number of technical deficiencies, of which the most important was that the extraction of cellulose from natural products was both energy intensive and polluting compared with the polyolefins [4]. Furthermore the modification of cellulose by acetylation to give technologically acceptable products sharply reduced the environmental biodegradability of the base polymer [18]. Consequently it is difficult to achieve an acceptable balance between the required technological performance and ultimate biodegradability. More recent experience has shown that this is a significant problem with other hydro-biodegradable polymers of biological origin and indeed it is in the nature of hydrophilic natural polymers such as cellulose and starch to be rapidly bioassimilated by hydrolytic microorganisms and any attempt to improve technological properties by chemical or physical modification interferes with nature's intention. The hydrophobic polymers such as natural rubber and the synthetic polyolefins stand at the other end of a spectrum of technological and ecological properties (Fig.1) [23]. Synthetic plastics have achieved a central position in the distribution of consumer goods because of their combination of flexibility, toughness and excellent barrier properties, which has made them the materials of choice for packaging applications. The polyolefins have been found to be particularly important in blown film and injection moulding technologies because of their ease of conversion and low cost. The present-day efficient distribution of perishable foodstuffs is a direct consequence of the low cost of polyolefin packaging and its resistance to water and water-borne microorganisms during use [1].

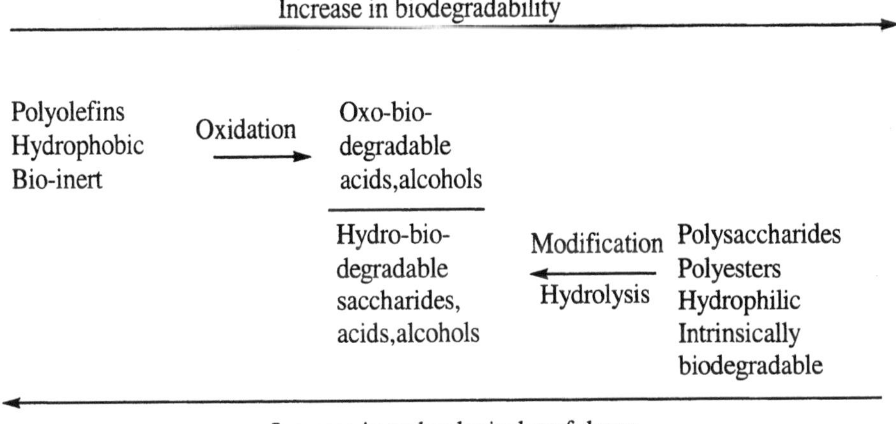

Fig. 1. Alternative approaches to environmentally acceptable polymers [23]

In agriculture, the new technology of plasticulture, based on polyethylene, has led to a revolution in the growing of soft fruits and vegetables. Polyolefins do not hydrolyse under any practical conditions but, as already indicated, they do oxidise rapidly in the environment unless protected by antioxidants and particularly in the presence of the oxygenase enzymes they are bioassimilated. A major advantage of the polyolefins is then that oxo-biodegradation can be degraded in a controlled way involving both prooxidant and antioxidants. This will be discussed in more detail in Chapter 3.

During the 1980s, trade associations such as the British Plastics Federation, the Industry Committee for Packging and the Environment (INCPEN) in the UK and the Council for Solid Waste Solutions in the USA positively campaigned against degradable plastics, primarily on the grounds that that induced degradability would interfere with recycling [24-26]. Consequently the subsequent development of polyolefins with induced degradability was developed by polymer additive companies in association with universities and the outcome of these development will be discussed in later chapters in this book.

3.2 BIODEGRADABLE POLYMERS BASED ON RENEWABLE RESOURCES

The search for biodegradable polymers based on renewable resources, which began in the 1980s, has been compared to the search for the holy grail [27]. It has been embraced with enthusiasm by the 'green' movement [28] and has in turn influenced the scientific community and subsequently the even the sceptical polymer manufacturing industry itself. The following are the stated objectives of this search [27];

1. To replace polyolefins (PO) and polyvinyl chloride (PVC) regardless of which properties are looked for.
2. To match the production costs of PO and PVC.
3. The reduce the full 'metabolic burning time' to two composting cycles (30 days) in a technical composting unit.

The renewability concept, which is frequently confused with sustainability (see Section 4), has now been embraced by the polymer industries. The first fully bio-synthetic/biodegradable polymer, poly-(3-hydroxy-butyrate) (PHB), although originally discovered in 1925 [29], was developed on a semi-technical scale by ICI in the 1980s by microbial fermentation of sugar. The development of this and related polyesters, the poly(alkanoates) (PHAs) was described by Hamond and Liggat in the first edition of this book [30] and has stimulated a search for other bio-based biodegradable polymers. Poly(lactic acid) (PLA), a hydro-biodegradable polymer synthesised from corn sugar by conventional abiotc chemistry, is now made on a small commercial scale. Its properties have been intensively studied and the mechanical behaviour of PLA is reported to be similar to that of polyethylene. However, cost and performance still lack the economic attractions of the polyolefins. The PHAs similarly have so far not had an easy technological development as they have been passed from one company to another. Progress toward the development of sustainable PHAs that satisfy the consumer market still lies in the future and scientific approaches to this objective will be discussed in Chapters 9 and 10.

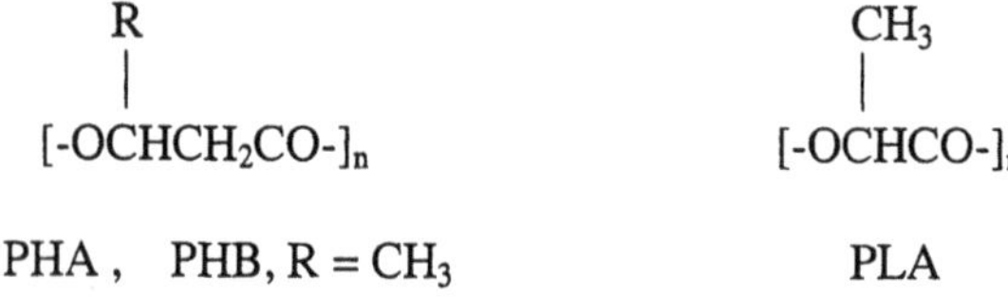

Starch, normally obtained from cereals, is of little value to the plastics industry in its natural form. However, by the process of extrusion cooking, it can be made into a plastic which, when plasticised, approaches the properties of the commodity plastics (see Chapter 6}. The ideological argument for using starch as a base material for plastics is that it hydro-biodegrades rapidly in the environment. Corn starch, is a relatively cheap commodity and producers in the more affluent counties would like to find an outlet for current excess capacity. However, the fact that it is renewable and cheap does not necessarily mean that it is sustainable and if the requirements of the above paradigm are to be met, then the agricultural production of starch for the plastics industry must soon come in conflict with the production of food (Section 4). The production of commodity plastics from a temporary excess of food is not then a long-term sustainable policy for an already overpopulated world. This does not mean that polymers based on carbohydrates will not find niche applications and, if the vast amounts of waste cellulose could be utilised by means of 'green' chemistry, cellulose-derived product could make a substantial impact on the packaging industry.

Nature's cellulose-based litter abounds on land and in the sea. Most of this is in combination with lignin and in principle there is the possibility of utilising these abundant materials as the basis of a genuinely sustainable polymer industry. However, useful products are not achieved at present without the input of fossil fuel energy (see Section 4). Renewability alone is then not a sufficient criterion of ecological acceptability. Packaging polymers are required not only to be compatible with the natural biocycle but they must also fulfil their intended function for the benefit of society [1] and at the same time be economic in the use of fossil resources [4]. The ideal renewable polymer has so proved to be difficult to achieve for these purely practical reasons.

3 Custom design of biodegradable polymers

3.1 END-OF-LIFE DESIGN

It is now recognised for the reasons discussed above that the design of biodegradable polymers depends on the end-of-life environment as well as on service requirements. Typical examples are items intended to end up primarily in sewage [18], which should be substantially converted to CO_2 and associated cell biomass during the time it is in the sewage plant. Some, but not all bio-based polymers (modified starch and aliphatic polyesters) satisfy this requirement and are suitable for the manufacture of short-lived personal hygiene products such as diapers, etc.

Agricultural products such as mulching films and tunnels have quite different service requirements and end-of-life requirements from domestic packaging (Chapter 14). Garden waste sacks and some food packaging may appear in municipal compost where a longer biodegradation time is not only reasonable but is also ecologically desirable (see Section 4). Agricultural mulch, silage films and fertilizer sacks and animal feed bags, either deliberately or inadvertently remain on the land as litter. To fulfil their design purpose, all these products require a safety period in use before they begin to biodegrade. There are also very important applications for biodegradable polymers in the body either in controlled drug release or in sutures and related applications where very specific durability requirements apply.

The long-term effects of man-made materials in the environment are as important, if not more important than their initial impact as litter. The use of degradable materials in consumer products, in agriculture and in prostheses should not lead to the generation of toxic or otherwise environmentally unacceptable chemicals in the human environment. For example the polyolefins that contain only carbon and hydrogen are converted by peroxidation to low molar mass carboxylic acids, hydroxy acids and alcohols that are nutrients for microorganisms. They thus appear ultimately as carbon dioxide and water (Chapter 3). Chlorinated polymers and other polymers containing hetero-atoms other than oxygen and nitrogen by contrast must be viewed with some suspicion. Low molar mass organo-chlorine compounds represent a particular threat to the environment due to their persistence and it is critically important that new polymers that are intended to biodegrade rapidly in the environment should be carefully assessed by standardisation organisations for eco-toxicity effects before being allowed into general application in waste and litter control. This will be discussed in Chapter 14.

3.2 DESIGN FOR SERVICE-LIFE

The applications of biodegradable polymers fall into two distinct categories. In the first, biodegradability is an essential part of the function of the product. Examples are temporary sutures in the body during surgery or the therapeutic controlled release of drugs (Chapter 10). In both, cost is relatively unimportant provided that the artefacts fulfil their intended purpose. Similarly in agriculture where very thin films of photo-biodegradable polyolefins (mulching films) are used to increase soil temperature, ensuring earlier harvest. A major ecological benefit of plastics mulch is to reduce the use of irrigation water and fertilisers, an increasingly important objective in parts of the world where water is scarce and becoming scarcer [19]. An important requirement is that no significant quantity of plastics residues must persist in the soil in subsequent seasons since these interfere with root growth and reduce productivity. The technological design and use of biodegradable polymers in agriculture is motivated by economics since the need to remove films from the land is eliminated [18-20,31,32] (see Chapter 14).

The second use of biodegradable polymers is in applications such as packaging where their use brings social benefit but does not bring overt economic gains to the manufacturer or user [33]. Biodegradable plastics do not add to the technological performance of a packaging material and manufacturers are reluctant to invest in new materials that do not bring cost-benefit. Consequently few developments in degradable packaging have established a position the marketplace during the past 15 years. Ethylene-carbon monoxide co-polymer (E-CO) is used in six-pack collars and has made a significant contribution in the protection of birds and animals from entrapment by carelessly discarded packaging [16,34] (see Chapter 13).

4 Life-cycle assessment

During the 1970s there was a popular but rather naive belief in industry that the incorporation of biodegradable materials such as starch into synthetic polymers such as polyethylene would transform these relatively intractable materials into environmentally biodegradable products. This was subsequently shown not to be the case but it resulted in serious over claiming by industrial companies wishing to take advantage of public enthusiasm for "environmentally friendly" packaging materials. This was categorised as "deceptive" by legislators. A timely and valuable investigation was carried out into "green marketing" by a Working Group of the Attorneys General of the USA. Their conclusions were summarised in "The 'Green Report' in 1990 [35], which subsequently provided a basis for standards for degradable polymers. The salient conclusions as to the purpose of such standards may be summarised as follows.

1. To protect the environment
2. To provide a 'level playing field" for business
3. To clarify competing claims for the benefit of the consumer

The basis proposed for the 'level playing field' is that "environmental claims must be uniform and supported by competent and reliable scientific evidence". The use of the term "environmentally friendly" was particularly criticised as being a vague term in normal use that was not based on objective criteria. It was concluded that this term should not be used unless product was first subjected to life-cycle assessment. The definition of LCA in the 'Green Report' given below is now the basis of LCA techniques as currently practiced.

"Product life-cycle assessment involves consideration of environmental effects at every stage in the products life-cycle, including the natural resources and energy consumed and the waste created in the manufacture, distribution and disposal of a product and its packaging......Such assessments will only provide useful comparative information about how to reduce environmental problems associated with products if they are conducted using uniform and consistent assumptions"

The term "environmentally friendly" has more recently been superseded by its modern equivalents "renewable" and "sustainable". As already discussed, it is sometimes assumed that polymers from renewable resources are by definition 'sustainable'. One definition of "sustainable" suggests that the development of new products for the benefit of society should not have an unacceptable effect on resource depletion and environmental pollution. However, 'acceptable' is a relative term and invites comparison of one material with another by life-cycle assessment (LCA) [4]. Companies engaged in the development degradable polymers from renewable resources have initiated life-cycle assessment comparisons of their products with the commodity synthetic polymers; notably polyethylene. As will be seen in the following Sections, these have not so far shown unambiguously that bio-based polymers are more environmentally sustainable than the present range of commodity polymers [36]. This results from the same reason that led to the "Green Report"; namely lack of consistency and uniformity of the assumptions made. In some cases they actually contradict one another.

4.1 ENERGY BALANCE DURING MANUFACTURE AND DISPOSAL

Energy input data for the manufacture of polyethylene quoted in the literature vary between ca. 65 GJ/t and 80 GJ/t [37]. . However, Dinkel et al. on behalf of Carbotech, in an LCA comparison of starch polymers with that of low density polyethylene [38], assumed a value of 92 GJ/t producing an immediate bias toward starch polymers.

Published LCAs of degradable polymers [14,38-42] assume that PE is disposed of only in landfill or by incineration. There is no recognition that PE can be "recovered" from the waste stream by composting (see Section 4.2), by pyrolysis to give monomers and fuels or by incineration with energy recovery. In practice, since the calorific value of PE (43 GJ/t) is almost identical to that of the oil from which it was manufactured, the carbon content of the plastic is ecologically neutral. The total non-recoverable energy used in the manufacture of PE is thus 21.6 GJ/t [12], not 65 GJ/t. This compares with the energy used in the manufacture of starch (Mater-Bi) products, which vary between 25.4 GJ/t and 52.5 GJ/t, depending on the co-agent in the formulations. Although the carbon energy input is assumed to be zero, since it is biosynthesised and returned to the carbon cycle by biodegradation, this does not apply to blends with fossil-based additives or polymers (co-agents) [36]. If biopolymers are used as a source of fuel in waste-to-energy incineration, the energy produced is considerably less from polysaccharides than that from PE. These data cast some doubt on the ecological benefits of bio-based polymers and emphasises the importance of using the same assumptions to produce the "level playing field".

4.2 LAND RESOURCE UTILISATION

None of the LCA studies have so far considered land utilisation in the ecological balance. At present bio-based polymers such as PHA, PLA and starch are produced from food crops. This does not present a problem in the short-term if the polymers are to be used in specialised 'niche' applications on the basis of a temporary surplus of food crops but it cannot be used as the basis of long-term sustainable development of bio-based plastics to replace polyolefins in packaging (Section 3.2). For example the anticipated scale of production of PLA during the present decade (not more than 500,000 tonnes/annum worldwide) is less than 1% of the worldwide production of polyolefins and, if the production of food-based biodegradable polymers was to increase toward the level of the fossil-based polymers, there would be serious competition between polymer and food production [18]. However, if, as discussed for cellulosic materials, bio-based feedstocks could be based on biological wastes or on crops grown on marginal land, then the situation would be changed. This represents a major challenge to the bioengineering industry and will be discussed further in later Chapters.

4.3 BIODEGRADATION TIME-SCALE

The final criterion of sustainability is the ultimate return of the materials to the carbon cycle. However, the assumption that fossil-based polymers do not biodegrade in the environment is not valid since polyolefins can be returned to the soil where they contribute to the fertility of the earth. It does a disservice to nature's versatility to draw a sharp distinction between natural and synthetic organic polymers. There is, for example, no intrinsic difference between the biodegradation of natural and synthetic rubbers. Both oxo-biodegrade in exactly the same way when first manufactured and both become highly resistant to biodegradation when formulated with antioxidants, for example in motor car tyres. The non-biodegradability of fully formulated engineering rubbers, whether natural or synthetic, as in the case of the commodity synthetic plastics, has much more to do with the presence of antioxidants, than with the inherent biodegradability of the polymers themselves. The process of oxo-biodegradation will be discussed in more detail in Chapter 3, but it is important to note here that some of nature's products, notably sequoia wood, are very resistant to degradation in a biotic environment [18]. This again is due to protection by powerful antioxidants and antibacterials (e.g. tannic acids).

5 Biodegradation test methods and standards

Many environmentalists are suspicious of all man-made materials that do not biodegrade rapidly in the environment after they have served their primary purpose. The reason for this is understandable. During the latter half of the twentieth century, it became evident that many industrial and agricultural chemicals have undesirable effects on the environment that could not have been anticipated when they were first introduced. However, this is often also applied to materials that are essentially inert because of their physical form. Polymers fall into this category. Polyvinyl chloride (PVC) has been denigrated because it contains chlorine. In fact PVC is physiologically inert, although vinyl chloride from which it is manufactured is highly toxic. No one knows at present what products are formed from PVC if and when it is induced to biodegrade. It must be assumed, until it is demonstrated otherwise, that the chlorine is only 'safe' in the environment in the form of the commercial undegraded high polymer.

Other carbon-chain polymers, particularly the polyolefins, when used in packaging, again present no toxicity hazard as high molar mass materials. However hydrocarbon polymers, unlike the halogenated polymers, when biodegraded or incinerated, produce only carbon dioxide and water. Even when they are pyrolysed they produce small molecule consisting mainly of a mixture of paraffins (methane, ethane etc) and olefins (ethene, propene, etc.). This process is safe when carried out under controlled conditions in a petrochemical plant [4]. In the environment, polyolefins are broken down by oxygen to a mixture of alcohols, carboxylic acids, hydroxy-carboxylic acids and the lower molar mass homologues are bioassimilated very rapidly *in situ* (see Chapter 3).

In spite of the scientific evidence, the popular perception that "Natural" is "good" and "Synthetic" is "bad" has led to a climate of fear of man-made materials. This has influenced the standards organisations and has resulted in the unscientific requirement that all synthetic polymers must be converted rapidly to carbon dioxide to be considered 'biodegradable'. At the same time, "natural" materials are exempted from this requirement because they are considered to be "by definition biodegradable" [43]. Biometric tests, such as the Sturm test for the measurement of the carbon dioxide produced during biodegradation of water-soluble chemicals are very convenient and easy to use in simple laboratory equipment. They were originally developed as test methods for hydrophylic polymers in sewage systems at ambient temperatures. This type of test is very convenient for measuring the biodegradability of hydro-biodegradable polymers, but it quite inappropriate for hydrophobic polymers degrading by oxo-biodegradation.

Lignocellulose, an abundant oxo-biodegradable aromatic polymer containing C-C and C-O bonds (see Chapter 3), does not biodegrade rapidly in an aqueous environment so it is unreasonable to require that man-made carbon-chain oxo-biodegradable polymers should do so. Furthermore, the requirement that polymers undergo rapid mineralisation in compost [44] is inconsistent with the concept of 'reclamation'. The European 'Waste Framework Directive' 1991 defines "recovery" as follows [45];

"Recycling/reclamation of organic substances.....use as fuel to generate energy and spreading on land resulting in benefit to agriculture or ecological improvement, including composting and other biological processes"

As discussed in Section 4, carbon dioxide generation is considered by LCA to be a negative parameter. CO_2 is an ecological deficit and does not improve the environment. Indeed it is much more beneficial to agriculture and the environment to retain the carbon in the soil as an available nutrient as happens naturally with lignocellulose. Composting standards [44] are much more concerned with the easy disposal of biodegradable plastics than with recovering their value as biomass. The usefulness of plastics as soil improvers would be lost before the compost ever reached the soil. The mechanism of biodegradation of lignocellulose and other naturally occurring carbon-chain polymers will be discussed in Chapter 3 but it should be noted here that similar microoganisms are involved in the biodegradation of the polyolefins, providing a model from which to develop composting test methods to alleviate popular concerns about the biodegradation of man-made polymers. The bioassimilation of carbon-chain polymers will be discussed in more detail in Chapter 14 when the applications of degradable polyolefins will be discussed.

6 Conclusions

During the first half of the twentieth century, the emphasis in the developing plastics industries was to make polymers as resistant as possible to environmental degradation. This tended to obscure the fact that most polymers as manufactured are relatively unstable materials when exposed to the outdoor environment. Consequently, when it became clear that synthetic packaging materials posed a threat to the environment, popular folklore categorised the commodity polymers as "indestructible" and "non-biodegradable".

A scientific understanding of polymer degradation, based on fundamental research on the abiotic environmental degradation of naturally occurring *cis-polyisoprene* rubber provided a rational explanation of why and how natural rubber and later the synthetic hydrocarbon polymers biodegraded and this in turn provided an explanation of how antioxidants inhibited this process in technological applications. It was recognised in the 1960s and 1970s that there was no fundamental difference between natural rubber and its synthetic analogue or in principle between natural rubber and the polyolefin plastics. In the absence of antioxidants, they all peroxidise and biodegrade at predictable rates. An extension of these studies was to use the scientific knowledge gained to develop degradable polyolefins with controlled rates of biodegradation.

However, the popular belief that "natural is good" and "synthetic is bad" has led to the subsequent obfuscation and neglect of this fundamental knowledge and the use of the term "sustainable" has come in recent years to mean polymers made from renewable resources, preferably by biological processes. The use of renewable resources is believed to reduce the exploitation of fossil resources and reduce the "greenhouse" effect. A deeper reflection on these objectives has raised considerable doubt about the premises on which they are based. All chemical processes require an energy input and today the source of this energy is fossil fuels, so that the lower the energy input and the less the associated pollution, the 'greener' is the product [19]. Life cycle assessment provides in principle a means of comparing the sustainability of biodegradable polymers. So far, however, LCA has not been applied in a "uniform and consistent" manner in comparing bio-based polymers with degradable synthetic polymers and there still remains considerable doubt as to whether bioplastics can be considered to be more sustainable than the biodegradable forms of the commodity polymers.

A second consequence of the false "natural-good, synthetic-bad" paradigm is that biodegradable plastics are considered to be good only if they are converted rapidly to carbon dioxide and water. This unscientific criterion automatically excludes carbon-chain polymers such as the polyolefins and many wood-based products from consideration as biodegradable materials. Wood chips, often containing preservatives, have been used for many decades as 'mulches' to reduce weed growth and to condition the soil. Similarly, polyolefins have been used for twenty years in the agricultural environment without detriment to the fertility of the soil. It will be seen in later Chapters that renewable and non-renewable polymers will both play a part in the range of polymers available to industry in the twenty first century, depending on their applications, environmental requirements and cost. However, in order that both types of polymer are able to fulfil their potential, it is of the utmost importance that science-based standards are developed that allow all biodegradable polymers to achieve their full and often complementary potentials. If the use of fossil resources for the production of energy were to cease tomorrow, fossil carbon resources would last at least 300 years and fossil-based CO_2 effluent would become a much less important consideration. Such a development would allow time for the development of sustainable bio-based carbon feedstocks for synthetic polymers.

7. Acknowledgements

I am grateful to Dr.Martin Patel of Utrecht University for very helpful discussions on Life-cycle Assessment and for giving me access to previously unpublished results of his own research.

References

1. Scott, G. (1999) *Polymers and the Environment,* Royal Society of Chemistry, Chapter 2.
2. Mosthaf, H. (1990) *Plast. Verarbeiter* **41**, 50.
3. Guillet,J. (1995) in G.Scott and D.Gilead, (eds), *Degradable Polymers: Principles and Applications*, 1st edition, Chapter 12.
4. Scott,G. (1999) *Polymers and the Environment*, Royal Society of Chemistry, (1999), Chapter 4.
5. Scott,G. (1999) Environmentally Degradable Polymers in Waste Management, in Z.F.Said and E.Chiellini (eds.) *Selected Papers from ICS-UNIDO International Workshop on EnvironmentallyDegradable Polymers: Polymeric Materials and the Environment*, Doha, Pub. University of Qatar.
6. Scott,G. (May 1999) *Wastes Management*, 38-39.
7. Plastics Recycling Foundation (1989) *Plastics recycling, A strategic vision.*
8. Council for solid waste solutions (1989) *From soda bottle to swimming pool.*

9. Council for solid waste solutions (1989) *The solid waste management problem.*
10. Sadrmohghegh,C., Scott,G and Setudeh,E. (1985) *Polym. Plast. Tech.Eng.,* **24**, 149.
11. Scott,G. (1990) Recycling of Plastics: A challenge to the polymer industries in A.V.Patsis (ed.), *Stabilisation and Controlled Degradation of Polymers*, Luzerne, p.215
12. Bousted ,I. and Hancock,G.F. (1981) *Energy and Packaging*, Ellis Horwood Publishers
13. Brandrup,J. (1998) *M ✦ll und Afall*, 8, 492.
14. Heyde, M. (1998) *Polym. Deg. Stab.,* **59**, 3-6.
15. Scott, G.(1993) in *Atmospheric Oxidation and Antioxidants,* 2nd Edition, Vol. II, ed. G.Scott, Elsevier, p.392..
16. Harlan,G. and Kmiec,C. (1995) in G.Scott and D.Gilead (eds.), *Degradable Polymers; Principles and Applications,* 1st edition, Chapman & Hall, Chapter 8.
17. Scott, G. (1995) in G.Scott and D.Gilead (eds.), *Degradable Polymers; Principles and Applications,* 1st edition, Chapman & Hall, Chapter 9.
18. Scott,G. (1999) *Polymers and the Environment*, Royal Society of Chemistry, Chapter 5.
19. Scott,G. (2000) *Polym. Deg. Stab.,* **68**, 1-7
20. Scott, G. (1997) *Trends in Polymer Science*, **5**, 361-368.
21. Li,S and Vert,M. (1995) in G.Scott and D.Gilead,(eds.), *Degradable Polymers: Principles and Applications*, 1st edition, Chapter 4.
22 Griffin, G.J.L. (1994) in *Chemistry and Technology of Biodegradable Polymers*, Ed. G.J.L. Griffin, Chapman and Hall, Chapter 3.
23. Scott,G. and Wiles, D.M. (2001) *Biomacromolecules*, **2**, 615-622.
24. SPI (1988) *Plastics and Degradability*
25. INCPEN, *Degradability and Plastics Packaging*, Discussion paper No. 4.
26. Klemchuk,P.P. (1990) *Polym. Deg. Stab.,* **27,** 183.
27. Tomka, I (1999) *Selected Papers from the International Workshop on Environmentally Degradable Plastics,* Smolenice, Slovak Republic, October 4-8, ICS-UNIDO, Trieste
28. Sadun, A.G., Webster, T.F. and Commoner, B. (1990).*Breaking down the degradable plastics scam,* Greenpeace, Washington.
29. Lemoigne,M. (1925) *Ann. Inst. Plast.*, **39**, 144.
30. Hammond,T. and Liggat,J.J. (1995) in G.Scott and D.Gilead (eds), *Degradable Polymers: Principles and Applications*, 1st edition, Chapter 5.
31. Fabbri, A. (1995) in G.Scott and D.Gilead, (eds), *Degradable Polymers: Principles and Applications,* 1st edition, Chapman & Hall, Chapter 11.
32. Gilead, D (1995) in G.Scott and D.Gilead, (eds), *Degradable Polymers: Principles and Applications,* 1st edition, Chapman & Hall, Chapter 10.
33. Scott, G. and Gilead, D. (1995) in G.Scott and D.Gilead, (eds), *Degradable Polymers: Principles and Applications,* 1st edition, Chapman & Hall, Chapter 13.
34. *Proceedings of the Second International Conference on Marine* Debris, Shomura, R.S. and Godfrey, M.L., Eds. US Department of Commerce (1990).
35. *The Green Report', Report of a task force set up by the Attorneys General of the* USA to investigate 'Green marketing' (1990).
36. Patel, M. (2002) *Proceedings of the Bioplastics Conference*, York, Europoint.
37. Patel, M. (2002} *Eco-profiles of plastics and related intermediates*, Association of Plastics Manufacturers in Europe, Brussels.

38. Dinkel,F., Pohl, C., Ros, M. and Waldeck, B., (Carbotech) (1996) *Ökobilanz stärkehaltiger Kunststoffe,* BUWAL, Bern, Switzerland.
39. Estermann, R. and Schwarszwälder (1998) for Novamont, Composto. Oltern Switzerland.
40. Conn R.E. (2000) for Cargill-Dow, Presentation at Süddeutches Kuntstoffzentrum, Würzurg, Germany.
41. Patel, M. (1999) *Closing Carbon Cycles*, PhD Thesis of the University of Utrecht.
42. BIFA (2001) *Interim Report on Starch loose-fill* for Flo-Pak. GMBH, Germany
43. **CEN TC 249 WG9, N18** (1999) *Characterisation of biodegradability*, CEN.
44. **EN 13432** (2000) *Packaging – Requirements for packaging recoverable through composting and biodegradation – Test scheme and evaluation criteria for the final acceptance of packaging*, Comit■ Europ■en de Normalisation (CEN).
45. *Waste Framework Directive* (1991) European Union.

2

AN OVERVIEW OF BIODEGRADABLE POLYMERS AND BIODEGRADATION OF POLYMERS

SAMUEL J. HUANG
Institute of Materials Science
University of Connecticut
Storrs CT 06269-3136
USA

1 INTRODUCTION

It is now widely recognized that, along with the importance of synthetic polymers possessing long-term stability, there is also a need for polymers that break down in a controlled manner. Biodegradable macromolecules can be tailored specifically for controlled degradation under the inherent environmental stress in biological systems either unaided or by enzyme-assisted mechanisms. Medical applications of these materials have led to significant developments, such as the controlled release of drugs, fertilizers and pesticides, absorbable surgical implants, skin grafts and bone plates. Many studies of the mechanisms of biodegradation of synthetic polymers were motivated by this and will be initially discussed. Recent interest in polymer waste management of packaging materials has and incentive to the research and will be discussed in a later section.

2 BIOMEDICAL POLYMERS

Bioerodable materials in current use are limited to applications that do not require long-term strength retention. It is acknowledged by the medical profession that problems exist with the current practices of bone fracture fixation. Two serious problems are osteoporosis due to stress shielding [1-3] and necessary second operations for device removal after bone healing. To alleviate these problems, polymers of α-hydroxy acids such as lactic and glycolic acid are being explored. They have shown potential utility as biocompatible, fully resorbable implant devices. The biocompatibility of poly(α-hydroxy acids) has been known for some time from *in vivo* acute and subacute tissue reaction [4], as well as *in vitro* cytotoxicity response [5]. Sutures of these materials have been in use now for many years.

G. *Scott (ed.), Degradable Polymers, 2nd Edition,* 17-26.

Many variables have been revealed which affect degradability. Key factors elucidated by Huang and coworkers include segment mobility, surface area, morphology, molecular weight, hydrophilic/hydrophobic interactions and the availability of hydrolysable links [6-7]. A complete understanding of behaviour is critical because biomaterial requirements are very severe [8-11]. Early work by this group examined ways to mimic biopolymer degradation by incorporation of amino acids in synthetic polymer backbones. It was learned, for example, that functional group recognition by specific enzymes enhances degradability [6, 12-15].

Detailed studies have been performed on poly(glycolic acid) (PGA) hydrolysis [16-17]. This work has shown interrelations of factors affecting degradation. Chu and Louis have quantified simultaneous changes in pH, tensile strength, glycolic acid concentration and degree of crystallinity. This complete information gives degradation mechanistic details [18]. Huggman and Casey have explored the effect of carboxylic acid end group concentration on PGA degradation rate. By capping the acid end group they were able to retard hydrolysis, but the results were not as good as expected [19].

2.1 FACTORS AFFECTING DEGRADABILITY *IN VIVO*

For a polymer to be degradable, main chain hydrolysable groups must be both present and accessible. Poly(lactic acid) differs from poly(glycolic acid) by one CH, group per repeat unit. This makes it more hydrophobic and increases the steric hindrance, so it is more hydrolytically stable. With semicrystalline polymers, the tightly packed crystalline regions are less accessible to degradants which must diffuse in to be effective. Therefore the amorphous regions degrade first, leaving the more crystalline regions. Since the amorphous chains tie the crystallites together, when they are hydrolysed, catastrophic mechanical failure occurs even before weight loss [20-21]. During hydrolytic testing of a semicrystalline copolyester suture there was only 1.7% weight loss for 66.1% strength loss [21].

In the synthesis of hydrolytically stable, crystalline poly(L-lactic acid), (PLLA), the choice of catalyst and reaction conditions are important to minimize the extent of racimization [22-23], which could lead to a decrease in properties. Also maximizing the molecular weight while minimizing the unreacted monomer has been shown to be important for the synthesis of poly(L-lactic acid) with longer *in vivo* strength retention [24].

Research on semicrystalline poly(ε-caprolactone) has confirmed that morphology affects degradability [25-30]. Spherulite morphology was clearly visible after hydrolytic attack, since the amorphous regions were preferentially degraded. The morphology of poly(α-hydroxy acids) similarly determines their properties. By varying the stereoisomer ratio of D- or L-lactate units, poly(lactic acid) (PLA) can be made semicrystalline or fully amorphous [31-34]. Variation in the comonomer ratio of a lactide and glycolide in copolymers is another way to control morphology [31, 35-36]. Hydrolysis studies of copolymers of lactide and glycolide show that the amorphous regions are also preferentially etched [21, 37-38], as was the case with poly(ε-caprolactone). As one would expect, more crystalline poly(α-hydroxy acids) have greater resistance to degradation [34-35]. Materials with more diverse degradation characteristics have also been obtained from copolymers of ε-caprolactone with lactide [39-41], and with glycolide [42-43].

2.2 EFFECT OF CROSS-LINKING

From arguments such as the proceeding ones, it might be concluded that cross-linked polymers should degrade more rapidly than linear counterparts, since they are less crystalline. All factors being equal, (regarding the basic chemical nature of the system), this is not so. Some explanation for this may be as follows.

With linear polymers, one hydrolysis event breaks a chain into two chains. The accessible amorphous tie chains between crystallites thus support a disproportionate amount of the load. Therefore after a low percentage of weight loss and a low percentage of chain cleavage, the material fails. In a cross-linked network, one hydrolysis event does not cause two new chains to form. The cross-link density, strength and weight will decrease, but the strength loss profile will more closely match the weight loss profile, since all chains bear the same proportions of the load, in contrast to semicrystalline polymers. Thus with random scission of a network structure, there is a more gradual strength loss without catastrophic failure. A tightly cross-linked network will decrease the diffusion of potential degradants. If these are large entities, such as enzymes, the degradation will be limited to the surface. Decreased diffusion will also increase the hydrolysis resistance of cross-linked systems. Cross-linking also immobilizes segments which disallows conformations necessary for enzyme binding [44].

Early work done with gelatin showed that cross-linking with hexamethylene diisocyanate (HDI) decreased biodegradation [13, 45-46]. Also `tanning' of collagen tubes with glutaraldehyde created covalent cross-links adding structural and functional stability and increasing *in vivo* residence times [47].

In some cases cross-linking can increase degradability. Work by Dickenson *et al.* on *in vivo* and *in vitro* degradation of hydrogels from poly(α-amino acids) serves to reinforce the dependence of degradability on many factors. Dodecane diamine, used as the cross-linking agent, was hypothesized to increase hydrophobicity and to favour diffusion of lipophilic enzymes which aid degradation [48-49].

Poly(ε-caprolactone) has been cured with benzoyl peroxide to form a network structure. Degradability was inversely proportional to cross-link density. After degradation by *Cryptococcus laurentii* the surface was highly porous, the structure sponge-like. Uncross-linked PCL showed crystalline spherulites remaining after the same bioregion process [25-29]. One cross-linked test sample looked spongy but maintained the original dimensions even after 70% weight loss. Porous morphology resulting after degradation is a desirable feature [50], since it has been shown to stimulate new tissue growth [51-55]. Materials made from linear polymers have purposely contained removable inclusions which leave a porous structure [52, 56-57]. To obtain microporous morphologies, incompatible systems have been blended [52-54, 57].

High modulus is necessary for bone plate applications. This can be obtained with the advantage of composite technology. Bone itself is a composite material [58]. New work is evaluating composites containing the same mineral as bone, hydroxyapatite [59-60]. The usage of both biostable [1-3] and completely bioabsorbable [31-32] composite implants is being explored, as well as biostable fibres in a degradable matrix, for bone plates [61-65] and artificial ligaments [63, 66-70].

Most biodegradable composites being tested employ a thermoplastic matrix. There are inherent difficulties in interfacial wetting between linear thermoplastics and reinforcing fibres which limit optimum stress transfer between the fibre and matrix. This also deleteriously affects hydrolytic stability [63]. While the use of solid polymer without fibre has been considered [24, 71-73], problems such as premature cracking can occur. Short fibre composites have been made and tested [63], but properties are not optimal.

One good way to successfully fabricate continuous fibre composites uses a resin that can be cured in stages. After the first cure stage the resin should be able to evenly coat fibres with good wetting. In the second stage of fabrication the impregnated continuous fibre weaves, mats or layers are pressurized and heated in mould of the final shape, and the resin hardens to its final reacted form. The unsaturated polyesters of lactic and glycolic acid which are being studied in this work have the potential to fulfill criteria for a good matrix resin suitable for a biodegradable composite.

Efforts have been directed to make cross-linkable copolymers of poly(α-hydroxy acids) by incorporating olefin groups. Variation in the relative amount of D- and L-lactic acid in the polymer is one way to control morphology. Adjusting monomer ratio in copolymers with glycolic acid is a second way. By varying the amount of cross-linkable olefin groups, a third type of degradation control is obtained.

Advantage was taken of the fact that the ring opening reaction of lactones can be initiated by alcohols [74]. Large polyols have been used in this way to synthesize triblock copolymers by initiating the reaction of glycolide with hydroxyl-terminated poly-(glycolide-co-trimethylene carbonate) [75], and poly(ethylene oxide) [76]. The hydroxyl-terminated initiators used in this work were 2-butene-1,4-diol and 2-butyne-1,4-diol. Polymers from 2-butyn-1,4-diol were discoloured, so further work on these is not included in this report. Lactone polymers initiated by alcohols terminate in hydroxyl groups. These were chain extended with unsaturated carboxylic acids, anhydrides, and acid chlorides. Diisocyanates were also used in cross-linking reactions with star oligomers initiated with polyols such as pentaerythritol and D-mannitol.

The usual synthetic route to useful high molecular weight poly(α-hydroxy acids) involves three steps [77]. In the first step lactic acid or glycolic acid is condensed to polymer, but due to the problems of water removal, the molecular weight is limited to the thousands at best. Next low molecular weight material is depolymerized to cyclic dimer under vacuum at high temperature [78]. To obtain high molecular weight poly(lactic acid), the dimer is purified to get rid of lactyllolactates and water. The lactide dimer is polymerized via a ring-opening mechanism. Molecular weights of 5 x 10^6 have been obtained [79-81].

Each lactide dimer can have three stereoisomeric possibilities; DD, LL, and DL. Mixtures of dimers can form DD/LL cocrystallized pairs (racemic lactide, m.p. 128°C) and DL-lactide crystals (meso-lactide m. p. 43°C). Poly(m-lactide) contains more syndiotactic dyads with poly(r-lactide) contains more isotactic dyads. The most atactic poly(lactic acid) is derived from D,L-lactic acid (DLLA) [77].

In the synthesis of low molecular weight polyesters for network formation it is generally convenient and less expensive to start from D,L-lactic acid, and glycolic acid, rather than from their respective cyclic dimers [78].

3 BIODEGRADABLE POLYMERS IN POLYMER WASTE MANAGEMENT

Polymer wastes are now approximately 10-11% of total wastes [79]. Polymer waste management requires sound complementary practices of conservation, recycling, incineration and biodegradation-bioconversion. Since biodegradation is potentially the most environmentally friendly of all these practices there is increasing activity in the area of biodegradable polymers as packaging materials [80-86].

Among the most studied polymers are biopolymers, modified biopolymers and blends containing biopolymers. Many microbial systems, especially bacteria, use poly-*R*-3-hydroxyalkanoates as energy storage materials, of which poly-*R*-3-hydroxybutylate (PHB) is the most common [87-93]. Since processing of the crystalline PHB has been proven to be rather difficult, various copolymers of PHB with hydroxyalkanates of different size and functionality have been investigated. Effects of stereochemistry, morphology and composition on the hydrolysis and biodegradation have been studied [94-98]. Data to date show that poly-3-hydroxyalkanoates are hydrolysed slowly and biodegradation proceeds rapidly in the presence of suitable micro-organisms. The effects of morphology on the degradation rates follow the trends observed for PCL and PLA degradations described in the previous section.

Modified biopolymers are logical materials for controllable lifetime materials in the biological environments. In general hydrophobic modifications retard biodegradation [99-103]. Detailed mechanisms of degradation are the subjects of active research. For example, it is not clear about the timing of deacetylation and chain cleavage of biodegradation of cellulose acetates.

Starch derivatives have received less attention as compared with cellulose acetates. Among the promising packaging materials are long chain esters of starch [104]. Again, the details of degradation processes remain to be clarified.

Blends of starch with biodegradable polymers are now commercially available as packaging materials [105-107]. The materials are degraded in composts. Since complex structures and morphologies exist in these materials, it will take some time before the biodegradation processes are fully understood.

4 CONCLUSION

Initial studies of biodegradation mechanisms were motivated by biomedical applications of biodegradable polymers. As the applications of these materials increase with time [108-127], understanding of the mechanisms will grow rapidly. Among these Huang and coworkers have found that curtinase is the depolymerase in soil microorganisms that degrades PCL (128). Polymers with carbon main chains require oxidation followed by hydrolysis in bio- and environmental degradations. This is discussed in Chapters 3 and 13 of this book. In recent years polymer waste management through biodegradation and bioconversion has become more important as one of the many means to reduce the problems of polymer waste management.

The biodegradation of synthetic and biopolymers is a complex process. Biodegradation proceeds via hydrolysis and oxidation. The stereoconfiguration, balance of hydrophobicity and hydrophilicity, and conformational flexibility contribute to the biodegradability of synthetic polymers presence of hydrolysable and/or oxidizable linkages in the polymer main chain, the presence of suitable substituents, correct. On the other hand, the morphology of polymer samples greatly affects their rates of biodegradation. The biodegradation of hydrolysable polymers proceeds in a diffuse manner, with the amorphous regions degrading prior to the degradation of the crystalline and cross-linked regions. Cross-linked polymers containing one or more hydrolysable functional groups (amides, enamine, enol-ketone, ester, urea and urethane) have been synthesized and found to be biodegradable at various rates. They show great promise for application in biomedical and agricultural areas, and in packaging technology.

References

1. Szivek, J. A., Weatherly, G. C., Pilliar, R. M. and Cameron, H. U. (1981) *J. Biomed. Res.,* **15**, 853.
2. Brown, S. A. and Mayor, M. B. (1978) *J. Biomed. Res.,* **12**, 67.
3. Woo, S. L.-Y., Akeson, W. H., Levenetz, B. *et al.* (1974) *J. Biomed. Mater. Res.,* **8**, 321.
4. Gourlay, S. J., Rice, R. M., Hegyeli, A. F. *et al.* (1978) *J. Biomed Mater. Res.,* **12**, 219.
5. Rice, R. M., Hegyeli, A. F., Gourlay, S. J. *et al.* (1978) *J. Biomed. Mat. Res.,* **12**, 43.
6. Huang, S. J., Bitritto, M., Leong, K. W. *et al.* (1978) in Adv. Chem. Ser. No. 169, *Stabilization and Degradation of Polymers* (eds D. L. Allara and W. L. Hawkins,) American Chemical Society, p. 205.
7. Huang, S. J. and Roby, M. S. (1986) *J. Bioact. Compat. Polym.,* **1**, 61.
8. Gilding, D. K. (1982) in *Biocompatibility of Clinical Implant Materials*, Vol II (ed. D. F. Williams), CRC Press, Boca Raton, FL.
9. Heller, J. (1983) in *Initiation of Polymerization* (ed. F. E. Bailey), ACS Symp. Ser. No. 212, p. 373.
10. Huang, S. J. (1985) in *Encyclopedia of Polymer Science and Engineering,* Vol. 2, 2nd edn, (eds Mark, Bikales, Overberger and Menges), pp. 220-244.
11. Williams, D. F. (1982) *J. Mater. Sci.*, **17**, 1233.
12. Huang, S. J., Bansleben, D. A. and Knox, J. R. (1979) *J. Appl. Polym. Sci.,* **23**, 429.
13. Huang, S. J. Bell, J. P., Knox, J. R. *et al.* (1976) in *Proc. 3rd Int. Biodeg. Symp.,* (eds J. M. Sharpley and A. M. Kaplan), Applied Science Publishers, London, p. 731.
14. Bitritto, M. M., Bell, J. R, Brenkle, G. M. *et al.* (1979) *J. Appl. Polym. Sci., Appl. Polym. Symp.,* **35**, 405.
15. Bell, J. P., Huang, S. J. and Knox, J. R. (1974) TR 75-48-CEMEL, US Army, Natick Laboratory, Natick, MA.
16. Williams, D. J. (1981) *J. Biomed. Mater. Res.,* **14**, 329.
17. Chu, C. C. (1981) *J. Biomed. Mater. Res.,* **15**, 19.
18. Chu, C. C. and Louis, M. (1985) *J. Appl. Polym. Sci.,* **30**, 3133.
19. Huffman, K. R. and Casey, D. J. (1985) *J. Polym. Sci., Polym. Chem. Ed.,* **23**, 1939.
20. Chu, C.C. (1985) Polym., **26**, 591.
21. Fredericks, R. J., Melveger, A. J. and Dolegiewitz, L. J. (1984) *J. Polym. Sci., Polym. Phys. Ed.,* **22**, 57.

22. Kricheldorf, H. R. and Serra, A. (1985) *Polym. Bull.,* **14**, 497.
23. Dunsing, R. and Kricheldorf, H. R. (1985) *Polym. Bull.,* **14**, 491.
24. Tune, D. C. (1986) *Polym. Prepr.,* **27**(1) ,431.
25. Jarrett, P., Cook, W. J., Bell, J. P. *et al.* (1981) *Polym. Prepr.,* **22**(2), 351.
26. Cook, W. J., Cameron, J. A., Bell, J. P. and Huang, S J. (1981) *J. Polym. Sci., Polym. Lett. Ed.,* **19**, 159.
27. Jarrett, R., Huang, S. J., Bell, J. P. *et al.* (1982) *Org. Coat. Plast. Chem. Proc.,* **47**, 45.
28. Jarrett, R., Benedict, C., Bell, J. P. *et al.* (1983) *Polym. Prepr.,* **24**(1), 32.
29. Jarrett, R., Benedict, C. V., Bell, J. P. *et al.* (1984) in *Polymers as Biomaterials,* (eds S. W. Shalaby, A. S. Hoffman, B. D. Ratner, and T. A. Horbett), Plenum Press, New York, p. 181.
30. Pitt, C. G., Chaslow, F. I., Hibionada, Y. M. *et al.* (1981) *J. Appl. Polym. Sci.,* **26**, 3779.
31. Christel, P., Chabot, F., Leray, J. L. *et al.* (1982) *Biomat.,* **1980**, 271.
32. Vert, M., Chabot, F., Leray, J. and Christel, P. (1981) *Makromol. Chem. Suppl.,* **5**, 30.
33. Kulkarni, R. K., Moore, E. G., Hegycli, A. F. and Leonard, F. (1971) *J. Biomed Mater. Res.,* **5**, 169.
34. Miller, R. A., Brady, J. M. and Cutright, D. E. (1977) *J. Biomed Mater. Res.,* **11**, 711.
35. Bilding, D. K. and Reed, A. M. (1979) *Polym.,* **20**, 1459.
36. Reed, A. M. and Gilding, D. K. (1981) *Polym.,* **22**, 494.
37. Carter, B. K. and Wilkes, G. L. (1984) in *Polymers as Biomaterials,* (eds S. W. Shalaby, A. S. Hoffman, B. D. Ratner and T. A. Horbett) Plenum Press, New York, p. 67.
38. Mohajer, Y., Wilkes, G. L. and Orler, B. (1984) *Polym. Sci. Eng.,* **24**(5), 319.
39. Pitt, C. G., Jeffcoat, A. R., Zweidinger, R. A. and Schindler, A. (1979) *J. Biomed Mater. Res.,* **13**, 497.
40. Feng, X. D., Song, C. X. and Chen, W. Y. (1983) *J. Polym. Sci., Polym. Lett. Ed.,* **21**, 593.
41. Song, C. X. and Feng, X. D. (1984) *Macromol.,* **17**, 2764.
42. Shalaby, S. W. and Jamiolkowski, D. D. (1985) *Polym. Prepr.,* **26**(2), 200.
43. Kricheldorf, H. R., Mang, T. and Jonte, J. M. (1984) *Macromol.,* **17**, 2173.
44. Pitt, C. G. (1984) *J. Contr. Rel.,* **1**, 3.
45. Huang, S. J. and Leong, K.-W. (1979) *Polym. Prepr.,* **20**, 552.
46. Huang, S. J., Bell, J. R., Knox, J. R. *et al.* (1977) in ACS Symp. Ser. No. 49, *Textile and Paper Chemistry and Technology* (ed. J. C. Arthur), p. 73.
47. Chvapil, M., Owen, J. A., Clark, D. S. *et al.* (1977) *J. Biomed Mater. Res.,* **11**, 297.
48. Dickenson, H. R., Hiltner, A., Gibbons, D. F. and Anderson, J. M. (1981) *J. Biomed Mater. Res.,* **15**, 577.
49. Dickenson, H. R. and Hiltner, A. (1981) *J. Biomed Mater. Res.,* **15**, 591.
50. Bruins, P. F. and Ashman, A. (1983) in *Proc., ANTEC 83, Soc. Plast. Eng.,* pp. 22-24.
51. Cameron, H. U., Macnab, I. and Pilliar, R. M. (1977) *J. Biomed Res.,* **11**, 179.
52. Gogolewski, S. and Pennings, A. J. (1982) *Makromol. Chem., Rapid Commun.,* **3**, 839.
53. Gogolewski, S., Pennings, A. L., Lommen, E. *et al.* (1983) *Makromol. Chem., Rapid Commun.,* **4**, 213.

54. Gogolewski, S. and Pennings, A. J. (1983) *Makromol. Chem., Rapid Commun,* **4**, 675.
55. Yannis, I. V., Burke, J. F., Chen, E. *et al.* (1982) *IUPAC Prepr.*, Amherst Mass., 336.
56. Leenslag, J.-W., Pennings, A. J., Veth, R. P. H. *et. al.* (1984) *Makromol, Chem., Rapid Commun.*, **5**, 815.
57. Leenslag, J. W., Gogolewski, S. and Pennings, A. J. (1984) *J. Appl. Polym. Sci.*, **29**, 2829.
58. Currey, J. D. (1983) in *Handbook of Composites*, Vol. 4, (eds A. Kelly and Y. N. Rabotnov), Elsevier Science Publishers, Amsterdam, sect. 3.2.4.
59. Hyon, S.Y., Jamshidi, K., Ikada, Y. *et al.* (1985) *Kobunshi Ronbun.*, **42**(11), 771.
60. Tencer, A. F., Mooney, Brown, K. L. and Silva, P. A. (1985) *J. Biomed. Mater. Res.*, **19**(8), 957.
61. Corcoran, S. F., Koroluk, J. M., Parsons, J. R. *et al.* (1980) *Current Concepts of Bone Fracture Fixation*, (ed. H. K. Uhthoff), Springer Verlag, Berlin, p. 136.
62. Alexander, H., Coreoran, S., Parsons, J. R. and Weiss, A. B. (1981) *Bioeng.*, **9**, 115.
63. Parsons, J. R., Alexander, H. and Weiss, A. B. (1983) *Biocompatible Polymers, Metals and Composites*, (ed. M. Szycher), Technomic Pub. Co., pp. 873-905.
64. Alexander, H., Parsons, J. R., Strauchler, I. D. and Weiss, A. B. (1983) US Patent No. 4,411,027.
65. Alexander, H., Weiss, A. B., Parsons, J. R. *et al.* (1979) *Trans. ORS, 25th Ann. ORS*, **4**, 27.
66. Witvoet, J. and Christel, P. (1985) *Clin. Orth. Rel. Res.*, **196**, 143.
67. Bercovy, M., Gontallier, D., Voisin, M. C. *et al.* (1985) *Clin. Orth. Rel. Res.*, **196**, 159.
68. Alexander, H., Weiss, A. B., Parsons, J. R. *et al.* (1979) *Trans. N. E Bioeng. Conf.*, **7**, 400.
69. Aragona, J., Parsons, J. R., Alexander, H. and Weiss, A. B. (1981) *Clin. Orthop.*, **160**, 268.
70. Alexander, H., Strauchler, I., Weiss, A. B. *et al.* (1978) *Trans. 4th Ann. Mtg. Soc. Biomat.*, **123**.
71. Getter, L., Cutright, D. E., Bhaskar, S. N. and Augsburg, J. K. (1972) *J. Oral Surg.*, **30**, 344.
72. Cutright, D. E. and Hunsuck, E. E. (1972) *Oral Surg.*, 33(1), 28.
73. Christel, P., Chabot, F. and Vert, M. (1984) in *2nd World Congr. Biomat., 10th Ann. Mtg. Soc. Biomat.*, 279.
74. Schindler, A. and Pitt, C. G. (1984) *Polym. Prepr.*, **25**(1), 257.
75. Casey, D. J. and Roby, M. S. (1984) US Pat No. 4,429,080.
76. Casey, D. J. and Roby, M. S. (1984) US Pat No. 4,452,973.
77. Chabot, F., Vert, M., Chapelle, S. and Granger, P. (1983) *Polym.*, **24**, 53.
78. Huang, S. J, Edelman, P. G. and Cameron, J. A. (1987) in *Advances in Biomedical Polymers,* (ed. C. G. Gebelein), Plenum, pp. 101-109.
79. Huang, S. J. (1990) *Polym. Mater. Sci. Eng.*, **63**, 633.
80. Huang, S. J. (1985) *Encyclopedia of Polymer Science and Engineering,* **2**, 2nd edn, John Wiley & Sons, New York, pp. 220-243.
81. Huang, S. J. (1989) in *Comprehensive Polymer Science,* Vol. 6 (eds G. Allen and J. C. Bevington), Pergamon Press, London, pp. 567-607.

82. Kaplan, D. L., Mayer, J. M., Ball, D. *et al.* (1993) in *Biodegradable Polymers and Packaging* (eds C. Ching, D. Kaplan and E. Thomas), Technomics Publishing, Lancaster-Basel, pp. 1-44.
83. Swift, G. (1993) *Acc. Chem. Res.*, **26**, 105.
84. Lenz, R. (1993) in *Advances in Polymer Series*, Vol. 107 (eds N. A. Peppas and R. S. Langer), Springer-Vierlanger, pp. 1-40.
85. Albertsson, A.-C. and Karlsson, S. (1993) in *Comprehensive Polymer Science,* First Supplement (eds G. A. Allen, S. L. Agarwal and S. Russo), Pergamon Press, London, p. 285.
86. Huang, J-C., Shetty, A. S. and Wang, M.-S. (1990) *Adv. Polym. Technol.*, **10**, 23.
87. Dawes, E. A. and Senior, P. J. (1973) *Adv. Micro. Physiol.*, **10**, 135.
88. Doi, Y. (1990) *Microbial Polyesters,* VCH, New York.
89. Anderson, A. J. and Dawes, E. A. (1990) *Microbiol. Rev.*, **54**, 450.
90. Homes, P. A., Collins, S. H. and Wright, L. F. (1984) US Pat. No. 4,477,654.
91. Steinbüchel, A. (1991) *Biomaterials: Novel Materials from Biological Sources,* Stockton Press, New York, pp. 123-214.
92. Brandly, H., Gross, R. A., Lenz, W. R. and Fuller, R. C. (1990) *Adv. Biochem. Eng. Biotech.*, **41**, 77.
93. Marchessault, R. H., Blulim, T. L., Deslandes, Y. *et al.* (1988) *Makromol. Chem. Makromol. Symp.*, **19**, 235.
94. Doi, Y., Kanesawa, Y., Kunioka, M. and Saito, T. (1990) *Macromolecules,* **23**, 26.
95. Kawaguchi, Y. and Doi, Y. (1992) *Macromolecules,* **25**, 2324.
96. Jendrossek, D., Knoke, L, Habibian, R. B. *et al.* (1993) *J. Environ. Polym Deg.*, **1**(1)53.
97. Nishida, H. and Tokiwa, Y. (1993) *J. Environ. Polym. Deg.*, **1**(1), 65.
98. Nishida, H. and Tokiwa, Y. (1993) *J. Environ. Polym. Deg.*, **1**(3), 235.
99. Buchanan, C. M., Gardner, R. M. and Komarck, R. J. (1993) *J. Appl. Polym. Sci.*, **47**, 1709.
100. Gu, J.-D., Eberiel, D. T., McCarthy, S. P. and Gross, R. A. (1993) *J. Environ. Polym. Deg.*, **1**(2), 143.
101. Chum, H. L. (ed.) (1989) *Biobased Materials,* Report SERI/TR-234-3610, Solar Energy Research Institute, Colorado, pp. 1-12.
102. Rossall, B. (1974) *Ind. Biodeterioration Bull.*, **10**, 95.
103. Arthur, J. Jr., (1985) in *Celluslose and Its Derivatives* (eds J. F. Kennedy and G. O. Phillips), Ellis Horwood Chichester.
104. Assempour, H., Koenig, M. F. and Huang, S. J. in *Unconventional and Nonfood Uses of Agricultural Biopolymers* (eds M. L. Fishman, R. B. Friedman and S. J. Huang), ACS Symposium Series, in press.
105. Corti, A., Vallini, G., Pera, A. *et al.* (1992) in *Biodegradable Polymers and Plastics,* (eds M. Vert, J. Feijeu, A. Albertsson *et al.),* Royal Society of Chemistry, pp. 245-248.
106. Bastioli, C., Bellotti, V., Del Giudice, L. and Gilli, G. (1992) in *Biodegradable Polymers and Plastics,* (eds M. Vert, J. Feijen, A. Albertsson *et al.),* Royal Society of Chemistry, pp. 101-111.
107. Bastioli, C., Bellotti, V., Del Giudice, L. and Gilli, G. (1993) *J. Environ. Polym. Deg.*, **1**(3),181.

108. Szivek, J. A., Weatherly, G. C., Pilliar, R. M. and Cameron, H. U. (1981) *J. Biomed. Res.,* **15**, 853.
109. Brown, S. A. and Mayor, M. B. (1978) *J. Biomed Res.,* **12**, 67.
110. Woo, S. L-Y., Akeson, W. H., Levenetz, B. *et al.* (1974) *J. Biomed. Mater. Res.,* **8**, 321.
111. Gourlay, S. J., Rice, R. M., Hegycli, A. F. *et al.* (1978) *J. Biomed Mater. Res.,* **12**, 219.
112. Rice, R. M., Hegyeli, A. F., Gourlay, S. J. *et al.* (1978) *J. Biomed. Mater. Res.,* **12**, 43.
113. Huang, S. J., Bitritto, M., Leong, K. W. *et al.* (1978) in Adv. Chem. Ser. No. 169, *Stabilization and Degradation of Polymers,* (eds D. L. Allara and W. L. Hawkins), American Chemical Society, p. 205.
114. Huang, S. J. and Roby, M. S. (1986) *J. Bioact. Compat. Polym.*, **1**, 61.
115. Gilding, D. K. (1982) in *Biocompatibility of Clinical Implant Materials*, Vol. II, (ed. D. F. Williams), CRC Press, Boca Raton.
116. Heller, J. (1983) *Initiation of Polymerization,* (ed. F. E. Bailey), ACS Symp. Ser. No. 212, p. 373.
117. Huang, S. J. (1985) in *Encyclopedia of Polymer Science and Engineering,* (eds.) Mark, Bikales, Overberger and Menges, V. 2, 2nd Ed., 1985, pp. 220-244.
118. Williams, D. F. (1982) *J. Mater. Sci.,* **17**, 1233.
119. Huang, S. J., Bansleben, D. A. and Knox, J. R. (1979) *J. Appl. Polym. Sci.*, **23**, 429.
120. Huang, S. J., Bell, J. P., Knox, J. R. *et al.* (1976) *Proc 3rd Int. Biodeg. Symp.,* (eds J. M. Sharpley and A. M. Kaplan), Applied Science Publishers, London, p. 73 1.
121. Bitritto, M. M., Bell, J. P., Brenkle, G. M. *et al.* (1979) *J. Appl. Polym. Sci, Appl. Polym. Symp.,* **35**, 405.
122. Bell, J. P., Huang, S. J. and Knox, J. R. (1974) TR 75-48-CEMEL, US Army, Natick Laboratory, Natick, Mass.
123. Williams, D. J. (1980) *J. Biomed Mater. Res.*, **14**, 329.
124. Chu, C. C. (1981) *J. Biomed Mater. Res.*, **15**, 19.
125. Chu, C. C. and Louie, M. (1985) *J. Appl. Polym Sci.*, **30**, 3133.
126. Huffman, K. R. and Casey, D. J. (1985) *J. Polym. Sci., Polym. Chem. Ed.,* **23**, 1939.
127. Chu, C. C. (1985) *Polym.*, **26**, 591.
128. Murphy, C. A., Cameron, J. A., Huang, S.J., and Vinopal, R. T., (1996), Appl. Environ. Microbiol., 456.

3

DEGRADATION AND STABILIZATION OF CARBON-CHAIN POLYMERS

GERALD SCOTT
Aston University
Birmingham B4 7ET, UK

1. Natural and synthetic carbon-chain polymers

Carbon-chain polymers contain a continuous sequence of carbon atoms that are not interrupted by hetero-atoms such as oxygen, nitrogen or sulphur. Although this concept is normally associated with synthetic polymers made by addition polymerisation such as polyethylene, polypropylene, polystyrene and the synthetic rubbers, many naturally occurring polymers contain the same uninterrupted sequence of carbon atoms [1]. The most studied and best understood of these is natural *cis*-poly(isoprene) (NR), synthesised by the rubber tree *Hevea Braziliensis.* The identical molecule is nowadays also synthesised in industrial chemical plants by addition polymerisation from *iso*prene.

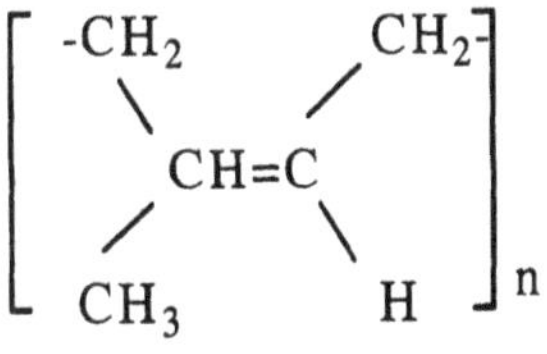

$$\left[\begin{array}{ccc} -CH_2 & & CH_2- \\ & CH{=}C & \\ CH_3 & & H \end{array}\right]_n$$

cis-poly(isoprene), NR

Natural rubber was one of the first important industrial polymers and it was recognised even before it reached the industrialised countries that that it very rapidly lost its initial useful properties in a process that, by analogy with human physical degradation, was given the name ‘ageing’. A particularly unique and desirable attribute of rubber was its rebound resilience and this was rapidly lost in a tropical environment. Furthermore, rubber latex products were attacked by microorganisms, which led to more general loss of mechanical properties and to eventual bioassimilation in the soil environment.

G. Scott (ed.), *Degradable Polymers, 2nd Edition*, 27-50.

The rapid ageing of rubber was a cause of great concern to rubber technologists in the 19th century and it was found empirically that both the deterioration of mechanical properties and subsequent biodeterioration could be inhibited by small amounts of aromatic amines and by some constituents of the vulcanisation system. These were given the name antioxygenes [2,3] or antioxidants [4] by scientists and technologists concerned with the mechanisms of oxidative deterioration of materials. The work of Bolland, Bateman and their co-workers at the British Rubber Producers Research Association in the UK [5,6] and Bevilaqua [7] in the USA was seminal in identifying the chemical processes that lead to oxidative scission of the rubber macromolecule causing loss of useful mechanical properties of the rubbers. Crucial to these investigations was the recognition of the importance of hydroperoxides which, due to the weak O-O bond, are the cause both thermo- and photo-instability in carbon-chain molecules. The early literature on the effects of peroxidation in hydrocarbon polymers was reviewed by the author in reference 8 and the environmental aspects of polymer degradation in references 9 and 10.

The synthetic hydrocarbon plastics and in particular the polyolefins and polystyrene are based on relatively cheap petroleum feedstocks. Although more environmentally stable than the polydiene rubbers, they are nevertheless much less resistant to the environment than might have been expected on the basis of their formal structures [9]. There are several reasons for this, which are now well documented [10,11]. The first is the presence of a small amount of unsaturation in the polymers as they are manufactured. It will be seen below that these markedly increase the peroxidisability of otherwise saturated polymers out of all proportion to their concentration. The second reason for the environmental instability of hydrocarbon polymers is that the processing operations (extrusion, injection moulding, etc.) are very damaging to the polymer due to the high shearing forces induced in the polymer molecules (equation 1). In the highly viscous state of the molten polymer, the chemical bonds of the polymer chains are broken to give free radicals, which immediately react with the oxygen that is always present in commercial operations to give peroxyl radicals and hydroperoxides. The latter play a key role in subsequent polymer degradation, since they are initiators for further peroxidation [12]. The third reason, which will be discussed below is the presence in the polymer of intrinsic impurities (e.g. olefinic unsaturation or carbonyl groups) introduced either during manufacture or during the conversion of polymers to industrial materials. This chemistry can be turned to an advantage in the accelerated degradation of polymers to lower molar mass materials, leading to their bioassimilation in the environment.

$$\text{P-P}' \xrightarrow{\text{Shear}} \text{P}\cdot + \text{P}'\cdot \xrightarrow{2\text{O}_2} \text{POO}^{\cdot} + \text{P}'\text{OO}^{\cdot} \xrightarrow{2\text{PH}} \text{POOH} + \text{P}'\text{OOH} + 2\text{P}^{\cdot} \qquad (1)$$

P, P′ are carbon chain segments; PH is any hydrocarbon polymer

2 Hydroperoxides and the peroxidation chain mechanism

Hydroperoxides are of fundamental importance to polymer degradation. Not only are the free radicals produced by their dissociation (reaction 2) the main initiators of the peroxidation chain reaction, [12,13], but $\text{PO}^{\cdot}$ is also the source of the ultimate low molar mass degradation products that are readily bioassimilated by microorganisms.

$$\mathrm{POOH} \xrightarrow{\Delta,h\nu} \mathrm{PO^{\cdot}} + {}^{\cdot}\mathrm{OH} \xrightarrow{2\mathrm{PH}} 2\mathrm{P^{\cdot}} + \mathrm{POH} + \mathrm{H_2O} \qquad (2)$$

The peroxidation chain sequence (3) and (4) continues as long as oxygen is present in the system.

$$\mathrm{P^{\cdot}} + \mathrm{O_2} \rightarrow \mathrm{POO^{\cdot}} \qquad (3)$$

$$\mathrm{POO^{\cdot}} + \mathrm{PH} \rightarrow \mathrm{POOH} + \mathrm{P^{\cdot}} \qquad (4)$$

Reaction (4) is the rate-controlling step and consequently, termination of the chain reaction occurs by reaction of $ROO^{\cdot}$ with other radical species in the system. The kinetics of the peroxidation chain reaction have been discussed in many reviews and the reader is directed to the following for further information [5,6,8].

The rate of peroxidation of hydrocarbon polymers in the absence of added initiators increases by two orders of magnitude, from the relatively stable unbranched polyethylenes through branched polypropylene to the polyunsaturated rubbers [11]:

$$\underset{\text{PE}}{-(\mathrm{CH_2CH_2})_n-} < \underset{\text{PP}}{-(\overset{\overset{\large \mathrm{CH_3}}{|}}{\mathrm{C}}\mathrm{HCH_2})_n-} < \underset{cis\text{-PB}}{-(\mathrm{CH_2CH{=}CHCH_2})_n-} < \underset{cis\text{-PI}}{-(\mathrm{CH_2}\overset{\overset{\large \mathrm{CH_3}}{|}}{\mathrm{C}}\mathrm{{=}CHCH_2})_n}$$

This order reflects the increasing ease of abstraction of the weakest P-H bonds (namely the methylenic hydrogens) in reaction (4) [14]. Consequently blends of saturated polymers with rubbers or co-polymers of saturated and unsaturated polymers peroxidise more rapidly than the saturated polymers themselves and this kind of modification has sometimes been used to increase the rate of bioassimilation of polymers through environmental peroxidation.

In the absence of antioxidants and stabilisers, the concentration of hydroperoxides increases rapidly in an 'autooxidising' polymer until the rate of decomposition of hydroperoxides is equal to the rate of their formation [8]. The rate at which this state is reached in the environment is normally determined by the influence of external factors that promote the decomposition of hydroperoxides.

2.1 PROMOTERS OF HYDROPEROXIDE DECOMPOSITION

Hydroperoxides are relatively stable at ambient temperatures in the absence of promoters of decomposition, the most important of which are UV light and transition metal ions. Activation of hydroperoxide decomposition by UV light, reaction (2), is the main cause of polyolefin peroxidation in sunlight. Consequently peroxidation and physical degradation of hydrocarbon polymers out-of-doors is very much faster in sunlight than in the dark at the same temperature, although the relative rates depend very much on the presence or absence of photosensitisers. Traces of transition metal ions have traditionally been a threat to the oxidative stability of polymers. In particular, vulcanising ingredients and pigments that are added to polydiene rubbers before cross-linking have to be purified of metallic prooxidants. In the modern synthesis of polymers from vinyl monomers using organometallic catalysts, the metal ions are not normally removed and this can also give rise to oxidative instability during service [15]. Transition metal catalysts react with polymer hydroperoxides as follows;

$$M^{n+} + POOH \rightarrow M^{(n+1)+} + PO^{\cdot} + OH^{-} \quad (5)$$

$$M^{(n+1)+} + POOH \rightarrow M^{n+} + POO^{\cdot} + H^{+} \quad (6)$$

$$POO^{\cdot} + PH \rightarrow POOH + P^{\cdot} \xrightarrow{O_2 + PH} POOH \quad (7)$$

$$PO^{\cdot} + PH \rightarrow POH + P^{\cdot} \xrightarrow{O_2/PH} POOH + P^{\cdot} \quad (8)$$

The reaction sequence (5)-(8) leads to the rapid accumulation of hydroperoxides and the attainment of a stationary peroxide concentration. Soluble Cu, Mn, Fe, Cr and Co compounds are the most effective promoters of peroxidation whereas Ni, Ce, V, Ti and Zn salts are less effective [16]. Transition metal ions are also photo-initiators of peroxidation but the order of activity is generally different. Indeed Cu^{2+}, which is an effective thermal prooxidant, is actually a photoantioxidant in some polymers [17]. Of greater practical importance, prooxidant transition metal ions such as Fe^{3+} can be deactivated as prooxidants and actually inverted to antioxidants by complexing [18]. This is important and practically useful when utilising transition metal ions to rapidly degrade carbon-chain polymers in the environment after discard. As will be seen later, some micro-organisms produce peroxidase enzymes that act similarly, although very much faster, than the metal complex systems discussed above.

2.2 PRODUCTS OF HYDROPEROXIDE DECOMPOSITION

The most important physical effect of peroxidation in hydrocarbon polymers is the reduction of the molar mass of the polymer leading to deterioration in mechanical properties. Peroxidation, whether abiotically or biologically initiated, leads to the formation of hydrophilic chemical species such as carboxylic acids and alcohols particularly in the surface layers of the polymer [13]. This makes them much more accessible to microorganisms which are able to colonise the polymer surface and utilise the low molar mass oxygenated species as nutrients in the absence of other sources of carbon [19].

Peroxidation of *cis*-poly(isoprene) leads in part to the formation of vicinal hydroperoxides as shown in Scheme 1. This process has been known for many years [5-7] but its full significance for rubber biodegradation has been recognised only recently. The low molar mass peroxidation products are rapidly biassimilated under environmental conditions. In addition, however, carbon dioxide and water, the normal end points of biodegradation, are also formed in significant quantities during abiotic oxidation and peroxidation of hydrocarbon polymers to CO_2 is an essential part of the natural biocycling process.

```
-CH2   CH2CH2   CH2CH2                         -CH2   CH2CH2  ·CHCH2
   \   /    \    /    \    /       PO·            \    /      \   /   \   /
    C=CH     C=CH       C=CH     ——————>           C=CH       C=CH    C=CH  + POH
   /        /          /                          /          /       /
 CH3      CH3        CH3                        CH3        CH3     CH3

                                       ↙ O2

-CH2   CH2CH2   CHCH2                          -CH2   CH2CH2  CHCH2
   \   /    \   //    \    /  O2 + PH             \   /      \  //    \   /
    C=CH     C-CH      C=CH ——————>   HOOC-CH        C-CH      C=CH  + P·
   /         /\        /                    /  \      /\        /
 CH3     ·O-O  CH3   CH3                  CH3  O - O   CH3   CH3

                                                      |
                                              O2/PH   |
                                                      ↓

   Biodegradable products    CH3COCH2CH2CHO  + CH3COOH  +  HCOOH
                                             (+ CO2 + H2O )
```

Scheme 1. Peroxidation of cis-PI to biodegradable products [7]

Synthetic *cis*-PI (IR) has an identical chemical structure to NR and of course it peroxidises by the same mechanism. It will be seen later that the biodegradation rate of IR is also very similar to that of NR confirming that the same abio- and bio-oxidation processes occur together in the environment with synergistic interaction between environmental abiotic and biotic processes.

Abiotic peroxidation of the polyolefins (reactions (3) and (4)) gives rise to some vicinal hydroperoxides and this process is particularly favoured in the poly-α-olefins, such as polypropylene due to the susceptibility of the *tertiary* carbon atom to hydrogen abstraction via a hydrogen-bonded intermediate (Scheme 2). A major proportion of the peroxidic products are hydrogen-bonded vicinal hydroperoxides that break down to small biodegradable molecules such as carboxylic acids, alcohols and ketones [20] as well as longer chain oxygen-modified breakdown products, which oxo-biodegrade more slowly. The decomposition of the vicinal hydroperoxides is also facilitated by internal hydrogen-bonding and the low molar mass products of this self-induced degradation are small biodegradable molecules such as acetic and formic acids.

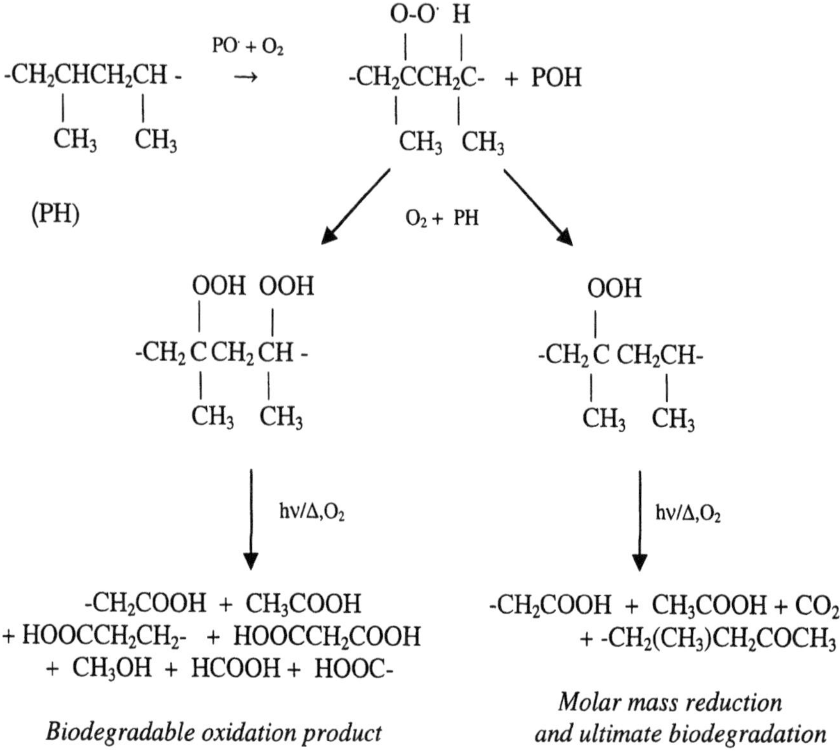

Scheme 2 Formation and breakdown of hydroperoxides in polypropylene

In the case of the polyolefins, random chain scission is initially the dominant process. This is shown typically for polypropylene in Scheme 2. However some low molar mass oxidation products are formed via vicinal hydroperoxides in both PP and PE [20]. The alkoxyl radicals formed by decomposition of the hydroperoxides contain weak carbon-carbon bonds in the α positions to the hydroperoxide groups, which lead to the formation of low molecular weight aldehydes and alcohols that rapidly oxidise further to carboxylic acids. These are biodegradable species, similar to products formed by hydrolysis of aliphatic polyesters and, as in the case of cis-PI, they are rapidly bioassimilated to give cell biomass (see below).

3. Microbial degradation of carbon-chain polymers

3.1 CIS-POLYISOPRENE

The free radical chain reactions (3), (4), initiated by hydroperoxides in the presence of transition metal ions and oxygen (5)-(8), continues in hydrocarbon polymers so long as oxygen is present in the system with the accumulation of abiotically stable carboxylic acids. However, in microbially active environments, carboxylic acids are bioassimilated by micoorganisms so that the carboxylic acids are in dynamic equilibrium in the system. It has been demonstrated experimentally [21,22] that pure strains of bacteria (in particular *actinomycetes*) and fungi cause up to 55% loss of mass of rubber sheets in 70 days. The actinomycete, *Nocardia* (sp. Strain 835A) was found by Tsuchii and co-workers [23] to be particularly effective in degrading NR rubber gloves in the absence of any other source of carbon. Weight losses of 75% were achieved in two weeks and the same strain in laboratory fermenters led to complete degradation of NR in 45 days [24].

More recently, Ikram and co-workers [25] have shown that in normal soils at 25°C, NR gloves showed 54% loss of thickness after 4 weeks. And 94% mass loss after 48 weeks. Commercial nitrile and neoprene rubbers showed insignificant loss in this time and plasticised PVC showed a smaller mass loss (11.6%) due entirely to biodegradation of the plasticiser and not to the biodegradation of the polymer itself. Bacterial populations on the NR gloves (12317/mg) were higher than for fungi (441.47/mg), which were in turn significantly higher than actinomycetes (297.02/mg). Nevertheless, Heisey and Papadatos [26] isolated 10 actinomycetes (seven strains of *Streptomycetes*, two strains of *Amycolatopsis* and one strain of *Nocardia*) from soil that reduced the mass of NR gloves from 10-18% in 6 weeks.

Ikram has subsequently shown [27] that mass loss of NR is highly dependent upon the nutrients in the soil; particularly nitrogen and phosphorus. After 24 weeks, NR in the high N (100mg/l), P (150mg/l) system had lost 61.5% of its mass whereas in the low N (10mg/l), P (15mg/l) system, only 23.6% mass was lost. Control (unfertilised) soil produced least mass loss (17.3%). Microbial growth measured on the rubber pieces were in decreasing order as expected.

Table 1. Effect of added soil nutrients on the mass loss of rubber and plastic films: (%) after 40 weeks in soil. [27].

Polymer	Nutrient treatment		
	High*	Low*	Control*
NR	-82.4	-38.5	-29.7
Neoprene	+0.3	-13.0	-1.1
Nitrile	-4.3	-3.2	-3.5
Plasticised PVC	-26.1	-13.4	-11.1

* Nutrients added: High 100 mg/l N and 150 mg/l P; Low 10 mg/l N. 15 mg/l P; Control nil

Steinbüchel and co-workers [28,29], using rubbers as the sole source of carbon, found that NR and IR (synthetic polyisoprene rubber) biograde at a similar rate in the presence of *Pseudomonas aeruginosa*. NR gloves were 26% mineralised in 6 week compared with 21% for IR gloves. This slight difference may well have been due to the difference in the antioxidants used in the formulation (see Section 4) since Berekaa et al. [30] in a similar study showed that removal of antioxidants by extraction markedly increased the rate of microbial growth. It is clear, however that, contrary to the views of some environmentalists [31], there is no intrinsic difference between natural and synthetic polymers .

It has been pointed out that some actinomyctes can utilise CO_2 as a source of carbon [32]. It is therefore necessary to equate microbial growth and associated formation of protein to loss of weight of the substrate [26]. Table 2, taken from the work of Heisey shows that there is indeed a broad correlation between mass loss and protein formation. *Nocardia* [23] and *P. aeruginosa* were shown to break the cis-PI chain by an oxidative mechanism since aldehyde groups were found to accumulate during microbial degradation. This is always the first product formed during the abiotic peroxidation of *cis*-PI and the evidence suggests that the bacteria initiate a radical-chain peroxidation, which is inhibited by antioxidants. This will be discussed further in the context of poyolefin biodegradation.

Chlorinated polymers are much more resistant to abiotic peroxidation than pure hydrocarbon polymers [11] and nitrile rubbers, although susceptible to peroxidation are normally highly stabilised by extraction-resistant antioxidants [9].

Table 2. Mass changes of NR strips and protein concentration produced by rubber metabolising microorganisms [26]

Isolate	Mass change of rubber strips (%)	Protein concentration* (mg/g of rubber)
Control	1 ± 1	1 ± 0
1	1 ± 0	2 ± 0
2	0 ± 2	2 ± 0
3	-8 ± 1	26 ± 11
4	-9 ± 1	35 ± 3
5	-11 ± 2	27 ± 9
6	-11 ± 0	29 ± 3
7	-12 ± 3	27 ± 4
8	-12 ± 1	28 ± 7
9	-13 ± 1	32 ± 2
10	-14 ± 2	40 ± 3
11	-16 ± 4	39 ± 4
12	-16 ±4	46 ± 9
13	-16 ± 2	45 ± 3
14	-18 ± 2	46 ± 2

* Total in the culture broth and on the rubber strips

Fully formulated tyre rubbers, in contrast to latex rubbers used in domestic products, are highly resistant to peroxidadation and hence biodegradation. Tyres survive almost unchanged in the outdoor environment when discarded and although secondary uses are found for a small proportion of these in agriculture to weigh down silos and in docks as bumpers, where do not biodegrade at a measurable rate and have to be ultimately disposed of by some other means. It will be seem (Section 4.1) that this has much more to do with the effectiveness of the antioxidant systems used in automobile tyres [9] than with the inherent biodegradability of the rubber molecule.

3.2. LIGNIN AND LIGNOCELLULOSE

Lignin is another polymer that, like *cis*-polyisoprene, bridges the gap between natural and synthetic polymers. Lignin is a cross-linked polymer containing benzene rings (see below). It is formed in chemical association with cellulose (lingocellulose) in the cell walls of plants [33]. The aromatic structures contain alkoxy and hydrocarbon substituents that link the basic unit below into a macromolecular structure through carbon-carbon and carbon-oxygen bonds. Both the chemical and physical properties of lignin resemble those of the synthetic phenol-formaldehyde (PF) resins and it acts as an adhesive for cellulose fibre in a manner that anticipates the synthetic fibreglass composites in modern polymer technology [34]. Like the PF resins, lignocellulose is strong and tough and provides physical protection to the growing plant. In addition it provides chemical and biological protection to wood, straw, husks, etc. A further similarity to PF resins is the relative resistance of lignin to peroxidation as a result of the presence in the polymer of many antioxidant-active phenolic groups which act as protective agents against abiotic peroxidation and biological attack by peroxidase enzymes (see below).

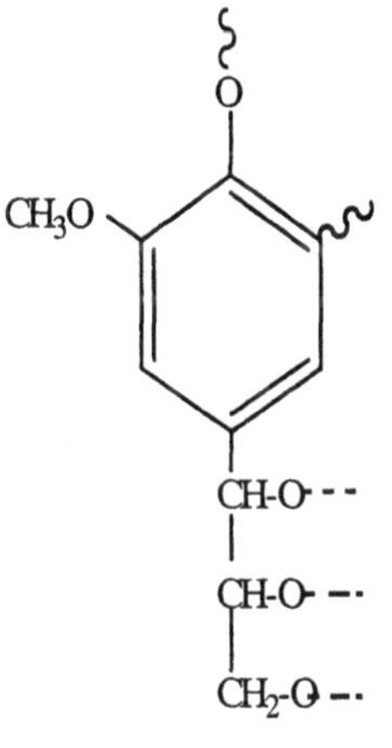

O

CH_3O

CH-O- - -

CH-O- - -

CH_2-O - -

Lignin monomeric unit

~~~ indicates potential sites through which dehydropolymerisation and cross-linking may occur.

----- indicates sites through which attachment to cellulose may occur.

Cellulose is almost always found in natural products in combination with lignin (25-30% in most woods). The crystalline structure of the former provides reinforce- ment and tensile strength for an otherwise rather weak material [34]. Lignin, on the other hand, by chemical attachment to cellulose through ether linkages acts as the polymer matrix, providing both impact strength and resistance to the environment. Relatively small amounts of lignin inhibit attack by the hydrolytic microorganisms that degrade pure
~~~

cellulose, for example in compost [35]. Lignocellulose, due to its physical (hydrophobic) and chemical inertness, does not readily degrade either abiotically or biotically and when it does biodegrade, the lignin tends to accumulate [36]. It biodegrades slowly under composting conditions and in grass, hay and straw, lignin was found to biodegrade to the extent of 17-53% in 100 days [37]. In laboratory incubation studies, thermophilic composting of grass straw showed 45% degradation in 45 days [38] but the process tends to slow down at more extended times presumably due to the increasing ratio of lignin to cellulose. Janssen [39] has estimated by measuring carbon-labelled CO_2 formation that the time for complete conversion of straw to carbon dioxide is about ten years and as will be seen in Section 3.3 a considerable proportion of the lignin is converted to humus.

There has been intense interest in recent years in the selective removal of lignin from wood pulp during papermaking. Since lignin cannot biodegrade by a hydrolytic process, the lignin component of lignocellulose biodegrades slowly by oxidative attack due to extra-cellular peroxidases formed from actinomycetes such as *Streptomyces viridosporus* [40-42]. A number of peroxidases have been isolated that remove lignin from lignocellulose without affecting cellulose itself. Manganese peroxidase (MnP) in particular has been implicated as an important enzyme formed by white rot fungi during the delignification of Kraft pulps [43-45].

Unlike the peroxidation of the hydrocarbon polymers, the oxidation of lignin occurs by a stoichiometric process and not a chain reaction. Because phenols are antioxidants, the phenoxyl radicals formed are too stable to participate in a peroxidation chain reaction (see Section 4), and the aromatic system must be converted to quinonoid compounds (and ultimately humus) during bioassimilation. Both abiotic transition metal ion catalysed peroxidation and biological oxidation are involved in the conversion of lignin to humus. It will be seen later (Section 3.5) that oxidising enzymes are also responsible for the oxo-biodegradion of the polyolefins by a similar mechanism.

3.1 TANNINS

Tannins are derived from gallic acid by dehydropolymerisation (see below). The role of redox reactions involving iron, hydrogen peroxides and hydroperoxides is well understood [46]. However, exo-cellular peroxidases (ferriprotoporphorins) are the biological equivalents, which act many times faster than the inorganic system [47]. Both carbon-carbon and carbon-oxygen linkages are formed in the dehydropolymerisation of simple phenols and these are also present in lignin and the tannins.

OH HO OH COOH → COOH HO OH OH HO OH COOH OH —ROO.→ RO', 'OH Peroxidase HOOC HO OH O O HO OH COOH

Gallic acid Polyphenolic dehydrodimers Polyhydroxy polyquinones

3.3 HUMUS

In the words of S.A.Waksman "humus serves as a reservoir and stabiliser for organic life on this planet...it is the storehouse of important chemical elements for plant growth" [48]. Humus is the further oxidation product of lignocellulose and its brown-black colour is due to the predominance of quinonoid polymers that can be partially extracted by organic solvents. It is found not only in fertile soils to the extent of 1-3% but also in 'brown' coals which are intermediate in degree of oxidation between lignin and anthracite, the ultimate dehydro-polymerisation product of lignin, which consists mainly of condensed aromatic rings of graphite.

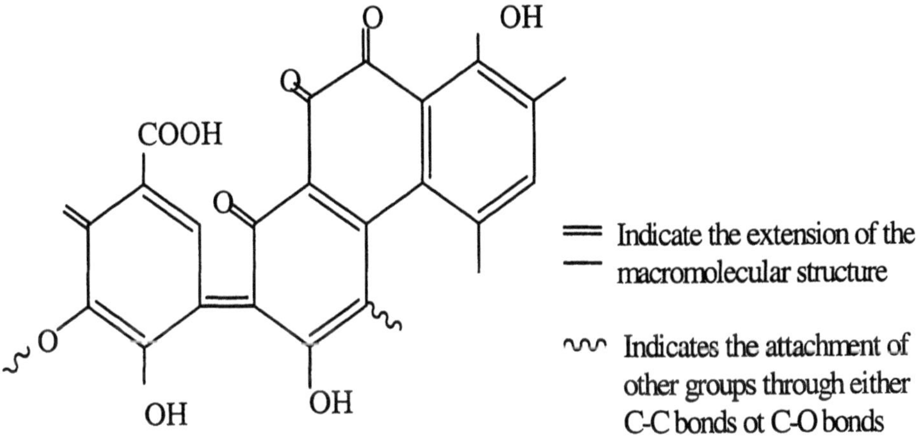

COOH
OH
OH
OH

═ — Indicate the extension of the macromolecular structure

∿ Indicates the attachment of other groups through either C-C bonds ot C-O bonds

It was seen in the previous sections that humus is formed from lignin and the tannins by further oxidation by a variety of peroxidase (polyphenoloxidase) enzymes and during the oxidation process the concentration of carboxyl groups is considerably reduced and the carbon-oxygen ratio increases. However although lignin and tannic acids are the major source of humus, they are almost certainly not the only source. The peroxidases, which can generate hydroxyl radicals [47], can thus convert aromatic rings present in proteins to phenols and hence to humus. Humus also contains substantial amount of nitrogen, chemically attached to the polyphenol-quinone molecules and this is slowly released as fertiliser.

Because of the presence of both carbonyl groups and hydroxyl groups, humus is able to chelate metal ions and is a source of trace elements for plant growth as well as available carbon and nitrogen compounds formed in the breakdown of proteins that act as nutrients for the growing plants. During the digestion of humus there is a slow but steady liberation of carbon dioxide as a consequence of the breakdown of aliphatic carbon sequences but it important to note that aromatic molecules do not convert rapidly to carbon dioxide. Guillet has noted [49] that when plants are grown on photooxidised polystyrene, approximately 50% of the carbon is absorbed directly by the plant without being liberated to the atmosphere as CO_2. The rate of ultimate mineralisation of both polymers is very similar.

3.4 POLYOLEFINS

Although commercial polyolefins are more oxidatively stable than the polydiene rubbers, they behave very similarly to ligneous materials on exposure to the environment and like the commercial rubber products, their resistance is due to the antioxidant and photostabiliser packages that have been developed over the past 50 years. As discussed in Section 2, the polyolefins are peroxidised in the outdoor environment to biodegradable products similar to those formed from the *cis*-PI. It has been known for many years [50] that, although commercial carbon-chain polymers are normally resistant to biodegradation, when polyethylene that contains transition metal ions is peroxidised at composting temperatures, it can be used as a carbon source by thermophilic fungi at 40°C and 50°C. It was also observed that fragmented polyethylene in conjunction with fertilisers actually had a beneficial effect on the growth of vegetables [50]. The potential for biologically recycling hydrocarbon polymers by accelerated peroxidation after use was appreciated and advocated as a way of utilising waste polyolefin packaging materials. This technology has also been used since the mid 1970s in the manufacture of protective agricultural films with controlled stability that subsequently biodegrade in contact with soils. However, environmentalists find it difficult to accept that polymers that are not 'natural' that is they are produced by synthesis from oil can ever biodegrade [31]. This has encouraged synthetic chemists to search for new polymers that are derived from or simulate the behaviour of natural products. The problems associated with this approach will be discussed in Chapter 14 where the applications of degradable polymers are discussed.

Lee and co-workers [51], in a study of the behaviour of starch-polyethylene blends (6% starch + transition metal ion prodegradants selected from Fe, Zn, Ni and Mn) in compost, examined the effect of a number of lignin-degrading micro-organisms. The polymer films were first peroxidised either thermally at 70°C in an air oven for up to 20 days or by long wave UV irradiation for up to 8 weeks before being exposed to three cellulose-degrading bacteria (*Streptomyces viridosporus, Streptomyces badius* and *Streptomyces setonii*) and one lignocellulose-degrading fungus (*Panerochaete chrysosporium*). Using a starch-agar assay, it was found that *S. setonii and P. chrysosporium* were unable to utilise cornstarch but the former did biodegrade polyethylene. Mass-loss measurements were inconclusive due to the difficulty on removing microflora but GPC showed a reduction of polydispersity in the case of the samples incubated with *Streptomyces* spp., indicating the selective removal of lower molar mass species. The authors confirmed previous findings that prior peroxidation is an essential prerequisite to the biodegradation of polyethylene.

More detailed experiments, similar to those described above, have been used to simulate the effect of environmental exposure on the chemical, physical and biological changes occurring in commercial degradable polyethylenes during service and on exposure to the environment [19]. PE films, after peroxidation at composting temperatures or after being subjected to photooxidation, were incubated with bacteria and fungi that had been isolated from soils that were adapted to the presence of partially degraded polyethylene. The peroxidised samples were used as the sole source of carbon for a period of six months. Three different kinds of degradable polymer were used.

(a) Photodegradable (photolytic) polymers made by copolymerisation of ethylene with carbon monoxide (E-CO). Union carbide technology [52].

(b) Conventional polyethylene containing a transition metal prooxidant blended with starch (E-St). Griffin technology [53].

(c) Photo-biodegradable (oxo-biodegradable) polymers based on conventional polyethylene containing a photo-sensitive transition metal ion complex antioxidant. (S-G). Scott-Gilead technology [54]

Although E-CO initially photodegraded to fragments more rapidly than S-G and E-St, photodegradation of the transition metal ion catalysed systems continued to a much lower molar mass. After fragmentation, the peroxidised polymers were incubated in the absence of any other source of carbon with three microorganisms isolated from soil in the vicinity of discarded polyethylene. Two were bacteria (*Nocardia asteroides* and *Rhodococcus rhodochrous*) and one was a fungus (*Cladosporium cladosporioides*). It recently has been shown by A-M. Delort using epifluorescence spectroscopy [55] that biofilm formation is very rapid on the surface of peroxidised polyethylene (TDPA™) films. This technique shown typically in fig.1(a) for *Nocardia asteroides* allows the direct measurement of the percentage surface area colonized. This is followed by disintegration of the surface of the polymer by exo-enzymes from the colonised bacteria. This is illustrated for *Rhodococcus rhodocrous* after one month of incubation in the absence of any other source of carbon in Fig. 1(b).

Three parameters were used to measure the extent of biodegradation. The first was mass loss as measured by the decrease in the thickness of the films during incubation. *Nocardia asteroides* was particularly effective in bioassimilating all thermally oxidised polymers (Table 3), whereas *Rhodococcus rhodochrous* bioassimilated photooxidised S-G but had little effect on photooxidised E-CO. It was seen in Section 3.1 that *Nocardia* was also very effective in bioassimilating *cis*-PI under similar conditions. The fungus, *Cladosporium* was least effective in reducing the mass of polyethylene samples but, surprisingly, it did degrade the photooxidised starch-filled polymer.

The second parameter was bioerosion of the surface of the films. This was assessed by removing the biomass by sterilisation, followed by scanning electron microscope examination (Table 3, SEM). The results broadly correlated with the mass-loss measurements, although in some cases, bioassimilation was greater by this measure than by mass loss.

(a) (b)

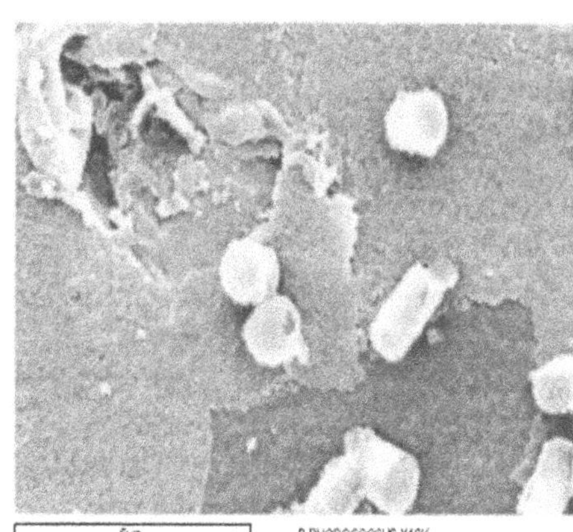

Fig. 1 (a) Colonization of Nocardia asteroides (15 min) on peroxidised degradable TDPA™ polyethylene by epifluorescence spectroscopy, (b) bioerosion of the surface of degradable TDPA™ by SEM after 1 month. (Reproduced with permission from Dr.Anne-Marie Delort and co-workers, Clermont-Ferrand University [55])

The third measure of biodegradation was molar mass reduction. In all cases M_w decreased most rapidly during both photooxidative and thermooxidative treatment (Table 3) and it was found that bioassimilation occurs readily at M_w up to ~40,000. This suggests that the predominant process occurring initially is the removal of the low molar mass oxidation products from the surface of the polymer (Scheme 2). Bioerosion was confirmed by the fact that M_w decreased relatively slowly during the biodegradation process.

It was concluded from this study that peroxidation is the rate-determining step in the oxo-biodegradation of the polyolefins. This is very rapid in sunlight but, when the polymer contains transition metal prooxidants, it also continues in soil after burial in the case of PBD and E-St). Table 4 shows typically the stages of bioassimilation of oxo-biodegradable PE. A major advantage of the polyolefins is the abilty to delay the time to the commencement of peroxidation (induction time) by the addition of antioxidant and stabilisers during the conversion process. High molecular weight polyolefins are inert to microbial degradation and biodegradation cannot begin until enough small molecules have been formed in the polymer surface to permit biofilm formation.

It will be seen in Section 4 that the rate of the abiotic and hence the biodegradation process can be readily controlled by antioxidants whereas no comparable control process has yet been developed for hydro-biodegradable polymers. A second conclusion was that starch plays no part in the biodegradation of a polyethylene matrix until the latter has been extensively peroxidised in the presence of transition metal ions. Similar conclusions have been reached by Wool [56] who showed that in the absence of PE degradation in starch-PE blends, biodegradation is controlled by the rate of migration of microorganisms through the hydrophilic polyethylene matrix. This becomes appreciable only when the starch is the major component

Table 3. Effect of Nocardia asteroides on the physical characteristics of commercial PE films after photo- and thermooxidation [19]

		Control		Nutrients only		*Nocardia* only			
Polymer	Abiotic Treatment	OD[a]	M_w ($x10^{-3}$)	OD[a]	M_w ($x10^{-3}$)	OD[a]	M_w ($x10^{-3}$)	SEM[b]	Δth[c] %
E-CO	Untreated	0.05	290	0.05	299	0.04	328	0	0
	Photoox.	0.12	16	0.10	14	0.11	12	2+	0
	Thermoox.	0.45	21	0.40	19	0.20	17	2+	-20
PBD	Untreated	0	248	0	280	0	282	0	0
	Photoox.	0.35	40	0.30	32	0.25	19	2+	0
	Thermoox.	1.05	16	0.90	16	0.75	15	3+	-20
E-St	Untreated	0.10	206	0.11	285	0.11	289	0	0
	Photoox.	1.15	16	1.08	15	0.70	12	4+	-27
	Thermoox.	1.90	nd	1.95	nd	1.55	nd	2	-6

[a] Optical density at 1715 cm^{-1}. [b] Visual rating of surface erosion; 0 = no erosion...... 4 = severely eroded. [c] Change in thickness by change in absorbance at 1375 cm

Table 4. Time-scale for biodegradation of peroxidised polyolefins

Photooxidation (SEPAP, 60°C)	100 h
or	
Thermooxidation (60°C)	300 h
Biofilm formation	0.25 h
Surface disintegration (*R.rhodocrous*)	1 month
Mass loss (6 months)	15-20%

Note: in compost and in soil, thermooxidation and biooxidation occur synergistically.

David and co-workers [57] confirmed the bioerosion of polyethylene films, peroxidised in the presence of cobalt acetylacetonate (Coacac) at 40°-70 °C both in compost and in a liquid medium by respirometry in the presence of an extract of microorganisms obtained from compost. The latter provides a method of measuring bioassimilation complementary to that described above. Table 5 shows PE degradation as a percentage of that theoretically required for total conversion to carbon dioxide and water.

Chiellini et al. [58] extracted thermally peroxidised polyethylene with acetone and measured the rate of mineralization of the solvent free extracts in forest soil. This is compared with cellulose and a number of low molar mass control hydrocarbons in Fig. 2. Surprisingly, the peroxidation products were converted to carbon dioxide and water more rapidly than cellulose. The extracted polyethylene degraded at a similar rate to the pure hydrocarbons and it is evident from this work that the rate controlling process in the overall sequence of degradation reactions is the initial peroxidation of the polymer. It has been demonstrated [19] that the exposure of peroxidised PE to an abiotic water-leaching environment did not remove the peroxidation products from the polymer, whereas bioassimilation began immediately (see Fig. 2)

Table 5. Extent of biodegradation of thermally peroxidised LDPE films by GPC and respirometry [57].

Concentration[1] (%)	Microorganisms[2]	M_n	M_w	M_w/M_n	Degradation (by oxygen consumption[3]) (%)
0.05	Extract 1	630	1900	3	17
0.05	Extract 2	700	1980	2.8	14
0.006	Extract 1	840	2130	2.5	32
0.006	Extract 2	950	2250	2.4	27

[1] Concentration of oxidised PE in the medium [2] Extracts from two different composts [3] Based on the difference between oxygen consumption with and without the polymer samples at plateau

Albertsson, Karlsson and their co-workers [20] have investigated the chemical constitution of the low molecular weight products formed by the peroxidation of photodegradable PE. Table 6 shows the types of product formed, many of them in very small amount but, as was seen above (Fig.1(a) and Fig. 2), sufficient to allow immediate colonisation of microorganisms on the surface of the polymer. Furthermore, these workers have shown that the products of polyethylene peroxidation are identified in different proportions depending on whether microorganisms were present in the system or not (Table 6). This indicates that some per-oxidation products (e.g. carboxylic acids and ketones) are bioassimilated more rapidly than others (short-chain paraffins) and although over a longer time these species are known to be biodegradable [59].

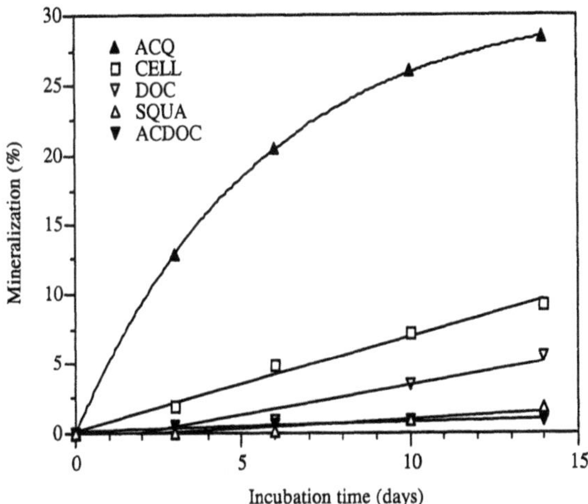

Fig. 2 Mineralization of acetone extract of PE (EPI TDPA™) after peroxidation (ACQ) compared with cellulose (CELL) and low molar mass hydrocarbon analogues of polyolefins. DOC is docosane ($C_{22}H_{46}$), SQUA is 2,6,10,15,-19,23-hexamethyl tetracosane, squalane and ACDOC is docosanoic acid ($C_{22}H_{44}O_2$).
(Published with kind permission from Professor Emo Chiellini and co-workers)

Table 6. Biodegradable products formed from polyethylene in abiotic and biotic *environments [20].*

Degradation product	Abiotic	Biotic
Carboxylic acids (C_6-C_{11})	+	-
Aliphatic alcohols (C_{20}-C_{26})	-	+
Ketones (C_{10}-C_{12})	+	-
Aliphatic hydrocarbons (C_{20}-C_{26})	-	+

Pandey and Singh [60] have recently shown that polypropylene biodegrades much more rapidly than polyethylene by mass loss in compost. This was carried out on solvent extracted polymers to remove antioxidants and PP lost over 60% mass in 6 months whereas LDPE lost about 10% in the same time. Ethylene-propylene co-polymers biodegraded at rates intermediate between PP and PE. As expected, prior UV irradiation (photooxidation) increased both the rate and extent of the bioassimilation. This is fully in accord with the rates of environmental peroxidation of these molecules [61] and it has been shown [62] that PP acts as a sensitiser for the peroxidation of LDPE.

4. Antioxidant control of peroxidation and biooxidation

4.1 MECHANISMS OF ANTIOXIDANT ACTION

Antioxidant is the general scientific description used for the inhibition of polymer peroxidation in both abiotic and biotic systems [63]. It embraces more specific technological terms such as antidegradants (thermoantioxidants), antifatigue agents (mechanooxidation inhibitors) and light stabilisers (photoantioxidants). It was seen in Section 2 that peroxidation of carbon-chain polymers occurs by a free radical chain reaction (reactions 4 and 5) and the intrinsic initiators are hydroperoxides that dissociate to radicals under the influence of heat and light.

The chemical mechanisms involved in the action of antioxidants have been discussed in standard texts [9,63-66] and the reader is directed to these and the references they contain for more detailed information. Two complementary antioxidant mechanisms are frequently used synergistically in polyolefins. The first is the kinetic chain-breaking hydrogen donor, (CB-D) mechanism in which chain-propagating peroxyl radicals ($POO^{\cdot}$) are preferentially reduced to hydroperoxide by the antioxidant (AH).

$$POO^{\cdot} + AH \rightarrow POOH + A^{\cdot} \qquad (10)$$

The relatively stable radicals ($A^{\cdot}$) produced (e.g. phenoxyl from phenols and amin-oxyl from aromatic amines) cannot continue the kinetic chain and disappear from the system by coupling with other or the same free radicals. It should be noted that this process is stoichiometric and hydroperoxides are produced in each inhibiting step (reaction 10).

The second antioxidant mechanism, the peroxidolytic or peroxide decomposing process (PD) removes the hydroperoxides (POOH) that are the main source of initiating radicals by reactions that do not produce free radicals.

$$POOH \rightarrow \text{Non-radical products} \qquad (11)$$

CB-D and PD antioxidants thus have complementary functions and when used together they reinforce one another (i.e. synergise).

The metal dithiocarbamates (MDRC) are an important group of catalytic PD antioxidants that have been used in rubbers and some plastics [63,67] for many years. Some are very effective light stabilisers (photoantioxidants) due to their UV stability. In particular the Ni and Co and Cu complexes slowly release sulphur acids which destroy hydroperoxides over a long period [63,67] FeDRC and MnDRC are effective thermal antioxidants but are rapidly thermolysed in compost or photolysed in sunlight to give the corresponding transition metal ion, which then behaves as prooxidants by reactions 6-9. The effectiveness of the MRDCs as light stabilisers depends crucially on the photo-stability of the metal complexes [47,65,67-69].

```
        S     S
      //  \  /  \
R2NC     M      CNR2       MDRC
      \  /  \  //
        S     S
```

M = Zn, Ca, etc., thermal antioxidant
M = Ni, Co, Cu, photoantioxidant
M = Fe, Mn, Ce, thermal antioxidants; photo-prooxidants

4.2 THE REQUIREMENT FOR ANTIOXIDANTS IN DEGRADABLE POLYMERS

Many applications of degradable polymers require stability against the environment from weeks, as in the case of much retail packaging, to many months or even years for specialised applications such as protective films in agriculture. The applications of degradable polyolefins will be discussed in more detail in later chapters, but it is important to note here that the oxo-biodegradable polyolefins, because of their ability to be stabilised against the environment have inherent advantages over polymers that are randomly attacked by microorganisms. The behaviour of the ideal degradable polymer is shown in Fig. 3 [65].

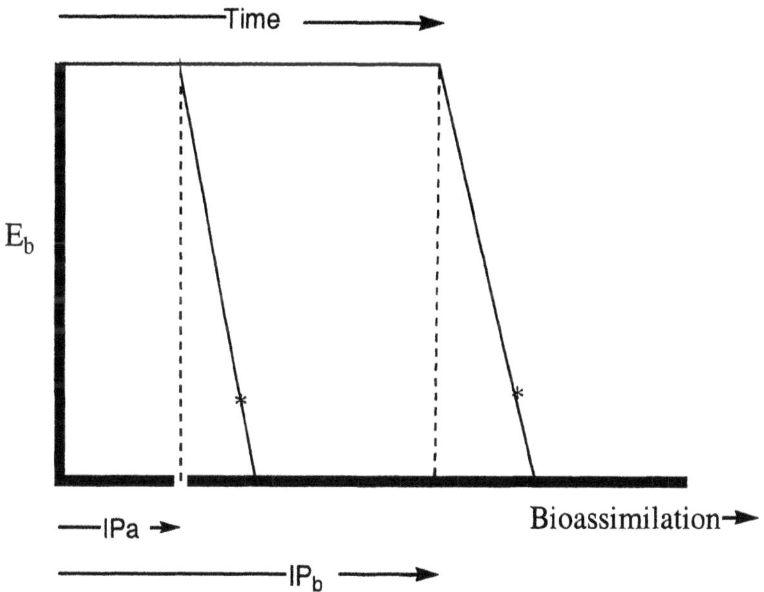

Fig. 3. Ideal degradable polymer [65].
$IP_{a,b}$ are the induction periods of two different formulations of the same polymer, during which no peroxidation or change of properties occurs.

Typical prooxidant transition metal compounds (e.g. iron, cobalt or manganese stearates) are used commercially to induce peroxidation in degradable plastics. However, such prooxidants alone have no practical utility in commercial products unless the prooxidants are deactivated during polymer fabrication, since oxidative degradation begins during processing [12], resulting in unserviceable products, particularly when exposed to the environment. Figure 4 shows the behaviour of a commercial degradable polyethylene film used in landfill covers and packaging applications (EPI's TDPA™) at composting temperatures and in a weatherometer. It is clear that at ambient temperatures, the shelf life of the polymer is adequate for the intended purpose, whereas at composting temperatures 60-70°C or in landfill, the polymer peroxidises rapidly as measured by the formation of carboxylic acids. Commercial hindered phenol antioxidants are relatively weak inhibitors of peroxidation under conditions where peroxyl radicals are rapidly generated (e.g. in the presence of prooxidant transition metal ions or in UV light and the induction time is relatively short under these conditions [68]. However, if the objective outlined in Fig. 3 is to be achieved, the induction time must be controllable [69-72]. This is especially true for agricultural applications of degradable plastics such a mulching films where the economics of the use of degradable plastics depend on retaining the film intact until just before harvest [63,64] followed by a rapid loss of properties so that the film disintegrates and can then be ploughed into the soil where it continues to degrade in the presence of oxygen, catalysed by transition metal ions and peroxidase enzymes (Section 3.5) associated oxidation products (C=O FTIR absorbance at 1715 cm^{-1}).

The transition metal dithiolates, of which the dithiocarbamates (MRDC) above are typical, combine antioxidant and prooxidant functions in the same molecule [18,65,69,70,73-76]. The MRDCs act as processing stabilisers during manufacture and as antioxidants during storage but they are very sensitive to light and rapidly 'invert' from UV stabilisers to prooxidants in the outdoor environment. By varying the concentration of the additive, both the induction time and the rate of loss of mechanical properties at the end of the induction time (figure 3) can be varied by the use of combinations of MRDC. This is the basis of the Scott-Gilead (S-G) photodegradable polyolefins, which are particularly useful in agricultural applications where relatively long induction times are often required.

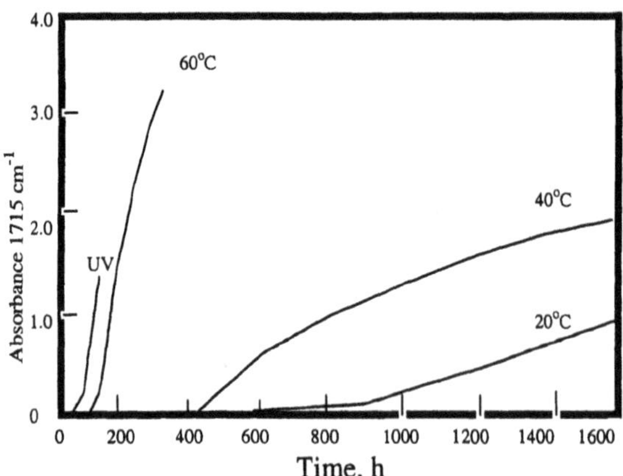

Fig. 4. Formation of oxidation products in degradable PE (TDPA™) during exposure to heat (20-60°C) and SEPAP accelerated weathering (UV) procedures.
I am grateful to Professor Jacques Lemaire and his co-workers for permission to publish this previously unpublished data.

Most commercial light stabilisers are not transition metal ion complexing agents. They are not intended to give a sharp end to the peroxidation induction period in order to avoid disastrous failure of polymer components. Thus for example, the UV absorbers (hydroxybenzophenones and hydroxybenzotriazoles) are destroyed relatively slowly by light. The hindered amine light stabilisers (HALS) are catalytic CB antioxidants [10,77] are depleted slowly in hydrocarbon polymers and do not show the required rapid 'inversion' behaviour characterisitic of the transition metal thiolates. The applications of S-G and TDPA™ technology in packaging and agriculture will be discussed in Chapter 14.

5 Conclusions

The above discussion allows us to conclude that the bioassimilation of synthetic carbon-chain polymers has much in common with that of their natural analogues (notably natural rubber, resins and lignin). In all cases, nature uses abiotic oxidation chemistry together with biotic chemistry, very often together. The degradation products formed by oxo-biodegradation are of benefit to the agricultural environment as biomass and ultimately in the form of humus.

Carbon is retained in the soil during oxo-biodegradation in a form accessible to growing plants, rather than by being eliminated to the environment as carbon dioxide as is the case with hydro-biodegradable polymers (e.g. pure cellulose and starch). The time-scale for complete oxo-bioassimilation of the synthetic polyolefins is very similar to that for their natural analogues such as *cis*(polyisoprene) and related plant exudates and lignocellulose, the structural material of plants..

Time control of biodegradation of the synthetic carbon-chain polymers is achieved by antioxidants that behave similarly to naturally occurring antioxidants present in lignin and tannin. The abiotic processes that lead to the peroxidation of the hydrocarbon polymers involve prooxidant transition metal ions that are analogous to the oxygenase and peroxidase enzymes used by nature to oxo-biodegrade rubber and lignin. Consequently the oxidation products (notably the carboxylic acids) are rapidly absorbed as nutrients by biological organisms.

Acknowledgements

I am indebted to Professor Emo Chiellini, Dr. Graham Swift and their co-workers for permission to reproduce Fig. 2 and to my collaborators, Professor Jacques Lemaire and Dr. Anne-Marie Delort and their co-workers for permission to publish Figs. 1 and 4. I also thank Dr.A.Ikram, Dr. R.P.Singhe and Professor A.Steinbűchel for helpful discussions.

References

1. G.Scott, (1999) *Polymers and the Environment*, Royal Society of Chemistry, Chapter 1.
2. Moureau, C. and Dufraisse, C. (1922) *Compte Rend..*, **174**, 258.
3. Moureau, C. and Dufraisse (1926-7) *Chem. Rev.*, **3**, 113.

4. Lowry, C.D., Egloff, G., Morrell, J.C. and Dryer C.G. (1933).*Ind. Eng. Chem.*,**25**, 804.
5. Bolland, J.L. (1949), *Quart. Rev.* **3**, 1.
6. Batemann,L. (1954), Quart. Rev., **8**, 147.
7. Bevilaqua, E.M. (1957) *Rubb. Chem Tecnol.*, **30**, 667.
8. Scott, G (1965) *Atmospheric Oxidation and Antioxidants*, Elsevier, Chapters 2,4 and 8
9. Scott, G. (1999) *Polymers and the Environment*, Royal Society of Chemistry, Chapter 3
10. Scott, G. (1993) *Atmospheric Oxidation and Antioxidants, 2nd Edition*, Vol. II, Ed. G.Scott, Elsevier, Chapter 8.
11. Grassie, N. and Scott, G. (1985) *Polymer Degradation and Stabilisation*, Cambridge University Press, Chapter 4.
12. Scott, G. (1993) *Atmospheric Oxidation and Antioxidants, 2nd Edition*, Vol. II, Ed. G.Scott, Elsevier, Chapter 3
13. Scott, G. (1995) in *Degradable Polymers: Principles and Applications*, 1st Edition, Eds. G.Scott and D.Gilead, Chapman & Hall, Chapter 1.
14. Al-Malaika, S. (1993) *Atmospheric Oxidation and Antioxidants, 2nd Edition*, Vol. I, Ed. G.Scott, Elsevier, Chapter 2.
15. *Advances in Polyolefins*, (1987), Eds. R.B.Seymore and T.Cheng, Plenum Press.
16. Osawa,Z. (1993) in *Atmospheric Oxidation and Antioxidants*, 2nd Edition, Vol. II, Ed. G.Scott, p.338.
17. Rasti, F and Scott, G, (1980) *Europ. Polym. J.*, **16**, 1153.
18. Amin, M.U. and Scott, G. (1974) *Europ. Polym. J.*, **10**, 1019-1028
19. Arnaud, R., Dabin, P., Lemaire, J., Al-Malaika, S., Chohan, S., Coker, M., , Scott, G., Fauve, A. and Maaroufi, A. (1994) *Polym. Deg. Stab.*, **46**, 211-224.
20. Albertsson, A-C, Barenstedt, C. Karlsson, S., and Lindberg T., (1995) *Polymer,* **36**, 3075-3083
21. Shaposnikov, V.N., Rabotnova, L.L., Yarmola, G.A., Kutznetsova, V.M. and Mozokhina-Porshnyakova (1952), *Microbiologiya*, **21**, 146-154.
22. Low, F.C., Tan, A.M. and John, C.K. (1992) *J.Nat. Rubb.Res.*, **7**, 195-205.
23. Tsuchii, A, Suzuki, T. and Takeda, K. (1985) *App. Environ. Microbial.* **50**, 965-970.
24. Kajikawa, S., Tsuchii, A. and Takeda, K. (1991) *Nippon Nogeikagaku Kaishi*, **65**, 981-986.
25. Ikram, A., Alias, O. and Napi, D. (2000) J. Rubb. Res., **3**, 104-114.
26. Heisey, R.M. and Papadatos,, S. (1995) *App. Environ. Microbiol.* **61**, 3092-3097.
27. Ikram, A., Alias, O., Bahri, A.R.S., Fauzi, M.S. and Napi, D. (2001) *J .Rubb, Res.*, **4**, 102-117.
28. Linos, A., Berekaa, M.M., Reichelt, R., Keller, U., Schmidt, J., Flemming, H-C., Kroppenstedt R.M. and Steinbűchel, A. (2000) *Appl. Environ. Microbiol.*, **66**. 1639-1645.
29. Linos, A., Reichelt, R., Keller, U. and Steinbűchel, A. (2000) *FEMS Microbiol. Lett.*, **182**, 155-161.
30. Berekaa, M.M., Linosa, R., Reichelt, R., Keller, U., and Steinbűchel, A. (2000) *FEMS Microbiology Lett.*, **184**, 199-206.
31. Sadun, A.G., Webster, T.F. and Commoner, B. (1990) *Breaking down the degradable plastics scam,* Greenpeace, Washington D.C.
32. Lechevalier, M.P., Prauser, H.. Labeda, D.P. and Ruan, J-S. (1986) *Int. J. Syst. Bacteriol.*, **36**, 29-37.
33. Adler. G. (1977) *Wood Sci. Technol.* **11**, 169-217
34. Scott, G., (1999) *Polymers and the Environment*, Royal Society of Chemistry, pp. 2-4.

35. Haug, R.T. (1993) *The practical handbook of compost engineering*, Lewis publishers, Boca Raton.
36. Wehmer, C, *Ber.* **48**, 130-134.
37. Hammouda, G.H.H. and Adams, W.A. (1989) in *Compost: Production, quality and use*, Eds. M. De Bertoldi, M.P. Ferranti, P. L'Hermite and F. Zucconi, Elsevier App. Sci., pp 245-253.
38. Horwath, W.R., Elliot, L.F. and Churchill, D.B, (1995) *Compost Science and Utilization*, **3**, 22-30.
39. Jansson, S.L. (1963) in *The use of isotopes in soil organic matter studies*, Report of the FAO/IAEA Technical Meeting, September 9-14, Pergamon Press, Oxford.
40. Adhi, T.P., Korus, R. A, and Crawford, D.L. (1989) *Appl. Environ. Microbiol*, **55**, 1165-1168.
41. Adhi, T.P., Korus, R.A., Pometto, A.L. and Crawford, D.L. (1987), *Appl. Biochem. Biotechnol.* **18**, 291-301
42. Ramachandra, M., Crawford, D.L. and Hertel, G. (1988) *Appl. Environ. Microbiol.* **54**, 3057-3063.
43. Hirai, H., Kondo, R. and Sakai, K. (1994) *Mokzua* Gakkaishii, **40** 980-986.
44. Katagiri, N., Tsutsumi, Y. and Nishida, T. (1995) *Appl. Environm Microbiol.*, **61**, 617-622.
45. Paice, M.G., Reid, I.D.,Bourbonnais, R., Archibald, F.S. and Jurasek, L. (1993) *Appl. Environm. Microbiol.* **59**, 260-265.
46. Scott, G. (1965) *Atmospheric Oxidation and Antioxidants*, Elsevier, 125-132.
47. S.A.Waksman, *Humus*, Baillièri, Tindal and Cox, 1938, p.xii.
48. Metodiewa, D and Dunford, B. (1993) *Atmospheric Oxodation and Antioxidants*, Vol. III, Ed. G, Scott, Elsevier, Chapter 11.
49. J.Guillet in *Degradable Polymers: Principles and Applications*, 1st Edition, Eds. G.Scott and D.Gilead, Chapman and Hall, p.240.
50. Eggins, H.O.W., Mills, J., Holt, A. and Scott, G. (1971) *Microbial Aspects of Pollution*, Eds. G.Sykes and F.A.Skinner, Academic Press, 270-277.
51. Lee, B., Poletto, A.L., Fratzke, A. and Bailey, T.B. (1991) *Am. Soc.Mmicrobiol.*, **57**, 678-685.
52. Harlan, G. and Kmiec, C. (1995) in *Degradable Polymers: Principles and Applications, First Edition, Eds.*, G.Scott and D.Gilead, Chapman & Hall, Chapter 8.
53. Griffin, G.J.L. (1994) in *Chemistry and Technology of Biodegradable Polymers*, Ed.. G.J.L. Griffin, Blackie Academic & Professional, Chapter 3
54. Scott,G. (1995) in *DegradablePolymers: Principles and Applications*, *First Edition*, Eds. G.Scott and D.Gilead, Chapman and Hall, Chapter 9.
55. Delort, A-M., Personal communication.
56. Wool, R.P. (1995) in *Degradable Polymers: Principles and Applications*, *First Edition*, Eds. G.Scott and D.Gilead, Chapman & Hall, Chapter 7.
57. Weiland, M., Daro, A. and David, C. (1995) *Polym. Deg. Stab.*, **48**, 275-289.
58. Chiellini, E., Personal communication.
59. Potts, J.E., Clendinning, R.A., Ackart, W.B. and Niegiscg, W. D. (1976) in *Polymers and Ecological Problems*, Ed. J.E. Guillet, Plenum Press,
60. Pandey, J.K. and Singh, R.P. (2001), *Biomacromolecules*, **2**, 880-885.
61. Scott, G. (1993) in *Atmospheric Oxidation and Antioxidants*, Vol.II, Elsevier, p. 387-389.
62. Sadrmohaghegh, C., Scott, G. and Setudeh, E. (1980) *Polym. Deg. Stab.*, **3**, 469.

63. Scott, G. (1997) *Antioxidants in science, technology, medicine and nutrition,* Albion Chemical Science Series, Chapters 3 and 4.
64. Scott, G. (1993) in *Atmospheric Oxidation and Antioxidants, Vol. I*, Ed. G. Scott, Elsevier, Chapter 1
65. Scott, G. (1999) *Polymers and the Environment*, Royal Society of Chemistry, Chapter 5.
66. Al-Malaika, S., Chakraborty, K.B. and Scott, G. (1983) in *Developments in Polymer Stabilisation-6,* Ed. G.Scott, App. Sci. Pub., Chapter 3.
67. Al-Malaika, S, Chakraborty, K.B. and Scott, G. (1983) in *Developments in Polymer Stabilisation-6*, Ed. G.Scott, Appl. Sci. Publishers, Chapter 3.
68. Scott. G. (1999) in *Degradability, Renewability and Recycling – Key functions for future materials*, Eds., A-C. Albertsson, E. Chiellini, J. Feijen, G. Scott and M. Vert, Macromolecular Symposia, **144**, Wiley-VCH.
69. Gilead, D. and Scott, G. in *Development in Polymer Stabilisation-5,* Eds. G.Scott and D. Gilead, App. Sci. Pub.,, Chapter 4.
70. Gilead, D. (1995) in *Degradable Polymers: Principles and Applications*, 1st Edition, Eds. G.Scott and D. Gilead, Chapman & Hall, Chapter 10.
71. Fabbri, A.. (1995) in *Degradable Polymers: Principles and Applications*, 1st Edition, Eds. G.Scott and D. Gilead, Chapman & Hall, Chapter 10.
72. Scott, G. (1997), *Trends in Polymer Science,***5**, 361-368.
73. Scott, G. (2000) in Proceedings of the 6th Arab Conference on Materials Science, *J. Appl. Polym Sci., in press.*
74. Scott, G. (1990), *Polym. Deg. Stab.,* **29**, 135-154.
75. Al-.Malaika, S., Marogi, A. and Scott, G. (1986) *J. Appl. Polym. Sci.,* **31**, 685.
76. Al-Malaika, S., Marogi, A. and Scott, G. (1987) *J. Appl. Polym.. Sci.,***33**, 1445.
77. Scott, G. (1993) in *Atmospheric Oxidation and Antioxidants*, 2nd Edition, Vol. I,Ed. G.Scott, Elsevier, Chapter 4

4

TECHNIQUES AND MECHANISMS OF POLYMER DEGRADATION

SIGBRITT KARLSSON AND ANN-CHRISTINE ALBERTSSON
Department of Polymer Technology
The Royal Institute of Technology (KTH)
SE-100 44 STOCKHOLM, Sweden.

1 Introduction

The lifetime of a polymer is dependent not only on the weak links in the material but also on its surroundings. Degradation must be carefully monitored in order to evaluate the usefulness and service-life of polymers in different applications. As environmental concerns have increased, new demands are placed on the prediction of long-term properties of synthetic and natural polymers. When polymeric materials are exposed to a complex environment (*e.g.* outdoors) we may describe this as environmental degradation. Environmental degradation occurs due to a combination of factors of which photo-oxidation, thermo-oxidation, humidity, erosion by the weather and chemical action due to pollutants and micro- and macro-organisms are the most important. A combination of these may be cumulative, synergistic or antagonistic.

Reliable techniques are required to predict the susceptibility of polymers to degradation. The analyses should be able to monitor chemical, physical and mechanical changes brought about during processing, use and waste disposal. Degradation tests may be long term but usually accelerated tests are more convenient. It is also important to monitor how the additives diffuse and/or migrate in order to relate the formation of degradation products to the degradation rate.

2 Life-time of inert and degradable polymers

In principle, all organic polymers are degradable, differing only in degradation mode and time. In practice, nearly all bulk polymers are inert materials either by themselves or by the addition of compounds rendering them insensitive towards ageing. Interest in

G. Scott (ed.), Degradable Polymers, 2nd Edition, 51-69.

environmental issues is growing and the demand for new materials with low environmental burden is high. The design of materials which continue to have strength and functionality while in service, but which degrade after use is a rather novel concept in the material sciences area. The new concept of "degradability" is now often demanded even at the development stage of new plastics products. At the same time, an additional demand is now widely proposed that polymeric materials should use renewable resources instead of petrochemical sources. A third requirement is to save materials by improved product design and by recycling. The integrated product policy (IPP) [information in *e.g.* *environ*REPORT, No. 2 January 2000] discussed and proposed within EU inflicts new strategies for manufacturer of polymers and plastic products. IPP states the all products should be materials- and energy-efficient, contain no substances that could harm health or the environment, have minimal effects on the environment in general. In addition, the products should be manufactured wherever possible from renewable materials. They should also be re-useable or recycable, or a source of energy when the products reach their useful life. The products should also give rise to less waste containing smaller amounts of substances harmful to health or the environment (EU Commission [information in *e.g.* *environ*REPORT, No. 2 January 2000]).

The development of a range of biodegradable polymer products, with a predetermined lifetime, are of interest in at least four main fields. These are 1) packaging materials, 2) mulching films in agriculture, 3) disposable items in different expendable items (table ware such as cups, spoons *etc.*), and in 4) medical applications, where an *in vivo* degradation must occur through natural metabolisms in the human or animal body.

Several interesting solutions have been proposed to make polyethylene degrade faster. Normally commodity plastics may take 25-50 years to biodegrade in the environment.This can be accelerated by the incorporation of chromophoric groups that increase the susceptibility of polyethylene towards photo-oxidation. Metal salts are useful and potent additives that give rise to radical formation within the polymer chains, which again increase the rate of photooxidation of these materials. The knowledge that biopolymers usually degrade quickly by microbiological action has led to the use of agricultural products as additives for synthetic polymers with the intention of producing biodegradable polyethylene. Starch is the most commonly used biopolymer in this context.

Polymers with hydrolysable linkages in the backbone are very useful in a range of degradable materials. For disposable table-wares as cups or expendable packages many of them are still too expensive and do not exhibit the desirable combination of mechanical and chemical properties. Well-known synthetic hydrolysable polymers are polyesters [1], polycarbonates [2], polyanhydrides [2], polyamides [2] and poly(amino acids) [2]. Hydrolysable biopolymers may be cheaper than synthetically produced polymers (*e.g.* aliphatic polyesters such as polylactides) and many scientists today are looking for new possibilities using such traditional natural polymers as polysaccharides, proteins and lipids. Special interest is focused on poly(β-hydroxybutyrate) and its copolymers [3,4] (see Chapters 9 and 10). Well-known natural products such as Pullulan (a bacterial polysaccharide produced by *Aerobasidium pullulans*), cellulose acetate and starch, as well as synthetic polyvinyl alcohol are important degradable materials.

3 Degradation mechanisms

Several of the more common commodity polymers like the polyolefins are susceptible to photo-oxidation. For a polymer like polyethylene, photo-oxidation leads to increasing amounts of carbonyl compounds. In-chain ketone groups act as sensitisers by UV light absorption. Through the well-known Norrish type I and II degradations; radicals, end-vinyl and ketone groups are formed. Other products often observed in photo-oxidised low-density polyethylene (LDPE) are esters [5]. Scheme 1 shows one mechanism for abiotic ester formation. By Norrish type I cleavage the radical formed can react with an alkoxyl radical on the polyethylene (PE) chain.

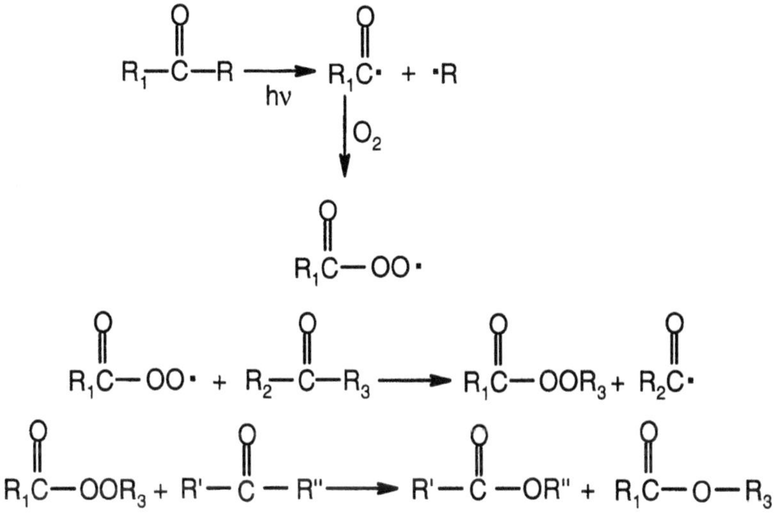

$$R_1-\overset{O}{\overset{\|}{C}}-R \xrightarrow[h\nu]{} R_1\overset{O}{\overset{\|}{C}}\cdot + \cdot R$$

$$R_1\overset{O}{\overset{\|}{C}}\cdot \xrightarrow{O_2} R_1\overset{O}{\overset{\|}{C}}-OO\cdot$$

$$R_1\overset{O}{\overset{\|}{C}}-OO\cdot + R_2-\overset{O}{\overset{\|}{C}}-R_3 \longrightarrow R_1\overset{O}{\overset{\|}{C}}-OOR_3 + R_2\overset{O}{\overset{\|}{C}}\cdot$$

$$R_1\overset{O}{\overset{\|}{C}}-OOR_3 + R'-\overset{O}{\overset{\|}{C}}-R'' \longrightarrow R'-\overset{O}{\overset{\|}{C}}-OR'' + R_1\overset{O}{\overset{\|}{C}}-O-R_3$$

Scheme 1. Abiotic ester formation in polyethylene
(see also G. Scott, Polymer Age 6 (3), 54 (1975), Biological Recycling of Polymers)

This mechanism can be compared with other proposed mechanisms for ester formation in biological systems (Scheme 2), in which several enzymes and coenzymes in concerted reactions rearrange the straight hydrocarbon chain to the ester compound. The mechanism were supported by isolation of cetylpalmitate from *Acinetobacter calcoaceticus* culture medium.

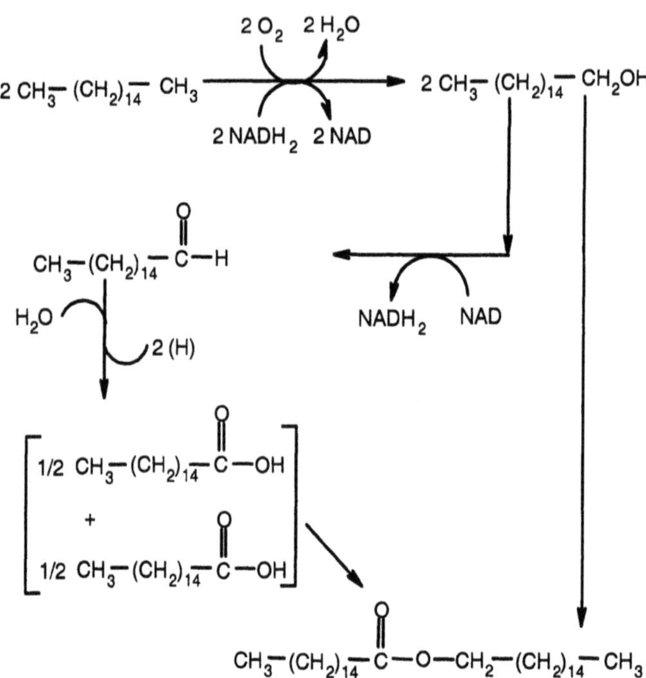

Scheme 2. Biotic ester formation (cetylpalmitate) by Acinetobacter calcoaceticus grown on hexadecane according to H.G. Schlegel, Allgemeine Mikrobiologie, 4th. ed., Georg Thiem Verlag, Stuttgart, 1976, p. 356.

By IR-spectroscopy a decrease was observed in the keto-carbonyl groups (ester groups at 1740 cm^{-1}, aldehyde groups at 1720 cm^{-1} and ketone groups at 1715 cm^{-1}) during microbial assimilation (*i.e.* biotic step) as a result of release of short-chain carboxylic as degradation products [5-8].

Thermo-oxidation increases the degradation rates considerably. During this type of degradation both molecular reduction and molecular enlargement reactions occur [9]. In LDPE with pro-oxidant and starch it was shown that oxidation at 100°C initiates carbonyl formation after 5 days compared to pure LDPE and LDPE with only starch which are unaffected under the same conditions [10].

Hydrolysis readily diminishes the lifetime of synthetic and natural polyesters. Scheme 3 shows the alkaline hydrolysis of poly(β-hydroxybutyrate) (PHB), the natural polyester belonging to the group polyhydroxyalkanoates (PHA), produced by micro-organisms which has application in above all medical materials [4]. The degradation product is often β-hydroxybutyric acid and when this reaction occurs in the human system, predominantly in the mitochondria, the compound is sometimes referred to as a ketone body (medical term, which is not related to the chemical term ketone). β-hydroxybutyrate is, together with other compounds, the fuel of respiration and a quantitatively important source of energy for micro-organisms.

Scheme 3 Mechanism of hydrolysis of poly(β-hydroxybutyrate)

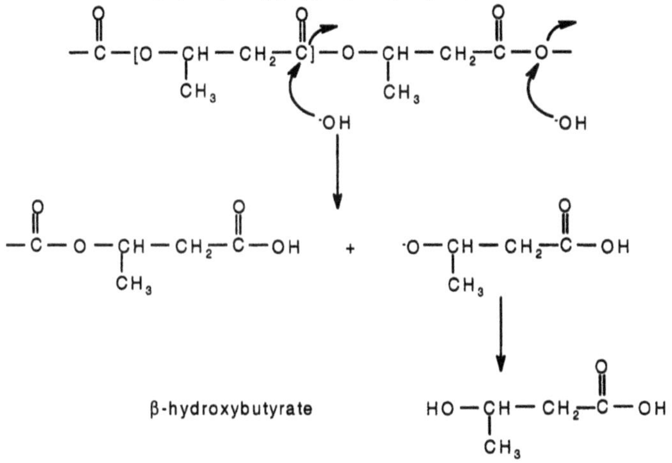

β-hydroxybutyrate

Scheme 4 shows another kind of biological hydrolysis, the putrefactive degradation of casein [11]. The later reaction is also a hydrolysis, but in contrast to the hydrolysis of PHB the resulting degradation products (amino acids) are used in the anabolic cycle forming new proteins.

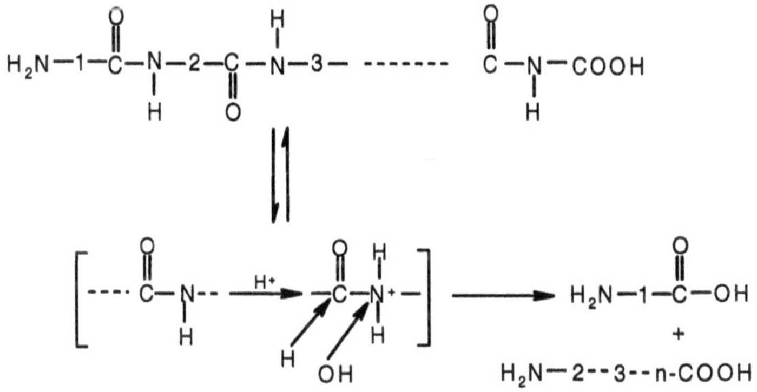

Scheme 4. Enzymatic hydrolysis of casein

4 Polymer analyses and characterisation

Scheme 5 demonstrates the physical, chemical and mechanical analyses which give information on functional group changes during degradation, Fourier Transform Infrared Spectroscopy(FTIR), changes in crystal dimensions, X-Ray Diffraction (XRD), stress-strain properties by Instron, or Dynamic Mechanical Analysis (DMA), thermal properties such as Tg, degree of crystallinity Differential Scanning Calorimetry (DSC) and DMA and morphology Scanning Electron Microscopy (SEM) Transmission Electron Microscopy (TEM). *Scheme 6* instead summaries the methods for obtaining results regarding changes in the polymeric chains as molar masses, Size Exclusion Chromatography (SEC) and Matrix-Assisted Laser Desorption/Ionisation (MALDI) and/or formation of low molecular weight compounds, Gas Chromatography (GC), Liquid Chromatography (LC), GC coupled to Mass Spectrometry (MS). The latter requires usually specialised extractions which may be traditional as Soxhlet or liquid-liquid extraction but more useful are Solid-Phase Extraction (SPE) or Solid-Phase MicroExtraction (SPME) which both allows the sampling of low levels of low molecular weight compounds. Other extraction techniques require supercritical gases such as CO_2 in Supercritical Fluid Extraction (SFE), ultrasonication as in ultrasonication extraction, or microwaves as in Microwave Assisted Extraction (MAE).

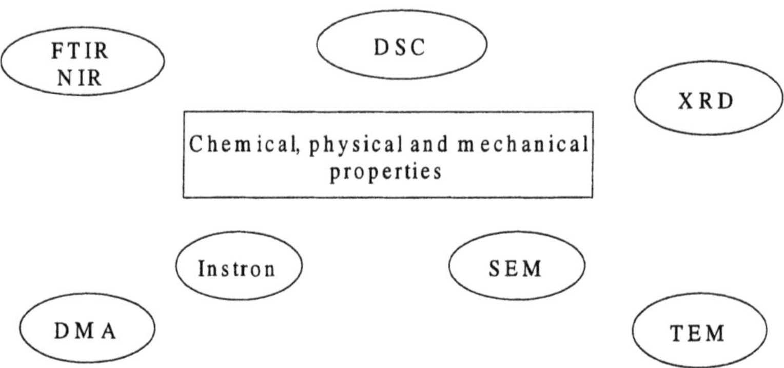

Scheme 5. Physical, chemical and mechanical characterisation techniques useful in polymer degradation studies

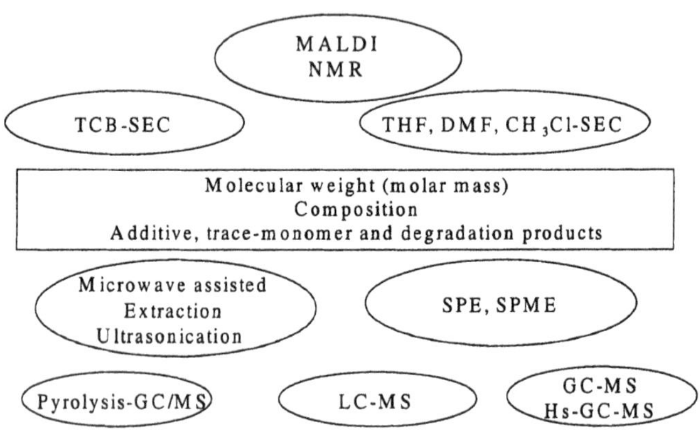

Scheme 6. Analysis techniques to obtain detailed understanding of changes in molar mass, composition and identification of low molecular weight compounds.

As a measure of oxidation in polyolefins, increasing amounts of carbonyl compounds absorbing in the region of 1640-1740 cm^{-1} can be observed. *Figure 1* shows one such example where the susceptibility to biodegradation of LDPE, containing a masterbatch of corn starch, linear LDPE and styrene butadiene copolymer, is compared with pure LDPE. It was shown that this masterbatch renders LDPE more susceptible to biodegradation compared to pure LDPE [12]. Besides change in the functional groups of the repeating chains also the crystal dimensions may also be changed during degradation. Melt-extruded poly(hydroxybutyrate-co-hydroxyvalerate) (PHB/HV) demonstrated causes changes in the atomic positioning of the oxygen atoms in the helix structure due to torsion of the main chain bonds when the processing temperature and screw speed were altered [13].

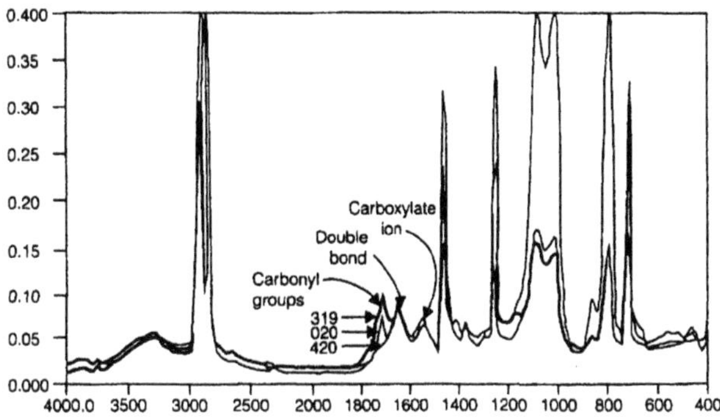

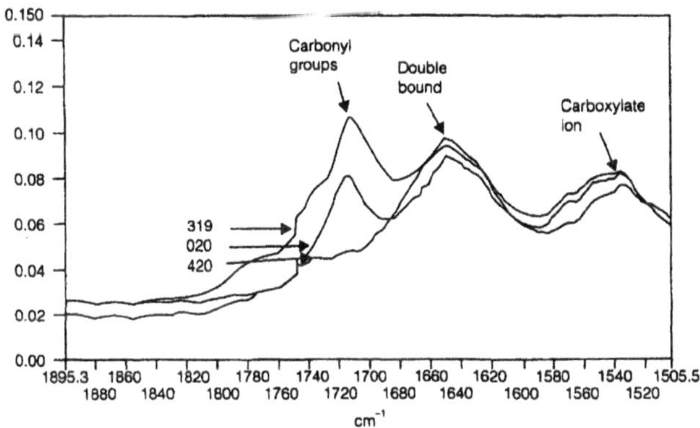

Fig. 1. ATR-FTIR of LDPE containing 15% masterbatch of a corn starch, linear LDPE, styrene-butadiene copolymer and Maganese Stearate. Samples were heated at 100°C for 6 days before biodegradation (319 and 020) and some were incubated without initial heating (420). Biodegradation was performed in a mixture of fungi at 20°C [12].

When degradation is caused by hydrolysis (chemical and/or biological) weight losses are generally reported. This is of course very easy to measure but gives only an initial rough guide to the percentage degradation. If weight loss data is coupled with SEC, it is possible to distinguish between surface erosion accompanied by a significant weight loss, as was found in biologically hydrolysed polycaprolactone (PCL), and random chain scission with little surface erosion and minor weight loss as for chemically hydrolysed PHB/HV. Biodegraded PHB/HV instead demonstrates significant weight loss with no or minor change in molecular weight. Weight loss data should, however, be used with caution as a measure of degradation, since we may have severe degradation resulting in rapidly decreasing properties without significant weight loss.

Morphology and crystallinity changes, melting temperature (T_m) and glass transition temperature (T_g) values are also helpful when describing the changes in long-term properties. Scanning electron microscopy (SEM) and differential scanning calorimetry (DSC) are techniques, which give rather straightforward answers although they are not very easy to interpret. The first heating should always be used when performing thermal analysis by DSC of degraded polymers, as the first heating reflects the undisturbed polymer. Often it is possible to follow the preferential ageing of the amorphous regions leaving the crystalline regions rather unaffected, at least during the early stages of degradation. Morphology changes were demonstrated which differentiated between physical/chemical (abiotic) and biological (biotic) degradation of LDPE with starch and pro-oxidant [14]. A decrease in lamellar thickness (*l*) was demonstrated for biotically (fungal) degraded samples as compared with the abiotically aged ones, which showed constant or increasing value of *l*. The degree of crystallinity showed that prolonged exposure to fungi resulted in a decreasing value as compared to the increasing value in abiotically exposed samples [14].

Using liquid scintillation counting, (which analysis the $^{14}CO_2$ emitted from a polymer during degradation), it is possible to follow small changes in the extent of abiotic degradation even in inert polymers such as PE [15]. The degradation curve is characterised by a straight-line progression in the first 100 days of observation before declining [16]. It was demonstrated that initially preheated LDPE-starch-pro-oxidant had a ten times higher rate of degradation in biological environments as compared with the similarly preheated pure LDPE [12].

4.1 MOLECULAR WEIGHT (MOLAR MASS) CHANGES BY SEC AND MALDI

Molecular weight changes give necessary understanding of whether. the degradation mode is random chain scission which results in very slow changes in M_n and M_w or if the samples are degraded from chain ends leading to rapid change in M_n/M_w. During thermo-oxidation of environmentally adaptable polyethylene, a stabilising effect of starch added together with a pro-oxidant system in LDPE was observed during monitoring of molecular weight changes (M_n , M_w and M_z) [17]. During thermo-oxidation multi-modality developed and the polydispersity index reached values higher than two, which generally indicates random (internal) chain scission. By SEC, multi-modality was also observed in another study where PCL were degraded at different temperatures by chemical hydrolysis and biological hydrolysis [18}. The recently developed matrix-assisted laser desorption/ionisation (MALDI) technique coupled with MS allowed a further understanding of this phenomenon. From MALDI-time-of-flight (TOF) spectra it was observed that the many peaks observed in SEC chromatograms of PCL differed by 10 repeat units from each other, which means that the two, three or four peaks observed in some of the degrading environments had molecular weights which differed by 10 repeating units CL [18]. A comparison of SEC/MALDI and SEC/NMR allowed the observation of multi-modality in SEC-chromatograms and this was related to degradation time and formation of low molecular weight compounds [19]. Thus, for certain polymer samples, MALDI offers yet another means to reveal the changes at molecular level (observe chain ends and changes in the chains) during degradation. Unfortunately it is not applicable to polyolefins, although recently a new method has been developed in which an organic cation is covalently bonded to polyethylene to produce the necessary ionisation for MALDI-TOF [20]. In addition, the synthetic polymers to be analysed should have low polydispersity (<1.1) and a molecular weight <10 000 Da. Otherwise a large discrepancy between SEC-measurements and MALDI is observed. The available MALDI-matrices by which the polymers are mixed limits the technique even further.

4.2 EARLY OXIDATIVE CHANGES IN POLYMERS MEASURED BY CHEMILUMINESCENCE

Degradative changes occurring in unlabelled polymers (*e.g.* in oxidation reactions) can be analysed using chemiluminescence (CL) [21, 22]. This method monitors degradation at a much earlier stage than FTIR or DSC. It is very sensitive (single photon counting) and the intensity of emission is dependent on the extent of oxidation of the polymer. It is generally accepted that the emission comes from the phosphorescence of an excited ketone [23].

Fig.2 shows two examples of CL analysis on aged polymers. CL was recorded from thermally aged samples of starch-pro-oxidant filled LDPE, compared with pure LDPE and LDPE with only starch [10]. The large content of unsaturation (oxidisable sites in the SBS-phase), which is part of the pro-oxidant formulation, is the explanation for the rapid initial oxidation in the filled LDPE, yielding hydroperoxides, compared with pure LDPE.

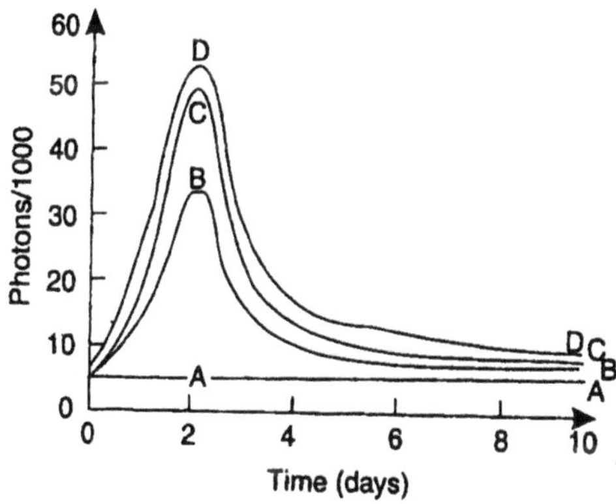

Figure 2. Chemiluminescence, measured in a nitrogen atmosphere of different LDPE samples. Thermo-oxidized (100°C) A = LDPE and LDPE+7.7% starch: B = LDPE + 7.7% MB: C = LDPE + 15% MB and D = LDPE = 20% MB [10].

Hydroperoxide degrades into an alcohol, an excited carbonyl group and a molecule of oxygen. The source of the luminescence is the hydroperoxide decomposition and the intensity reflects the hydroperoxide concentration. Fig. 3 shows how the addition of a surfactant influences the intensity of the emitted light from LDPE films initially irradiated by UV light for 0, 40 and 300 days, respectively, and thereafter mixed in a salt solution containing *Pseudomonas aeruginosa* [24]. The LDPE containing surfactant emits more light than pure LDPE after 490 days of biodegradation. This is an indication of the greater oxidative degradation in LDPE with surfactant aged with *Pseudomonas aeruginosa* compared with the pure LDPE.

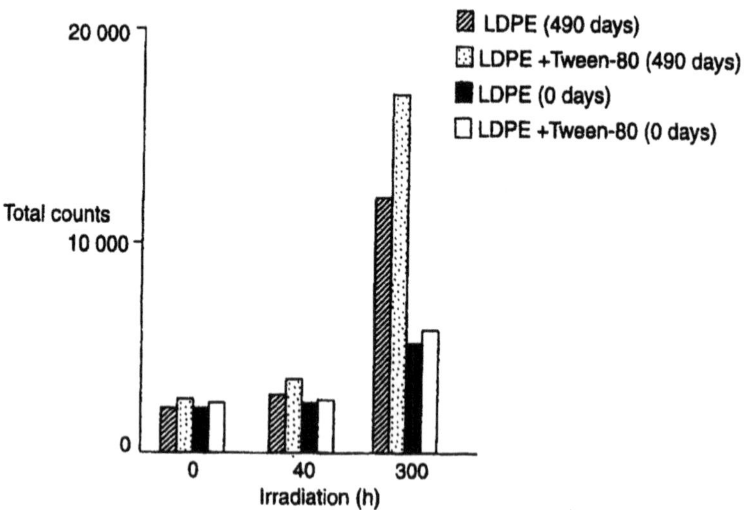

Figure 3. Chemiluminescence measured in a nitrogen atmosphere of biodegraded (Pseudomonas aeruginosa). LDPE with and without Tween-80 biodegraded and undegraded, all samples pre-irradiated for 0, 40 and 300 days before ageing [24].

4.3 CHROMATOGRAPHIC ANALYSES OF THE LOW MOLECULAR WEIGHT COMPOUNDS IN POLYMERS

There are several different types of low molecular weight compounds that can be present in polymers. It is possible to detect traces of monomers, solvents and initiators, different additives and, of course, degradation products. The low molecular weight compounds can interact with the environment and create problems and change the properties of the materials. The long-term properties are affected, e.g. leaching of additives will shorten the lifetime of the plastics and can, of course, damage materials in the vicinity of the polymer. Pharmaceuticals and foods should not contain any traces of low molecular weight compounds from polymers.

There is considerable interest in the possibility of analysing polymers to identify all the possible low molecular weight compounds present. The environmental concern about the use of plastics and the demand for degradable polymers demands specific and skilful analytical techniques to monitor these products. It is necessary to be able to predict what type of degradation products may be formed, for example during recycling or in the human body.

Gas chromatography (GC) and liquid chromatography (LC) together with mass spectroscopy (MS) are the methods most suitable to detect low molecular weight compounds. Often it is necessary to develop special extraction procedures when analysing polymers. The materials are very complex and the degradation environment complicates the analyses even further. Volatile compounds can be analysed using head-space (HS) gases coupled with GC. The method is convenient, as no sample preparation has to be done. The powder or film is placed in a glass vial, which is sealed and the gas above the polymer is subsequently analysed after thermostatting. Fig. 4 demonstrates the versatility of HS-GC which gives the degradation products of thermo-oxidised LDPE containing iron dimethyl dithiocarbamate and carbon black. The material is used in agriculture for mulch films (see Chapter 14). Typical degradation products detected were acetaldehyde, methanol, acetone, butanol among others not identified.

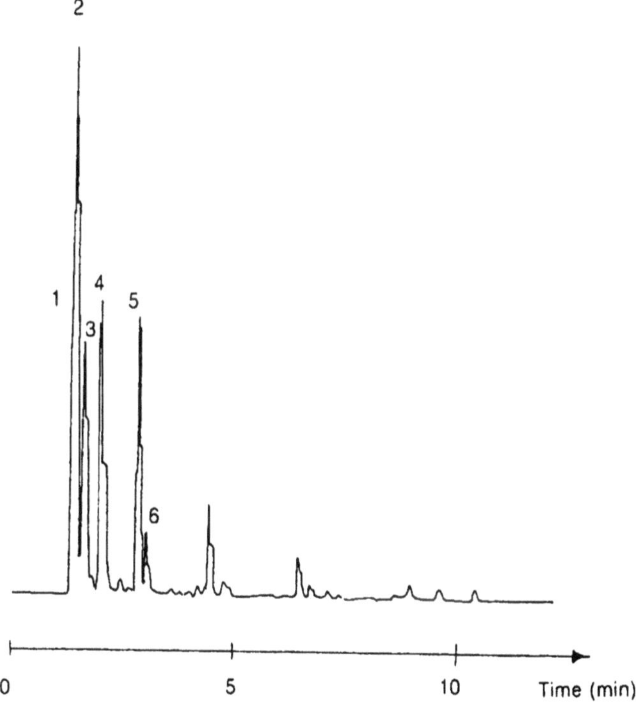

Fig. 4. Volatile degradation products of photo-oxidized LDPE with iron dimethyl dithiocarbamate and carbon black monitored by HS-GC. 1 = acetaldehyde, 2 = methanol, 3 = acetone, 4 = 1-butanol, 5 = butanol, 6 = 3-pentanol.

Fig. 5 presents the degradation products obtained by thermo-oxidation of nylon 66 where products such as cyclopentanone and copper-acetate were formed during the ageing. The copper-acetate originates from additives and the formation of cyclopentanone is described by the mechanism given in Scheme 7. Lately, SPME has provided an even better sampling method for collecting volatile species from polymers [25]. In comparison to classical head-space chromatography, the detection level in SPME is lower and a range of absorbents (from non-polar to polar) are commercially available [25].

It is possible to postulate degradation mechanisms and then verify them by GC or LC coupled with mass spectrometry (MS). Many of the polymers formed through step-wise polymerisation degrade by a simple scission from the chain ends whereby mainly monomers are formed as degradation products. When hydrolysed, poly(1,5-dioxepan-2-one) (PDXO) and copolymers of PDXO with the homopolymer of poly(lactic acid) produce 2-hydroxy-ethoxypropanoic acid (the linear form of DXO) and lactic acid [26, 27]. On the other hand polymers such as polyethylene produce several hundreds of degradation products during degradation [10,28].

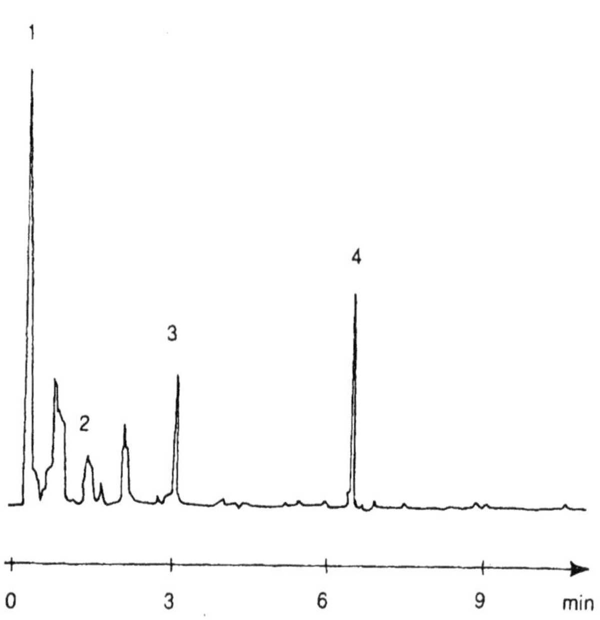

Fig. 5. Volatile degradation products of thermo-oxidised nylon 66 monitored by HS-GC. 1 = acetaldehyde, 2 = Cu-acetate, 3 = cyclopentanone, 4 = aniline

Natural polymers usually produce very complex degradation product patterns, which make the identification of the products difficult. Casein, used in self-levelling concrete, can be attacked by several microorganisms and, when the degradation is performed by the anaerobic *Clostridia,* the mechanisms are denoted as putrefactive degradation; a microbiological term used to describe the anaerobic degradation of proteins forming malodorous amines and carboxylic acids. Several very malodorous compounds are produced as in the case of the "sick building" syndrome during the early 1980s [29]. GC and LC identified volatile organic acids, mono-, di- and polyamines [11,30]. Casein degrades through an enzyme-mediated hydrolysis reaction, producing amino acids as a first step. The amino acids are further deaminated or decarboxylated forming organic acids or amines (see Scheme 4).

The debate about degradable polymers has often focused on the use of natural polymers and it is often believed that all biopolymers are degradable. This is not quite true, even in nature there are materials with very long degradation time, *e.g.* lignin and bone. Nature combines polymers with short degradation times with materials having long degradation times in an energy and material optimised process as in cellulose, hemicellulose and lignin in the wood cell. In wood, three different polymers have vastly different degradation times which when combined constitute a material with excellent mechanical and physical strength.

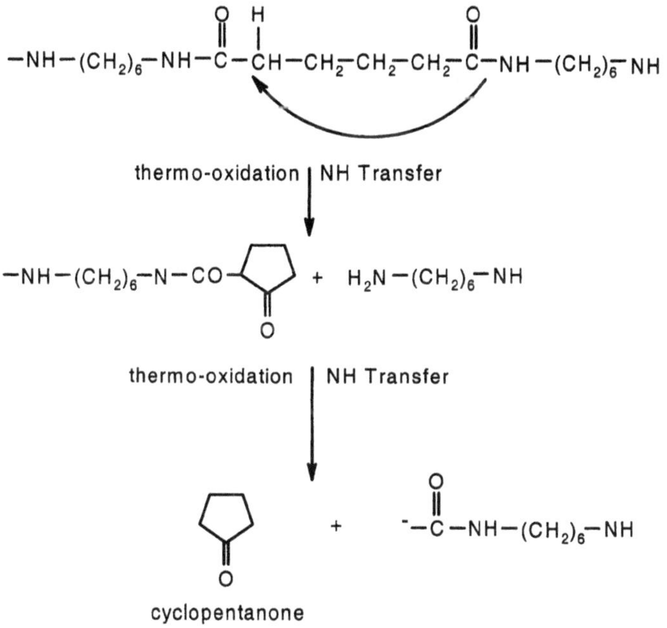

$-NH-(CH_2)_6-NH-C(=O)-CH(H)-CH_2-CH_2-CH_2-C(=O)-NH-(CH_2)_6-NH$

thermo-oxidation | NH Transfer

$-NH-(CH_2)_6-N-CO-$(cyclopentanone ring) + $H_2N-(CH_2)_6-NH$

thermo-oxidation | NH Transfer

(cyclopentanone ring) + $^{\bullet}-C(=O)-NH-(CH_2)_6-NH$

cyclopentanone

Scheme 7. Mechanism of thermo-oxidation of nylon 66

Several hundreds of degradation products have been detected and identified in environmentally degraded polyethylenes. Thermo-oxidised LDPE with starch and prooxidants form, after 6 weeks, ketones, alcohols, aldehydes, lactones and carboxylic acids and often in homologous series [28, 31]. A confirmation of the biodegradation mechanism of PE was presented which showed that samples of biotically aged LDPE contain a smaller number of carboxylic acids than abiotically aged LDPE. This indicates that these compounds are assimilated by the microorganisms [14]. A continuation of this work gave intermediate and final degradation products of six different polyethylenes with enhanced degradability exposed to different abiotic environments [32]. From the degradation product pattern a zip depolymerisation through a cyclic transition state mechanism was proposed for the formation of dicarboxlic acids and ketoacids [32].

The identification of degradation products can be very tedious. The sampling step is vital as a mistake will result in the loss of some degradation products or those new ones are formed by rearrangements of the original ones. Traditional liquid-liquid extractions are less useful nowadays, and many very good new techniques have been introduced. SPME allows extraction of degradation products in a much lower amounts. A comparison between SPME and HS-GC revealed that homologous series of carboxylic acids, ketones and furanones were identified after SPME while only lower carboxylic acids (up to C_6) were identified after head-space sampling [25]. Microwave-assisted extraction (MAE) and ultrasound assisted extraction are new extraction techniques, which are very adequate for sampling of low molecular weight compounds in polymers. A careful optimisation of mixtures of polar and nonpolar solvents are needed for each type of polymer extracted by MAE [33]. Ultrasonication was developed to analyse the migration of aromatic antioxidants and UV stabilisers in LDPE. A fast and total recovery of these additives were achieved after 10-15 minutes of ultrasonication in chloroform at 60°C while Soxhlet extraction usually requires 24 hours or more [34}.

4.4 CHEMOMETRY OF COMPLEX DATA OBTAINED IN DEGRADATION STUDIES

The introduction of chemometrics to handle the often very high number of data obtained after analyses of polymer degradation offers a tool, which simplifies the evaluations. Multivariate data analysis improves the possibility of obtaining a description of the series of events that take place during degradation. Multivariate data analysis is a method to extract information from data tables (or many results). Principal Component Analysis (PCA) is one projection technique by which a mathematical tool is provided to explain variance in a single data table X (or Y). PCA shows how observations (or different results) are related. (It can also tell if an experiment failed by in a two-dimensional figure showing one point far away from the others). A variety of data-analytical objectives may be analysed. Fig. 6 shows a score plot, PCA, obtained from a Pyrolysis-GC/MS data table, describing the changes in PCL degraded in biotic and abiotic environments. Samples degraded by mesophilic microorganisms has a positive PC1 while thermophilic and compost samples have zero or negative PC1 value. In addition, we observe that the environments with micro-organisms give a large positive influence on the PC2. By showing this two-dimensional projection a simple picture gives the relation between the pyrolysis products formed and degradation mode (abiotic and biotic hydrolysis by different micro-organisms). Thus, multivariate models allow us to distinguish between biological and abiotic degradation mechanisms.

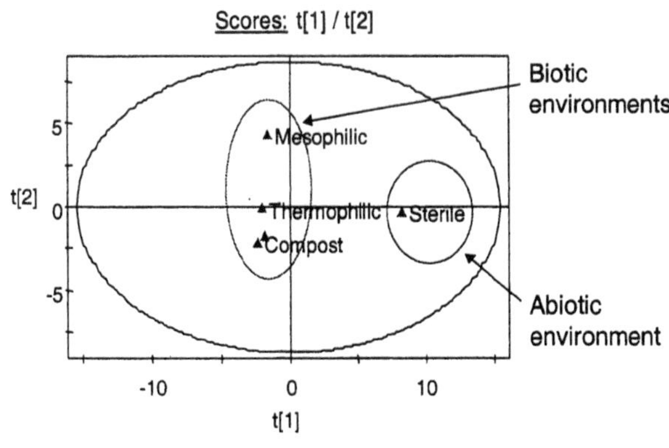

Fig. 6 Score plots. PCA of pyrolysis-GC/MS data from PCL degraded in biotic and abiotic environment at different temperatures.

Similar models were constructed for glass-fibre reinforced polyester composites, which were degraded for 20 years at 40°C and 60°C. By PCA it was not possible to find a pattern relating the degradation products identified by head-space-GC-MS with temperature and degradation time. By partial least square (PLS) it was instead revealed that two different degradation mechanisms were operating at the two temperatures [35].

Additives are also subject to changes during degradation. Many additives are aromatic by nature and must therefore be analysed by LC rather than GC. The migration and release of additives from polymers means that the stabilising effect is reduced in the material. LC was used to monitor the migration of Chimassorb 944 from LDPE in simulated land-fills. Quantification was carried out by UV-spectroscopy [36]. It was shown that a total release of this additive could be expected only after more than 10 years and that the levels were well below toxicity (0.9 mg/l compared to toxicity level for *e.g.* fish >9 mg/l) [36]. Transformation products of a primary phenolic and a secondary phosphite antioxidant in abiotic and biotic environments revealed that these consisted of esters, acids, dealkylated cinnamate, quinonoid products and 2, 4-di-tert-butylphenol [37] formed in varying amounts in the different degradation environments.

5 Correlation of different analysis methods

It is not enough to just follow one parameter and perform one analysis (or one kind of analysis) when describing degradation. Several events take place at the same time and/or at different time-scales. Generally, the more complex the degrading environment, the more difficult the evaluation of the test results becomes. Outdoors, hydrolysis, biodegradation by micro- or macroorganisms, thermo-oxidation and mechanical erosion might be expected in a synergistic mode. In designing new degradable polymers, a good picture is obtained by analysing during degradation, the molecular weight changes (SEC), functional group changes (FTIR) and identifying by GC-MS the low molecular weight products that are formed. From these kinds of analysis it was deduced that a more effective degradation, without the formation of aromatic degradation products, was obtained by using nitrile-butadiene rubber (NBR) instead of SBR rubber in LDPE with manganese stearate [38].

Generally, biological degradation implies that, not only microbial degradation can occur but also degradation due to chemical hydrolysis, thermolysis and sometimes also photolysis depending on the experimental conditions. Bacteria, fungi and algae, which are the most often used micro-organisms degrade an organic compound in quite different ways. Some organisms have only inter-cellular enzymes, which means that a compound must be transferred into the cell before degradation takes place. Others excrete series of extracellular enzymes (*e.g.* several fungi) which of course facilitates the needed contact between enzyme and the substrate) [39].

The test period is another factor to be aware of. Standard biodegradation tests have usually been adopted from the traditional biological oxygen demand tests (BOD) which uses cellulose degrading micro-organisms and requires a maximum test period of about one month. Using such a test in a standard mode, could prove to be wrong. This type of test does not take into account, on the one hand the very many degradation reactions occurring in nature which take more than one month, and on the other hand the fact that most degradation products of nature are used again and again in the catabolic cycles; that is the building of new biopolymers. Thus, the design of the test, specifying micro-organisms used and period of degradation, must be done based on the kind of polymer and its application. Even if the analysis results indicate that the polymer has not degraded more than 60% wt after one month, we still can talk about biodegradation [40, 41].

6 Conclusions

Polymer degradation encompasses a series of events during which physical, chemical and mechanical properties change. These changes are sometimes very slow which make the analysis of them difficult. In order to fully characterise the long-term properties of polymers a variety of techniques must be used. Several of them may give similar results but often the techniques complement each other. Infra red spectroscopy gives a change in functional groups but does not take into account the often very highly heterogeneous nature of polymer degradation as a very small area of a polymeric film or small amount of polymer powder are analysed. Mechanical testing, on the other hand, gives a more general picture of the degradation in terms of stress-strain properties and if the degradation may be heterogeneous the result is still a summary of all changes taken place in the materials. Different analyses give more or less clues to allow postulation of degradation mechanisms. In particular, chemical polymer characterisation by chromatography is useful in that respect. Size Exclusion Chromatography (SEC) and Matrix-Assisted Laser Desorption Ionisation (MALDI) give information changes in molecular weights and molar masses. Gas or Liquid Chromatography (GC and LC) coupled with Mass Spectrometry (MS) allows detection and identification of the presence of low molecular weight compounds (trace monomers and catalysts, additives, degradation products *etc.*). GC (LC)-MS needs well-adapted separation/extractions before it is possible to identify series of degradation products and other low molecular weight compounds. Such extraction methods which are useful in polymer characterisation are solid-phase extraction (SPE), solid-phase microextraction (SPME), ultrasonication, microwave assisted extraction (MAE), superfluid critical extraction (SFE) while traditional liquid-liquid extraction (*e.g.* Soxhlet) are more time-consuming and requires large amounts of organic solvents.

References

1. Albertsson, A-C. and Ljungquist, O. (1988) Degradable polyesters as biomaterials, *Acta Polymerica* **39**, 95-104.
2. Albertsson, A-C. and Karlsson, S. (1992) Biodegradable polymers, in S.L. Aggarwal and S. Russo (eds.), *Comprehensive Polymer Science*, First supplement, Pergamon Press, Oxford, pp. 285-297.
3. Doi, Y. (1990) *Microbial Polyesters*, VCH Verlag, Weinheim, New York.
4. Karlsson, S., Sares, C., Renstad, R. and Albertsson, A-C. (1994) GC, LC and GC-MS identification of degradation products in accelerated aged PHA, *J. Chrom.* **669**, 97-102.
5. Albertsson, A-C., Andersson, S.O. and Karlsson, S. (1987) The mechanism of biodegradation of polyethylene, *Polym. Degrad. Stab.* **18**, 73-87.
6. Albertsson, A-C. and Karlsson, S. (1990) The influence of biotic and abiotic environments on the degradation of polyethylene, *Prog. Polym. Sci.* **15**, 177-192.
7. Albertsson, A-C. and Karlsson, S. (1990) Polyethylene degradation and degradation products, in J.E. Glass and G. Swift (eds.), *Agricultural and Synthetic Polymers*, ACS Symp. Series. No. 433, ACS, Washington DC, pp. 60-64.
8. Albertsson, A-C. and Karlsson, S. (1990) Biodegradation and test methods for environmental and biomedical applications of Polymers in S. Barenberg *et al.* (eds.), *Degradable Materials*, CRC Press, Boca Raton, pp. 263-286.
9. Holmström, A. and Sörvik, E.M. (1978) Thermooxidative degradation of polyethylene .1. .2. Structural-changes occurring in low-density polyethylene, high-density polyethylene, and tetratetracontane heated in air, *J. Polym. Sci., Polym. Chem.* **16**, 2555-2586.
10. Albertsson, A-C., Barenstedt, C. and Karlsson, S. (1992) Susceptibility of enhanced environmentally degradable polyethylene to thermal and photooxidation, *Polym. Degrad. Stab.* **37**, 163-171.
11. Karlsson, S., Banhidi, Z.G. and Albertsson, A.C. (1988) Detection by high performance liquid chromatography of polyamines formed by clostridial putrefaction of caseins. *J. Chrom.* **442**, 267-277.
12. Albertsson, A-C., Barenstedt, C. and Karlsson, S. (1993) Increased biodegradation of LDPE-matrix in starch-filled LDPE-materials, *J. Environ. Polym. Degrad.* **1**, 241-245.
13. Renstad, R., Karlsson, S., Albertsson, A-C., Werner, P.-E. and Westdahl, M. (1997) Influence of processing parameters on the mass crystallinity of poly(3-hydroxybutyrate-co-3-hydroxyvalerate), *Polym. Int.* **43**, 201-209.

14. Albertsson, A-C., Barenstedt, C., Karlsson, S. and Lindberg, T. (1995) Degradation product pattern and morphology changes as means to differentiate abiotically and biotically aged degradable polyethylene, *Polymer* **36**, 3075-3083.
15. Albertsson, A-C. (1978) Biodegradation of synthetic polymers II. Limited microbial conversion of ^{14}C in polyethylene to $^{14}CO_2$ by some soil fungi, *J. Appl. Polym. Sci.* **22**, 3419-3433.
16. Albertsson, A-C. (1980) The shape of the biodegradation curve for low and high density polyethylenes in prolonged series of experiments, *Europ. Polym. J.* **16**, 623-630.
17. Erlandsson, B., Karsson, S., and Albertsson, A-C. (1997) The mode of action of corn starch and a prooxidant system in LDPE: Influence of thermooxidation and UV-irradiation on the molecular weight changes, *Polym .Degrad. Stab.* **55**, 237-245.
18. Eldsäter, C., Erlandsson, B., Renstad, R., Albertsson, A-C. and Karlsson, S. (2000) The biodegradation of amorphous and crystalline regions in film-blown poly(ε-caprolactone), *Polymer* **41**, 1297-1304.
19. Gallet, G., Carroccio, S, Rizzarelli, P and Karlsson, S. (2000) Thermal degradation of poly(ethylene oxide-propylene oxide-ethylene oxide) triblock copolymer: comparative study by SEC/NMR, SEC/MALDI-TOF-MS and SPME/GC-MS *Polymer* **43**, 1081-1094.
20. Bauer, B.I., Wallace, W.E., Fanconi, B.M. and Guttman, C.M. (2001) "Covalent cationization method" for the analysis of polyethylene by mass spectrometry, *Polymer* **42**, 9949-9953.
21. Billingham, N.C., O'Keefe, E.S. and Theu, E.T.H. in *Proceeding 12th Ann. Conf. Adv. in Stabilization and Controlled Polymer Degradation* (ed. P. Klemchuk), Lucerne, Switzerland (1990), p. 1.
22. Billingham, N.C., O'Keefe, E.S. and Theu, E.T.H.: *Polym. Mat. Sci. Eng*. **58** (1988), 431.
23. Zlatkevich, L. (ed.) *Luminescence Technique in Solid-State Polymer Research*, Marcel Dekker, New York (1989), Chapter 3.
24. Albertsson, A-C., Sares, C. and Karlsson, S. (1993) Increased biodegradation of LDPE with nonionic surfactant *Acta Polymerica* **44**, 243-246.
25. Hakkarainen, M., Albertsson, A-C. and Karlsson, S. (1997) Solid phase microextraction (SPME) as an effective means to isolate degradation products in polymers, *J. Environ. Polym. Degrad.* **5**, 67-73.
26. Karlsson, S., Hakkarainen, M. and Albertsson, A-C. (1994) Identification by GC-MS and HS-GC-MS of in vitro degradation products of homo- and copolymers of L- and D,L-lactide and 1,5-dioxepan-2-one, *A. Chrom. J.* **688**, 251-259.
27. Albertsson, A-C. and Löfgren, S. (1992) Copolymers of 1,5-dioxepan-2-one and L- or D,L-lactide, synthesis and characterization, *Makromol. Chem., Makromol. Symp.* **53**, 221-231.
28. Albertsson, A-C., Barenstedt, C. and Karlsson, S. (1994) Degradation of enhanced environmentally degradable polyethylene in biological aqueous media. mechanisms during the first stages, *J. Appl. Polym. Sci.* **51**, 1097-1105.
29. Karlsson, S. and Albertsson, A-C. (1990) The biodegradation of a biopolymeric additive in building materials, *Materials and Structures* **23**, 352-357.

30. Karlsson, S., Banhidi, Z.G. and Albertsson, A.C. (1989) Gaschromatographic detection of volatile amines formed in indoor air due to putrefactive degradation of casein-containing building materials, *Materials and Structures* **22**, 163-169.
31. Albertsson, A-C., Barenstedt, C. and Karlsson, S. (1994) Abiotic degradation products from enhanced environmentally degradable polyethylene, *Acta Polymerica*, **45**, 97-103.
32. Karlsson, S., Hakkarainen, M. and Albertsson, A-C. (1997) Dicarboxylic acids and ketoacids formed in degradable polyethylenes by zip depolymerization through a cyclic transition state, *Macromolecules* **30**, 7721-7728.
33. Camacho, W. and Karlsson, S. (2000) Quality-determination of recycled plastic packaging waste by identification of contaminants by CC-MS after microwave assisted extraction (MAE) *Polym. Degrad.Stab.* **71**, 123-134.
34. Haider, N. and Karlsson, S. (1999) A rapid ultrasonic extraction technique to identify and quantify additives in poly(ethylene) *Analyst* **124**, 797-800.
35. Hakkarainen, M., Gallet, G. and Karlsson, S. (1999) Prediction by multivariate data analysis of long-term properties of glassfiber reinforced polyester composites *Polym. Degrad. Stab.* **64**, 91-99.
36. Haider,N. and Karlsson, S. (1999) Migration and release profile of Chimassorb 944 from low-density polyethylene film (LDPE) in stimulated landfills *Polym. Degrad. Stab.* **64**, 321-328.
37. Haider, N. and Karlsson, S.: *J. Appl. Polym. Sci.* (2001), in press.
38. Khabbaz, F., and Albertsson, A-C. (2000) Great advantages in using a natural rubber instead of a synthetic SBR in a pro-oxidant system for degradable LDPE, *Biomacromolecules* **1**, 665-673.
39. Albertsson, A.-C. (1999) Biodegradation of polymers in historical perspective versus modern polymer chemistry, in S.H. Hamid, *et al.* (eds.), *Handbook of Polymer Degradation*, Marcel Dekker, New York, pp. 417-435.
40. Albertsson, A.-C. (1993) Degradable polymers, *J. Macromol. Sci., Pure Appl. Chem.* **A30**, 757-765.
41. Hakkarainen, H., Khabbaz, F., Albertsson, A.-C. Biodegradation of polyethylene followed by assimilation of degradation products, in A. Steinbüchel (ed.) *Biopolymers Vol. 9*, in press.

5

BIODEGRADATION OF ALIPHATIC POLYESTERS

*SUMING LI AND MICHEL VERT
Centre de Recherche sur les Biopolymères Artificiels
UMR CNRS 5473, Faculté de Pharmacie
15 avenue Charles Flahault, 34060 Montpellier, France

1 Introduction

Synthetic polymers appeared about sixty years ago and immediately medical people realized that this new class of materials was of interest for therapeutic applications. For example, isotonic aqueous solutions of polyvinylpyrrolidone (PVP) were used as plasma expander during World War II and although this compound was far from ideal, it stayed in use for years before substitutes appeared [1]. Since then, many polymers have been evaluated as candidate biomaterials [2]. However, only a number of them have reached the stage of clinical applications and commercial availability.

Polymeric biomaterials can be divided in two main classes according to desired lifetimes: namely biostable and biodegradable. Biostable materials have been used for a long time for both permanent and time-limited applications. Nevertheless, it was gradually realized that biodegradable systems could be of great interest for temporary therapeutic applications in surgery, in pharmacology and in tissue engineering. The first biodegradable synthetic polymer was poly(glycolic acid). When invented in 1954 [3], this polymer was discarded because of its poor thermal and hydrolytic stabilities which did not allow it to be used as a regular plastic material. Two decades later it became the first biodegradable suture material not related to natural polymers [4-5]. Since then, a number of other synthetic polymers have been investigated and are currently referred to as biodegradable, bioabsorbable, bioresorbable and bioerodible in the literature. All these terms could have been used to qualify differences in degradation characteristics or

G. Scott (ed.), Degradable Polymers, 2nd Edition, 71-131.

mechanisms. Authors tend to use these words competitively, creating much confusion, which is unacceptable as interest in environmental biodegradation increases [6].

Before discussing the various aspects of the biodegradation of aliphatic polyesters, let us first make a few semantic comments. Heller defined the term **bioerosion** as the conversion of an initially water insoluble material to a soluble one without involving necessarily major chemical degradation [7]. Langer and Peppas, on the other hand, made initially no distinction between **bioerodible** and **biodegradable** [8], but later discussed these terms with respect to the control of drug release rate [9]. **Biodegradable polymer** was widely used for polymers which undergo in vivo degradation [10]. Williams defined **biodegradation** as a biological breakdown of polymeric material [11] as opposed to simple hydrolytic breakdown [12]. This definition was also defended by Gilbert et al. [13]. For Graham and Wood, **biodegradable** systems are not only able to degrade in vivo but also to form soluble products easily removed from the implantation site and excreted from the body [14]. Griffin proposed a terminology based on degradation pathways [15]. **Direct biodegradation** was used to reflect enzymatic scission and metabolization of macromolecules. **Indirect biodegradation** was reserved to oxidative cleavage of polymer chains followed by metabolization, whereas **macrobiological biodegradation** was taken as involving mechanical degradation by animals. For Holland et al., **biodegradation** reflects hydrolytic, enzymatic or bacteriological degradation processes occurring within a polymer matrix. Accordingly, biodegradation does not necessarily imply that the physical form of the polymer is altered. **Bioerosion** was proposed to reflect physical loss from the polymer matrix brought about by a variety of physical and chemical processes which are not always specified [16]. For Guillet et al., **biodegradable** corresponds to the ability of being chemically transformed by the action of biological enzymes or microorganisms into products which themselves are capable of further biodegradation [17]. A similar definition was proposed by Albertsson and Karlsson who considered **biodegradation** as reflecting transformation and deterioration of polymers solely by living organisms (including microorganisms and/or exogenous enzymes) [18].

In order to elucidate the confusing situation, a series of definitions were adopted at the Second International Scientific Workshop on Biodegradable Polymers and Plastics (Montpellier, France) [19]:

Polymer degradation is a deleterious change in the properties of a polymer due to a change in the chemical structure

A biodegradable polymer is a polymer in which the degradation is mediated at least partially by a biological system.

A bioabsorbable polymer is a polymer that can be assimilated by a bio-system

Erosion describes processes of dissolution or wearing away of a polymer from the surface.

These definitions will be used in this review regardless of whether degradation occurs in living bodies or in the outdoor environment. The prefix **bio-** will be considered as reflecting phenomena which result from the contact with living elements such as tissues, cells, body fluids or microorganisms. Accordingly, water, oxygen and enzymes are also regarded as biological elements. All these terms, nevertheless, do not specify the fate of degradation by-products. The term **bioresorbable** was introduced to qualify compounds for which elimination of degradation by-products through natural pathways (metabolization or kidney filtration) had been proven [20]. Therefore, bioresorption implies total elimination of the initial foreign material with no residual side effects. In the same way, **bioassimilation** will be used to design materials which can be degraded and further assimilated in the natural environment. Most hydro-biodegradable polymers contain hydrolyzable linkages such as amide, ester, urea and urethane along the polymer chains [21]. However, the flexible ester-containing polymers, and in particular aliphatic polyesters appear the most attractive because of their variable biodegradability and versatile physical, chemical and biological properties. The synthesis of aliphatic polyesters can be realized either by ring-opening polymerization of heterocyclic monomers bearing at least one ester bond in the ring (1) or by step-growth polycondensation of hydroxyacids (2) or of diols and diacids (3):

$$\text{R}\overset{\text{O}}{\frown}\text{C=O} \longrightarrow \text{-[-O-R-CO-]}_n\text{-} \quad (1)$$

$$\text{HO-R-COOH} \longrightarrow \text{H-[O-R-CO-]}_n\text{-OH} \quad (2)$$

$$\text{HO-R-OH + HOOC-R'-COOH} \longrightarrow \text{H-[-O-R-O-CO-R'-CO-]}_n\text{–OH} \quad (3)$$

Ring-opening polymerization generally yields polymers with high molecular weight (MW), whereas direct polycondensation results in low MW.

The main members of the aliphatic polyester family are presented in Table 1. All these polymers have been tentatively investigated for temporary therapeutic applications during the last twenty years. However, several of them are now on the market, mainly in the form of sutures, such as Dexon® (PGA), Vicryl® (90/10 GA/L-LA copolymer), PDS® (PDO), Maxon® (67.5/32.5 GA/1,3-dioxane-2-one copolymer), Monocryl® (GA/CL copolymer), Polysorb® (GA/L-LA copolymer), dental devices for guided tissue regeneration such as Antrisorb® (PLA_{50}), Resolut® (GA/DL-LA copolymer), Guidor® (PLA_{50}) membranes, and orthopedic fixation devices such as Phusiline® (L-LA/DL-LA copolymer), Sysorb® (PLA_{50}), Endofix® (PLA_{100} or GA/1,3-dioxane-2-one copolymer), Bioscrew® (PLA_{100}) interference screws. This situation results mostly from the fact that there is a huge gap between laboratory experimentation and industrialization. Indeed, many prerequisites must be fulfilled before the stage of clinical application is realised [22]. A list of these prerequisites is given in Table 2. It is worth noting that these prerequisites have to be adapted according to each application. So far, no similar attention has been paid to the case of environmental biodegradation of polymers.

In the aliphatic polyester family, polymers derived from lactic acid enantiomers and glycolic acid have been widely investigated and seem to be the most promising, at least for biomedical applications. High molecular weight LA/GA polymers are obtained by ring-opening polymerization of cyclic diesters, i.e., L-lactide, D-lactide, DL-lactide and glycolide [23]. In the case of LA-containing polymer chains, chirality of LA units provides a worthwhile means to adjust bioresorption rates as well as physical and mechanical characteristics [24].

Table 1 Aliphatic Polyesters

Polymer and acronym	Structure
Poly(glycolic acid) (PGA)	$-[-O-CH_2-CO-]_n-$
Poly(lactic acid) (PLA)	$-[-O-{}^{*}CH(CH_3)-CO-]_n-$
Poly(ε-caprolactone) (PCL)	$-[-O-(CH_2)_5-CO-]_n-$
Poly(valerolactone) (PVL)	$-[-O-(CH_2)_4-CO-]_n-$
Poly(ε-decalactone) (PDL)	$-[-O-{}^{*}CH((CH_2)_3CH_3)-(CH_2)_4-CO-]_n-$
Poly(1,4-dioxane-2,3-one)	$-[-O-(CH_2)_2-O-CO-CO-]_n-$
Poly(1,3-dioxane-2-one)	$-[-O-(CH_2)_3-O-CO-]_n-$
Poly(*para*-dioxanone) (PDO)	$-[-O-(CH_2)_2-O-CH_2-CO-]_n-$
Poly(hydroxybutyrate) (PHB)	$-[-O-{}^{*}CH(CH_3)-CH_2-CO-]_n-$
Poly(hydroxyvalerate) (PHV)	$-[-O-{}^{*}CH(CH_2-CH_3)-CH_2-CO-]_n-$
Poly(β-malic acid) (PMLA)	$-[-O-{}^{*}CH(COOH)-CH_2-CO-]_n-$

Table 2 Criteria for marketable biodegradable polymers

1. Biocompatibility,	including : - polymer - leachable : • oligomers • residual monomers • degradation products - shape - surface properties
2. Biofunctionality,	including : - physical properties - mechanical properties - biological properties
3. Stability,	including : - processing - sterilization - storage
4. Bioresorption,	including : - degradability - controlled degradation rate - resorption of degradation products

The various LA/GA polymers are presented in Table 3. For the sake of simplicity, polymers are identified in this paper by using acronyms PLAxGAy where x is the percentage of L-LA units present in the monomer feed, y is that of GA units, (100-x-y) being the percentage of D-LA units whenever these two moieties are present in the feed. This nomenclature may appear unusual with respect to literature. However, it has the advantage of reflecting clearly the chemical and configurational compositions of the polymers, the average polymer chain composition being generally very close to that of the feed [25].

In this review, we will comment on biodegradation of various aliphatic polyesters. Efforts will be focused on clarifying mechanisms of biodegradation in general before considering various degradation characteristics of each compound. The discussion will be largely based on recent advances in the field of LA/GA polymers. However, information will also be extracted from papers dealing with other problems such as drug delivery and bone surgery. Convergences and discrepancies will be underlined when it is reasonably possible.

Table 3 LA/GA-derived homo- and copolymers

Polymer and acronym	Structure
Poly(glycolic acid) PGA	$-[-O-CH_2-CO-]_n-$
Poly(L-lactic acid) PLA_{100}	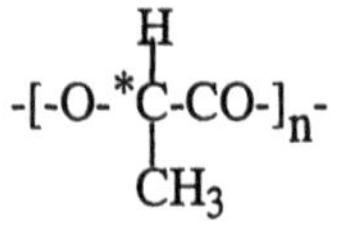 $-[-O-{}^{*}\overset{H}{\underset{CH_3}{C}}-CO-]_n-$
L-LA/D-LA stereocopolymer PLA_X $\{ x = 100\, n / (n+p) \}$	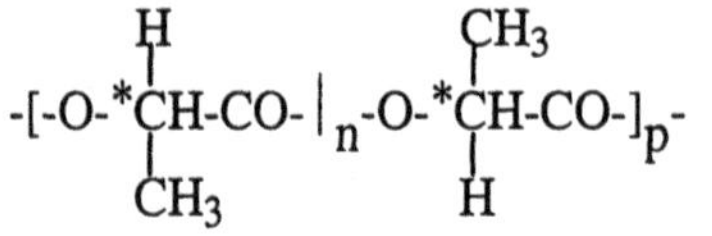 $-[-O-{}^{*}\overset{H}{\underset{CH_3}{C}}H-CO-\vert_n-O-{}^{*}\overset{CH_3}{\underset{H}{C}}H-CO-]_p-$
L-LA/GA copolymer $PLA_XGA_{(100-x)}$ $\{ x = 100\, n / (n+q) \}$	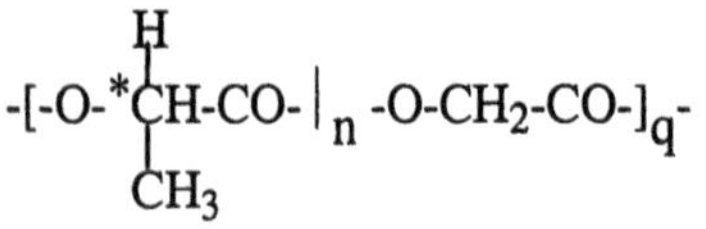 $-[-O-{}^{*}\overset{H}{\underset{CH_3}{C}}H-CO-\vert_n-O-CH_2-CO-]_q-$
L-LA/D-LA/GA terpolymer PLA_XGA_Y $\{ x = 100\, n / (n+p+q) \}$ $\{ Y = 100\, q / (n+p+q) \}$	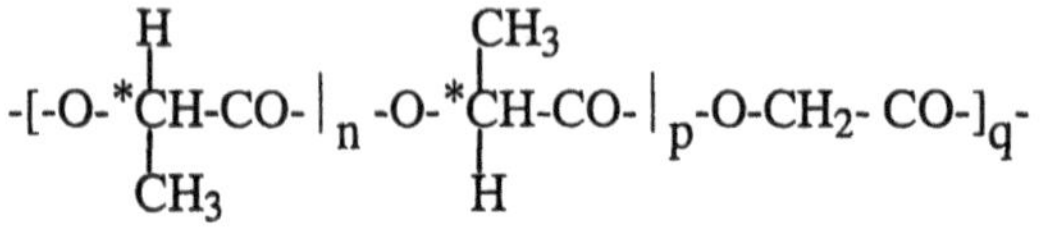 $-[-O-{}^{*}\overset{H}{\underset{CH_3}{C}}H-CO-\vert_n-O-{}^{*}\overset{CH_3}{\underset{H}{C}}H-CO-\vert_p-O-CH_2-CO-]_q-$

2 Biodegradation mecchanisms

According to the literature, degradation of polymeric materials in a living environment can result from either enzymatically or chemically mediated cleavages. The two mechanisms can act separately or simultaneously. Although environments are different *in vivo* from those outdoors, there is no fundamental difference between the biodegradation of a polymer by animal cells and by microorganisms. Both involve water, enzymes, metabolites, ions, etc. which interact with the material [26]. Under these conditions, it is possible to distinguish enzymatic, hydrolytic and microbial degradation.

2.1 ENZYMATIC DEGRADATION

It is now well understood that biopolymers such as proteins, polysaccharides, polynucleotides and bacterial poly(β-alkanoates) (PHA) degrade enzymatically [10,13,27], in agreement with the two main characteristics of living systems, *i.e.* biodegradation and biorecycling. The situation is totally different in the case of synthetic polymers. In fact, there has been much debate about the involvement of enzymes in the *in vivo* degradation of PLA, PGA and PCL homo- and copolymers. Some authors argue in favor of substantial enzymatic degradation [28-30], while most people relegate the enzymatic involvement to a secondary role [31-34]. The differences between parenteral and outdoor conditions further increased the confusion. In some cases, enzymatic degradation was shown from differences between the behaviors of samples in the presence and in the absence of living organisms. In the case of *in vitro* studies, comparison was generally made between data in the presence and in the absence of enzymes. However, the observation of such differences is not conclusive because there are many other factors capable of interfering with polymer degradation when experimental conditions are not similar [22]. Therefore, the demonstration of enzymatic degradation (biodegradation) must be based on the concordance of data using different analytical methods and from a careful monitoring of the generation and fate of the degradation products.

Various enzymes have been investigated in attempts to clarify their effect on the degradation of PLAGA polymers. Williams and Mort examined the role of fifteen enzymes in the in vitro degradation of Dexon® sutures and found that four of them increased the hydrolytic rate [29]. Herrmann et al. suggested that tissue esterases largely affected PGA degradation [30]. Proteinase K, an enzyme secreted by the fungus Tritirachium album Limber, was shown to be able to strongly accelerate the degradation rate of PLA stereo-copolymers [35-42], L-LA units being preferentially degraded as compared to D-LA [37-42]. Enzymes such as tissue esterases, pronase and bromelain also affect PLA degradation [35]. In contrast, many other enzymes seem to be inactive [35].

Pitt *et al.* investigated the *in vivo* degradation of a series of elastomeric homo- and copolymers of PCL and PVL crosslinked with bis-caprolactone. These compounds were subject to bioerosion involving immediate attack at the surface [43-44]. This finding was attributed to the rubbery nature of the materials, polymeric chains having enough freedom to take on chain comformations convenient for enzymatic attack. The authors also noted that for degradable polymers in the glassy state, this conformation could hardly be achieved and thus small (if any) enzymatic degradation could occur. Michizuki *et al.* examined the enzymatic degradation of PCL fibres by lipase [45]. Scanning electron microscopy (SEM) photographes showed that the enzyme preferentially attacked amorphous regions rather than crystalline ones. As enzymatic degradation proceeded, the diameter of the fibres became gradually smaller. The enzymatic degradation of PCL by various lipases was confirmed by other authors [41,46].

It should be noted that enzymes can be inactive on high MW material and become active at the later stages of degradation when the chain fragments become small and soluble in surrounding fluids. Once formed during degradation, tiny crystalline particles can be phagocytosed and undergo intracellular degradation [47]. Microspheres can also be easily phagocytosed [48].

In the case of PHB and PHBHV copolymers, it has been shown that biodegradation occurred in aerobic and anaerobic microbially active environments (aerobic and anaerobic sewage sludges and compost, estuarine sediments, soil, riverwater and seawater). According to Cox, the degradation rate depended on moisture level, nutrient supply, temperature and pH [49]. Biodegradation appeared to proceed by colonization of the polymer surface by bacteria or fungi, which secreted an extra cellular depolymerase capable of degrading the polymer in the vicinity of the cell. The soluble degradation products were then absorbed through the cell wall and metabolized. Accordingly, enzymatic activity resulted in surface erosion, the thickness of PHB injection moulded bars gradually decreasing with time in an aerobic 1:1 sewage/riverwater environment. In addition, the MW of the residual PHB in soil did not change during biodegradation [49]. Doi et al. and Gilmore et al. confirmed the enzymatic degradation of these polymers although hydrolysis also contributed [50-54]. The rate of enzymatic degradation by PHA depolymerase appeared faster than that of simple hydrolytic degradation by two or three orders of magnitude [51-52]. On the other hand, blends of PHB with PHBHV copolymer, poly(β-propiolactone) and poly(ethylene adipate) degraded enzymatically faster than each polymer component, thus showing synergistic effects between the two polymers [53].

In conclusion, it is now generally admitted that in the case of glassy aliphatic polyesters like PLA, PGA and their copolymers, enzyme involvement is unlikely at the early stages of degradation *in vivo* or under outdoor conditions. Nevertheless, enzymes contribute at the later stages, especially when soluble by-products are released. In contrast, for rubbery polymers like crosslinked PCL, enzymes seem to be active from the very beginning via surface erosion phenomena [43-44]. Actually, enzymatic degradation of aliphatic polyesters should not be claimed unless weight loss and dimensional changes without MW decrease are shown, and non-enzymatic degradation is excluded. Under in vitro conditions, however, a number of specific enzymes do accelerate the degradation of PLA, PGA, PCL and PHBHV polymers.

2.2 HYDROLYTIC DEGRADATION

From the molecular viewpoint, ester hydrolysis is a well known reaction in organic chemistry. The hydrolytic reaction can be catalyzed by both acids or bases. Also, the reaction product, RCOOH, is able to accelerate ester hydrolysis by autocatalysis. In the case of aliphatic polyesters, chain cleavage at the ester bond level is autocatalyzed by carboxyl end groups initially present or generated by the degradation reaction.

Pitt *et al.* studied the *in vivo* degradation of films of PLA_{50}, PCL and corresponding copolymers. The authors suggested that the first stage of degradation was confined to a MW decrease due to random hydrolytic ester cleavage autocatalyzed by the carboxyl endgroups, the second stage being characterized by the onset of weight loss and a decrease in the rate of chain scission [55-56].

The kinetics of the autocatalyzed hydrolytic degradation proposed by Pitt et al. were derived according to the following equations :

$$d[E] / dt = -d[COOH] / dt = - k\,[COOH] \bullet [H_2O] \bullet [E] \qquad \text{(i)}$$

where [COOH], $[H_2O]$ and [E] represent respectively carboxyl endgroup, water and ester concentrations in the polymer matrix. By using the following relationships :

$$[COOH] = W / (\overline{M}n \bullet V) = \rho / \overline{M}n$$

$$[COOH] = [E] / (\overline{DP}n - 1)$$

$$\overline{M}n = m \bullet \overline{DP}n$$

where W is the polymer matrix weight, V its volume, ρ its density, $\overline{M}n$ the number average molecular weight, $\overline{DP}n$ the number average degree of polymerization and m the repeat unit mass, one obtains eq.ii.

$$d(1/\overline{DP}n) / dt = k (\rho/m) [H_2O] (\overline{DP}n - 1) \overline{DP}n^{-2} \qquad \text{(ii)}$$

Integration of eq.ii leads to eq.iii :

$$Ln \{(1-\overline{DP}n) / (1-\overline{DP}n_0)\} = k't \qquad \text{(iii)}$$

where $k' = k (\rho / m) [H_2O]$, and $\overline{DP}n_0$ is the $\overline{DP}n$ at time zero. This kinetic expression is valid before the onset of weight loss. If $\overline{DP}n >> 1$, eq.iii can be simplified to eq.iv :

$$Ln (\overline{DP}n / \overline{DP}n_0) = Ln (\overline{M}n / \overline{M}n_0) = - k't \qquad \text{(iv)}$$

According to this relationship, semilog plots of $\overline{DP}n$ or of $\overline{M}n$ versus hydrolysis time should be linear prior to the onset of weight loss, a feature which was observed experimentally [55-56].

From the macroscopic viewpoint, degradation of aliphatic polyesters has been regarded as homogeneous, although surface erosion has occasionally been claimed. Ginde and Gupta investigated the *in vitro* degradation of PGA pellets and fibres in aqueous media at different pH values [57]. The authors found that pellets showed considerable surface degradation, whereas fibres showed little surface changes. Singh *et al.* studied a drug delivery system based on PLA_{50} microcapsules and concluded that erosion-based degradation occurred by the hydrolytic cleavage of ester bonds in the polymer backbone at the surface of microcapsules leading to the formation of lactic acid monomers [58]. Kimura *et al.* investigated the *in vitro* and *in vivo* degradations of fibers deriving from a copoly(ester-ether) composed of PLA_{100} and polyoxypropylene blocks [59]. Surface erosion was observed in both cases and the authors concluded that hydrolysis was limited to the surface.

By contrast, many authors have argued either explicitly or implicitly in favor of autocatalyzed bulk degradation. Hutchinson investigated the *in vitro* and *in vivo* releases of polypeptides ($633 < \overline{M}_n < 22{,}000$ daltons) from PLAGA copolymer matrices containing from 25 to 100% DL-LA. The degradation process of the polymer matrix was considered as homogeneous [60]. Sanders *et al.* studied a $PLA_{22}GA_{56}$ microsphere-based delivery system and observed a homogeneous (bulk) rather than heterogeneous (surface) degradation [61]. This conclusion was derived from the biological response and from changes in microsphere aspects *in vivo*. Kenley *et al.* examined a series of PLAGA copolymers representing a range of monomer ratios and MW, with the goal of studying polymer degradation kinetics *in vivo* and *in vitro* [62]. Hydrolysis was supposed to proceed throughout the bulk of the polymer structure because the onset of weight loss lagged behind MW decrease. Schakenraad *et al.* and Helder *et al.* investigated a series of glycine/DL-LA copolymers *in vivo* and *in vitro* [63-64]. The degradation mechanism was ascribed to bulk hydrolysis in both cases, MW decreasing continuously with the ageing time. Cohen *et al.* used a PLAGA (75/25) copolymer for long term delivery of high MW, water soluble proteins [65]. At all times, MW distribution displayed a unimodal pattern, suggesting homogeneous degradation. St. Pierre and Chiellini reviewed the degradability of synthetic polymers for pharmaceutical and medical uses. These authors concluded that hydrolysis of PLAGA polymers was a bulk process with random cleavage of ester functions [66]. In another review concerning the controlled release of bioactive agents from PLAGA polymers, Lewis concluded that degradation of aliphatic polyesters occurred in the bulk [67].

Chemical degradation of PHB and PHBHV copolymers has been also the subject of debate. Holland et al. investigated comparatively the hydrolytic degradation of PHB and of a series of PHBHV copolymers under various conditions [68]. The observed increase in surface energy was assigned to the presence of increasing amounts of hydroxyl and carboxyl groups at the surface as a consequence of ester hydrolysis in agreement with a "surface erosion process". However, surface erosion rapidly appeared in competition with a "bulk erosional process" which resulted from the diffusion of products of chain scission from the matrix [68]. Knowles and Hastings studied hydrolytic degradation of a PHBHV (93/7) copolymer in buffered physiological saline at various pH and found that neutral and acidic solutions produced a diffuse "surface degradation", whereas the alkaline solution appeared more agressive, with site-specific attacks causing deep points of "surface erosion" [69].

In the late 80s, the discovery of a faster degradation inside large-size PLAGA specimens greatly changed the understanding of the hydrolytic degradation of PLAGA polymers [70-79]. The heterogeneous degradation was assigned to diffusion-reaction phenomena as summarized in the following schema (Fig. 1). Typically, the polymer matrix is initially homogeneous in the sense that the average MW is the same throughout the matrix. Once placed in an aqueous medium, water penetrates into the specimen leading to hydrolytic cleavage of ester bonds (step 1). At the very beginning, degradation occurs in the bulk and is macroscopically homogeneous. Each ester bond cleavage forms a new carboxyl endgroup which accelerates the hydrolytic reaction of the remaining ester bonds by autocatalysis [80]. However, the autocatalytic effect does not work at the surface because of two factors. First, as the aqueous medium is always buffered, *in vitro* as *in vivo*, the carboxyl endgroups present at the surface are neutralized and lose their catalytic potence. Second, when soluble oligomeric compounds are generated in the bulk, the soluble oligomers which are close to the surface can escape from the matrix before total degradation, while those located inside can hardly diffuse out of the matrix. Therefore, autocatalysis is larger in the bulk than at the surface, thus leading to a surface/interior differentiation (step 2).

As the degradatrion proceeds, more and more carboxyl endgroups are formed inside to accelerate the internal degradation and enhance the surface/interior differentiation (step 3). Bimodal MW distributions are observable due to the presence of two populations of macromolecules degrading at different rates (Fig. 2). Finally, hollow structures are formed when the internal material, which is totally transformed to soluble oligmers, dissolves in the aqueous medium (step 4) (see Fig. 3). Hollow structures were observed for amorphous polymers like PLA_{50}, $PLA_{62.5}$, PLA_{75}, and $PLA_{37.5}GA_{25}$ [72,73,79]. In contrast, in the case of crystallizable polyesters like $PLA_{87.5}$, PLA_{96}, PLA_{100}, $PLA_{75}GA_{25}$ and $PLA_{85}GA_{15}$, no hollow structures were obtained due to the crystallization of degradation products (Fig. 4), although degradation was faster inside than outside [73-79].

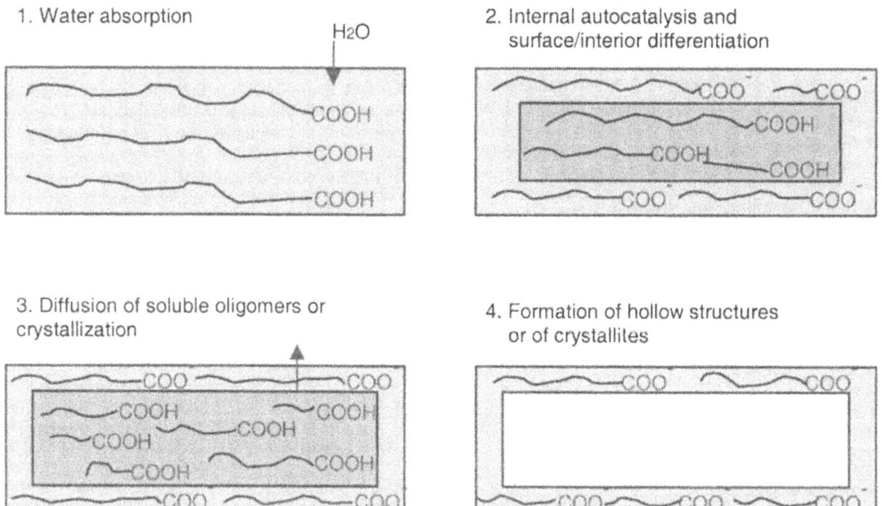

Fig.1. Schematic presentation of the faster internal degradation mechanism due to auto- catalysis (Source : from Ref. 72)

The faster internal degradation of PLA and PLAGA polymers is now regarded as a general phenomenon. It was confirmed by many authors [81-88] for PLACL copolymers [89], lactide/glycine copolymers [63-64], lactide/1,5-dioxepan-2-one copolymers [90], poly(trimethylene carbonate) [91] and poly(*para*-dioxanone) [92]. According to diffusion/reaction phenomena, small devices should degrade slower (*i.e.* at the surface rate) than large ones since thinness facilitates neutralization of carboxyl endgroups and elimination of soluble oligomeric compounds, thus minimizing the autocatalytic effect. This trend was confirmed by comparing the degradation rates shown by submillimetric microparticles and films and millimeric plates made from the same PLA_{50} [93]. As expected, the plates degraded faster than the microparticles and films.

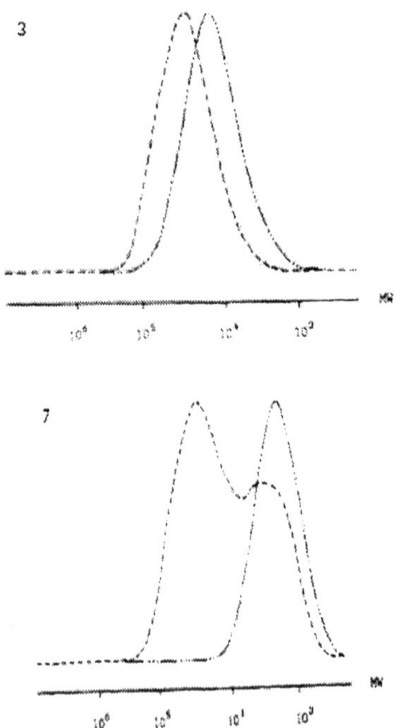

Fig. 2. SEC chromatograms of PLA_{50} after 3 and 7 weeks in vitro degradation : - - - - surface ; —···—··· interior (Source : from Ref. 72)

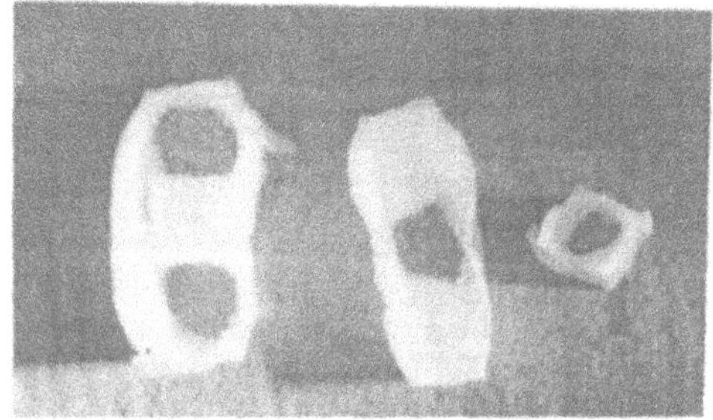

Fig. 3. Hollow structure of a PLA_{50} specimen after 10 weeks in vitro degradation

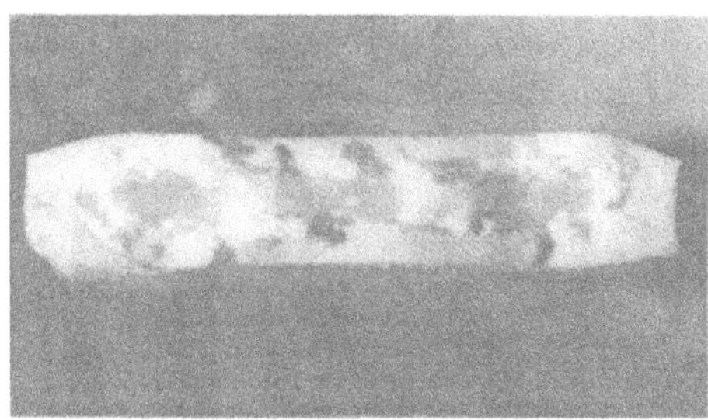

Figure 4. Cross section of an initially amorphous PLA_{96} specimen after 40 weeks in vitro degradation (Source : from Ref. 75)

In the case of PCL, hydrolytic degradation is very slow. However, the degradation mechanism is the same as that of PLAGA polymers. Typically, degradation of PCL begins with random hydrolytic cleavage of the ester linkages autocatalyzed by the carboxyl endgroups. The MW decreases continuously, but there is no weight loss during this first stage of degradation. The second phase is characterized by the onset of weight loss due to the formation of low MW fragments small enough to diffuse out of the bulk [56,89,94]. No faster internal degradation has been reported for the PCL homopolymer, so far, probably because of the combination of high crystallinity, hydrophobicity and low degradation rate.

Therefore, the hydrolytic degradation of aliphatic polyesters is a complex process involving four main phenomena, namely water absorption, ester cleavage, diffusion of soluble oligomers and solubilization of fragments [95]. These phenomena depend on many factors such as matrix morphology, chemical composition and configurational structure, MW, size, distribution of chemically reactive compounds within the matrix, and composition of the degradation media [96-97]. Considering the interdependence of degradation and drug release in controlled drug delivery systems, all the factors listed above can affect the characteristics of drug release, as will be shown later on [97].

2.3 MICROBIAL DEGRADATION

More and more attention is at present being paid to aliphatic polyesters of the PLA, PGA and PCL-type with respect to the problems of solid waste accumulation, delivery of chemicals to plants, and seed protection. Whether these polymers can biodegrade in the natural environment or whether the by-products, generated by abiotic hydrolysis, are bioassimilated by fungi or bacteria, are questions currently being investigated, particularly in the context of composting.

Torres *et al.* studied the ability of some microorganisms to use lactic acids, PLA oligomers and polymers, and PLAGA copolymers as sole carbon and energy sources under controlled or natural conditions [98-100]. Among the 14 filamentous fungal strains tested, two strains of *Fusarium moniliforme* and one strain of *Penicillium roqueforti* were identified as able to totally assimilate L- and DL-lactic acids as well as PLA_{50} oligomers [98-99]. In contrast, PLA_{100} oligomers appeared biostable because of crystallinity. A synergistic effect was observed when both microorganisms were present in the same culture medium containing PLA_{50} oligomers.

High MW polymers were also considered via degradation tests in soil and in selected culture media. *Fusarium moniliforme* filaments were found to grow on the surface of a $PLA_{37.5}GA_{25}$ copolymer and through the bulk. On the other hand, five strains of different filamentous fungi were isolated from the soil as capable of bioassimilating PLA_{50} soluble oligomers in mixed cultures. Based on the results obtained from various measurements, the authors proposed a mechanism including abiotic hydrolysis of macromolecules followed by bioassimilation of soluble oligomers [100]. Therefore, when a PLA material is placed in a degradation medium in the presence of microorganisms, only abiotic hydrolytic degradation occurs unless the few enzymes mentioned above as capable of degrading high molar mass PLA are present. Whether the degradation is homogeneous or heterogeneous depends on the size of the material and on the physical state of the medium (liquid or moisture). In a liquid medium, degradation is heterogeneous due to larger internal autocatalysis for large-size devices, while in the case of a humid environment, degradation has to be homogeneous since the soluble oligomers cannot diffuse out of the matrix. Anyhow, sooner or later, low MW assimilable oligomers are formed throughout the polymer matrix. The bioassimilation process will then take place. At this time, fungal filaments will penetrate the partially degraded mass to take advantage of the internal oligomers, thus turning the abiotic degradation to a biotic one. Under natural conditions, this should happen after a long lag period. However, PLA polymers will be definitely degraded and assimilated since the ultimate degradation products (L- and D-lactic acids) have been shown to be metabolized by some common microorganisms [100]. In partircular, it is now well known that PLA polymers degrade completely and rather rapidly in a compost where the temperature is usually between 50 and 60°C.

Jarrett *et al.* investigated the microbial degradation of crosslinked PCL films and of PCL single crystals in the presence of a yeast, *Cryptococcus laurentii*, and a fungus, *Fusarium.* Data were compared with those obtained for the chemical degradation in a 40% methylamine aqueous solution [101]. These microorganisms produced both endo- and exoenzymes and a cofactor (surfactant) which worked together in the degradation of the polymer. In the absence of the cofactor, the access of enzyme to the hydrophobic polymer was limited. Lefebvre *et al.* examined the biodegradation of hydroxy- or methoxy-terminated PCL samples by mixed cultures of microorganisms issued from a suspension of a compost, and by a pure culture of an actinomycete isolated from compost, too [102]. The authors suggested that the initiation of degradation took place in the vicinity of the chain ends. Methylation of the chain ends did not affect the biodegradability. Akahori and Osawa investigated the biodegradability of PCL/paper composites by outdoor exposure and outdoor soil burial [103]. All the samples crumbled within 6 months. The MW of residual PCL was found to decrease slowly, suggesting a contribution from hydrolytic degradation.

3 Biodegradable aliphatic polyesters

Let us now consider the various factors which can affect the degradation of aliphatic polyesters. For the sake of clarity, each type will be discussed separately.

3.1 PLAGA HOMOPOLYMERS AND COPOLYMERS

As mentioned above, degradation of PLAGA polymers is rather complex. It depends not only on polymer characteristics such as morphology, chemical composition and configurational structures, MW and MW distribution, size and porosity, chemically reactive additives, but also on experimental parameters such as pH, ionic strength, temperature and implantation site [16, 96-97].

3.1.1 *Morphology*

Polymer morphology plays a critical role in degradation phenomena. It is now well known that degradation of semicrystalline polymers occurs in two stages. The first stage consists in water-diffusion into the amorphous regions with random hydrolytic scission of ester bonds. The second stage starts when most of the amorphous regions are degraded. Hydrolytic attack then progresses within crystalline domains [104-106]. This phenomenon was first observed by Fischer et al. who investigated the structure of solution-grown crystals of L-LA/D-LA stereocopolymers by means of chemical reactions [104]. It was found that disordered regions of single crystals were selectively degraded by methanolic sodium hydroxide, oligomeric compounds with very narrow MW distributions being produced as shown by size-exclusion chromatography (SEC). In the case of a $PLA_{92.5}$ stereocopolymer, for example, the authors obtained a trimodal MW distribution with three distinct peaks corresponding respectively to onefold, twofold and threefold the MW of the crystalline lamellae thickness.

Carter and Wilkes carried out morphological studies on Vicryl® pellets by chemical etching and observed that the amorphous phase was preferentially degraded and removed [105]. Fredericks et al. studied structure and morphology changes of Vicryl® sutures in vitro [106]. The results showed that hydrolytic attack was initiated in the amorphous areas of the polymer, resulting in an augmentation of crystallinity. As hydrolysis advanced, crystalline areas were attacked and eventually removed. Leeslag et al. confirmed the predominant degradation of amorphous regions in the case of PLA100 plates, an increase in crystallinity being detected [107].

In a series of publications, Chu described the degradation of Dexon® suture material as resulting from bulk degradation with selective attack in amorphous zones [108-111]. In vitro tests showed that degradation occurred in two stages: the first from day 0 to day 21, and the second from day 21 to day 49. Maximum crystallinity was detected around the transition period [108]. This finding seems to disagree with the steady increase in crystallinity of PGA fibres and pellets reported by Ginde and Gupta [57]. However, one must be cautious with these results obtained from differential scanning calorimetry (DSC) since polymer morphology can be altered by heating. Chu suggested that diffusion of oligomers was probably a significant component of the degradation process during the later stage of degradation [108]. A study of the hydrolytic degradation of Dexon® sutures by SEM also showed that hydrolysis occurred initially in the amorphous regions sandwiched between two crystalline ones [109]. The annealed specimens exhibited a faster decrease in mechanical properties than the unannealed ones when both were subjected to hydrolysis [110-111], in agreement with the work of the present authors [74] and with the work of Nakamura et al. [112] with PLA_{100}. Faster decrease of mechanical properties, however, does not imply faster degradation. Miller et al. observed that fast-annealed PGA degraded more rapidly than slow-annealed PGA [113], the half-lives being of 0.85 and 5 months respectively. The difference was related to the higher crystallinity of the latter.

Ginde and Gupta examined the influence of polymer morphology on the chemical degradation of PGA fibres and pellets [57]. The degradation rate was found to be much faster for pellets than for fibres and regarded as in agreement with the presence of long range order in fibres. Pellets of commercial origin were slightly more crystalline (38%) when compared to melt-spun fibres (35%), whereas small angle X-ray diffraction showed the lack of orientation in the case of pellets. Therefore, chain orientation also plays an important role in the hydrolytic degradation [57]. Leeslag et al. examined the hydrolyzability of melt-spun PLA_{100} fibres and found that looser fibrillar structure of PLA100 fibres increased the degradation rate, fibres spun in the presence of camphor losing 50% of their tensile strength after 32 days whereas those spun in the presence of toluene, which had a more compact fibrillar structure, lost only 20% of the initial tensile strength after 70 days in vitro [114].

Our systematic investigation on the degradation of PLAGA polymers agreed well with the concept of preferential degradation in the amorphous zones of semicrystalline PLA_{100} [74]. In particular, bimodal MW distributions were obtained with two distinct peaks corresponding respectively to one and two transverse lengths of crystalline lamellae. As degradation advanced, the SEC peak corresponding to one transverse length increased at the expenses of the other. Finally, when all the amorphous regions and the crystalline lamellae surfaces were totally degraded, a monomodal MW distribution was obtained [74]. Moreover, in vitro degradation of quenched PLA_{100} and PLA_{96} was investigated with the aim of examining the influence of initial morphology on degradation [74-75]. X-ray diffraction and DSC data showed that the initially amorphous PLA_{100} and PLA_{96} crystallized during degradation at 37°C (Fig. 5).

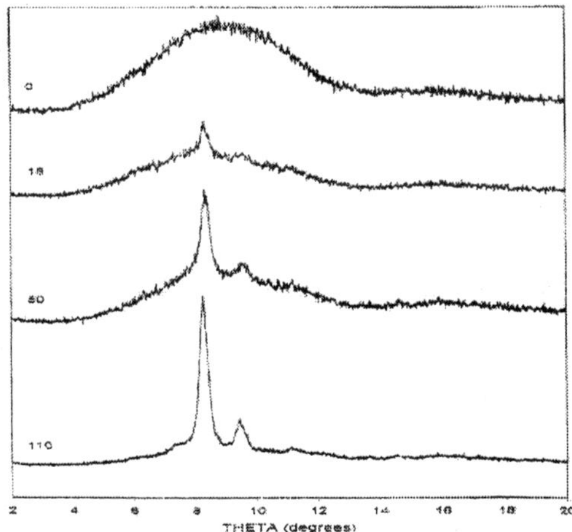

Fig. 5 X-ray diffraction spectra of PLA_{100} after 0, 18, 50 and 110 weeks in vitro degradation (Source : from Ref. 96)

Low temperature crystallization was attributed to the larger mobility of low MW chains and to the plasticizing effect of absorbed water which is known to lead to a reduction of the glass transition temperature of these polymers [115]. Surprisingly, initially amorphous PLA_{96} crystallized faster than PLA_{100} (Fig. 6). After 50 weeks, for example, crystallinity of PLA_{96} attained 50%, while that of PLA_{100} was only 20%. PLA_{100} normally crystallizes faster than PLA_{96} because of higher isotacticity. The faster crystallization of PLA_{96} confirmed, in fact, the previous conclusion that crystallization resulted from degradation by-products or short chains. PLA_{96} degraded faster than PLA_{100}, leading to more short chains and to a faster crystallization. The formed crystalline zones are more resistant to further degradation and as a consequence, a very narrow peak corresponding to low MW crystalline zones was detected in SEC chromatograms [74-75].

Fig. 6. Crystallinity changes of initially amorphous PLA_{96} and PLA_{100} with in vitro degradation (Source : from Ref. 96)

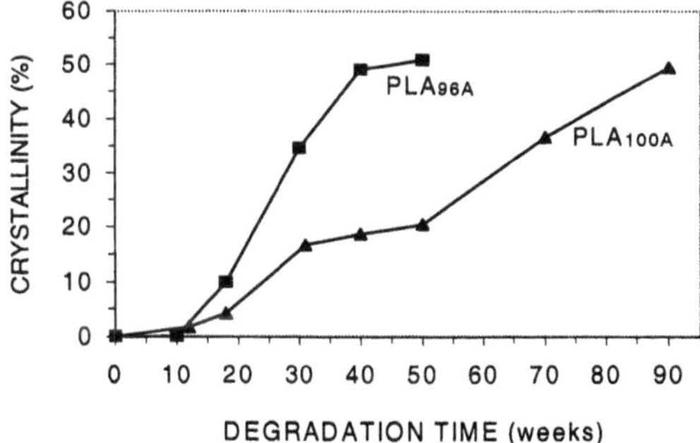

The fate of intrinsically amorphous PLAGA polymers appeared more complex depending on their chain structures. Two classes can be distinguished: the first class consists of polymers which can never crystallize, and the second one of polymers whose degradation by-products can crystallize. $PLA_{37.5}GA_{25}$ is a good example of the first class. This terpolymer cannot crystallize even after degradation due to the high irregularity of its chain structure. PLA_{50}, $PLA_{62.5}$, PLA_{75}, $PLA_{87.5}$, $PLA_{75}GA_{25}$ and $PLA_{85}GA_{15}$ belong to the second class. These polymers are crystallizable because of the presence of stereoregular blocks which, once released by degradation, are able to crystallize. For example, a crystalline oligomeric stereocomplex composed of L-LA and D-LA-rich fragments was obtained at the later stages of PLA_{50} degradation, as shown in Fig. 7 [78,116]. These stereoregular fragments were initially present within PLA_{50} chains obtained by ring-opening polymerization of DL-lactide, during which chain growth proceeds by addition of LL or DD pairs [25]. In consequence, PLA_{50} contains non-negligible isotactic sequences provided transesterification reactions are limited. At the later steges of degradation, the released L-LA and D-LA fragments were susceptible to forming an oligomeric stereocomplex. $PLA_{62.5}$ also led to slightly crystalline residues of the stereocomplex-type [79]. In contrast, PLA_{75} and $PLA_{87.5}$ degradation residues had the same crystalline structure as PLA_{100} since insufficient D-LA fragments were available [79]. Partial crystallization of degradation residues was also observed with $PLA_{75}GA_{25}$ and $PLA_{85}GA_{15}$ [73,75]. However, the structure of the crystalline residues is slightly different from that of PLA_{100}, showing that GA units are also included in the crystallized zones. Similar results have been reported for the hydrolytic degradation of $PLA_{73}GA_{27}$ [117].

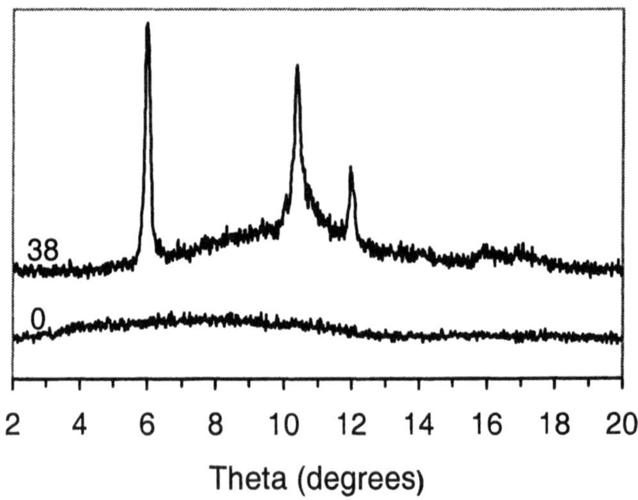

Fig. 7. X-ray diffraction spectra of PLA_{50} after 0 and 27 weeks in vitro degradation : formation of stereocomplex (Source : from Ref. 116)4

A schematic representation of polymer constitution/degradation relationships is given in Fig. 8. The triangle can be divided in two types of zones: C zones are composed of intrinsically crystalline polymers and A zones of intrinsically amorphous polymers. For polymers in C zones, if they are prepared in the amorphous state by quenching, degradation will be characterized by crystallization of degradation by-products. If they are prepared in semicrystalline state by annealing, degradation will be characterized by selective degradation of amorphous regions. In both cases, no hollow structures will be obtained for large-size devices in spite of the faster internal degradation. In the case of amorphous polymers, three subzones can be distinguished : A1, A2 and A3. For A1 polymers, degradation will lead to hollow structures which remain amorphous due to the high irregularity of polymer chain structures. For A2 polymers, degradation will also lead to hollow structures which will partially crystallize, the crystalline structure depending on the initial composition. For A3 polymers, no hollow structures can be obtained, both the surface and interior will crystallize. It is noteworthy that degradation of small-size amorphous polymers will not lead to hollow structures, but morphological changes are comparable to large-size ones [116].

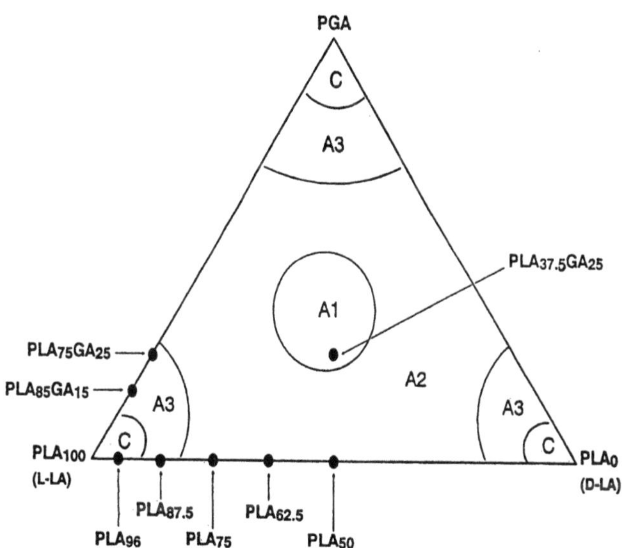

Fig. 8. Schematic presentation of the degradation behaviors of various PLAGA polymers (Source : from Ref. 96)

3.1.2 *Chemical composition and configurational structure*

It is now well known that the composition of the polymer chains (i.e. the contents of L-LA, D-LA and/or GA units) strongly influences the degradation rate of aliphatic esters [16, 96]. As L-lactic acid is available from renewable resources, PLA_{100} was largely preferred to PLA_0 for investigations. PLA_X stereocopolymers have variable degradation rates. PLA_{50} degrades the most rapidly, whereas PLA_{100} is the most stable member of the PLAGA polymer family [96]. PLA_X is intrinsically semicrystalline if x is greater than 90, the degradation rate being faster when the same material is quenched and thus amorphous. The copolymerization of LA with GA considerably enhances the degradation rate [113, 118-119]. Gilding and Reed reported that most of the copolymer composition range is taken up by amorphous polymers. For L-LA/GA polymers, the 25 to 75% GA composition range corresponds to amorphous compounds, whereas in the DL-LA/GA series, this range extends from 0 to 70% [120].

Cutright et al. carried out a systematic in vivo degradation study on a series of L-LA/GA homo- and copolymers. The degradation rate of these polymers was found to increase as follows: PGA < PLA_{100} < $PLA_{75}GA_{25}$ < $PLA_{75}GA_{25}$ < $PLA_{25}GA_{75}$ [118], showing that degradation rates can be altered by varying the proportion of LA to GA. Later, Miller et al. confirmed this point but found a slightly different and more logical order: PLA_{100} << PGA < $PLA_{75}GA_{25}$ < $PLA_{25}GA_{75}$ < $PLA_{50}GA_{50}$ [113]. The half-life of these polymers decreased from 5 months for PGA to 1 week for $PLA_{50}GA_{50}$ and rapidly increased to 6.1 months for PLA_{100}. Reed and Gilding investigated comparatively in vitro the degradation of PLA_{100}, PGA and $PLA_{50}GA_{50}$ [119]. They found that PGA sutures were totally degraded within 10 to 12 weeks, $PLA_{50}GA_{50}$ discs within 2 to 4 weeks, whereas PLA_{100} discs lost only 10-15% of weight over a period of 16 weeks. Kaetsu et al. developed implantable luteinizing hormone releasing hormone (LHRH) agonist-LA/GA polymer composites and observed that in vivo degradation and release rates could be altered and controlled over a wide range by changing MW, stereo-isomerism, crystallinity, monomer side-chain length and copolymer composition [121]. PLA_{50} was found to degrade faster than PLA_{100} and, PLA_{100} was found to degrade faster than PLA_0 [121]. Zhu et al. carried out an in vivo degradation study of several DL-LA/GA homo- and copolymers and found that all corresponding threads were completely absorbed by 60 days [122]. Ogawa et al. studied the in vivo degradation of 1x10x10 mm plates made of DL-LA/GA polymers and found that both the lag time and the half-life decreased with the increase in GA content [123]. Amarpreet and Hubbell prepared a series of 66 terpolymers of DL-lactide, glycolide and ε-caprolactone [124]. The half-lives of these polymers were found to vary from a few weeks to several months.

Hyon et al. showed very little change in tensile strength and weight loss for high MW PLA_{100} fibres immersed in phosphate-buffered saline at 37°C for 6 months [125]. Tunc et al. evaluated a high MW PLA_{100} osteosynthesis device and observed a linear relationship between the stability and the reduction in MW [126-127] ; six weeks after implantation, 71% of the initial tensile strength were still present. Eitenmüller et al. investigated a selection of polymers and copolymers of L-LA, DL-LA and GA for use as resorbable implants in orthopaedic surgery [128]. Both in vivo and in vitro investigations showed comparable high values for the bending stability of high MW PLA_{100} and PLA_{95} over a period of 8 weeks. Degradation of $PLA_{90}GA_{10}$, $PLA_{87.5}$ and PLA_{75} was complete within one year and complete degradation of PLA_{100} was suggested to take about 2 years. Nakamura et al. studied the in vivo degradation of PLA_{100}, PLA_{50}, $PLA_{45}GA_{10}$ and $PLA_{40}GA_{20}$ [112]. The results showed that PLA_{50} degraded less rapidly than the two copolymers and that PLA_{100} was very stable since no weight loss was observed after 6 months for both the crystallized or the amorphous forms of PLA_{100}.

It is obvious from literature data that there are discrepancies between the degradation rates of similar PLAGA polymers, which can be attributed to the fact that polymeric materials identified by the same name were actually different compounds in agreement with particularities of synthetic polymers. According to our systematic investigation (Table 4), the half-life, in terms of weight loss, of initially amorphous PLA_{100} was 110 weeks and that of crystallized PLA_{100} was even longer [74]. The introduction of 4% of D-LA units to L-LA chains enhanced considerably the degradation of PLA_{100} since PLA_{96} had a half-life of about 90 weeks [75]. PLA_{50} degraded much more rapidly than PLA_{100} and PLA_{96} with a half-life of 10 weeks [72]. Copolymerization with GA strongly enhanced the degradation, $PLA_{37.5}GA_{25}$ having a half-life of 3 weeks (Table 4).

It is noteworthy that the degradation rate of a polymer is closely related to the rate of water absorption or swelling. Semicrystalline polymers are intrinsically less inclined to water diffusion and swelling than amorphous polymers. For example, PLA_{100} and PLA_{96} absorbed respectively 7% and 12% after 10 weeks in a phosphate buffer, while PLA_{50} absorbed 150% and largely swelled [72,74,75]. In the cases of PLAGA copolymers, $PLA_{75}GA_{25}$ absorbed only 3% after 3 weeks, whereas $PLA_{37.5}GA_{25}$ absorbed 23% in the meantime [73]. Again, the rate and ratio of water absorption depend on many factors such as initial morphology, chemical composition and configurational structure, MW and MW distribution, polymerization conditions, etc.

Table 4 In vitro degradation rates of PLAGA polymers

Polymer	$\overline{M}_w$ (E-3)	$\overline{M}_w / \overline{M}_n$	Half-life (weeks)
PLA_{100}	130.0	1.8	110
PLA_{96}	100.0	1.7	90
$PLA_{87.5}$	190.4	1.9	80
PLA_{75}	48.6	1.6	22
$PLA_{62.5}$	67.6	1.9	23

PLA_{50}	65.0	1.6	10
$PLA_{85}GA_{15}$	112.0	1.8	20
$PLA_{75}GA_{25}$	111.0	1.8	10
$PLA_{37.5}GA_{25}$	51.0	1.6	3

It is noteworthy that the degradation rate of a polymer is closely related to the rate of water absorption or swelling. Semicrystalline polymers are intrinsically less inclined to water diffusion and swelling than amorphous polymers. For example, PLA_{100} and PLA_{96} absorbed respectively 7% and 12% after 10 weeks in a phosphate buffer, while PLA_{50} absorbed 150% and largely swelled [72,74,75]. In the cases of PLAGA copolymers, $PLA_{75}GA_{25}$ absorbed only 3% after 3 weeks, whereas $PLA_{37.5}GA_{25}$ absorbed 23% in the meantime [73]. Again, the rate and ratio of water absorption depend on many factors such as initial morphology, chemical composition and configurational structure, MW and MW distribution, polymerization conditions, etc.

Little attention has been paid to chemical composition changes of the copolymers during degradation. Fredericks *et al.* claimed that there was no preferential attack on either the LA or GA units of Vicryl® polymer chains despite the difference in hydrophilicity [106]. In contrast, Hutchinson and Furr reported that the LA/GA ratio increased with degradation of PLAGA copolymers as GA blocks degraded preferentially [60]. Degradation-related composition changes were observed in the case of $PLA_{75}GA_{25}$ and $PLA_{85}GA_{15}$, as shown in Fig. 9. The L-LA-enriched segments crystallized under *in vitro* degradation conditions to form crystalline zones [73-75]. The crystallization limited in return the preferential degradation of GA units which were included in crystalline zones, as confirmed by Grijmpa *et al.* [117]. In the case of $PLA_{37.5}GA_{25}$, no significant preferential degradation of GA units was observed probably due to the fact that $PLA_{37.5}GA_{25}$ degraded too fast [73].

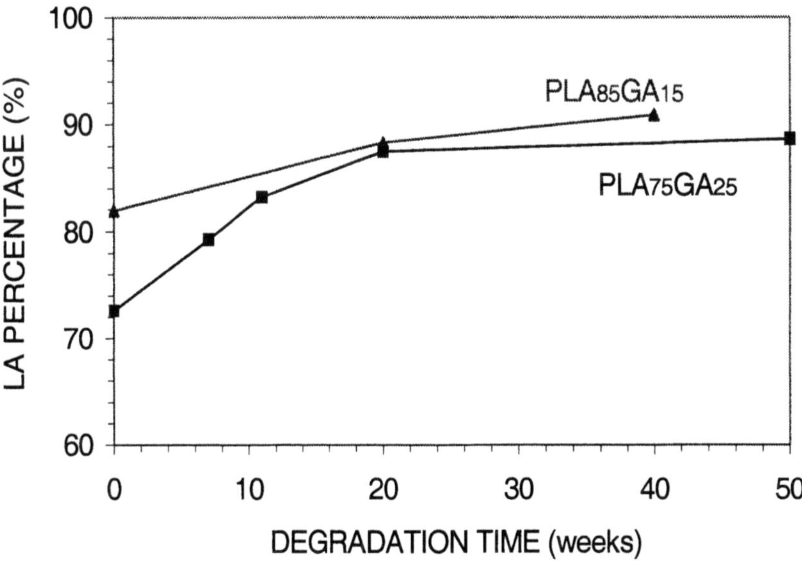

Fig. 9. Evolution of lactic acid (LA) percentage of $PLA_{85}GA_{15}$ and $PLA_{75}GA_{25}$ with in vitro degradation.

In conclusion, the degradation rate of PLAGA polymers can be modified by varying chemical and configurational structures. Copolymers and stereocopolymers degrade generally faster than homopolymers. Chemical composition changes can be observed for PLAGA copolymers.

3.1.3 *MW and MW distribution*

MW and MW distribution are important factors in polymer degradation because of the autocatalytic character of aliphatic polyester hydrolysis. Vert et al. recommended the elimination of low MW products by the dissolution-precipitation method [23], a means to minimize degradation on processing. This procedure allowed the same authors to process totally bioresorbable composites composed of a semicrystalline PLA matrix reinforced by PGA fibers [129]. Leeslag et al. also showed that the purification of PLA_{100} led to a material more resistant to hydrolytic degradation [107]. Pitt et al. found that PLA_{50} films with an initial $\overline{M}_n$ of 14,000 was absorbed by week 28, compared with 60 weeks for those with an initial $\overline{M}_n$ of 49,000 [56]. Chawla and Chang investigated the in vivo degradation of 4 PLA_{50} samples having different high MW and found that, at the end of a 48 week implantation period, samples with lower MW degraded faster [130]. Kaetsu et al. studied in vivo degradation of 4 PLA_{50} needle samples with relatively low MW (1,100 to 2,200 determined by terminal group titration) [121]. In this case, the higher the initial MW, the slower the degradation rate. Ogawa et al. observed that both the lag time and the half-life of PLA_{50} plates increased with MW during in vivo degradation [123]. Fukuzaki et al. studied the degradation of different PLA_X with relatively low MW [81]. The authors proposed three types of PLA_{50} degradation profiles according to MW, namely parabolic, linear and S types. The parabolic type observed for $\overline{M}_n$ =1,500 daltons can be explained by the dissolution of the soluble fraction of initial oligomers, whereas the S-type ($\overline{M}_n$ =3,500 daltons) reflects a lag-time due to higher MW.

The effect of MW distribution, or polydispersity ($\overline{M}_w/\overline{M}_n$) on PLAGA polymer degradation has so far been investigated only qualitatively. Nevertheless, the presence of low MW chains and/or monomers does accelerate the degradation. The elimination of low MW and residual monomers is thus very important if one wants a relatively stable polymer. Nakamura et al. observed an important increase in degradation rate of PLA stereocopolymers in the presence of DL-lactide and glycolide [112]. Mauduit et al. showed that oligomers accelerated the rate of degradation of PLA_{50} films [131]. It may be concluded then that the lower the MW, the faster is the degradation rate, in agreement with the presence of more carboxylic acid catalyzing groups. The incorporation of low MW oligomers or monomers also increases the degradation rate of PLAGA polymers

3.1.4 *Size and porosity*

The size of polymer samples is also an important factor. Kwong et al. observed a rapid disintegration of insulin-PLA_{100} pellet matrix 28 days after implantation as compared with microbeads which remained intact. The faster degradation of pellets was attributed to porosity [132]. Ginde and Gupta examined the influence of macroscopic dimensions on the chemical degradation of PGA fibres and pellets [57]. The degradation rate was found to be much faster in pellets than in fibres. This feature was assigned to the lack of long range order in pellets. Visscher et al. investigated the effect of particle size on *in vitro* and *in vivo* degradation of three $PLA_{25}GA_{50}$ microsphere samples whose average diameters were respectively less than 45-75, 75-106 and 106-177 mm. The largest microcapsules exhibited a slightly greater trend to undergo both in vivo and in vitro degradation relative to other groups [133]. However, it was concluded that over the examined size range, there were minimal differences in the degradation properties of polymeric matrices and thus of microcapsules. In contrast, Törmälä et al. evaluated the strength retention of self-reinforced PGA rods with different diameters (1.5, 2.0, 3.2 and 4.5 mm) in a phosphate buffer [134]. Small rods were found to lose strength more rapidly than large ones. Differences were related to surface area/weight ratios which controlled water diffusion and thus hydrolysis. Mauduit investigated the degradation of low MW and high MW PLA_{50} particles and films in vitro [135]. In contrast to large size devices, corresponding SEC chromatograms remained monomodal during the whole degradation period. Zhu et al. observed the same phenomenon for microspheres of PLA and two PLAGA copolymers [136]. Grizzi et al. found that degradation of PLA_{50} in forms of films, powder and microspheres was much slower compared to that of large-size specimens [93], showing conclusively that the smaller the polymer size, the slower the degradation rate. This unexpected behavior was well explained by the mechanism of faster internal degradation described above. It is of interest to keep in mind that literature is often misleading in comparing degradation rates of microspheres, fibres, films, pellets and other massive implants [57, 132].

The porosity of the polymer matrix is an influencing factor. Lam *et al.* evaluated the *in vitro* and *in vivo* degradation of PLA_{100} films. The authors observed a faster degradation of non-porous films as compared with porous ones [137]. Smith and Hunneyball prepared 1 - 10 μm PLA_{50} microspheres by solvent evaporation and microparticles by grinding the glassy solid formed by the cooling of molten PLA, both containing various amounts of prednisolone [138]. The microparticles exhibited a more prolonged drug release profile, indicating that the fusion process may have substantial advantages for thermostable drugs requiring long-term release. Eenink *et al.* developed hollow fibres for the delivery of hormones (tritium labeled-levonorgestrel). The release was found to be dependent on the membrane structure of the hollow fibre wall. A zero order release was achieved both *in vitro* and *in vivo* for as long as 6 months [139].

3.1.5 *Chemically reactive additives*

In the case of drug delivery systems, matrix/drug interactions can be critical for both matrix degradation and drug delivery profiles. For acidic drugs, one can expect faster hydrolysis of ester bonds because of acid catalysis. In contrast, in the case of basic drugs, two phenomena can be expected; base catalysis of ester bond cleavage, and neutralization of carboxyl endgroups of polymer chains, which minimizes or eliminates the autocatalytic effect. Thus, the degradation may be accelerated or slowed down by a base depending on the relative importance of the two effects.

Maulding *et al.* reported an acceleration of the degradation of PLA_{50} in the presence of a basic drug, thioridazine, which is a tertiary amine compound. Catalysis was attributed to the nucleophilic nature of the amino group [140-141]. Kishida *et al.* obtained similar results [142]. The authors found that the degradation rates of PLA_{100} matrix, which normally hydrolyzes slowly in aqueous media, could be increased by incorporating basic compounds, such as cinnarizine, thioridazine, indenolol and clonidine, the catalytic effect increasing with the loading. The degradation rate depended also on the apparent pK_a of the conjugate acid of each basic compound. The higher the pK_a, the faster the degradation according to the following order : indenolol > clonidine > thioridazine > cinnarizine [142]. Cha and Pitt examined the effect of four tertiary amines: methadone, naltrexone, promethazine and meperidine on the hydrolytic rate of PLA_{100} and of $PLA_{80}GA_{20}$ [143]. It was found that the catalytic effect of the amines increases with the loading, but the order of activity (meperidine > methadone > promethazine >> naltrexone) did not correlate with the pK_a nor with $\log(P_{oct})$ (octanol-water partition coefficient) of the amines, in contrast to data reported by Kishida *et al.* Cha and Pitt suggested that the steric accessibility of the unsolvated nitrogen might be the determining factor. Moreover, drug release was degradation-controlled ; faster degradation corresponding to shorter induction time prior to drug release [143]. The same authors also examined the case of PLACL copolymer (75-85 mol% L-LA units). In contrast to PLA_{100} and of $PLA_{80}GA_{20}$, Fickian diffusion was responsible for the drug release kinetics [144]. Fitzgerald and Corrigan investigated the mechanism governing the release of levamisole from $PLA_{25}GA_{50}$ microspheres and discs [145]. Drug release profiles were sigmoidal and fitted a degradation controlled release model. On the other hand, polymer degradation was dramatically accelerated with increasing additive content.

Other authors observed a decrease of the degradation rate or of the drug release rate due to interactions between polymer chain ends and drugs. Bodmeier and Chen evaluated degradable PLA pellets prepared by pressing without heat or solvent [146]. They detected a lag-time in drug release in the case of low MW PLA_{50} pellets. This was assigned to interactions between drugs (quinidine sulfate or propranolol hydrochloride) and carboxyl endgroups. Drug/polymer interactions were absent in the case of PLA_{100} pellets. The same authors suspected similar interactions in the case of microspheres and films made of high and low MW PLA_{50} blends containing quinidine [146]. Mauduit *et al.* investigated various gentamycin/PLA_{50} blends [131,135,147]. It was found that the base form of the drug was able to neutralize polymer chain ends on blending in acetone, whereas the sulfate form required an aqueous environment. In the case of gentamycin sulfate/low MW PLA_{50} systems, interactions between drug and chain ends stabilized the matrix, the free drug in excess being rapidly released [147]. Both phenomena acted against base catalysis.

The effects of sparingly soluble additives were also investigated. Verheyen *et al.* examined the physico-chemical properties of hydroxylapatite/PLA_{100} composites in solution tests [148]. Although the authors did not mention it, data showed that the higher the hydroxylapatite content, the slower the decrease of MW. Li and Vert investigated the hydrolytic behavior of PLA_{50}/coral blends [149]. It was found that coral, which is composed of calcium carbonate, significantly slowed down the degradation of PLA_{50} matrix and suppressed the faster internal degradation. Zhang *et al.* observed a decrease of degradation rate by 1.7 to 3.0 times with incorporation of salts such as $Mg(OH)_2$, $MgCO_3$, $CaCO_3$ and $ZnCO_3$ into $PLA_{25}GA_{50}$ films [150]. The stabilizing effect of hydroxylapatite, coral and other metal salts on the hydrolysis of polymer chains can be assigned to the neutralization of carboxyl end groups by the additives and/or by the degradation medium which penetrated the matrix due to the presence of the polymer/additive interfaces.

Li *et al.* investigated the mechanism of hydrolytic degradation of PLA_{50} matrix in the presence of a tertiary amine, namely caffeine, in order to elucidate the influence of this basic compound on the hydrolytic cleavage of polyester chains [116]. Caffeine was incorporated into PLA_{50} in various contents (0 to 20%) by blending in acetone followed by solvent evaporation. The resulting blends were processed to 1.5 mm thick plates and 0.3 mm thick films by compression moulding. Degradation was carried out in isoosmolar pH = 7.4 phosphate buffer at 37°C. The effects of caffeine on degradation characteristics were rather complex and largely depended on the loading. Low contents ($\leq$ 2%) of molecularly dispersed caffein accelerated considerably the degradation of matrices with respect to caffeine-free devices. However, the increase in degradation rate was not proportional to the caffeine content due to the combined effects of base/carboxyl endgroup interaction, crystallization and matrix-controlled or channeling-controlled diffusion of caffeine. In the early stages of degradation, the overall catalytic effect was larger for devices with low caffeine contents than for highly loaded ones where caffeine was in the crystalline state and thus less available for basic catalysis. During the later stages, however, the neutralization of carboxyl end groups became predominant and governed the degradation in the case of highly loaded devices. Generally, plates degraded slightly faster than films.

Therefore, the degradation of PLA polymers in the presence of basic compounds appeared rather complex because of the contribution of a number of parameters, i.e. base catalysis, neutralization of carboxyl endgroups, porosity, dimensions of devices, loading and morphology of added compounds. They all contribute to control the matrix degradation, and drug release profiles in the case of drug release systems. None of these factors can be considered separately if one wants to understand the effect of basic drugs on the properties of drug delivery systems.

3.1.6 *γ Irradiation*

Several authors examined the influence of γ irradiation on the degradation of PLAGA biomedical devices. The data are also of interest in the field of drug delivery. Gupta and Deshmukh claimed that irradiation in air or in a nitrogen atmosphere can generate chain scission and crosslinking simultaneously [151]. Chu *et al.* considered chain scission as predominant on the basis of the reduction in mechanical properties of Dexon® sutures upon γ irradiation [109]. It was speculated that the scission process depended on free radical chemistry. In addition, the authors observed formation of surface cracks during the hydro-lysis of Dexon® sutures whose number and regularity increased with increasing irradiation dosages. In another study, Chu showed that γ irradiation of Dexon® and Vicryl® fibres resulted in the earlier decrease of pH of the degradation medium [152]. All these observations are in agreement with a MW decrease on irradiation although no MW data are available.

Tsai *et al.* carried out effective sterilization of PLA/mitomycin C microcapsules for parenteral use by ^{60}Co γ irradiation [153]. The authors found that the sterilization did not affect microcapsule structure, release rate and drug stability. However, no MW measure-ments were performed. Spenlehauer *et al.* found that γ sterilization dramatically decreased the MW of PLA_{50}, $PLA_{45}GA_{10}$ and $PLA_{37.5}GA_{25}$ microspheres and this degradation continued on storage for GA-containing compounds [154]. γ sterilization also modified the release pattern of cisplatin-loaded microspheres. Birkinshaw *et al.* studied γ irradiation of compression moulded PLA_{50} samples in air with doses as high as 10 Mrads [155]. Substantial embrittlement occurred at higher doses and the irradiated material absorbed water at a slightly slower rate than without irradiation. The authors concluded that the primary effect of irradiation on hydrolytic degradation was associated with the initial reduction in MW, the degradation mechanism remaining the same. Volland *et al.* investigated the influence of γ sterilization on captopril-containing $PLA_{25}GA_{50}$ microspheres [156]. The MW of the polymer decreased with increasing irradiation dose, but the polydispersity remained unchanged, suggesting a random chain cleavage rather than an unzipping process.

3.1.7 *pH and ionic strength*

Chu examined the effect of pH on the degradation of Dexon® and Vicryl® sutures by using 3 different buffer solutions with pH = 5.25, 7.44 and 10.06, respectively [157-161]. For Vicryl® sutures, maximum retention of tensile properties occurred around pH 7.0. For Dexon® sutures, no major difference was observed between physiological and acidic media. In contrast, the alkaline buffer had a dramatic effect on Dexon® degradation [157-161]. Makino *et al.* observed that PLA_{100} degraded rapidly in a strongly alkaline solution or in solutions of high ionic strength. The effects of pH and ionic strength were interpreted in terms of electric potential distribution at the polymer-solution interface. Degradation was also found to be affected by salt concentration in buffer solutions, suggesting that the cleavage of polymer ester bonds was accelerated by conversion of the acidic degradation products into neutral salts [162].

Reed and Gilding considered that pH changes had no effect on the *in vitro* degradation of PLAGA polymers in citrate-phosphate buffers at pH 5 and pH 7 and boric acid-borax buffer at pH 9. Any sensitivity of PLAGA polymers to pH was actually dominated by combined effects of crystallinity and hydrophobicity, according these authors [119]. Kenley *et al.* studied the *in vitro* degradation of $PLA_{25}GA_{50}$ cylindrical samples in aqueous buffers with pH varying from 4.5 to 7.4 at 37°C [62]. They obtained superimposable total weight and MW profiles for all samples, suggesting pH-independence for the *in vitro* hydrolysis. Ginde and Gupta studied the *in vitro* degradation of PGA pellets and fibres in 4 different buffer solutions with pH 4.7, 7.0, 9.2 and 10.6 [57]. No major difference was found between the slightly acidic and neutral media. However, the two alkaline solutions considerably accelerated the degradation of PGA samples.

Li *et al.* compared the *in vitro* degradation of large-size PLAGA polymers in three different media; distilled water, isoosmolar phosphate buffer and saline [72-74]. There was no significant difference between the two isoosmolar media despite the pH fall observed in saline solution at the later stages of degradation. In contrast, the absence of ionic strength in distilled water promoted water absorption and thus the surface/centre differentiation in the early stages (Fig. 10). Furthermore, the phosphate buffer enhanced solubilization of degradation products in the later stages (Fig. 11), the carboxylate form $RCOO^-Na^+$ of organic acids being more hydrophilic than the carboxylic form RCOOH [73].

Therefore, one can reasonably conclude that alkaline and strong acidic media accelerate polymer degradation. The difference between slightly acidic and physiological media, however, is much less pronounced due to autocatalysis by carboxyl endgroups. The effects of pH and ionic strength on drug release profiles are rarely discussed in literature. Nevertheless, one can assume that for degradation controlled release, drug release should be enhanced if degradation is enhanced. In the case of diffusion controlled release, one should take into account the solubility of the drug in the external medium.

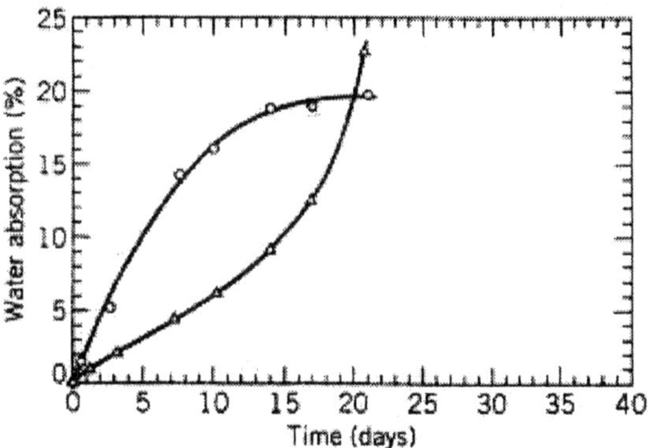

Fig. 10. Water absorption profiles of PLA37.5GA25 specimens with degradation at 37°C in a pH = 7.4 phosphate buffer (Δ) and in distilled water (∘) (Source : from Ref. 73)

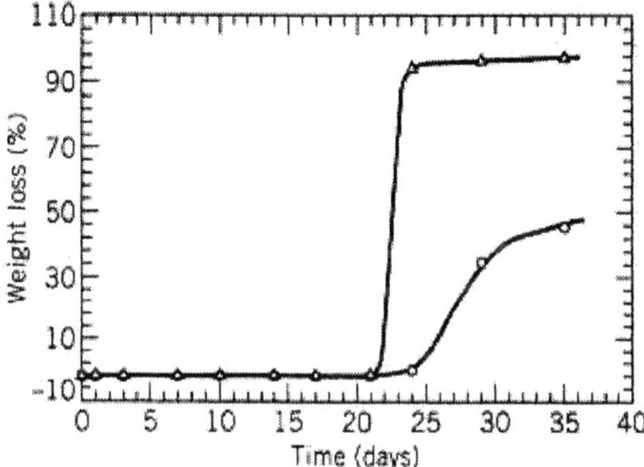

Figure 11. Weight loss changes of PLA37.5GA25 specimens with degradation at 37°C in a pH = 7.4 phosphate buffer (Δ) and in distilled water (∘) (Source : from Ref. 73)

3.1.8 *Comparison between in vitro and in vivo degradation*

Many authors have studied comparative matrix degradation behaviors *in vitro* and *in vivo*. Chegini *et al.* carried out a comparative SEM study on the degradation of Lactomer® ($PLA_{70}GA_{30}$) ligating clips *in vivo* by implantation in rabbits and *in vitro* by incubation in pH 7.3 phosphate buffered saline [163]. Lactomer® clips were found to show a greater change in *vivo* than *in vitro*. Törmälä *et al.* observed a faster loss of mechanical strength of self-reinforced PGA rods and PLA_{100} screws and plates *in vivo* than *in vitro* [134,164]. This difference in rates of strength loss was assigned to the effect of cellular enzymes or other biological or biochemical factors as well as to the mechanical stresses caused to the implants by physical movements of the rabbits. Albertsson and Karlsson also suggested that the higher *in vivo* degradation rate resulted probably from the dynamic environment of *in vivo* systems [18]. Therin *et al.* [76] and Spenlehauer *et al.* [154] also found a faster degradation *in vivo* than *in vitro*. Differences were assigned to mechanical stresses due to muscular movements of the rabbits [76], in agreement with data reported by Suuronen *et al.* [164] and Albertsson *et al.* [18].

Leeslag *et al.* investigated the *in vivo* degradation of high MW PLA_{100} devices used for fixation of mandibular fractures in sheep and dogs or subcutaneously implanted in rats. They also considered the *in vitro* degradation in pH=6.9 phosphate buffered saline [107]. These authors showed that except for dynamically loaded bone plates, there were no significant differences between *in vivo* and *in vitro* degradation, thus excluding the additional effect of enzymes. Kenley *et al.* also found no major difference between $PLA_{25}GA_{50}$ copolymer degradation kinetics *in vitro* in buffer solutions and *in vivo* after subcutaneous implantation in rats [62]. In contrast, Visscher *et al.* observed a much faster degradation of $PLA_{25}GA_{50}$ *in vitro* than *in vivo*, and attributed it to the totally aqueous environment and constant physical movement that the microspheres were subjected to *in vitro* [133].

Authors also compared the drug release behaviors *in vitro* and *in vivo*. Ikada *et al.* investigated degradation and release behaviors of a dideoxykanamycin B (DKB)-containing composite material prepared from hydroxyapatite and low MW PLA_{50} *in vitro* in pH=8 phosphate-buffered saline and *in vivo* by implantation into fenestrated tibias of rats [165]. Both the amount and the period of retention of DKB were much greater *in vivo* than *in vitro*, indicating a faster degradation *in vitro*. In contrast, Smith and Hunneyball observed a faster release of prednisolone from PLA_{50} microspheres *in vivo* than *in vitro*, which was assigned to the faster penetration by biological fluids and/or faster degradation of PLA *in vivo* [138]. On the other hand, Kwong *et al.* [132], Sanders *et al.* [61], and Heya *et al.* [166] obtained a good correlation between *in vivo* and *in vitro* drug release profiles.

Therefore the literature shows that in most cases, PLAGA polymers exhibited comparable *in vitro* and *in vivo* matrix degradation behaviors. Whenever differences are observed, effects of physical factors (temperature, stirring, pH) or physiological ones related to implantation sites (subcutaneous, intramuscular or in bony tissues) should be considered prior to any reference to enzymatic activities.

3.1.9 *Enzymes*

As mentioned above, the enzymatic degradation of PLAGA polymers can be affected by a number of enzymes [35-42]. Recently, McCarthy's group carried out a detailed study on the degradation of a series of PLA stereocopolymers by *proteinase K* [37-39]. It was observed that *proteinase K* preferentially degraded L-lactyl units as opposed to D-lactyl ones ; poly(D-lactide) being non-degradable. Fig. 12 shows weight loss data of three PLA stereocopolymers in the presence of *proteinase K.* PLA_{40} degraded the most rapidly with a weight loss of 96% after 168 hours. PLA_{25} exhibited a lower degradation rate, with about 45% of weight loss. In the case of PLA_{10}, very little weight loss was detected, less than 5% after 168 hours. On the other hand, the enzymatic degradation preferentially occurred in the amorphous regions of semicrystalline PLA polymers. Water uptake appeared to be another important factor in the enzymatic degradation as water could lead to swelling of the polymer and thus facilitate enzymatic attack [42].

The enzymatic degradation is a surface phenomenon. It consists of two steps : first, enzyme adsorbs on the surface of a polymer matrix through its bonding domain ; second, ester bonds are cleaved due to the effect of the calalytic domain of the enzyme. The surface morphology changes of PLA polymers with enzymatic degradation were examined by environmental scanning electron microscopy (ESEM). As shown in Fig. 13a, PLA_{25} had initially a smooth surface with the presence of some stripes resulting from the mold [42]. After only 24 hours in the degradation medium, numerous tiny pores of several μm appeared at the surface. After 168 hours, the film appeared greatly degraded with large pores (Fig. 13b).

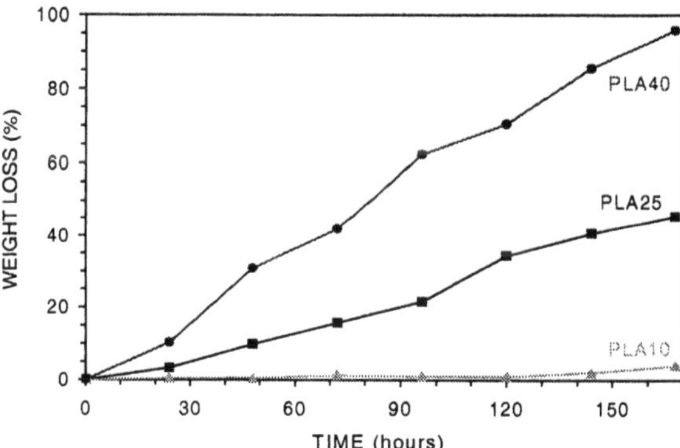

Fig. 12. Weight loss changes of PLA_{10}, PLA_{25} and PLA_{40} films with enzymatic degradation in the presence of proteinase K (Source : from Ref. 42)

The enzymatic degradation is a surface phenomenon. It consists of two steps : first, enzyme adsorbs on the surface of a polymer matrix through its bonding domain ; second, ester bonds are cleaved due to the effect of the calalytic domain of the enzyme. The surface morphology changes of PLA polymers with enzymaticdegradation were examined by environmental scanning electron microscopy (ESEM). As shown in Fig. 13a, PLA_{25} had initially a smooth surface with the presence of some stripes resulting from the mold [42]. After only 24 hours in the degradation medium, numerous tiny pores of several μm appeared at the surface. After 168 hours, the film appeared greatly degraded with large pores (Fig. 13b).

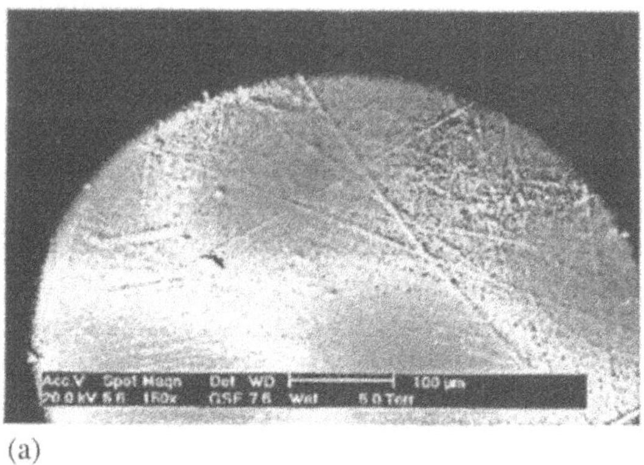

(a)

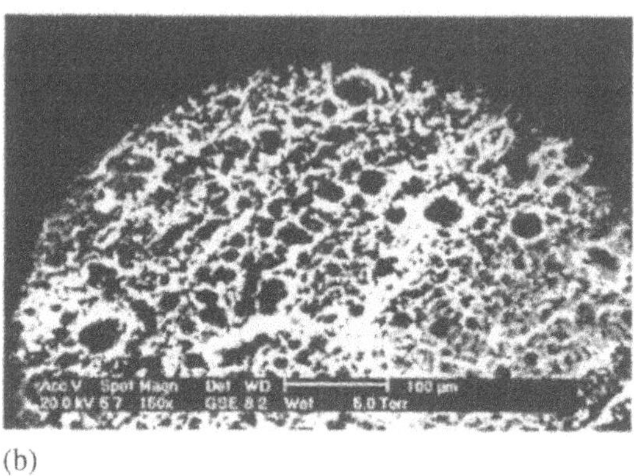

(b)

Fig. 13. ESEM micrographs of PLA_{25} after 0 (a) and 168 hours (b) enzymatic degradation in the presence of proteinase K (Source : from Ref. 42)

3.2 POLYCAPROLACTONE AND COPOLYMERS

PCL is an important member of the aliphatic polyester family. It is mostly investig-ated for use in drug delivery systems. The degradation of PCL and its copolymers is known to proceed in at least two distinct stages as in the case of PLA [56, 94]. The first stage of the degradation process involves non enzymatic, random hydrolytic ester cleavage, autocatalyzed by carboxyl end groups of polymer chains. The duration of the first stage is determined by the initial MW of polymer as well as its chemical structure. When the MW has decreased to about 5000, the second stage starts with the slowing down of the rate of chain scission and the beginning of weight loss because of the diffusion of oligomeric species from the bulk. The polymer becomes prone to fragmentation and, at this point, either enzymatic surface erosion or phagocytosis can contribute to the absorption process [56, 94]. Pitt and Gu evaluated the modification of the rate of PCL chain scission as films in water, alcohols, acidic and basic reagents [167]. It was observed that partial ethoxylation of the carboxyl end groups of PCL reduced the degradation rate in water, consistent with an autocatalytic mechanism. However, the discrepancy between calculated and experimental data suggested that the hydrolysis mechanism might not be exclusively autocatalytic and that uncatalyzed hydrolysis contributed also to the chain scission rate. On the other hand, Gabelnick found that PCL system with an initial Mn of 50000 required three years for total removal from the body, showing a remarkable slow degradation [168].

Fig.14 shows the MW changes of PCL with in vitro degradation. $\overline{M}_w$ decreased steadily from initial 58,700 to 7,000 after 200 weeks. After 133 weeks, the MW distribution became trimodal due to the selective degradation of amorphous zones and of the crystalllite edges. The MW of the three peaks were 2,600, 5,200 and 8,800, respectively, corresponding to one, two and three times the thickness of the crystallite lamellae.

3.2.1 *Morphology*

PCL behaves similarly to PGA in that residual crystallinity increases with time. Pitt et al. observed a steady increase in crystallinity of PCL films from 45% to nearly 80% after 120 weeks implantation, which was attributed to crystallization of tie segments made possible by the chain cleavage in the amorphous phase, facilitated by the low glass transition temperature of PCL (-60°C) [94]. Similar results were obtained by Pitt and Gu in an in vitro degradation study [167]. Jarrett et al. investigated the degradation mechanism of PCL films and single crystals in phosphate buffered solution with enzymes (cryptococcus, fusarium) and in 40% methylamine [101]. Interesting results were obtained by SEC which showed that, in the case of PCL films degrading in presence of enzymes, the high MW peak was simply decreasing in size with no discernable change in its position or distribution. In contrast, the chemical degradation showed a shift toward low MW along with appearance of single and double traverse length peaks which resulted from the selective degradation of amorphous regions, as detected earlier for PLA_{100} [74]. This finding was attributed to the inability of the enzyme system to diffuse into the polymer matrix, the 40% methylamine solution penetrating readily. Insofar as the biodegradation of PCL single crystals is concerned, SEC results showed a MW shift, reflecting the presence of an endo-enzyme [101].

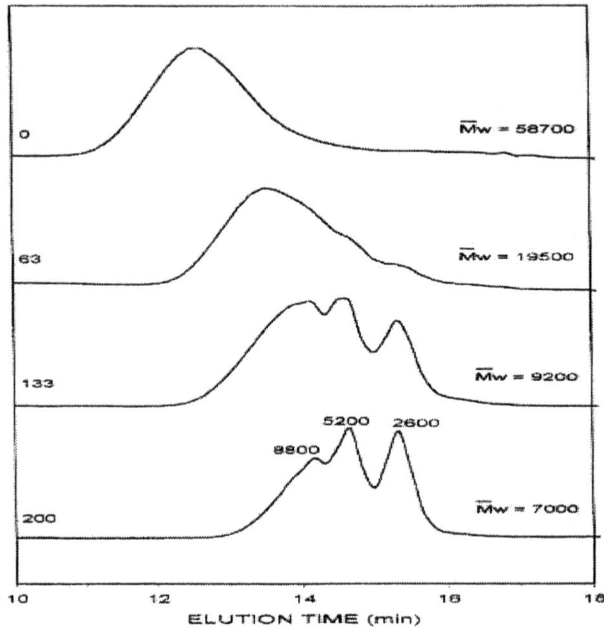

Figure 14. SEC chromatograms of PCL after 0, 63, 133 and 200 weeks in vitro degradation (Source : from Ref. 97)

3.2.2 *Copolymers with other lactones*

Pitt et al. investigated the *in vivo* degradation of PCL and of copolymers of ε-caprolactone with δ-valerolactone and DL-ε-decalactone in rabbits [56, 94]. The mechanisms of biodegradation of these polymers were found to be qualitatively similar despite a range of structures and morphologies. The degradation of the copolymers of PCL with DL-ε-decalactone and with δ-valerolactone showed that the copolymer with greater amount of comonomer degraded more rapidly due to a reduction in crystallinity [56]. The brittleness of implanted films and capsules was brought about by an increase in polymer crystallinity associated with chain cleavage in the amorphous phase and crystallization of the resulting unrestrained tie segments [94]. Subsequent studies by Woodward et al. support the proposition that intracellular degradation of PCL was the principal *in vivo* degradation pathway once the $\overline{M}_n$ of the polymer fell to 3000 [47], with a microbial degradation (*P. pullulans*) study by Fields et al. showing the necessity of low MW for enhanced polymer degradation [169].

Data on the biodegradation of a series of homo- and copolymers of ε-caprolactone, crosslinked to varying degrees with 2,2-bis(ε-caprolactone-4-yl)propane (BCP), were reported later by Pitt et al. as previously mentioned [44]. These elastomeric polymers were found to be bioabsorbed by an enzymatic surface erosion process. Concurrent with the surface erosion, the crosslinked polymers were subject to slower nonenzymatic hydrolysis of ester groups.

3.2.3 *Copolymers with lactic and glycolic acids*

Copolymers of PCL with PLA or PGA have also been considered. Song *et al.* studied the degradation of random and block copolymers of ε-caprolactone with DL-lactide and found that the random copolymers degraded faster than the parent homopolymers, block copolymers degrading at intermediate rates [170]. Li and Feng investigated the degradation behaviors of ABA, ABC and ACB triblock copolymers of ε-caprolactone (A), DL-lactide (B) and glycolide (C) in water at 37°C and 50°C [171]. The order of degradation rates of the different phases was found to be C>B>>A, which permitted the variation of degradation rates by changing copolymer composition. During the degradation process, the crystallinity of the A phase increased. The composition changed too, with increase in A and decrease in B and C content. Fukuzaki *et al.* synthesized a series of low MW L-LA/CL copolymers by direct condensation, and studied the *in vivo* degradation by subcutaneous implantation in the back of rats [172]. Pasty copolymers (30-70 mol% CL) degraded faster than solid (0-15 mol% CL) and waxy copolymers (85-100 mol% CL). Grijmpa et al. described the preparation of high MW CL/L-LA copolymers by ring-opening polymerization and evaluated their mechanical properties as implants [173].

Pitt et al. investigated the in vivo degradation of copolymers of ε-caprolactone with DL-lactide in rabbits [56]. Copolymers with 11, 23, 47 and 90 mol% of DL-lactide were found to degrade much more rapidly than corresponding homopolymers. Cha and Pitt investigated the in vitro degradation in phosphate buffer of blends of PCL, PLA_{100} and $PLA_{83}GA_{17}$ prepared by three different methods: compression moulding, coprecipitation and solvent evaporation of a methylene chloride-in-water emulsion of the polymers [174]. They found that the degradation rate was dependent on blending methods. For compression moulded blends, the degradation rate of $PLA_{83}GA_{17}$ decreased but that of PCL and PLA_{100} increased. However, there was no evidence of blend miscibility.

Li et al. investigated the degradation of a Zn metal initiated PLACL copolymer [89]. The degradation rate was very much enhanced as shown by the increased water absorption and weight loss rates, in comparison with the parent homopolymers PCL and PLA_{100}. Crystallinity increased from initial 14% to 52% at 63 weeks. The crystalline structure was of the PLA_{100}-type, showing a phase separation between the two components.

3.2.4 Enzymes

Poly(ε-caprolactone) (PCL) is one of the most promising synthetic polymers since it can degrade in an aqueous medium or in contact with microorganisms, and thus can be used to make compostable polymeric devices [101-103]. The enzymatic degradation of PCL polymers has also been investigated, especially in the presence of *lipase*-type enzymes [41, 45,46,172]. Three kinds of *lipase* were found to significantly accelerate the degradation of PCL, *i.e. Rhizopus delemer lipase* [172], *Rhizopus arrhizus lipase* [45], and *Pseudomonas lipase* [41,46]. Highly crystalline PCL was reported to totally degrade in 4 days, the crystallinity decreasing during degradation [46], in contrast to hydrolytic degradation which takes several years.

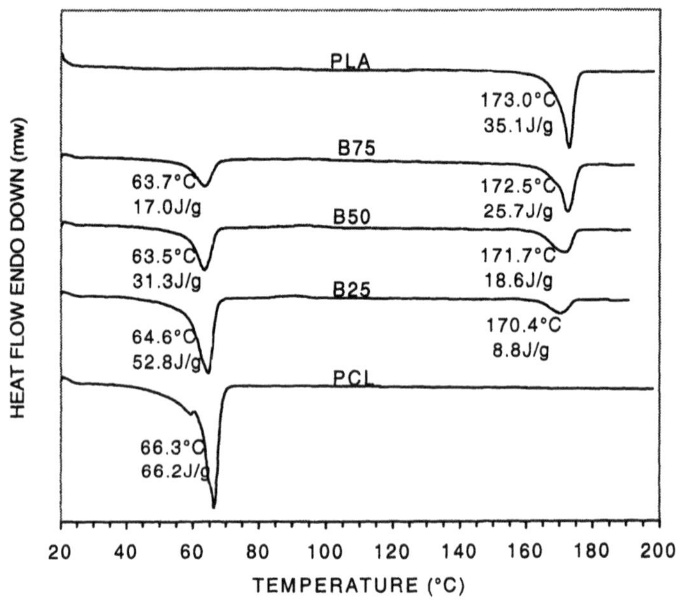

Fig. 15. DSC thermograms of PLA_{100} and PCL homopolymers and blends with PLA_{100}/PCL ratios of 75/25 (B75), 50/50 (B50) and 25/75 (B25) (Source : from Ref. 41)

The enzymatic degradation of PCL and its blends with PLA_{100} were examined in the presence of *pseudomonas lipase* or *proteinase K* [41]. The two polymers in the blends exhibited well separated crystalline domains as shown in Fig. 15. PLA_{100} and PCL homopolymers showed only a melting temperature (Tm) at 173.0°C and 66.3°C, respectively, the corresponding melting enthalpy (ΔHm) being 35.1 and 66.2 J/g. In contrast, the three blends showed two melting peaks belonging to the two components. The Tm values of both components were slightly lower than those of the pure polymers. The ΔHm values were almost proportional to the contents.

PCL films showed well defined spherulites of 50-100 μm whose boundaries could be clearly distinguished (Fig. 16a). Dramatic changes were observed in the presence of *pseudomonas lipase*. The surface was strongly eroded to leave sponge-like and fibrillar structures (Fig. 16b), showing that both amorphous and crystalline zones can be degraded. The boundaries between these zones became better defined because they were mainly composed of crystallite defects or amorphous material which could be preferentially degraded. The presence of both sponge-like and fibrillar structures might reflect different degrees of degradation, the former reflecting a more advanced degradation.

In the case of the blends, the selective degradation of PCL or PLA_{100} components revealed their inner morphology. After degradation by *proteinase K*, the surface of PLA_{100}/PCL(25/75) showed numerous pores of several micrometers (Fig. 17a). Spherulitic structures could also be distinguished. In the presence of *pseudomonas lipase*, the surface presented a microsphere-like pattern (Fig. 17b). In fact, PLA_{100} spherical microdomains were regularly dispersed within the PCL continuous matrix. *Proteinase K* selectively degraded the PLA_{100} microdomains, leaving a porous structure. In contrast, *pseudomonas lipase* selectively degraded the PCL continuous matrix, leaving PLA_{100} microdomains in the form of microspheres [41].

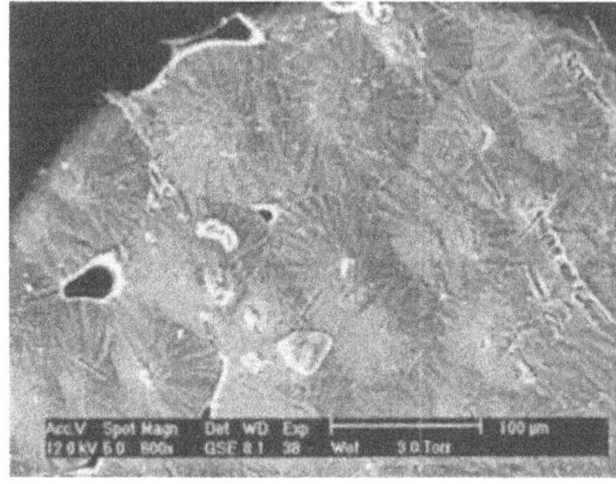

(a)

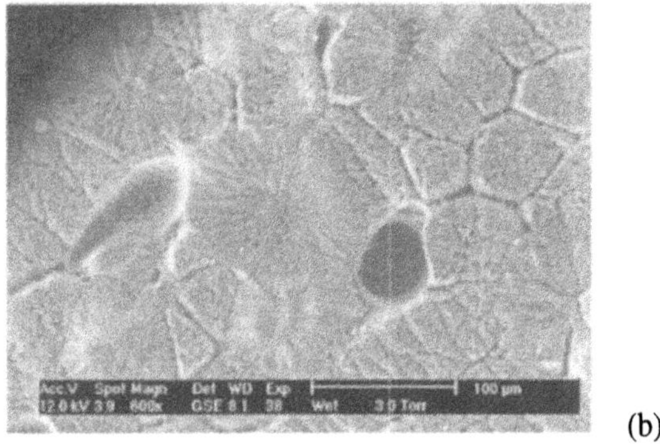

(b)

Fig. 16. ESEM micrographs of the surface of a solution cast PCL film before (a) and after 72 hours degradation in the presence of pseudomonas lipase (b) (Source : from Ref. 41)

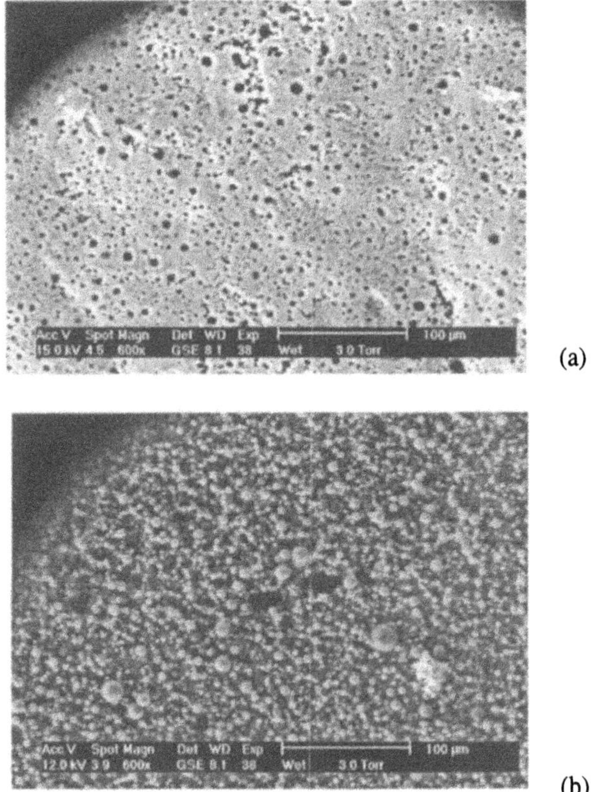

(a)

(b)

Fig. 17. ESEM micrographs of the surface of a PLA_{100}/PCL(25/75) blend film after 72 hours degradation in the presence of Proteinase K (a) and in the presence of pseudomonas lipase (b) (Source : from Ref. 41)

3.3 POLY(HYDROXYBUTYRATE) (PHB) AND COPOLYMERS (PHBHV)

PHB is a polymer obtained from many strains of bacteria. Von Korsatko claimed that PHB of various MW can be readily manufactured depending on the extraction method [175]. Grassie et al. and Tanahashi et al. described how PHB can be prepared synthetically [176-177]. Bleoembergen et al. reported the synthesis of PHBHV copolymers by coordination polymerization of β-lactones [178]. PHBHV copolymers from 0 to 30% HV contents are available as Biopol® on the market.

3.3.1 *Composition, temperature, physical form and MW*

Work by Gilding on a PHB sample with MW > 2 x 10^6 showed that thermal processing caused a decrease in MW, and in vitro hydrolysis showed no apparent degradation after 6 months [10]. Mergaert et al. reported data on the in vitro degradation of test pieces of PHB and PHBHV (90/10 and 80/20) copolymers in phosphate buffer at temperatures from 4° to 55°C and in sterilized freshwater and seawater at 15°C [179]. No weight loss was observed in any of the solutions, at any temperature after 98 days.

Holland et al. investigated the hydrolytic degradation of PHB together with a series of PHBHV copolymers under various conditions [68]. The results showed that alteration of the copolymer ratio by increasing the HV content enhanced the degradation rate probably because of a fall in crystallinity. Additionally, a change from physiological temperature to 70°C increased the degradation rate by some 30 to 100-fold. The physical form of the polymer matrix also had a pronounced effect on the degradation rate. Melt pressed discs and injection moulded samples degraded less rapidly than solvent cast films or cold compressed tablets. Differences were tentatively assigned to lower crystallinity and higher porosity in the latter cases. Moreover, the authors found that polymer MW was a dominant parameter, with large decrease in degradation rate occurring as polymer MW increased from 3.6 x 10^6 to 3 x 10^6. Finally, a comparison with Dexon®, Vicryl® and PDS® sutures showed that the degradation rate of PHB was much slower than any of these commercial sutures. Even in biological environments, the authors believed that the underlying hydrolytic mechanism is likely to prevail although some low esterase activity may be involved [68].

3.3.2 *pH and chemically reactive additives*

Holland et al. observed that increasing alkalinity speeded up the degradation rate, indicating ester hydrolysis as degradation mechanism [68]. A more complicated pattern emerged from work by Knowles and Hastings who studied the in vitro degradation of a PHBHV (93/7) copolymer in the form of tensile test bars in aqueous media with different pHs [69]. It was observed that degradation behaviors of this copolymer and the effect of pH did not follow an easily definable pattern. Alkaline media were found to facilitate weight loss, but the mechanical properties and MW changes were pH-independent. SEC analyses showed that the copolymer exhibited initially a bimodal MW distribution, i.e., two populations of different MW were present. As degradation proceeded, a drop in the high MW peak and the production of a pronounced low MW peak were detected. It was concluded that bimodal MW distribution and pH played a significant role in the degradation pattern. In addition, the authors suggested that hydrolysis occurred by different mechanistic pathways, depending on the pH of the surrounding medium. In acidic and neutral solutions, hydrolysis proceeded by a protonation process, whereas in alkaline media, hydroxyl ions were attached to the carbonyl carbons [69].

Yoshioka et al. examined the influence of basic additives on PHBHV polymer degradation [142, 180]. They found that the degradation of PHB and PHBHV copolymers could be accelerated by incorporating basic compounds and that the rate could be controlled by changing the loading and the basicity of basic compounds. On the other hand, the hydrolysis rate was governed by water solubility and polymer-water partition of incorporated bases, and corresponded with the diffusion rate of water in films.

3.3.3 *Selective degradation of amorphous zones*

Welland et al. investigated the selective degradation of solution-grown freeze-dried PHB single crystals with gaseous methylamine by using SEC to follow MW distribution changes [181]. In the course of degradation, two narrow peaks were observed in the chromatograms whose MW were equal to once and twice the value of the lamellae thickness as determined by small angle X-ray scattering, meaning that the etching was confined to the lamellae surface and did not progress significantly into the crystal interior. Slight shortening of the single and double traverse fragments observed during degradation and the persistence of the double traverse peak were explained by the presence of a small fraction of poorly accessible folds. These findings confirmed the selective degradation mechanism in the case of semicrystalline polyesters and were in good agreement with what had been reported in literature concerning the degradation of $PLA_{92.5}$ single crystals [104] and of semicrystalline PLA_{100} [74].

3.3.4 *Bacterial and enzymatic degradation*

The extent and nature of enzymatic involvement in PHB degradation were discussed by Holmes who reported an acceleration of degradation in the presence of certain bacteria [182]. Recently, Cox reported that the biodegradation rate of Biopol tended to be faster for PHBHV copolymers compared to PHB homopolymer in certain environments [49]. However, the biodegradation rate of Biopol copolymers of different HV contents appeared to be similar. According to Cox, the surface erosion mechanism was favoured by a large surface area because of the enzymatic involvment ; an increase in film thickness leading to a decrease in biodegradation rate. Moreover, the decrease in initial MW and crystallinity as well as the presence of low MW additives increased the biodegradation rate. In contrast, Doi et al. found in, a study of the biodegradation of microbial and synthetic poly-(hydroxyalkanoates) (PHA), that the rate of surface erosion of P(R-3HB) film by PHA depolymerase decreased with an increase in crystallinity, but was little influenced by the size of spherulites [50]. The rate of enzymatic degradation was strongly dependent on the molecular structure of the monomer units in poly(3-hydroxyalkanoates), the enzyme being unable to hydrolyze the sequences of (S)-3HB units.

Stinson and Merrick reported in vitro degradation of PHB by Pseudomonas Leimoignei [183] and Mergaert et al. studied the biodegradation of PHB and PHBHV copolymers in soil and in compost at different temperatures [179]. It was concluded that these polymers were degraded in all natural environments studied, and the copolymers tended to be degraded faster and depended on the environment and on temperature. In natural soils, the polymers were degraded by a variety of microorganisms such Gram-negative bacteria, Gram-positive bacilli and streptomycetes, as well as moulds.

Kemnitzer et al. investigated the relative degradability of a series of PHB stereocopolymers, having (R)-repeat unit contents from 50% to 100%, with a PHB depolymerase enzyme isolated from Penicillium funiculosum [27]. Two opposing effects on the degradation rate were studied: the increase due to the disruption of the crystalline phase and the decrease due to a stereochemical enzyme impediment with increasing (S)-HB content, the used enzyme catalyzing only the hydrolysis of (R)-PHB. It was shown that for stereocopolymers with (R)-HB contents greater than 81%, the degradation rate was lower than that of 100% (R)-PHB of similar MW. Therefore, the preference for (R)-HB repeat units dominated over crystalline morphology effects for the compositional range 81-100%. However, at lower (R)-HB contents, effects of crystalline morphology prevailed [27].

Gilmore et al. investigated the biodegradability of a PHBHV (80/20) copolymer blended with ester-substituted celluloses such as CAB® (Eastman Kodak EAB 500-1; Butyryl 48%, acetyl 6%, Hydroxyl 0.7%) and CAP® (Bayer cellit PR 900; propionyl 45%, acetyl 3.5%, hydroxyl 1.6%) [54]. It was found that both environmental (sewage) and enzymatic assays on the blend films showed a strong inhibiting effect of CAB or CAP on the copolymer degradation. Weight loss never exceeded 10% when the cellulose ester content was 50% or higher. Kinetics of degradation of blends containing 25% CAP or CAB were different, the blend with CAP degrading more rapidly in both the sewage and enzymatic solutions.

3.3.5 *In vivo degradation*

In contrast to the extensive investigations on the bacterial degradation of PHB, few data are available in the literature on the *in vivo* degradation of PHBHV polymers. Nevertheless, early work by Kronenthal showed loss of strength in vivo but no weight loss within a period of several months [184]. Korsatko et al. found that in vitro release of 7-hydroxyethyl-theophilline (HET) was three times faster than in vivo release (mice), suggesting that PHB was degraded nonenzymatically and probably because of simple hydrolysis [185].

Bissery et al. evaluated the distribution of PHB microspheres in mice [186]. Millar and Williams reported little degradation of monofilament fibres of PHB in rats [187]. Saito et al. found little change in the intraperitoneally or subcutaneously implanted

PHB films in the rat for several months [188]. No inflammatory activity was detected. The authors claimed that the final degradation product might be D(-)-3-hydroxybutyrate, a physiological compound always present in the human body as an energy source and that PHB might be used as a slow drug-releasing material buried inside the body [188].

3.4 POLYDIOXANONES

Polydioxanones can be synthesized from dioxane-diones, some of which are lactide/glycolide hybrids. This series of polymers will be considered in three sections: unsymmetrically substituted poly(1,4-dioxane- 2,5-diones); poly(1,3-dioxane-2-one) and poly(1,4-dioxane-2,3-dione); poly(para-dioxanone) and derivatives.

3.4.1 *Unsymetrically substituted poly(1,4-dioxane-2,5-diones)*

Unsymmetrically substituted poly(1,4-dioxane-2,5-diones) have been prepared by Augurt and co-workers [189-190]. The 3-methyl member of the series is a hybrid of half a lactide and half a glycolide molecule fused together. The "dimer" is then polymerized to give a polymer with chain structures similar to GA-LA-GA-LA sequences. The authors proposed that it was suitable for use as an absorbable surgical suture or as a bone pin. Copolymerization with lactide and glycolide yields a family of copolymers and the high GA-containing members may find use in sustained drug release systems.

3.4.2 *Poly(1,3-dioxane-2-one) and poly(1,4-dioxane-2,3-dione)*

Rosensaft and Webb copolymerized L-LA or 1,3-dioxane-2-one (trimethylene carbonate, TMC) with GA to form LA/GA/LA or TMC/GA/TMC triblock structures, with polymer chains containing up to 57.5 mol% TMC [191-192]. It was reported that TMC/GA copolymers showed a decrease in in vivo (rat) tensile strength as the TMC content increased from 48.5% to 57.4% and that these copolymers had a greater in vivo strength than PGA when in suture form. The authors suggested in addition that 1,4-dioxane-2,3-dione can be used instead of TMC, the major use of these dioxanones being for fabricating absorbable surgical sutures. Later Katz et al. described a new commercial suture material, Maxon®, which is a copolymer of TMC and glycolide (32.5% TMC) [193]. Sanz et al. studied comparatively the in vivo degradation of Maxon®, Vicryl® , PDS® and chromic Catgut® [194]. They observed that Maxon® and PDS® evoked smaller inflammatory response than Vicryl® and Catgut®. On the other hand, Maxon® and Vicryl® were statistically stronger than PDS® and Catgut®. It was observed moreover that PDS® and Maxon® continued to retain tensile strength during the late postoperative period.

3.4.3 *Poly(para-dioxanone) and derivatives*

The polymerization of para-dioxanone as well as that of its methyl and dimethyl homologues were described by Doddi et al. [195]. Poly(para-dioxanone) is primarily used as the absorbable suture material PDS® (manufactured by Ethicon Inc.) because of its good tensile properties with respect to PGA and its ability to form monofilaments [196]. PDS®

material has been investigated for arterial regeneration in rabbit [197] and for internal suspension and fixation of facial fractures clinically [198], for cerclage of the eyeball [199], for closure of abdominal wounds [200] and for orbital floor reconstruction [201] as well as for use in pediatric cardiovascular operations [202] and in orthopeadic surgery [203].

PDS® suture has a crystallinity of about 37%, thus the degradation mechanism was presumed to be similar to that of PGA with selective degradation of amorphous regions [195]. In vivo degradation work on PDS® sutures showed a slow linear cross sectional area profile loss for 5 months, followed by a complete loss during the sixth month [196,204]. A slow weight loss was detected by ^{14}C studies for the first twelve weeks, with major loss occurring between 12 and 18 weeks and complete degradation after 26 weeks. In another study it was found that approximately 25 weeks were required for the total degradation of polydioxanone Absolok clips in the pouch under the in vivo conditions [163]. The degradation rate of PDS® can be enhanced by copolymerization with GA over the 5% to 25% GA composition range [205]. The correlation between in vivo and in vitro results suggested that the degradation mechanism involved non-enzymatic hydrolysis of ester bonds (i.e. homogeneous degradation). However, ^{14}C studies showed that in vivo degradation products, unlike those of PLA and PGA, were principally removed in the urine (93%). This may suggest a different enzymatic degradation of oligomeric fragments from that occurring with PGA. The main degradation product was found to be 2-hydroxyacetic acid [205].

It seems that polydioxanones and, in particular poly(para-dioxanone) and its deriv-atives, are attractive biodegradable materials for various surgical applications.

3.5 CROSS-LINKED POLYESTERS

Cross-linked polyesters have been investigated mainly for applications in drug release systems. At least two types can be distinguished. In the first type, which can be described as polyester hydrogels, a diacid is reacted with a diol (or polyol), with unsaturation in either or both monomers [206-207]. Fumaric acid and polyethylene glycol (PEG) yield typically the following structure;

$$-[-CO-CH=CH-CO-O-(-CH_2-CH_2-O-)-CO-CH=CH-CO-O-]-$$

The resulting polyester is typically water soluble except when it is crosslinked through the double bonds, for example with N-vinylpyrrolidone. Hydrolysis of the ester linkages leads to water soluble fragments consisting of PEG and short chains of polyvinylpyrrolidone attached to fumaric acid.

Heller et al. and Baker et al. studied in vitro release of bovine serum albumin (BSA) from fumaric acid/PEG hydrogel and found that the release was controlled by matrix degradation rather than diffusion and lasted several months [206, 208]. The release rate could be regulated by constructing unsaturated polyesters containing varying proportion of esters activated by electron-withdrawing substituents vicinal to the ester function and/or by varying crosslink density. Inclusion of stronger or more activated acids (ketomalonic acid for example) enhanced the release rate. Han et al. synthesized three kinds of low MW unsaturated polyesters by the reaction of PCL diol or DL-lactide and glycolic acid with maleic anhydride or fumaric acid, which were further thermally crosslinked in the presence of radical initiator to prepare a matrix resin for biomedical composites [209]. Hydrolysis of the crosslinked polyesters was investigated in buffer solutions at pH 5.4, 7.4 and 10.0 in comparison with PLA_{50} and PLA_{100}. Weight loss data showed that the crosslinked material degraded much more rapidly than both PLA_{50} and PLA_{100} and that basic solutions enhanced the degradation. However, no mechanistic conclusion was given. Sawhney et al. synthesized copolymers having a poly(ethylene glycol) central block, extended with DL-LA or GA oligomers and terminated with acrylate groups, with the goal of obtaining a biodegradable hydrogel by photopolymerization [210]. These gels degraded upon the hydrolysis of the DL-LA or GA blocks into poly(ethylene glycol), lactic or glycolic acids and acrylic acid oligomers. The degradation rate could be tailored by appropriate choice of DL-LA or GA oligomers from less than one day to 4 months.

Another type of crosslinked polyester can be prepared by using a trifunctional alcohol or trifunctional acid (or both) to give a polymer network. For exemple, glycerol has been used to form crosslinked polyesters with citric acid and aspartic acid [211-212]. Polymer degradation was found to coincide with total release of a low MW drug (methyldopa).

3.6 WATER-SOLUBLE POLY(β-MALIC ACID) AND DERIVATIVES

Poly(β-malic acid) is an aliphatic polyester of the poly(hydroxyacid)-type which is water soluble regardless of pH. It is considered to be a promising carrier for polymeric prodrugs because of the presence of a carboxyl pendant group [213-216]. Malic acid like lactic acid is a chiral compound.Various racemic and optically active poly(β-malic acids), PMLAx, as well as their sodium salts PMLAxNa have been synthesized with different enantiomeric excess, x being the percentage of L-malic acid units in the main chains [217].

Braud et al. evaluated the in vitro degradation of low MW $PMLA_{100}Na$ in pH 7.5 phosphate buffer [218]. It was shown that the degradation rate obeyed first order kinetics at the first stage and that the ultimate degradation product was malic acid. The same authors found later that PMLAx can be easily obtained by hydrogenolysis of poly(benzyl β-malate) even for highly isotactic compounds and that the main chain degradation depends on pH and temperature but not on chain configurational structure [219]. Recently, they investigated the degradation of PMLAx by monitoring the formation of oligomers with aqueous SEC and high performance capillary electrophoresis (HPCE) [220]. It was suggested that the degradation pathway was random ester bond-scission by simple hydrolysis, in agreement with the rapid initial decrease of MW and the appearance of a whole series of detectable oligomers and a continuous change of oligomer chromatograms over a several month period.

4 Conclusions

Aliphatic polyesters at present constitute the most attractive class of synthetic polymers that can degrade in contact with living tissues or under outdoor conditions. In this family, polymers deriving from lactides, glycolide and ε-caprolactone, have been most widely investigated for applications in the fields of surgery, pharmacology, tissue engineering as well as environmental protection. Factors which can affect the biodegradation of these polymers have been gradually identified. This paper has tried to present the state-of-the-art on the basis of recent advances in this field. One of the first conclusions is that polymers bearing the same name can behave very differently for many reasons. Sources of differences can be found at any stage of the history of a polymer device from synthesis, through processing, sterilization and storage. It is only by considering all of them simultaneously that one can expect to succeed in the real control of degradation characteristics.

References

1. A. Hassig and K. Stampfli, "Plasma substitutes: past and present", Bibliotheca. Haemat., 33, 1-8, 1969.

2. D.V. Rosato, "Polymers, processes and properties of medical plastics: including markets and applications", in "Biocompatible Polymers, Metals and Composites", M. Szycher ed., Technomic. Publ. Co. Inc., Lancaster, chap. 45, 1019-1067, 1983.

3. E.L. Charles and N.Y. Buffalo, "Preparation of high molecular weight polyhydroxyacetic ester", U.S. Patent 2,668,162, 1954.

4. E.E. Schmitt and R.A. Polistina, "Surgical sutures", U.S. Patent 3,297,033, 1967.

5. E.J. Frazza and E.E. Schmitt, "A new absorbable suture", J. Biomed. Mater. Res. Symposium, 1, 43-58, 1971.

6. M. Vert, "Introductory remarks", in "Biodegradable Polymers and Plastics", M. Vert, J. Feijen, A. Albertsson, G. Scott and E. Chiellini eds., Royal Society of Chemistry, London, 1-3, 1992.

7. J. Heller, "Use of polymers in controlled drug release", in "Biocompatible Polymers, Metals and Composites", M. Szycher ed., Technomic. Publ. Co. Inc., Lancaster, chap. 24, 551-584, 1983.

8. R.S. Langer and N.A. Peppas, "Present and future applications of biomaterials in controlled drug delivery systems", Biomaterials, 2, 201-214, 1981.

9. R.S. Langer and N.A. Peppas, "Chemical and physical structure of polymers as carriers for controlled release of bioactive agents: a review", J. Macromol. Sci., REC. Macromol. Chem. Phys., C23, 61-126, 1983.

10. D.K. Gilding, "Biodegradable polymers", Biocompat. Clin. Impant. Mater., 2, 209-232, 1981.

11. D.F. Williams, "Biodegradation of surgical polymers", J. Mater. Sci., 17, 1233-1246, 1982.

12. C.G. Pitt, T.A. Marks and A. Schindler, "Biodegradable drug delivery systems based on aliphatic polyesters: application to contraceptives and narcotic antagonists", in "National Institute on Drug Abuse Research Monograph", R.E. Willette and G. Barnett eds., Naltrexone, Vol. 28, 232-253, 1981.

13. R.D. Gilbert, V. Stannett, C.G. Pitt and A. Schindler, "The design of biodegradable polymers: two approachs", in "Development in Polymer Degradation", N. Grassie ed., Vol. 4, Applied Science Publishers, London, 259-293, 1982.

14. N.B. Graham and D.A. Wood, "Hydrogels and biodegradable polymers for the controlled delivery of drugs", Polym. News, 8, 230-236, 1982.

15. G.J.L. Griffin, "Synthetic polymers and the living environment", Pure Appl. Chem., 52, 399-407, 1980.

16. S.J. Holland, B.J. Tighe and P.L. Gould, "Polymers for biodegradable medical devices. 1. The potential of polyesters as controlled macromolecular release systems", J. Control. Rel., 4, 155-180, 1986.

17. J.E. Guillet, H.X. Huber and J. Scott, "Studies of the biodegradation of synthetic plastics", in "Biodegradable Polymers and Plastics", M. Vert, J. Feijen, A. Albertsson, G. Scott and E. Chiellini eds., Royal Society of Chemistry, London, 73-92, 1992.

18. A.C. Albertsson and S. Karlsson, "Biodegradation and test methods for environmental and biomedical applications of polymers", in "Degradable Materials: Perspectives, Issues and Opportunities", S.A. Barenberg, J.L. Brash, R. Narayan and A.E. Redpath eds., CRC Press, Boca Raton, 263-286, 1990.

19. R.M. Ottenbrite, A.C. Albertsson and G. Scott, "Discussion on degradation terminology", in "Biodegradable Polymers and Plastics", M. Vert, J. Feijen, A. Albertsson, G. Scott and E. Chiellini eds., Royal Society of Chemistry, London, 73-92, 1992.

20. M. Vert, "Bioresorbable polyesters for bone surgery", Makromol. Chem., Suppl. 5, 30-41, 1981.

21. S.J. Huang, M. Bitritto, K.W. Leong, J. Pavlisko, M. Roby and J.R. Knox, "The effects of some structural variations on the biodegradability of step-growth polymers", Stabilization and Degradation of Polymers (Am. Chem. Soc.), 17, 209-214, 1978.

22. M. Vert, S. Li, G. Spenlehauer and P. Guerin, "Bioresorbability and biocompatibility of aliphatic polyesters", J. Mater. Sci., Materials in Medicine, 3, 432-446, 1992.

23. J. Leray, M. Vert and D. Blanquaert, "Nouveau matériau de prothèse osseuse et son application", French Patent Appl. 76 28163, 1976.

24. M. Vert, P. Christel, F. Chabot and J. Leray, "Bioresorbable plastic materials for bone surgery", in "Macromolecular Biomaterials", G.W. Hastings and P. Ducheyne eds., CRC press, Boca Raton, chap. 6, 119-142, 1984.

25. F. Chabot, M. Vert, S. Chapelle, P. Granger, "Configurational structures of lactic acid stereocopolymers as determined by C- H n.m.r.", Polymer, 24, 53-60, 1983.

26. M. Vert, A. Torres, S. Li, S. Roussos and H. Garreau, "The Complexity of the Biodegradation of Poly(2-hydroxy acid)-type Aliphatic Polyesters", in Y. Doi and K. Fukuda eds., *Biodegradable Plastics and Polymers*, Elsevier Sciences B.V., Amsterdam, 1994, pp. 11-23.

27. J.E. Kemnitzer, R. Gross and S.P. McCarthy, "Poly(β-hydroxybutyrate) stereoisomers : a model study of the effects of stereochemical and morphological variables on polymer biological degradability", in "Polymers as Biomaterials", *Macromolecules*, 25, 5227-5234 (1992).

28. D.F. Williams, "Enzyme-polymer interactions", J. Bioeng., 1, 279-294, 1977.

29. D.F. Williams and E. Mort, "Enzyme accelerated hydrolysis of poly(glycolic caid)", J. Bioeng., 1, 231-238, 1977.

30. J.B. Herrmann, R.J. Kelly and G.A. Higgins, "Polyglycolic acid sutures, laboratory and clinical evaluation of a new absorbable suture material", Arch. Surg., 100, 486-490, 1970.

31. T.N. Salthouse and B.F. Matlaga, "Polyglactin 910 suture absorption and the role of cellular enzymes", Surg. Gynecol. Obstet., 142, 544-550, 1975.

32. J.M. Schakenraad, M.J. Hardonk, J Feijen, I. Molenaar and P. Neuwenhuis, "Enzymatic activity toward poly(L-lactic acid) implants", J. Biomed. Mater. Res., 24, 529-545, 1990.

33. H. Younes, P.R. Nataf, D. Cohn, Y.J. Appelbaum, G. Pizov and G. Uretzky, "Biodegradable PELA block copolymers: in vitro degradation and tissue reaction", Biomat., Art. Cells, Art. Org., 16, 705-719, 1988.

34. G.E. Zaikov, "Quantitative aspects of polymer degradation in the living body", JMS-Rev. Macromol. Chem. Phys., C25, 551-597, 1985.

35. D.F. Williams, "Enzyme hydrolysis of polylactic acid", Eng. Med., 10, 5-7, 1981.

36. S.L. Ashley and J.W. McGinity, "Enzyme-mediated drug release from poly(DL-lactide) matrices", Congr. Int. Technol. Pharm., 5, 195-204, 1989.

37. M.S. Reeve, S.P. McCarthy, M.J. Downey and R.A. Gross, "Polylactide stereochemistry : effects on enzymatic degradability", *Macromolecules,* 27, 825-831 (1994).

38. R.T. MacDonald, S.P. McCarthy and R.A. Gross, "Enzymatic degradation of poly(lactide) : effects of chain stereochemistry and material crystallinity", *Macromolecules,* 29, 7356-7361 (1996).

39. S. Li and S.P. McCarthy, "Influence of crystallinity and stereochemistry on the enzymatic degradation of poly(lactide)s", *Macromolecules*, **32**, 4454-4456, 1999.

40. S. Li, M. Tenon, H. Garreau, C. Braud and M. Vert, "Enzymatic degradation of stereocopolymers derived from L-, DL- and meso-lactides", *Polymer Degradation and Stability*, **67**, 85-90, 2000.

41. L.J. Liu, S. Li, H. Garreau and M. Vert, "Selective enzymatic degradations of poly(L-lactide) and poly(ε-caprolactone) blend films", *Biomacromolecules*, **1**, 350-359, 2000.

42. S. Li, A. Girard, H. Garreau and M. Vert, "Enzymatic degradation of PLA stereocopolymers with predominant D-lactyl contents", *Polym. Degr. Stabl.*, **71**, 61-67, 2001.

43. A. Schindler and C.G. Pitt, "Biodegradable estomeric polyesters", Polym. Prepr., Amer. Chem. Soc., Div. Polym. Chem., 23, 111-112, 1982.

44. C.G. Pitt, R.W. Hendren, A. Schindler and S.C. Woodward, "The enzymatic surface erosion of aliphatic polyesters", J. Control. Rel., 1, 3-14, 1984.

45. M. Mochizuki, M. Hirano, Y. Kanmuri, K. Kudo and Y. Tokiwa, "Hydrolysis of polycaprolactone fibers by lipase : effects of draw ratio on enzymatic degradation"*J. Appl. Polym. Sci.*, 55, 289-296 (1995).

46. Gan Z., Yu D., Zhong Z., Liang Q. and Jing X., "Enzymatic degradations of poly(ε-caprolactone)/poly(DL-lactide) blends in phosphate buffer solution", Polymer, 1999, 40, 2859-2863.

47. S.C. Woodward, P.S. Brewer, F. Moatamed, A. Schindler and C.G. Pitt, "The intracellular degradation of poly(ε-caprolactone)", J. Biomed. Mater. Res., 19, 437-444, 1985.

48. Y. Tabata and Y. Ikada, "Macrophase phogocytocis of biodegradable microspheres composed of L-lactic acid/glycolic acid homo- and copolymers", J. Biomed. Mater. Res., 22, 837-858, 1988.

49. M.K. Cox, "The effect of material parameters on the properties and biodegradation of 'BIOPOL'", in "Biodegradable Polymers and Plastics", M. Vert, J. Feijen, A. Albertsson, G. Scott and E. Chiellini eds., Royal Society of Chemistry, London, 95-100, 1992.

50. Y. Doi, Y. Kumagai, N. Tanahashi and K. Mukai, "Sturctural effects on biodegradation of microbial and synthetic poly(hydroxyalkanoates)", in "Biodegradable Polymers and Plastics", M. Vert, J. Feijen, A. Albertsson, G. Scott and E. Chiellini eds., Royal Society of Chemistry, London, 139-148, 1992.

51. Y. Doi, Y. Kanesawa, M. Kunioka and T. Saito, "Biodegradation of microbial copolyesters: poly(3-hydroxybutyrate-co-3-hydroxyvalerate) and poly(3-hydroxybutyrate-co-4-hydroxyvalerate)", Macromolecules, 23, 26-31, 1990.

52. Y. Kanesawa and Y. Doi, "Hydrolytic degradation of microbial poly(3-hydroxybutyrate-co-3-hydroxyvalerate) fibres", Makromol. Chem., Rapid Commun., 11, 679-682, 1990.

53. Y. Kumagai and Y. Doi, "Enzymatic degradation of binary blends of microbial poly(3-hydroxyybutyrate) with enzymatically active polymers", Polym. Degr. Stabl., 37, 253-256, 1992.

54. D.F. Gilmore, N. Lotti, R.W. Lenz, R.C. Fuller and M. Scandola, "Biodegradability of blends of poly(hydroxybutyrate-co- hydroxyvalerate) with ester-substituted celluloses", in "Biodegradable Polymers and Plastics", M. Vert, J. Feijen, A. Albertsson, G. Scott and E. Chiellini eds., Royal Society of Chemistry, London, 251-254, 1992.

55. C.G. Pitt, "Non-microbial degradation of polyesters: mechanisms and modifications", in "Biodegradable Polymers and Plastics", M. Vert, J. Feijen, A. Albertsson, G. Scott and E. Chiellini eds., Royal Society of Chemistry, London, 7-19, 1992.

56. C.G. Pitt, M.M. Gratzel, G.L. Kimmel, J. Surles and A. Schindler, "Aliphatic polyesters. 2. The degradation of poly(DL-lactide), poly(ε-caprolactone) and their copolymers in vivo", Biomaterials, 2, 215-220, 1981.

57 R.M. Ginde and R.K. Gupta, "In vitro chemical degradation of poly(glycolic acid) pellets and fibres", J. Appl. Polym. Sci., 33, 2411-2429, 1987.

58. M. Singh, A. Singh and G.P. Talwar, "Controlled delivery of diphtheria toxoid using biodegradable poly(D,L-lactide) microcapsules", Pharm. Res., 8, 958-961, 1991.

59. Y. Kimura, Y. Matsuzaki, H. Yamane and T. Kitao, "Preparation of block copoly(ester-ether) comprising poly(lactide) and poly(oxypropylene) and degradation of its fibre in vitro and in vivo", Polymer, 30, 1342-1349, 1989.

60. F.G. Hutchinson and B.J.A. Furr, "Biodegradable polymers for the sustained release of polypeptides", Biochem. Soc. Trans., 13, 520-523, 1985.

61. L.M. Sanders, G.I. McRae, K.M. Vitale and B.A. Kell, "Controlled delivery of an LHRH analogue from biodegradable injectable microspheres", J. Control. Rel., 2, 187-195, 1985.

62. R.A. Kenley, M.O. Lee, T.R. Mahoney, II and L.M. Sanders, "Poly(lactide-co-glycolide) decomposition kinetics in vivo and in vitro", Macromolecules, 20, 2398-2403, 1987.

63. J.M. Schakenraad, P. Neuwenhuis, I. Molenaar, J. Helder, P.J. Dijkstra and J. Feijen, "In vivo and in vitro degradation of glycine/DL-lactic acid copolymers", J. Biomed. Mater. Res., 23, 1271-1288, 1989.

64. J. Helder, P.J. Dijkstra and J. Feijen, "In vitro degradation of glycine/DL-lactic acid copolymers", J. Biomed. Mater. Res., 24, 1005-1020, 1990.

65. S. Cohen, T. Yoshioka, M. Lucarelli, L.H. Hwang and R. Langer, "Controlled delivery systems for proteins based on poly(lactic/glycolic acid) microspheres", Pharm. Res., 8, 713-720, 1991.

66. T.St. Pierre and E. Chiellini, "Biodegradability of synthetic polymers for medical and pharmaceutical applications : Part 2 - Backbone hydrolysis", J. Bioact. Comp. Polym., 2, 4-30, 1987.

67. D.H. Lewis, "Controlled release of bioactive agents from lactide/ glycolide polymers", Drugs Pharm. Sci., 45 (Biodegrad. Polym. Drug Delivery Syst.), 1-41, 1990.

68. S.J. Holland, A.M. Jollly, M. Yasin and B.J. Tighe, "Polymers for biodegradable medical devices. II. Hydroxybutyrate-hydroxyvalerate copolymers: hydrolytic degradation studies", Biomaterials, 8, 289-295, 1987.

69. J.C. Knowles and G.W. Hastings, "In vitro degradation of a PHB/PHV copolymer and a new technique for monitoring early surface changes", Biomaterials, 12, 210-214, 1991.

70. S. Li, H. Garreau and M. Vert, "Bioresorbable polyesters of the glycolic/lactic type: in vitro investigations of the mechanism of degradation", in "Preprints of Kunming International Symposium on Polymeric Biomaterials", Kunming, China, May 3-7, 1988.

71. S. LI, "Etude de la degradation des poly(α-hydroxy acides) aliphatiques derives des acides lactique et glycolique en milieux aqueux modeles", Ph.D. thesis, University of Rouen, France, 1989.

72. S. Li, H. Garreau and M. Vert, "Structure-property relationships in the case of the degradation of massive aliphatic poly(α-hydroxy acids) in aqueous media. Part 1 : Poly(DL-lactic acid)", J. Mater. Sci.: Materials in Medicine, 1, 123-130, 1990.

73. S. Li, H. Garreau and M. Vert, "Structure-property relationships inthe case of the degradation of massive poly(α-hydroxy acids) in aqueous media. Part 2 : Degradation of lactide/glycolide copolymers : PLA37.5GA25 and PLA75GA25", J. Mater. Sci.: Materials in Medicine, 1, 131-139, 1990.

74. S. Li, H. Garreau and M. Vert, "Structure-property relationships inthe case of the degradation of massive poly(α-hydroxy acids) in aqueous media. Part 3 : Influence of the morphology of poly(L-lactic acid)", J. Mater. Sci.: Materials in Medicine, 1, 198-206, 1990.

75. M. Vert, S. Li and H. Garreau, "More about the degradation of LA/GA-derived matrices in aqueous media", J. Control. Rel., 16, 15-26, 1991.

76. M. Therin, P. Christel, S. Li, H. Garreau and M. Vert, "In vivo degradation of massive poly(α-hydroxy acids): validation of in vitro findings", Biomaterials, 13, 594-600, 1992.

77. M. Vert, S. Li and H. Garreau, "New insights on the degradation of bioresorbable polymeric devices based on lactic and glycolic acids", Clinical Materials, 10, 3-8, 1992.

78. S. Li and M. Vert, "Crystalline oligomeric stereocomplex as intermediate compound in racemic poly(DL-lactic acid) degradation", *Polym. Inter.*, **33**, 37-41, 1994.

79. S. Li and M. Vert, "Morphological changes resulting from the hydrolytic degradation of stereocopolymers derived from L- and DL-lactides", *Macromolecules*, **27**, 3107-3110, 1994.

80. K.R. Huffman and D.J. Casey, "Effects of carboxylic end groups on hydrolysis of polyglycolic acid", J. Polym. Sci.: Polym. Chem. Ed., 23, 1939-1954, 1985.

81. H. Fukuzaki, M. Yoshida, M. Asano and M. Kumakura, "Synthesis of copoly(D,L-lactic acid) with relatively low molecular weight and in vitro degradation", Eur. Polym. J., 25, 1019-1026, 1989.

82. E.A. Schmitt, D.R. Flanagan, and R.J. Linhardt, "Importance of distinct water environments in the hydrolysis of poly(dl-lactide-co-glycolide)", *Macromolecules*, 27, 743-748 (1994).

83. S.A.M. Ali, P.J. Doherty and D.F. Williams, "Mechanism of polymer degradation in implantable devices. 2. Poly(DL-lactic acid)" *J. Biomed. Mater. Res.*, 27, 1409-1418 (1993).

84. X. Zhang, U.P. Wyss, D. Pichora and M.F.A. Goosen, "An investigation of poly(lactic acid) degradation ", *J. Bioact. Compat. Polym.*, 9, 80-100 (1994).

85. H. Pistner, R. Gutwald, R. Ordung, J. Reuther and J. Müling, "Poly(L-lactide) : a long term degradation study in vivo. Part I : biological results", *Biomaterials*, 14, 671-677 (1993).

86. H. Pistner, H. Stallforth, R. Gutwald, J. Müling, J. Reuther and C. Michel, "Poly(L-lactide) : a long term degradation study in vivo. Part II : physical-mechanical behavior of implants", *Biomaterials*, 15, 439-450 (1994).

87. H. Pistner, D.R. Bendix, J. Müling, J. Reuther, J. Reuther, "Poly(L-lactide) : a long term degradation study in vivo. Part III : analytical characterization", *Biomaterials*, 14, 291-298 (1993).

88. T.G. Park, "Degradation of poly(lactic-co-glycolic acid) microspheres : effect of copolymer composition", *Biomaterials*, 16, 1123-1130 (1995).

89. S. Li, J.L. Espartero, P. Foch and M. Vert, "Structural characterization and hydrolytic degradation of a Zn metal initiated copolymer of L-lactide and ε-caprolactone", *J. Biomater. Sci.: Polym. Ed.*, **8**, 165-187. 1996

90. A. Löfgren and A. Albertsson, "Copolymers of 1,5-dioxepan-2-one and L- or DL-lactide : hydrolytic degradation behavior", *J. Appl. Polym. Sci.*, 52, 1327-1338 (1994).

91. A.C. Albertsson and M. Eklund, "Influence of molecular structure on the degradation mechanism of degradable polymers : in vitro degradation of poly(trimethylene

carbonate), poly(trimethylene carbonate-co-caprolactone), and poly(adipic anhydride)", *J. Appl. Polym. Sci.*, 57, 87-103 (1995).

92. H. Schliephake, D. Klosa and M. Rahlff, "Determination of the 3-D morphology of degradable biopolymer implants undergoing in vivo resorption ", *J. Biomed. Mater. Res.*, 27, 991-998 (1993).

93. I. Grizzi, H. Garreau, S. Li and M. Vert, "Biodegradation of devices based on poly(DL-lactic acid): size-dependence", *Biomaterials*, **16**, 305-311, 1995.

94. C.G. Pitt, F.I. Chasalow, Y.M. Hibionada, D.M. Klimas and A. Schindler, "Aliphatic polyesters. I. The degradation of poly(ε-caprolactone) in vivo", J. Appl. Polym. Sci., 26, 3779-3787, 1981.

95. M. Vert, "Bioresorption Synthetic Polymers and Their Operation Field", in G. Walenkamp ed., *Biomaterials in Surgery*, Georg Thieme Verlag, Stuttgart, 1998, pp. 97-101.

96. S. Li, "Hydrolytic degradation characteristics of aliphatic polyesters derived from lactic and glycolic acids", *J. Biomed. Mater. Res.: Appl. Biomat.*, 1999, **48**, 342-353.

97. S. Li and M. Vert, "Biodegradable polymers : polyesters ", in *The Encyclopedia of Controlled Drug Delivery*, E. Mathiowitz ed., John Wiley & Sons, 1999, pp. 71-93.

98. A. Torres, S. Roussos, S.Li and M. Vert, "Screening of microorganisms for biodegradation of poly(lactic acid) and lactic acid-containing polymers", *Applied and Environmental Microbiology*, **62**, 2393-2397, 1996.

99. A. Torres, S.M. Li, S. Roussos and M. Vert, "Degradation of L- and DL-lactic acid oligomers in the presence of Fusarium moniliforme and Pseudomonas putida", *J. Environ. Polym. Degr.*, **4**, 213-223, 1996.

100. A. Torres, S.M. Li, S. Roussos and M. Vert, "Poly(lactic acid) degradation in soil or under controlled conditions", *J. Appl. Polym. Sci.* **62**, 2295-2302, 1996.

101. P. Jarrett, C. Benedict, J.P. Bell, J.A. Cameron and S.J. Huang, "Mechanism of the biodegradation of polycaprolactone", in "Polymers as Biomaterials", S.W. Shalaby, A.S. Hoffman, B.D. Ratner and T.A. Horbett eds., Plenum Publ. Corp., 181-192, 1985.

102. F. Lefebvre, C. David and C. Vander Wauven, "Biodegradation of polycaprolactone by microorganisms from an industrial compost of household refuse", *Polym. Degr. Stab.*, 45, 347-353 (1994).

103. S.I. Akahori and Z. Osawa, "Preparation and biodegradation of polycaprolactone-paper composites", *Polym. Degr. Stab.*, 45, 261-265 (1994).

104. E.W. Fischer, H.J. Sterzel and G. Wegner, "Investigation of the structure of solution grown crystals of lactide copolymers by means of chemical reactions", Kolloid-Z. u. Z. Polymere, 251, 980-990, 1973.

105. B.K. Carter and G.L. Wilkes, "Some morphological investigations on an absorbable copolyester biomaterial based on glycolic and lactiic acid", in "Polymers as

Biomaterials", S.W. Shalaby, A.S. Hoffman, B.D. Ratner and T.A. Horbert eds., Plenum Press, New York, 67-92, 1984.

106. R.J. Fredericks, A.J. Melveger and L.J. Dolegiewtz, "Morphological and structural changes in a copolymer of glycolide and lactide occurring as a result of hydrolysis", J. Polym. Sci.: Polym. Phys. Ed., 22, 57-66, 1984.

107. J.W. Leeslag, A.J. Pennings, R.R.M. Bos, F.R. Rozema and G. Boering, "Bioresorbable materials of poly(L-lactide). VII. In vivo and in vitro degradation", Biomaterials, 8, 311-314, 1987.

108. C.C.Chu, "Hydrolytic degradation of poly(glycolic acid): tensile strength and crystallinity study", J. Appl. Polym. Sci., 26, 1727-1734, 1981.

109. C.C. Chu and N.D. Campbell, "Scanning electron microscopic study of the hydrolytic degradation of poly(glycolic acid) suture", J. Biomed. Mater. Res., 417-430, 1982.

110. A. Browning and C.C. Chu, "The effect of annealing treatments on the tensile properties and hydrolytic degradative properties of poly(glycolic acid) sutures", J. Biomed. Mater. Res., 20, 613-632, 1986.

111. A. Browning and C.C. Chu, "The effect of annealing treatments on the mechanical and degradative properties of poly(glycolic acid) sutures", in "Proc. ACS Division of Polymeric Materials: Science and Engineering", Vol. 53, Fall Meeting 1985, Am. Chem. Soc., Washington, DC, 510-514, 1985.

112. T. Nakamura, S. Hitomi, S. Watanabe, Y. Shimizu, K. Jamshidi, S.-H. Hyon and Y. Ikada, "Bioabsorption of polylactides with different molecular properties", J. Biomed. Mater. Res., 23, 1115-1130, 1989.

113. R.A. Miller, J.M. Brady and D.E. Cutright, "Dedradation rates of oral resorbable implants (polylactates and polyglycolates): rate modification with changes in PLA/PGA copolymer ratios", J. Biomed. Mater. Res., 11, 711-719, 1977.

114. J.W. Leeslag, S. Gogolewski and A.J. Pennings, "Bioresorbable materials of poly(L-lactide). V. Influence of secondary structure on the mechanical properties and hydrolyzability of poly(L-lactide) fibres produced by a dry-spinning method", J. Biomed. Mater. Res., 29, 2829-2842, 1984.

115. U. Siemann, "The influence of water on the glass transition of poly(DL-lactic acid)", Thermochemica Acta., 85, 513-516, 1985.

116. Li, S.; Girod-Holland, S.; Vert, M. Hydrolytic degradation of poly(DL-lactic acid) in the presence of caffeine base. J. Control. Rel. 40: 41-53; 1996.

117. D.W. Grijpma, A.J. Nijenhuis and A.J. Pennings, "Synthesis and hydrolytic degradation behaviour of high-molecular-weight L-lactide and glycolide copolymers", Polymer, 31, 2201-2206, 1990.

118. D.E. Cutright, B. Perez, J.D. Beasley, W.J. Larson and W.R. Posey, "Degration rates of polymers and copolymers of polylactic and polyglycolic acids", Oral Surg., 37, 142-152, 1974.

119. A.M. Reed and D.K. Gilding, "Biodegradable polymers for use in surgery - poly(glycolic)/poly(lactic acid) homo- and copolymers: 2. in vitro degradation", Polymer, 22, 494-498, 1981.

120. A.M. Reed and D.K. Gilding, "Biodegradable polymers for use in surgery - poly(glycolic)/poly(lactic acid) homo- and copolymers. 1", Polymer, 20, 1459-1464, 1979.

121. I. Kaetsu, M. Yoshida, M. Asano, H. Yamanaka, K. Imai, H. Yuasa, T. Mashimo, K. Suzuki, R. Katakai and M. Oya, "Biodegradable implant composites for local therapy", J. Control. Rel., 6, 249-263, 1987.

122. J.M. Zhu, Y.M. Shao, S.Z. Zhang and W.M. Sui, "Homopolymers and copolymers of glycolide and lactide", Journal of China Textile University (Eng. Ed.), 8, 57-61, 1991.

123. Y. Ogawa, H. Okada, M. Yamamoto and T. Shimamoto, "In vivo release profiles of leuprolide acetate from microcapsules prepared with polylactic acids or copoly(lactic/glycolic) acids and in vivo degradation of these polymers", Chem. Pharm. Bull., 36, 2576-2581, 1988.

124. S.S. Amarpreet and J.A. Hubbell, "Rapidly degraded terpolymers of dl-lactide, glycolide and e-caprolactone with increased hydrophilicity by copolymerization with polyethers", J. Biomed. Mater. Res., 24, 1937-1411, 1990.

125. S.H. Hyon, K. Jamshidi and Y. Ikada, "Melt spinning of poly(L-lactide) and hydrolysis of the fibre in vitro", in "Polymers as Biomaterials", Plenum Press, S.W. Shalaby ed., New York, NY, 51-65, 1984.

126. D.C. Tunc, M.W. Rohovsky, B. Jadhav, W.B. Lehman, A. Strongwater and F. Kummer, "Evaluation of body absorbable bone fixation devices", Polym. Mater. Sci. Eng., 53, 502-504, 1985.

127. D.C. Tunc and B. Jadhav, "Development of absorbable ultra-high-strength polylactide", Polym. Mater. Sci. Eng., 59, 383-387, 1988.

128. J. Eitenmüller, G. Muhr, K.L. Gerlach and T. Schmickal, "New semi-rigid and bioabsorbable osteosynthesis devices with a high molecular weight polylactide (an experimental investigation)", J. Bioact. Compat. Polym., 4, 215-241, 1989.

129. M. Vert, F. Chabot, J. Leray and P.Christel, "Nouvelles pièces d'ostéosynthèse, leur préparation et leur application", French Patent 78 29978, 1978.

130. A.S. Chawla and T.M.S. Chang, "In vivo degradation of poly(lactic acid) of different molecular weights", Biomed. Med. Dev. Art. Org., 13, 153-162, 1985-86.

131. J. Mauduit, N. Bukh and M. Vert, "Gentamycin/poly(lactic cid) blends aimed at sustained release local antibiotic therapy administered per-operatively: III. The case

of gentamycin sulfate in films of high and low molecular weight poly(DL-lactic acid)", J. Control. Rel., 25, 43-49, 1993.

132. A.K. Kwong, S. Chou, A.M. Sun, M.V. Sefton and M.F.A. Goosen, "In vitro and in vivo release of insulin from poly(lactic acid) microbeads and pellets", J. Control. Rel., 4, 47-62, 1986.

133. G.E. Visscher, J.E. Pearson, J.W. Fong, G.J. Argentieri, R.L. Robison and H.V. Maulding, "Effect of particle size on the in vitro and in vivo degradation rates of poly(DL-lactide-co-glycolide) microcapsules", J. Biomed. Mater. Res., 22, 733-746, 1988.

134. P. Törmälä, H.M. Mikkola, J. Vasenius, S. Vainionpaa and P. Rokkanen, "Strength retention of self-reinforced, absorbable polyglycolide rods in hydrolytic environment", Angew. Makromol. Chem., 185-186, 293-302, 1991.

135. J. Mauduit, N. Bukh and M. Vert, "Gentamycin/poly(lactic acid) blends aimed at sustained release local antibiotic therapy administered per-operatively: II. The case of gentamycin sulfate in high molecular weight poly(DL-lactic acid) and poly(L-lactic acid)", J. Control. Rel., 23, 221-230, 1993.

136. H. Zhu, Z.R. Shen, L.T. Wu and S.L. Yang, "In vitro degradation of polylactide and poly(lactide-co-glycolide) microspheres", J. Appl. Polym. Sci., 43, 2099-2106, 1991.

137. K.H. Lam, P. Nieuwenhuis, I. Molenaar, H. Esselbrugge, J .Feijen, P.J. Dijkstra and J.M. Schakenraad, "Biodegradation of porous versus non-porous poly(L-lactic acid) films",*J. Mater. Sci.: Mater. Med.*, 5, 181-189 (1994).

138. A. Smith and I.M. Hunneyball, "Evaluation of poly(lactic acid) as a biodegradable drug delivery system for parenteral administration", *Inter. J. Pharm.*, 30, 215-220 (1986).

139. M.J.D. Eenink, J. Feijen, J. Olijslager, J.H.M. Albers, J.C. Rieke and P.J. Greidanus, "biodegradable hollow fibers for the controlled release of hormones", *J. Control. Rel.*, 6, 225-247 (1987).

140. H.V. Maulding, T.R. Tice, D.R. Cowsar, J.W. Fong, J.E. Pearson and J.P.tNazareno, "Biodegradable microcapsules: acceleration of polymeric excipient hydrolytic rate by incorporation of a basic medicament", J. Control. Rel., 3, 103-117, 1986.

141. H.V. Maulding, "Prolonged delivery of peptides by microcapsules", *J. Control. Rel.*, 6, 167-176 (1987).

142. A. Kishida, S. Yoshioka, Y. Takeda and M. Uchiyama, "Formulation-assisted biodegradable polymer matrices", Chem. Pharm. Bull., 37, 1954-1956, 1989.

143. Y. Cha and C.C. Pitt, "A one-week subdermal delivery system for L-methadone based on biodegradable microcapsules", J. Control. Rel., 7, 69-78, 1988.

144. Y. Cha and C.C. Pitt, "The acceleration of degradation-controlled drug delivery from polyester microspheres", J. Control. Rel., 8, 259-265, 1989.

145. J.F. Fitzgerald and O.I. Corrigan, "Investigation of the mechanisms governing the release of levamisole from poly(DL-lactide-co-glycolide) delivery systems",*J. Control. Rel.*, 42, 125-132 (1996).

146. R. Bodmeier, K.H. Oh and H. Chen, "The effectof the addition of low molecular weight poly(DL-lactide) on drug release from biodegradable poly(DL(-lactide) drug delivery systems", *Inter. J. Pharm.*, 51, 1-8 (1989).

147. J. Mauduit, N. Bukh and M. Vert, "Gentamycin/poly(lactic acid) blends aimed at sustained release local antibiotic therapy administered per-operatively: I. The case of gentamycin base and gentamycin sulfate in poly(DL-lactic acid) oligomers", J. Control. Rel., 23, 209-220, 1993.

148. C.C.P.M. Verheyen, C.P.A.T. Klein, J.M.A. De Blieckhogervorst, J.G.C. Wolke, C.A. Van Blitterswijn, K. De Groot, *J. Mater. Sci.: Mater. Med.*, 4, 58-65 (1993).

149. S. Li and M. Vert, "Hydrolytic degradation of coral/poly(DL-lactic acid) bioresorbable material", *J. Biomat. Sci.: Polym. Ed.*, 7, 817-827 (1996).

150. Y. Zhang, S. Zale, L. Sawyer and H. Bernstein, "Effects of metal salts on poly(DL-lactide-co-glycolide) polymer hydrolysis", *J. Biomed. Mater. Res.*, 34, 531-538 (1997).

151. M.C. Gupta and V.G. Deshmukh, "Radiation effects on poly(lactic acid)", Polymer, 24, 827-830, 1983.

152. C.C. Chu, "Degradation phenomena of two linear aliphatic polyester fibres used in medicine and surgery", Polymer, 26, 591-594, 1985.

153. D.C. Tsai, S.A. Howard, T.F. Hogan, C.J. Malanga, S.J. Kandzari and J.K.H. Ma, "Preparation and in vitro evaluation of polylactic acid-mitomycin C microcapsules", *J. Microencapsulation*, 3, 181-193 (1986).

154. G. Spenlehauer, M. Vert, J.P. Benoit and A. Boddaert, "In vitro and in vivo degradation of poly(DL-lactide/glycolide) type microspheres made by solvent evaporation method", Biomaterials, 10, 557-563, 1989.

155. C. Birkinshaw, M. Buggy, G.G. Henn and E. Jones, "Irradiation of poly(DL-lactide)", Polym. Degr. Stabl., 38, 249-253, 1992.

156. C. Volland, M. Wolff and T. Kissel, "The influence of gamma-sterilization on captopril containing poly(DL-lactide-co-glycolide) microspheres ", *J. Control. Rel.*, 31, 293-305 (1992).

157. C.C. Chu, "An in vitro study of the effect of buffer on the degradation of poly(glycolic acid) sutures", J. Biomed. Mater. Res., 15, 19-27, 1981.

158. C.C Chu, "The in vitro degradation of poly(glycolic acid) sutures - effect of pH", J. Biomed. Mater. Res., 15, 795-804, 1981.

159. C.C. Chu, "A comparison of the effect of pH on the biodegradation of two synthetic bioabsorbable sutures", Ann. Surg., 195, 55-59, 1982.

160. C.C. Chu, "The effect of pH on the in vitro degradation of poly(glycolide/lactide) copolymer absorbable sutures", J. Biomed. Mater. Res., 16, 117-124, 1982.

161. C.C. Chu and G. Moncrief, "An in vitro evaluation of the stability of mechanical properties of surgical suture materials in various pH conditions", Ann. Surg., 198, 223-228, 1983.

162. K. Makino, H. Ohshima and T. Kondo, "Mechanism of hydrolytic degradation of poly(L-lactide) microcapsules: effects of pH, ionic strength and buffer concentration", J. Microencapsulation, 3, 203-212, 1986.

163. N. Chegini, D.L. Hay, J.A. von Fraunhofer and B.J. Masterson, "A comparative scanning electron microscopic study on degradation of absorbable ligating clips in vivo and in vitro", J. Biomed. Mater. Res., 22, 71-79, 1988.

164. R. Suuronen, T. Pohjonen, R. Taurio, P. Tormala, L Wessman, K. Ronkka and S. Vainionpaa, "Strength retention of self-reinforced poly-L-lactide screws and plates: an in vivo and in vitro study", J. Mater. Sci.: Materials in Medicine, 3, 426-431, 1992.

165. Y. Ikada, S.-H. Hyon, K. Jamshidi, S. Higashi, T. Yamamura, Y. Katutani and T. Kitsugi, "Release of antibiotic from composites of hydroxyapatite and poly(lactic acid)", J. Control. Rel., 2, 179-186, 1985.

166. T. Yeya, H. Okada, Y. Ogawa and H. Toguchi, "Factors influencing the profiles of TRH release from copoly(d,l-lactic/glycolic acid) microspheres", Inter. J. Pharm., 72, 199-205, 1991.

167. C.G. Pitt and Z.W. Gu, "Modification of the rates of chain cleavage of poly(ε-caprolactone) and related polyesters in the solid state", J. Control. Rel., 4, 283-292, 1987.

168. H.L. Gabelnick, "Biodegradable implants: alternative approaches", in "Advances in Humain Fertility and Reproductive Endocrinology: vol. 2, Long Acting Steroid Contraception", Raven Press, New York, NY, 149-173, 1983.

169. R.D. Fields, F. Rodriguez and R.K. Finn, "Microbial degradation of polyesters: polycaprolactone degraded by P. pullulans", J. Appl. Polym. Sci., 18, 3571-3579, 1974.

170. C.X.. Song, H.F. Sun and X.D. Feng, "Microspheres of biodegradable block copolymer for long acting controlled delivery of contraceptives", Polym. J., 19, 485-491, 1987.

171. Y.X. Li, "Synthesis and studies of the controlled drug release system of biodegradable polymers as carriers", Ph.D. Thesis, Peking University, China, 1988.

172. H. Fukuzaki, M. Yoshida, M. Asano and M. Kumakura, T. Mashimo, H. Yuasa, K. Imai and H. Yamanaka, "Synthesis of low molecular weight copoly(L-lactiic acid/ε-caprolactone) by direct copolycondensation in the absence of catalysts, and enzymatic degradation of the polymers", Polymer, 31, 2006-2014, 1990.

173. D.W. Grijmpa, G.J. Zondervan and A.J. Pennings, "High molecular weight copolymers of L-lactide and ε-caprolactone as biodegradable elastomeric implants materials", Polym. Bull. 25, 327-333, 1991.

174. Y. Cha and C.G. Pitt, "The biodegradability of polyester blends", Biomaterials, 11, 108-112, 1990.

175. W. Von Korsatko, B. Wabnegg, G. Braunegg, R.M. Lafferty and F. Strempfl, "Poly-D-(-)-3-hydroxybyttersäure (PHB) - ein biologisch abbaubarer Arzneistoffträger zur Liberations-verzögerung. 1. Mitt: Eintwicklung von parenteral applizierbaren Matrixtabletten zur Langzeitabgabe von Arzneistoffen", Pharm. Ind., 42, 525-527, 1983.

176. N. Grassie, E.J. Murray and P.A. Holmes, "The thermal degradation of poly(-(D)-b-hydroxy butyric acid). Part 1. Identification and quantitative analysis of products", Polym. Degr. Stab., 6, 47-61, 1984.

177. N. Tanahashi and Y. Doi, "Thermal properties and stereoregularity of poly(3-hydroxybutyrate) prepared from optically active b-butyrolactone with a zinc-based catalyst", Macromolecules, 24, 5732-5733, 1991.

178. S. Bleoembergen, D.A. Holden, T.L. Bluhm, G.K. Hamer and R.H. Marchessault, "Synthesis of crystalline b-hydroxybutyrate/ b-hydroxyvalerate copolyesters by coordination polymerization of b-lactones", Macromolecules, 20, 3086-3089, 1987.

179. J. Mergaert, A. Wouters, J. Swings and K. Kersters, "Microbial flora involved in the biodegradation of polyhydroxyalkanoates", in "Biodegradable Polymers and Plastics", M. Vert, J. Feijen, A. Albertsson, G. Scott and E. Chiellini eds., Royal Society of Chemistry, London, 95-100, 1992.

180. S. Yoshioka, A. Kishida, S. Izumikawa, Y. Aso and Y. Takeda, "Base-induced polymer hydrolysis in poly(b-hydroxybutyrate/ b-hydroxyvalerate) matrices", J. Control. Rel., 16, 341-348, 1991.

181. E.L. Welland, J. Stejny, A. Halter and A. Keller, "Selective degradation of chain folded single crystals of poly(β-hydroxybutyrate)", Polym. Commun., 30, 302-304, 1989.

182. P.A. Holmes, "Applications of PHB - a microbially produced biodegradable thermoplastic", Phys. Technol., 16, 32-36, 1985.

183. M.W. Stinson and J.M. Merrick, "Extracellular enzyme secretion by Pseudomonas lemoignei", J Bacteriol., 119, 152-161, 1974.

184. Kronenthal, R.L. (1974) Biodegradable polymers in medicine and surgery, in *Polymers in Medicine and Surgery*, (eds R.L. Kronenthal, Z. User and E. Martin), Plenum Press, New York, pp. 119-37.

185. Von Korsatko, W., Wabnegg, R., and Tillian, H.M. et al. Poly-D-(-)-3-hydroxybyttersäure (PHB) - ein biologisch abbaubarer Arzneistoffträger zur Liberations-verzögerung. 3. Mitt: Gewebsverträglishkeitsstudien parenteral applizierbarer poly-D-(-)-3-hydroxybyttersäure-tabletten in Gewebekultur und in vivo, Pharm. Ind., **46**, 952-954, 1984.

186. M.C. Bissery, F. Varelote and C. Thies, "In vitro and in vivo evaluation of CCNU-loaded microspheres prepared from poly((±)lactide) and poly(β-hydroxybutyrate)", in "Microspheres and Drug Therapy. Pharmaceutical and Medical Aspects", S.S. Davis, L. Illum, J.G. McVie and E. Tomlinson eds., Elsevier Science Publisher, Amsterdam, 217-227, 1984.

187. N.D. Miller and D.F. Williams, "On the biodegradation of poly(β-hydroxybutyrate) (PHB) homopolymer and poly(β-hydroxybutyrate/hydroxyvalerate) copolymers", Biomaterials, 8, 129-137, 1987.

188. T. Saito, K. Tomita, K. Juni and K. Ooba, "In vivo and in vitro degradation of poly(3-hydroxybutyrate) in rat", Biomaterials, 12, 309-312, 1991.

189. T.A. Augurt, M.N. Rosensaft and V.A. Perciaccante, "Surgical sutures of unsymmetrically substituted 1,4-dioxane-2,5-diones", U.S. Patent 3,960,152, 1976.

190. T.A. Augurt, M.N. Rosensaft and V.A. Perciaccante, "Polymers of unsymmetrically substituted 1,4-dioxane-2,5-diones", U.S. Patent 4,033,938, 1977.

191. M.N. Rosensaft and R.L. Webb, "Synthetic polyester surgical articles", U.S. Patent 4,243,775, January 6, 1981.

192. M.N. Rosensaft and R.L. Webb, "Synthetic polyester surgical articles", U.S. Patent 4,300,565, November 17, 1981.

193. A.R. Katz, D.P. Mukherjee, A.L. Kaganov and S. Gordon, "A new synthetic monofilament absorbable suture made from polytrimethylene carbonate", Surg. Gynecol. Obstet., 161, 213-222, 1985

194. L.E. Sanz, J.A. Patterson, R. Kamath, G. Willett, S.W. Ahmed and A.B. Butterfield, "Comparison of MAXON suture with VICRYL, chromic catgut and PDS sutures in fascial closure in rats", Obstet. Gynecol., 71, 418-422, 1988.

195. N. Doddi, C.C. Versfelt and D. Wasserman, "Synthetic absorbable surgical devices of polydioxanone", U.S. Patent 4,052,988, 1976.

196. J.A. Ray, N. Doddi, D. Regula, J.A. Williams and A. Melveger, "Polydioxanone (PDS), a novel monofilament synthetic absorbable suture", Surg. Gynecol. Obstet., 153, 497-507, 1981.

197. H.P. Greisler, J. Ellinger, T.H. Schwarcz, J. Golan, R.M. Raymond and U.K. Dae, "Arterial regeneration over polydioxanone prostheses in the rabbit", Arch. Surg., 122, 715-721, 1987.

198. J. Cornah and J. Wallace, "Polydioxanone (PDS): a new material for internal suspension and fixation", British Journal of Oral and Maxillofacial Surgery, 26, 250-254, 1988.

199. B. Biardzka and J. Kaluzny, "Experimental and clinical investigations on the suitability of polydioxanone threads for cerclage of the eyeball", Ophthalmologica (Basel), 197, 47-50, 1988.

200. D.J.J.R. Schoetz, J.A. Coller and M.C. Veidenheimer, "Closure of abdominal wounds with polydioxanone: a prospective study", Arch. Path. Lab. Med., 123, 72-74, 1988.

201. T. Lizuka, P. Mikkonen, P. Paukku and C. Lindqvist, "Reconstruction of orbital floor with polydioxanone plate", Inter. J. Oral Maxillofacial Surg., 20, 83-87, 1991.

202. J.L. Myers, D.B. Campbell and J.A. Waldhausen, "The use of absorbable monofilament polydioxanone suture in pediatric cardiovascular operations", Journal of Thoracic and Cardiovascular Surgery, 92, 771-775, 1986.

203. J.S. Miles, "Use of polydioxanone absorbable monofilament sutures in orthopeadic surgery", Orthopeadics (Thorofare), 9, 1533-1536, 1986.

204. Ethicon Inc., "Absorbable polymer-drug compositions", U.K. Patent 1,573,459, 1980.

205. C.J. Schaefer, P.M. Colombani and G.W. Geelhoel, "Absorbable ligating clips", Surg. Gynecol. Obstet., 154, 513-516, 1982.

206. J. Heller, R.F. Helwing, R.W Baker and M.W. Tuttle, "Controlled release of water soluble macromolecules from bioerodible \ hydrogels", Biomaterials, 4, 262-266, 1983.

207. J. Heller, "Water soluble polyesters", U.S. Patent 4,502,976, 1985.

208. R.W Baker, M.W. Tuttle and R.F. Helwing, "Novel erodible polymers for the delivery of macromolecules", Pharm. Technol., 26-30, Feb. 1984.

209. Y.K. Han, P.G. Edelman and S.J. Huang, "Synthesis and characterization of crosslinked polymers for biomedical composites", J. Macromol. Sci.-Chem., A25, 847-869, 1988.

210. A.S. Sawhney, C.P. Pathak and J.A. Hubbell, "Bioerodible hydrogels based on photopolymerized poly(ethylene glycol)-co-poly(α-hydroxyacid) diacrylate macromers", Macromolecules, 26, 581-587, 1993.

211. D. Pramanick and T.T. Ray, "Synthesis and biodegradation of copolyesters from citric acid and glycerol", Polym. Bull., 19, 365-370, 1988.

212. D. Pramanick and T.T. Ray, "Synthesis and biodegradation of polymers derived from aspartic acid", Biomaterials, 8, 407-410, 1987.

213. C. Braud, M. Vert and R.W. Lenz, "Polyelectrolytical properties of poly-β-malic acid and its partially benzylated derivatives", Proceedings of IUPAC 27th International Symposium on Macromolecules, Strasbourg, France, 1981 (Proceedings B5, Vol. II, 1086-1089).

214. C. Braud, C. Bunel, H. Garreau and M. Vert, "Evidence of the amphiphilic structure of partially hydrogenolyzed poly(β-malic acid benzyl ester)", Polym. Bull., 9, 198-203, 1983.

215. C. Braud and M. Vert, "Poly(β-malic acid) as a source of polyvalent drug carriers: possible effects of hydrophobic substituents in aqueous media", in "Polymers as Biomaterials", S.W. Shalaby, A.S. Hoffman, B.D. Ratner and T.A. Horbetts eds., Plenum Publ. Co., 1-15, 1984.

216. A. Caron, C. Braud, C. Bunel and M. Vert, "Blocky structure of copolymers obtained by Pd/C-catalyzed hydrogenolysis of benzyl protecting groups as shown by sequence-selective hydrolytic degradation in poly(β-malic acid) derivatives", Polymer, 31, 1797-1802, 1990.

217. P. Guerin, M. Vert, C. Braud and R.W. Lenz, "Optically active poly(β-malic acid)," Polym. Bull., 14, 187-193, 1985.

218. C. Braud, C. Bunel and M. Vert, "Poly(β-malic acid): a new polymeric drug carrier, evidence for degradation in vitro", Polym. Bull., 13, 293-299, 1985.

219. C. Braud, A. Caron, J. Francillette, P. Guerin and M. Vert, "Poly(β-malic acid) stereocopolymers: structural characteristics and degradation in aqueous media", Polym. Prepr., (Am. Chem. Soc., Div. Polym. Chem.), 29, 600-601, 1988.

220. C. Braud and M. Vert, "Degradation of poly(β-malic acid) - monitoring of oligomers formation by aqueous SEC and HPCE", Polym. Bull., 29, 177-183, 1992

6

STARCH -POLYMER COMPOSITES

CATIA BASTIOLI
Novamont S.p.A.
Via G. Fauser, 8
28100 Novara –Italy

1 Introduction

In nature, starch represents a link with the energy of the sun, which is partially captured during photosynthesis. Starch serves as a food reserve for plants and provides a mechanism by which non-photosynthesizing organisms, such as man, can utilize the energy supplied by the sun.

Today, starch is inexpensive and is available annually from corn and other crops, and is produced in excess of current market needs in the United States and Europe [1]. Starch is totally biodegradable in a wide variety of environments and could permit the development of totally degradable products for specific market demands. Degradation or incineration of starch products would recycle atmospheric CO_2 trapped by starch-producing plants and would not increase potential global warming [2].

All these reasons aroused a renewed interest in starch-based plastics in recent years. In the past, the study of starch esters and ethers [3,4-10] was abandoned due to the inadequate properties of these materials in comparison with cellulose derivatives for most applications. More recently, starch graft copolymers [2], starch plastic composites [11, 12], and starch itself [13-17], have been proposed as plastic materials.

Starch consists of two major components: amylose, a mostly linear alpha-D(1-4)-glucan and amylopectin, an alpha-D-(1-4) glucan which has alpha-D(1-6) linkages at the branch point. The linear amylose molecules of starch have a molecular weight of 0.2-2 million, while the branched amylopectin molecules have molecular weights as high as 100-400 million [18-19].

In nature starch is found as crystalline beads of about 15 μm-100 μm in diameter, in three crystalline modifications designated A (cereal), B (tuber), and C (smooth pea and various beans), all characterized by double helices: almost perfect left-handed, six-fold structures, as elucidated by X-ray diffraction experiments [18, 20, 21]. Starch beads may also show V crystallinity, characterized by a single helix when starch is in presence of fatty acids [22].

Crystalline starch beads in plastics can be used as fillers or can be transformed into thermoplastic starch which can be processed alone or in combination with specific synthetic polymers. To make starch thermoplastic, its crystalline structure has to be destroyed by pressure, heat, mechanical work and plasticizers such as water, glycerine or other polyols.

G. Scott (ed.), Degradable Polymers, 2nd Edition, 133-161.

This chapter reviews the main results obtained in the fields of starch-filled plastics and thermoplastic starch with a particular attention to the concept of gelatinization, destructurization, extrusion cooking, and the complexation of amylose by means of polymeric complexing agents with the formation of specific supramolecular structures. The behaviours of products now in the market are considered in terms of processability, physical-chemical and physical-mechanical properties and biodegradation rates.

2 Starch-filled plastics

Starch can be used as a natural filler in traditional plastics [11, 23-33] and particularly in polyolefins. When blended with starch beads, polyethylene films [34] biodeteriorate on exposure to a soil environment. The microbial consumption of the starch component, in fact, leads to increased porosity, void formation, and the loss of integrity of the plastic matrix. Generally [32, 35-38], starch is added at fairly low concentrations (6-15%); the overall disintegration of these materials is achieved by the use of transition metal compounds, soluble in the thermoplastic matrix, as pro-oxidant additives which catalyze the photo- and thermo-oxidative process [39-42]. An example of the contribution of starch in promoting the disintegration of photodegradable low density polyethylene (LDPE) is shown in Figs. 6. I and 6.2 [43, 44].

Starch-filled polyethylenes containing pro-oxidants are commonly used in agricultural mulch film, in bags and in six-pack yoke packaging. Commercial products based on this technology are sold by Ecostar and Archer Daniels Midland Companies [45, 46]. In the St Lawrence Starch [47, 48] technology, bought by Ecostar, regular corn starch is treated with a silane coupling agent to make fit compatible with hydrophobic polymers, and dried to less than 1% of water content. It is then mixed with the other additives such as an unsaturated fat or fatty acid autoxidant to form a master-batch which is added to a commodity polymer. The polymer can then be processed by convenient methods, including film blowing, injection moulding and blow moulding.

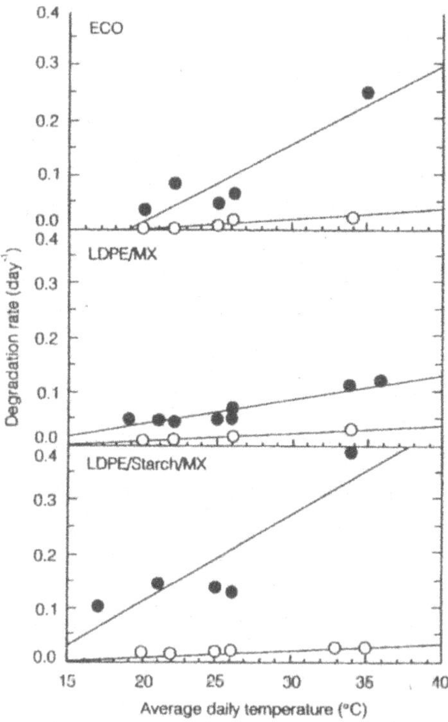

Figure 1 Average elongation at break versus total global solar radiation (45° South) for all locations [43]. ECO = ethylene-carbon monoxide copolymer (-1% CO), of Illinois Tool Works Co; LDPE/MX = LDPE film produced by Plastigone, containing metal compound pro-oxidant additives; LDPE/Starch/MX = LDPE film produced by ADM, containing 6% by weight of starch and metal compound pro-oxidant additives.

The temperature must be kept below 230 °C to prevent decomposition of the starch, and exposure of the master-batch to air must be minimized to avoid water absorption. Direct addition of starch and autoxidant without the master-batch step can also be used; as this requires some specific equipment, it is only practical for large volumes [42]. It is claimed that under appropriate conditions, the disintegration time of a buried carrier bag, containing an Ecostar additive to reach 6% starch, will be reduced from hundreds of years to 3-6 years [38]

However there is no evidence of a compliance of such materials with the norms of biodegradability and compostability already in place at international level. Moreover, the destabilization of polyethylene induced by the pro-oxidants may significantly affect its in use performances as a function of time

Within the field of starch-filled materials other systems were studied, some of which were completely biodegradable such as starch/poly(epsilon-caprolactone) [49], others partially biodegradable, such as starch/PVC/poly (epsilon-caprolactone) and its derivatives [50] or starch/modified polyesters [51]. In all these cases starch granules are used to increase the surface area available for attack by micro-organisms.

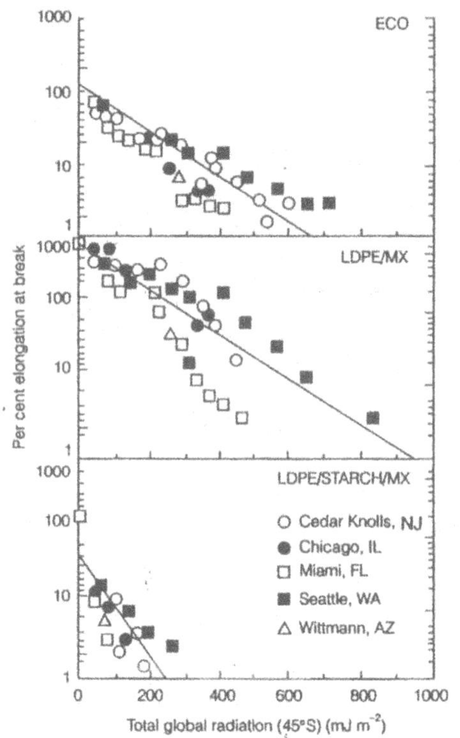

***Figure 2** The dependence of the empirical rate constant for degradation on the average ambient air temperature (• = data for degradable polymer; o = data for control polymer) [43].*

3 Thermoplastic starch

Starch can be gelatinized by extrusion cooking technology [51a-65]. As described by Conway in 1971, extrusion cooking and forming is characterized by sufficient work and heat being applied to a cereal-based product to cook or gelatinize completely all the ingredients. In general the main components of high pressure cooking extruders are feeders, compression screws, barrels, dies, and heating systems [51a]. The effects of processing conditions on the gelatinization of starch and on the texture of the extruded product have been studied by several researchers [52-69]. Gelatinized materials with different starch viscosity, water solubility and water absorption have been prepared by altering the moisture content of the raw product and the temperature or the pressure in the extruder. It was demonstrated that an extrusion-cooked starch can be solubilized without any formation of maltodextrins, and that the extent of solubilization depends on extrusion temperature, moisture content of the starch before extrusion and the amylose/amylopectin ratio. Mercier [68] analysed the properties of different types of starch and considered the influence of the following parameters: moisture content between 10.5 and 28%, barrel temperature between 70 and 250 °C, residence time between 20 seconds and 2 minutes, in a twin screw extruder. Corn starch, after extrusion cooking, gave a solubility lower than 35°/o, while potato starch solubility was up to 80% (Fig. 6.3).

Starch gelatinization is a difficult term to define clearly and it was used in the past to describe loss of crystallinity of starch granules, notwithstanding the process conditions applied [18]; namely, extrusion cooking, spray drying or heating of diluted starch slurries.The work carried out by Donovan in 1979 [70] and by Colonna and Mercier in 1985 [71] gave, however, a clear explanation of two different conditions for the loss of crystallinity of starch. Colonna reported that all starches exhibit a pure gelatinization phenomenon, which is the disorganization of the semicrystalline structure of the starch granules during heating in the presence of a water fraction > 0.9. For normal genotypes, gelatinization occurs in two stages. The first step, at around 60-70 °C, corresponds mainly to swelling of the granules, with limited leaching. Loss of birefringence, demonstrating that macromolecules are no longer oriented, occurs prior to any appreciable increase in viscosity. By contrast, differential scanning calorimetry (DSC) permits the determination of the gelatinization temperature more easily and precisely than microscopy and, additionally, the energy input needed to disorganize the crystalline structure of the granules. The second step, above 90 °C,

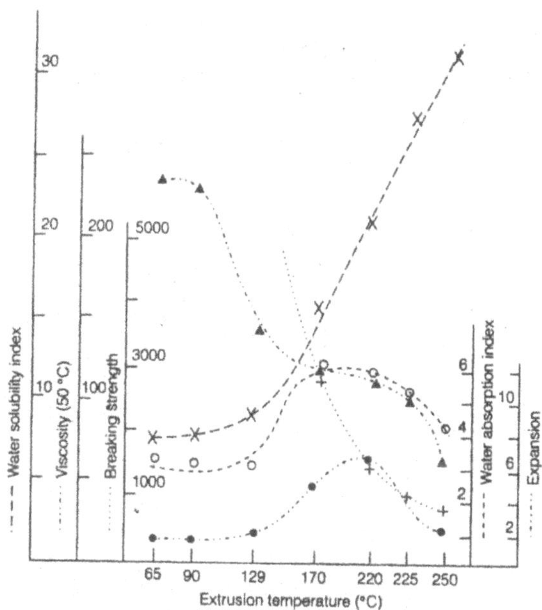

Figure 3 Effects of extrusion temperature on expansion (•), breaking strength (+), viscosity at 50 °C (A), water absorption index (o), and water-solubility index (x) of extruded products from corn grits. Initial moisture content before extrusion was 18.2% by weight [68].

implies the complete disappearance of granular integrity by excessive swelling and solubilization. Nevertheless this last transition is not detectable by DSC. Only at this stage the swollen granules can be destroyed by shear.

As observed by Donovan [70] and Colonna [71], at low water volume fractions ($V1 < 0.45$) loss of crystallinity occurred by two (pea and high amylose maize) or three (standard maize) crystalline melting steps, according to the Flory equation (Fig. 6.4):

$$1/Tm - 1/Tm = R/\Delta Hu \cdot Vu/V1 [V1 - X1\ V1]$$

where R is the gas constant, ΔHu the fusion enthalpy per repeating unit (anhydroglucose), Vu/V1 the ratio of the molar volume of the repeating unit to the molar volume of the diluent (water), Tm (K) the melting point of the crystalline polymer plus diluent, Tm (K) the true melting point of undiluted polymer crystallites, V1 the volume fraction of the diluent and X1 the Flory-Huggins interaction parameter. At high water volume fractions, melting of crystallites and swelling are co-operative processes. According to Colonna, during extrusion cooking and mainly under the conditions described by Mercier (water volume fraction < 0.28) [67] starch undergoes a real melting.

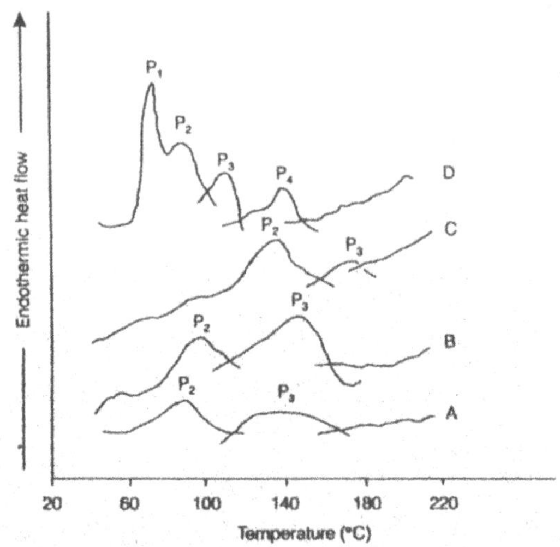

Figure 4 *Typical DSC curves for pea and maize starch reprinted from Phyto chemistry, 24-8, P. Colonna et al., "Gelatinization and Melting ...", p. 1670, copyright 1985 with kind permission from Elsevier Sci. Water volume fractions (V1): wrinkled pea (A), V1 = 0.35; smooth pea (B) V1 = 0.29; high amylose maize (C), V1 = 0.20 and normal maize (D), V1 = 0.55 [71].*

In the patent literature the term "destructurized starch " [72-90] refers to a form of thermoplastic starch described as molecularly dispersed in water [91]. Destructurization of starch is defined as melting and disordering of the molecular structure of the starch granules and as a molecular dispersion [91, 74]. The molecular structure of the starch granules is molten and consequently the granular structure disappears. This is achieved by heating the starch above the glass transition and the melting temperature of its components until they undergo endothermic transitions. In the melt stage both the crystalline and the granular structure of the starch are destroyed and the starch –water system forms a single phase in which no structure is discernible microscopically. The disappearance of the molecular structure of the starch granule may be determined using conventional light microscopy techniques [92].

If starch is heated above the glass transition and melting temperatures in presence of plasticizers the endothermic transition can be replaced by an exothermic transition. Destructurized starch, in simple terms, is a form of thermoplastic starch suitable for applications in the sector of plastics, with minimized defects tied to the granular structure of native starch [17,93,97].

Thermoplastic starch alone can be processed as a traditional plastic [68, 91, 98]; its sensitivity to humidity, however, makes it unsuitable for most of the applications (Fig. 6.5). Starch can be also made thermoplastic at water contents lower than 10%, in the presence of high boiling point plasticizers [14, 17], to avoid expansion phenomena at the die.

Another term which can be found in the literature is "Thermoplastically Processable Starch" (TPS), defined as a thermoplastic starch substantially water free.Thermoplastically Processable Starch is a modified native starch which is obtained without water, since instead of water use is made of a plasticizer or additive. The starch is thermoplastically processed together with the additive and the thermal transition taking place here is exothermic [100-105].

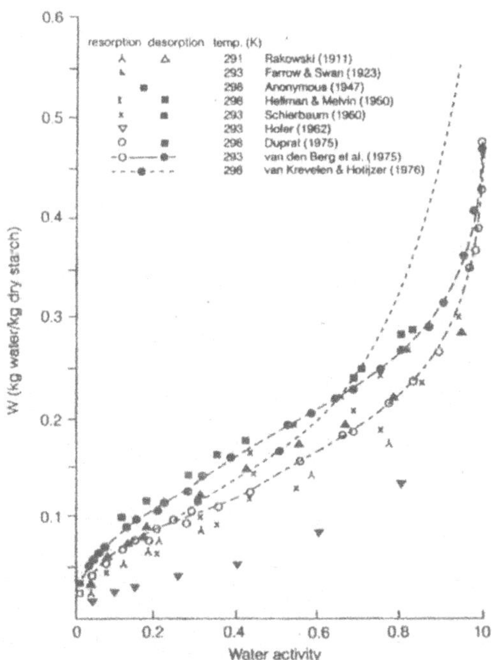

Figure 5 Some sorption isotherms of water vapour on native potato starch as reported in the literature [99].

Starch can be destructured in combination with different synthetic polymers to satisfy a broad spectrum of needs for the market. In this case it is possible to reach starch contents higher than 50%. Otey has studied EAA (ethylene-acrylic acid copolymer)/ thermoplastic starch composites since 1977 [106-116] and has demonstrated that the addition of ammonium hydroxide to EAA makes it compatible with starch. Urea, in these formulations, enhances the film tear propagation resistance and reduces ageing phenomena due to segmental motions in amorphous starch [117,118]. The films obtained with a content of plasticized starch of about 50% showed good tensile properties (Table 6.1) [111]. The sensitivity to environmental

Table 1 Influence of starch/EAA ratio and of partial replacement of EAA with PE or PVOH on the tensile strength and elongation of starch/EAA films [111,117]

Starch (phr)	EAA (phr)	PE (phr)	PVOH (phr)	Elongation (%)	UTS* (MPa)
10	90	–	–	260	23.9
30	70	–	–	150	22.2
40	60	–	–	92	26.7
40	40	20	–	66	23.9
40	25	25	–	85	21.7
40	20	40	–	34	20.1
40	55	–	5	97	32.0
40	40	–	20	59	39.7

*ultimate tensile strength.

changes and in particular the susceptibility to tear propagation precluded their use in most packaging applications [117]; moreover EAA is not biodegradable at all.

In 1989 studies on EAA-thermoplastic starch films, containing 40% by weight of EAA, processed at water contents lower than 2%, led to improved processability and film properties with elongation at break up to 200% [92]. By microscopic analysis it was possible to observe at least three different phases: one consisting of destructured starch, one consisting of the synthetic polymer alone, and a third one described as `interpenetrated', characterized by a strong interaction between the two components. As a confirmation, phase changes observed by DSC, nuclear magnetic resonance (NMR) [112,116,119-123], for starch-EAA-PE films showed at least Tour phases. DSC endotherms and extraction of free starch with hot water demonstrated the existence of a starch phase. DSC showed melting of an EAA phase and a LDPE phase but did not indicate the presents of EAA in amorphous regions of the PE. NMR, X-ray diffraction and extraction indicated the presents of an insoluble starch-EAA complex [123]. It was demonstrated that a portion of the starch forms complexes [121, 122] with EAA when EAA is salified by ammonium hydroxide or other salts during extrusion cooking, providing partial miscibility between the two polymers.

Rheological studies were performed on a product consisting of 60% of starch and natural additives and by 40% of EAA copolymer, containing 20% by mole of acrylic acid [124]. A strong non-Newtonian behaviour was shown by the viscosity curves at high shear rates; at intermediate shear rates the material seemed to approach a Newtonian plateau, while at low shear rates a viscosity upturn was observed, as shown by Fig.6, suggesting the presence of yield stress. Breaking-stretching data for the same material are also reported in the literature, together with those of LDPE [124] (Figs 7 and 8).

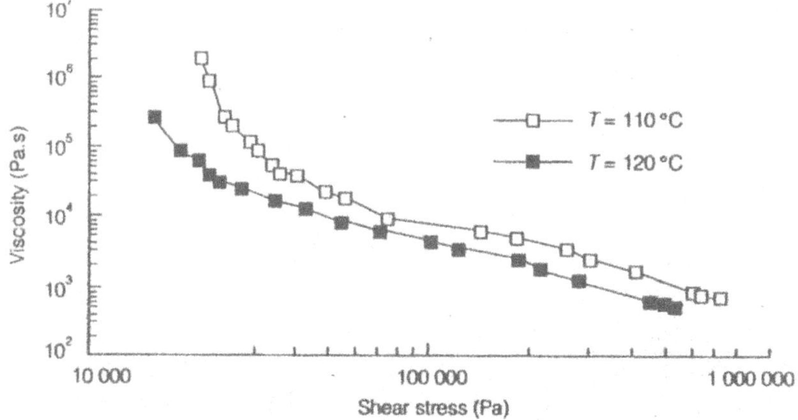

Fig. 6 Viscosity curve of a starch-based material containing 60% of thermoplastic starch and 40% of EAA [124].

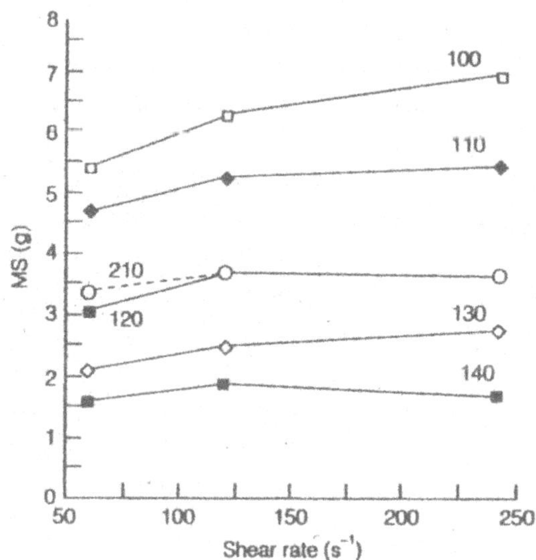

Fig. 7 Melt strength (MS) of thermoplastic starch/EAA 60/40 w/w composite material at different temperatures in comparison LDPE (---) [124].
Breaking-stretching (BSR) values are well above those of LDPE, while, at temperatures below T = 120°C, the melt strength becomes greater than that of polyethylene. Thus this starch-based polymer can be processed by all operations which involve elongational flow.

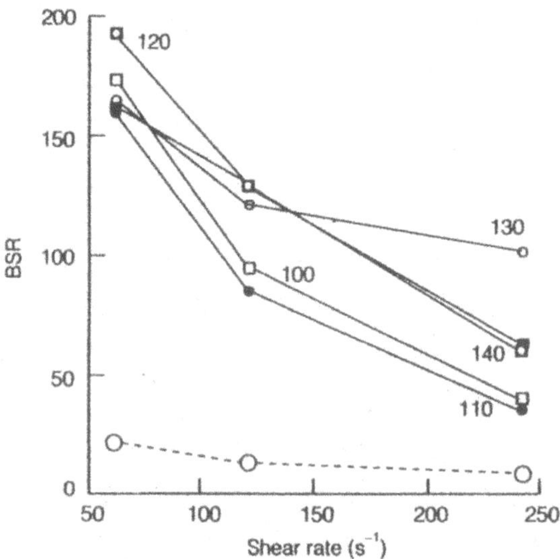

Fig. 8 Breaking-stretching (BSR) values of thermoplastic starch/EAA 60/40 w/w composite material at different temperatures in comparison with LDPE (---) [124].

Starch/vinyl alcohol copolymer systems [125-130], can generate a wide variety of morphologies and properties, depending on the processing conditions, the starch type and the copolymer composition. Different microstructures were observed, from droplet-like to layered, as a function of different hydrophilicity of the synthetic copolymer. Furthermore, for this type of composite, materiale containing starch with an amylose/amylopectin ratio > 20/80 w/w do not dissolve even with stirring in boiling water. Under these conditions a microdispersion, consisting of microsphere aggregates is produced, whose individua! particle diameter is under 1 um (Fig. 9). A droplet-like structure is also confirmed by transmission electron microscopic (TEM) analysis of film slices [126]. The droplet size is comparable with that of the microdispersion obtained by boiling.

For these products, high levels of melt elasticity is monitored by exit pressure data, whereas its recoverable fraction is almost negligible (low die swell) [128, 129]. The morphology of materials in film form, containing starch with an amylose/amylopectin ratio lower than 20/80 w/w, gradually looses the droplet-like form, generating layered structures (Fig. 6.10). In this case no microspheres are produced by boiling and the starch component becomes partially soluble. Fourier transform infrared (FTIR) second derivative spectra of materials with droplet-like structure, in the range of starch ring vibrations between 960 and 920 cm -1, gives an absorption peak at about 947 cm -1 (Fig. 11). This peck, observed also when starch is complexed with butanol, is

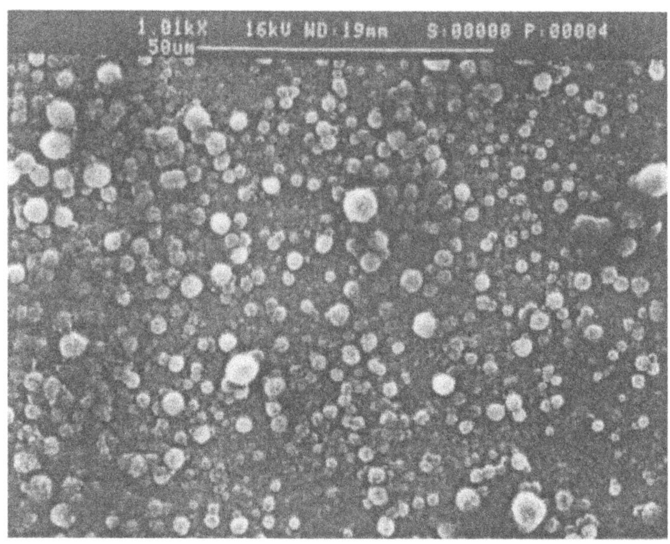

Fig. 9 Droplet-like structure of thermoplastic corn starch/EVOH blend in film from, after disagregation in boiling water [128].

attributed by Cael et al. [20] to ring vibrations, which result when amylose assumes a conformation known as the V form (a left-handed single helix). Therefore, the absorption at 947 cm-1 does not correspond to crystalline or gelatinized amylose, but to a complexed one (V-type complex), as in the presence of low molecular weight molecules such as butanol and fatty acide [20, 128]. Starch-based materials with an amylose content close to zero, even in the presence of vinyl alcohol copolymers, do not show any peak at 947 cm-1, demonstrating that vinyl alcohol copolymers, as well as butanol, leave the amylopectin conformation unchanged.

On the other hand, the V complex formed by starch, having an amylose/amylopectin ratio higher than 20% by weight, with ethylene-vinyl alcohol (EVOH) copolymers makes even amylopectin insoluble in boiling water (Table 2).The experimental evidence was accounted for by a model considering large invididual amylopectin molecules interconnected at several points per molecule as a result of hydrogen bonds and entanglements by chains of amylose/vinyl alcohol copolymer V complexes [128]. The biodegradation rate of starch in these materials is inversely proportional to the content of amylose/vinyl alcohol complex (Fig. 6.12). Furthermore FTIR second derivative spectra show the 947 cm ' peak increasing with biodegradation, which means

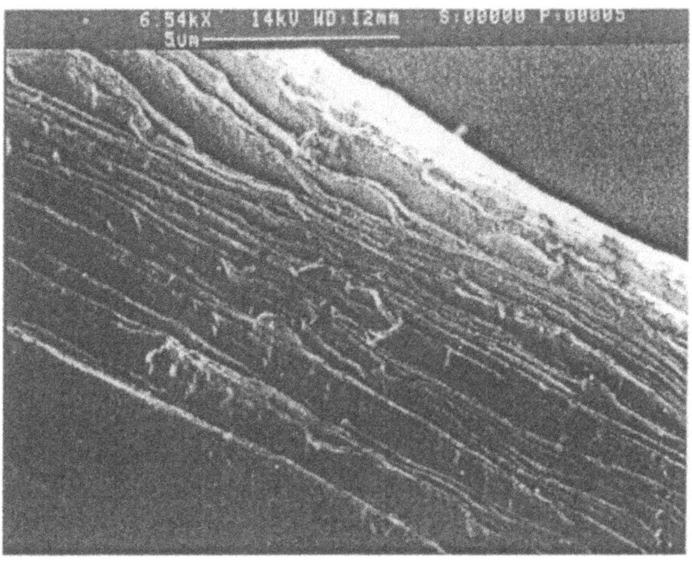

Fig. 10 Layared structure of thermoplastic waxy maize/EVOH film after 3 days of soil burial test [128].

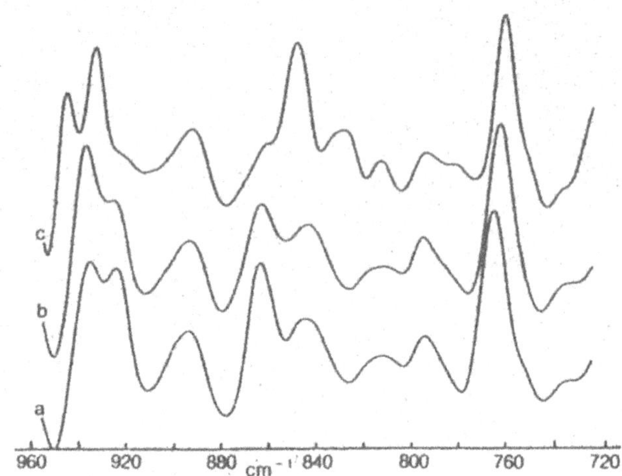

Fig. 11 FTIR second derivative spectral of corn starch. (a) cystalline, (b) gelatinized; (c) blended with EVOH [128].

Table 2 Insoluble residue of starch/EVOH 1:1 films after disagregation in boiling water as a function of starch composition [128
]

Amylose content (%)	Insoluble residue* (%)
5	58.3
10	67.5
15	75.3
20	92.1
25	97.5
28	96.8
70	97.1

*dry starch plus EVOH = 100%.

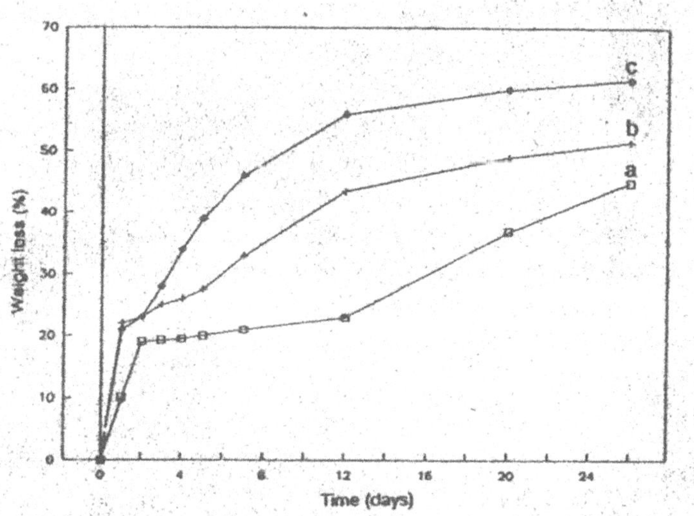

Fig. 12 Weight of corn starch/EVOH films in a soil burial test as a function of time and of amylose content. (a) 70% amylose; (b) 25% amylose; (c) 5% amylose [128].

a delayed microbial attack of complexed amylose relative to amylopectin [128]. In addition, water permeability of starch/EVOH films is a function of the V-type complex and can range from about 820 to 334 gr30 um/m2/24h [129].

A general study of shear flow characteristics was performed on a material containing about 60% of starch and natural additives and 40% of ethylene-vinyl alcohol copolymer 40/60 mol/mol [130]. A strong pseudoplastic behaviour at high shear stresses as well as yield stress at lower ones was detected (Fig. 13). The non-linear Bingham fluid model [131] well described its viscous behaviour over a wide range of shear rates. High levels of melt elasticity were detected from steady shearing testa, whereas its recoverable fraction was almost negligible, at least for a reasonable time scale. The peculiar viscous and elastic behaviour has been explained on the basis of the droplet-like morphology generated by the ability of starch to form V complexes in the presence of EVOH. Notwithstanding the peculiar rheological behaviour shown by starch/EVOH systems, traditional processing techniques such as film blowing can be easily applied.

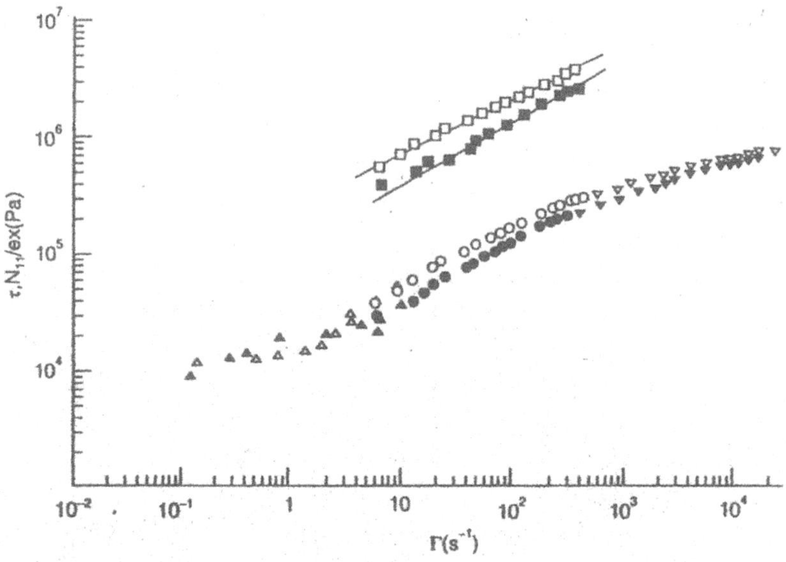

Fig. 13 Shear stress (○,●) and normal stress difference (N11/ex) (□,■) versus wall shear rate, of a thermoplastic starch/EVOH blend at 140 and 150°C [130].

The products based on starch/EVOH show mechanical properties good enough to meet the needs of specific industrial applications [132]. Their mouldability is comparable with that of traditional plastica such as polystyrene (PS) and acrylonitrile-butadiene-styrene copolymer (ABS). Nevertheless, they continue to behighly sensitive to low humidities, especially when in film form, with evident embrittlement.

In terms of biodegradation, ten months of aerobic biological treatment performed by a high sensitivity respirometric test, provoked the degradation of more than 90%, w/w of a product constituted by 60% of maize starch and natural additives and by 40% of ethylene-vinyl alcohol copolymer (EVOH) at 40% mol/mol of ethylene. Furthermore it has also been demonstrated that the synthetic component was degraded to about 80'% w/w, notwithstanding interrupting the test when CO2 evolution was stili relevant [126, 127]. A material with the same composition, containing an EVOH copolymer, characterized by a lower ethylene content (29% instead of 40% mol/mol) and, therefore, by a reduced ability to generate interpenetrated structures showed, in the Sturm test, an initial biodegradation rate significantly higher [126]. The SCAS test and biodegradation in lake water of a product constituted of 70% maize starch and natural additives and 30% EVOH support the hypothesis of a substantially different biodegradation mechanism for the two components [127]:

• the natural component, even if significantly shielded by the interpenetrated structure, appeared to be initially hydrolysed by extracellular enzymes;
• the synthetic component appeared to be biodegraded through surface adsorption of micro-organisms, assisted by the increase of available surface area during the hydrolysis of the natural component.

Other limited evidences for disappearance of ethylene-vinyl alcohol copolymers bave been produced by Roemesser [133] and Kaplan and coworkers [134]. The presence of starch improves the biodegradation rate of these synthetic polymers; a fundamental role is also played by size and distribution of ethylene blocks. The degradation rate is too slow to consider these materials as compostable [127] (Fig. 6.14). Studies to speed up the biodegradation process selecting specific micro-organisms are in progress.

Specific types of plasticizers were selected in order to avoid migration phenomena and physical ageing [132]. The possibility of speeding up the biodegradation process was considered by modifying the ethylene-vinyl alcohol copolymer by introducing carbonyl groups making it more sensitive to photodegradation [135]. The transparency of the material was also improved by adding additives such as boric acid, borax and other saline compounds [136]. Surface treatment by wax lamination or coextrusion was also considered [137]. With this kind of material it is possible to obtain finished parts by film blowing, injection moulding, blow moulding, thermoforming, etc. It is also possible to make foamed parts [138], particularly by an expansion process

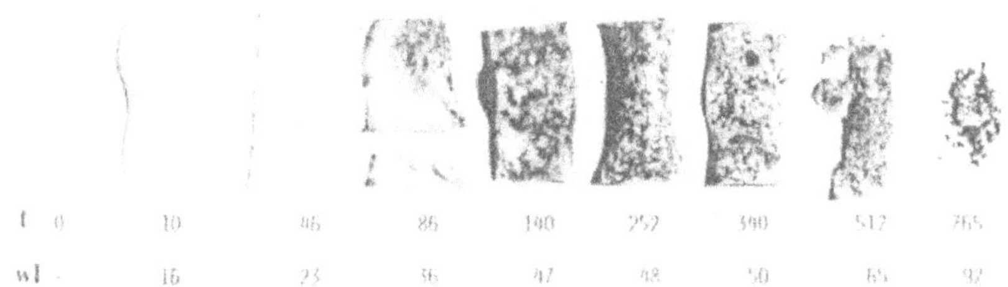

Fig. 14 Weight loss (wl) of starch complexed with EVOH under semicontinuos Activated Sludge (SCAS) test conditions [128] as a function of time (f)

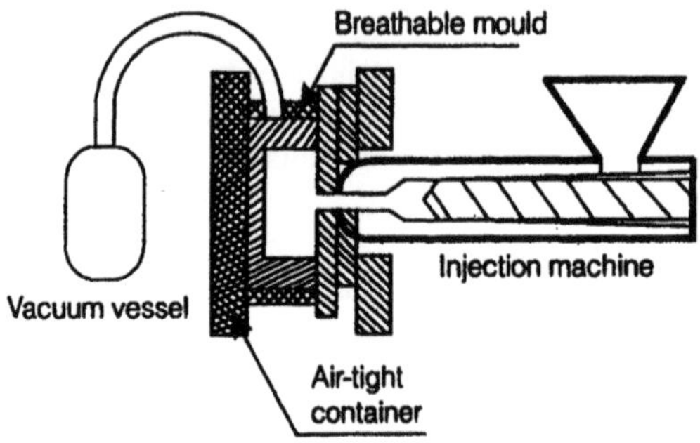

Fig. 15 Expansion-injection moulding of Mater-Bi [139]. Figure reprinted with kind permission of Elsevier Science as for Fig. 16.

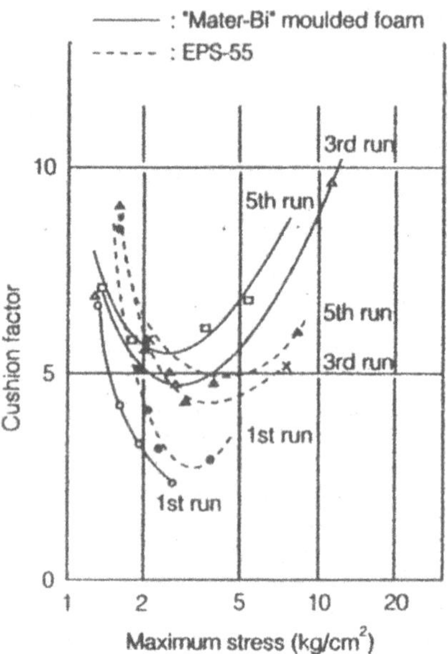

Fig. 16 Dynamic cushioning properties of Mater-Bi moulded foams[139].
(Reprinted from "Biodegradable Plastics and Polymers", Y. Yoshida and T. Uemura, p443-450, copyright 1994, with kind permission from Elsevier Science Ltd).

based on injection moulding technology. The technology consists of a breathable mould connected with a vacuum pump, applied to an ordinary injection moulding madrine (Fig. 15) [139]. Cushioning characteristics of these materials are close to expanded polystyrene (EPS-55) (Fig. 16); moreover, the foam density is of 0.040 g ml-'.

Starch can also be destructured in the presence of more hydrophobic polymers such as aliphatic polyesters [140]. It is known that aliphatic polyesters with low melting points are difficult to process by conventional techniques for thermoplastic materials, such as film blowing and blow moulding. With reference particularly to poly(epsilon-caprolactone) and its copolymers, films produced thereby are tacky as extruded, and rigid, and have low melt strength over 130 °C; moreover, due to the slow crystallization rate of such polymers, the crystallization process proceeds for a long time after production of the finished articles with an undesirable change of properties with time.

Novamont's Mater-Bi starch-based technology implies processing conditions able to almost completely destroy the crystallinity of amylose and amylopectine, which in the presence of macromolecules is able to form a complex with amylose such as specific polyesters. They can be of natural or synthetic origin, and are biodegradable. The complex formed by amylose with the complexing agent is generally crystalline and it is characterized by a single helix of amylose formed around the complexing agent. Unlike, amylose, amylopectine does not interact with the complexing agent and remains in its amorphous state The specification of the starch, i.e. the ratio between amylose and amylopectine, the nature of the additives, the processing conditions and the nature of the complexing agents allows engineering of various supramolecular structures with very different properties.

The scheme of a "droplet like" and a "layered" structure which can be produced as a result of Mater-Bi technology is reported in fig.6.17 and fig.6.18. The "droplet like" structure (fig.6.17) is constituted by a core of an almost amorphous amylopectine molecule screened by complexed amylose molecules which render amylopectine unsoluble.[141-143]

The layered structure is constituted by submicronic layers of amylopectine molecules intercalated by layers of complexing agent, being such layers compatibilized by complexed amylose (fig.6.18). The two structures and the many others derived from them explain the wide range of mechanical, physical-chemical, and rheological properties and the different biodegradation rates of Mater-Bi products.

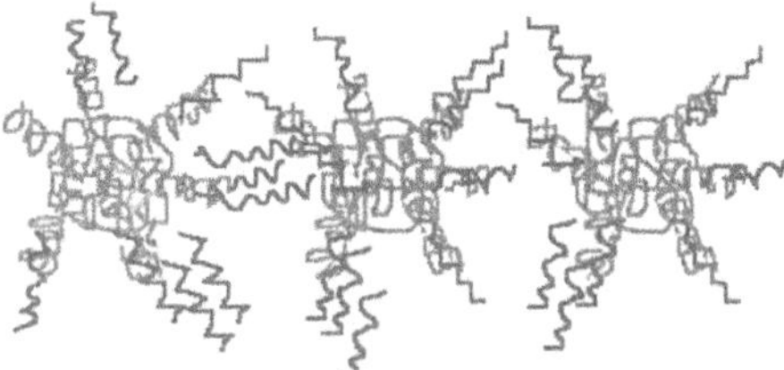

Fig. 17 Mater-Bi technology: Droplet-like structure. Green: amylopectine; red: amylose; blue complexing agent.

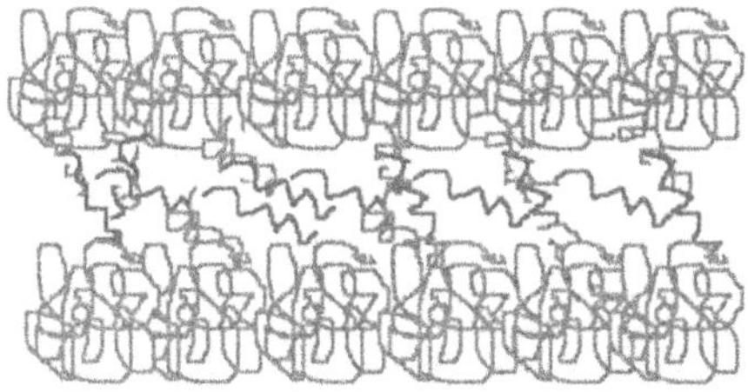

Fig. 18 Mater-Bi technology: Layered structure. Green: amylopectine; red: amylose; blu: complexing agent.

Blending of starch with aliphatic polyesters improves their processability and biodegradability. Particularly suitable polyesters are poly(epsilon caprolactone) and its copolymers, or polymers of higher melting point formed by the reaction of 1,4 butandiol with succinic acid or with sebacic acid, azelaic acid or poly(lactic acid), poly hydroxyalkanoates and aliphatic-aromatic polyesters.

The compatibilazation between starch and aliphatic polyesters can be promoted either by the processing conditions and or by the presence of compatibilizers between starch and aliphatic polyesters such as amylose/EVOH V-type complexes [128], starch-grafted polyesters, chain extenders like diisocyanates, epoxides, etc. are preferred. These types of materials are characterized by excellent compostability, mechanical properties and reduced sensitivity to water.

Thermoplastic starch can also be blended with polyolefins [144]. In this case about 50% of thermoplastically processable starch is mixed with 40% of polyethylene and 10% of ethyl acrylate-maleic anhydride copolymer. During this mixing process an esterification reaction takes place between the maleic anydride groups in the copolymer and the free hydroxyl groups in starch. Other studies bave been performed on polyamide/high amylose [74, 145, 146] and acrylic copolymers/high amylose starch systems [74, 146, 147]. The problem of partial biodegradability and too high a sensitivity to humidity persists.

Starch/cellulose derivative systems are also reported in other publications [136, 140, 148, 149], particularly, cellulose acetate and butyrate/starch blends in presence of glycerine and epoxidized soybean oil [148].

The combination of starch with a water soluble polymer such as polyvinyl alcohol (PVOH) and/or polyalkylene glycols has been widely considered since 1970 [150]. Recently, the system, thermoplastic starch/PVOH has been mainly studied for producing starch-based loose fillers as a substitute for expanded polystyrene [151-157]. As an example, Altieri and Lacourse developed a technology based on hydroxy propylated high amylose starch containing small amounts of PVOH for improving foam resiliency and density [151-155]. In this case loose fill was produced directly by a twin screw extruder. More recent patents. Recently more advanced processes and alloys have been developed which have resulted in foams with lower foam densities (8-6 kg/m3) and better performance [158-160].

Table .3 Some physical properties of Mater-Bi grades for film, in comparison with traditional plastics

TEST	PROCEDURE	UNIT	MATER-BI	LDPE
MFI	ASTM D 1338	g/10 min	2-8*	0.1-22*
Strength at Break	ASTM D 882	MPa	24-30	8-10
Elongation at Break	ASTM D 882	%	200-1000	150-600
Young's Modulus	ASTM D 882	MPa	100-400	100-200
Tear strength	ASTM D 1938			
• Primer		N mm^{-1}	30-90	60
• Propagation		N mm^{-1}	30-90	60

*150°C,5Kg ; †190°C, 2.16 Kg

Table .4 Some physical properties of Mater-Bi grades for injection molding, in comparison with traditional plastics.

TEST	PROCEDURE	UNIT	MATER-BI	PP†	PS‡
MFI	ASTM D 1238	g/10 min	20-10	0.3—40	1.2-25
Strength at Break	ASTM D 638	MPa	20-30	23	30-60
Elongation at Break	ASTM D 638	%	20-500	400-900	1-4.5
Young's Modulus	ASTM D 638	MPa	200-2000	1400-1800	3000-3500
IZOD (notched impact)	ASTM D 256	KJ m^{-2}	1- 80	3-10	2-3

*170°C,5Kg ; †23°C,2.16 kg; ‡ 200°C,5Kg

4 Starch-based materials on the market

The market of destructurized and complexed starch-based bioplastics accounts for about 25000ton/year, 75% of which is for packaging applications and including soluble foams for industrial packaging and films for bags and sacks. The market share of these products accounts for about 75-80% of the global market of bioplastics. [161]

Leading producers with well established products in the market are Novamont, , National Starch, main Novamont partner and licensee in the sector of loose-fills and of foamed sheets, and, finally, Biotec with a capacity of about 2000ton/year .

Following the recent start-up of its third line dedicated to the production of Mater-Bi film grades in Terni, Novamont's internal production capacity is of 20000 ton/year. The total capacity, including the network of licensees in the sector of loose fills, is of about 35000 ton/year.The technology for the production of starch based loose fills is licensed together with National Starch and Chemical Co.

The wide patent portfolio of Novamont covers the technologies of complexed starch developed bv Novamont and of destructurized starch developed by Warner Lambert and acquired by Novamont in 1997 after the exit of Warner Lambert from the market in 1993. Moreover, on August 2001 Novamont acquired the film technology of Biotec which included an exclusive license of the Biotec's patents on TPS in the sector of film [161].

In recent years companies such as Earthshell, Apack, Avebe dedicated significant efforts to the development of food containers through the "baking technology". Market tests are in place in USA and Europe to check their performances. [162]. Moreover very recently in the Netherlands Rodemburg built up a plant for the transformation of potato wastes generated by the industry of fried potatoes in a granulate to be used for the injection moulding of slow release devices. The claimed capacity is of 40000ton/year.

The price of starch-based bioplastics ranges from 1.25 to 4 Euro/Kg, with possibilities to compete even with traditional materials in some limited areas [162].

The properties achieved by starch-based bioplastics in certain applications and the commitment of the companies today dealing with this family of bioplastics give more confidence in the future possibilities of this market sector. Bioplastics from renewable origin, either biodegradable or non biodegradable, still constitute a niche market which requires high efforts in the areas of material and application development; the technical and economical breakthroughs achieved in the last three years, however, open new possibilities for such products in the mass markets.

Novamont today boasts a diversified portfolio of industrial tailor made materials for a wide range of applications which explains its position as market leader[161].
After more than 12 years of research and development, Mater-Bi products are able to fulfil specific in-use performances in different application sectors, and offer original solutions both from the technical and the environmental point of view.

Under the Mater-Bi trademark today Novamont produces a wide range of materials, divided into 5 families, according to the processing technologies: film , extrusion/thermoforming, injection moulding, foaming, tyres technology. Mater-Bi products are mainly used in specific applications where biodegradability is required; examples include composting bags and sacks (fig.6.19) fast food tableware (cups, cutlery, plates, straws, etc.), packaging (soluble and unsoluble foams for industrial packaging) (fig.6.20), film wrapping (fig.6.21), laminated paper, food containers (fig.6.22), agriculture (mulch film) (fig.6.23), nursery pots, plant labels, slow release devices etc.), hygiene (nappy backsheet, cotton swabs).

New sectors are also growing outside biodegradability, driven by the unique technical performance of some Mater-Bi products versus traditional materials, as in the case of breathable films with silky handle for nappies , chewable items for pets or biofillers for tyres. The new tyre Biotred GT3, launched by Goodyear in 2001 and recently adopted by BMW and Ford is an example of the high tech performances reached by Mater-Bi products [163].

Mater-Bi starch-based materials offer an ideal combination of properties showing:

- complete biodegradability and compostability according to existing standards (fig.6.24) [164-166];
- significant reduction of environmental impact, particularly with respect to CO_2 emissions and energy consumption, in comparison with traditional materials in specific uses [167-168] ;
- in use performances similar to traditional plastics;
- processability similar or improved in comparison with traditional plastic materials[161].
  Other properties of Mater-Bi films of last generation can be summarised as follows:
- soft, silky handle;
- wide range of permeability to water vapour (from 250 to 1000g/30um/m2/24h);
- wide range of mechanical properties from soft and tough materials to rigid ones, with no significant ageing after one year of storage [162].
- Antistatic behaviour;
- colourability with food contact approved pigments;
- compostability in a wide range of composting conditions: from home composting and static windrows to rotary fermenting reactors.

They are biodegradable and compostable according to the present European standards and are certified by AIB Vincotte in Belgium, by DINCERTCO in Germany and by IIP in Italy, according respectively to CEN EN13432, DIN 54900 and UNI 10785 standards.

After the acquisition of Enpac in 1998 and the subsequent agreement with Novamont, National Starch is licensing two technologies for the production of loose-fills: one from hydroxypropilated high amylose starch and a second from almost unmodified starch. The loose-fills' densities range from 6 to 10 kg /m3. The main licensees are Unisource, American Excelsior, Storopack and Flow Pack in USA.

Biotec, the German company which acquired in 1994 the patents of Fluntera, was acquired by EKI (Essem Kashoggi Industries) in 1998. Biotec, after the sale of the film business to Novamont is concentrated on foodserviceware products and on pharmaceutical products.

Figure 19 *Mater-Bi bags*

Figure 20 *Mater Loose Fills*

Figure 21: *Mater-Bi wrappings*

Figure 22 *Mater-Bi knitted net.*

Figure 23 *Mater-Bi mulch film*

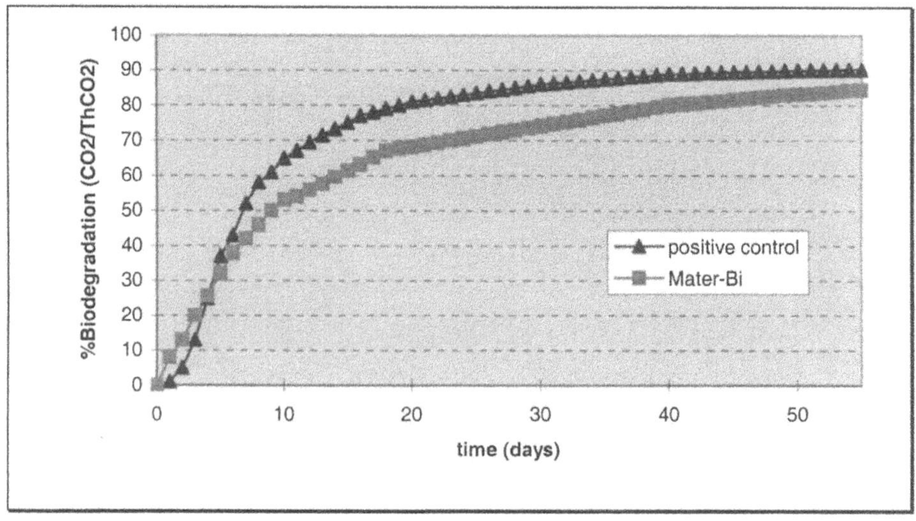

Figure 24 Aerobic biodegradation of Mater-Agro under controlled composting conditions (EN13432), in comparison with pure cellulose (test performed by OWS, Belgium)

5 CONCLUSIONS

Starch-based bioplastics constitute a new generation of materials able to significantly reduce the environmental impact in terms of energy consumption and green-house effect in specific applications, to perform as traditional plastics when in use, and to completely biodegrade within a composting cycle through the action of living organisms when engineered to be biodegradable.They offer a possible alternative to traditional materials when recycling is unpractical or not economical or when environmental impact has to be minimized.

After more than twelve years of research and development starch-based materials start to fulfil specific in-use performances in different application sectors. They are able to offer original solutions both from the technical and the environmental point of view.

Today some of the bioplastics available in the market are used in specific applications where biodegradability is required such as the sectors of composting (bags and sacks), fast food tableware (cups, cutlery, plates, straws etc.), packaging (soluble foams for industrial packaging, film wrapping, laminated paper, food containers), agriculture (mulch film, nursery pots, plant labels), hygiene (diaper back sheet, cotton swabs) slow release in agricultural and pharmaceutical sectors. Moreover new sectors are growing outside biodegradability, driven by improved technical performances versus traditional materials, as in the case of biofillers for tyres, chewable items for pets.

The price of bioplastics from renewable origin is decreasing and ranges from 1.25 to 4 Euro/Kg, with possibilities to compete even with traditional materials in some limited areas.

The world market for biodegradable plastics is still small, but it has grown significantly in the last few years reaching about 33000 ton/year in the year 2000; products totally or partially from renewable resources represent nearly 85-90% of this market[169].

References

1. *Corn Annual (1991)* Corn Refiners Assoc. Inc. Washington, DC, p. *20.*
2. Bagley, E. B., Fanta, G. F. and Burr, R. C. *et al. (1977) Polym. Eng. Sci.**17**, 311-6.*
3. Seiberlich, J. *(1941) Mod. Plastics,**18** (7), 64-5.*
4. Bagley, E. B., Fanta, G. F. and Doane, W. M. *et al. (1977) US* Pat. *40,26,849.*
5. Barach, B. and Shasha, S. *(1983) US* Pat. *43,82,813.*
6. Trinmell, D. and Shasha, B. S. *(1984) US* Pat. *44,39,488.*
7. Shasha, B. S., Doane, W. M. and Russel, C. R. *(1982) US* Pat, *43,44,857.*
8. Shasha, B. S. and Abbott, T. P. *(1982) US* Pat. *43,48,492.*
9. Fanta, G. F. and Stout, E. I. *(1984) US* Pat. *44,89,50.*
10. Rankin, J. C. *(1982) US* Pat. *43,30,443.*
11. Griffin, G. J. L. *(1977) US* Pat. *40,16,117.*
12 Otey, F. H., Westhoff, R. P. and Russell, C. R. *(1975)* in Symposium Papers, Technical Symposium, Nonwoven Product Technology, Miami Beach, FL, pp. *40-7.*
13. Stepto, F. T. and Tomka, I. *(1987) Chimia, **41**(3), 76-81.*
14. Tomka, I. and Troesch, A. *(1990) PTC* Int. Pat. Appl. WO *90/05161.*
15. Lacourse, N. L. *(1989)* Eur. Pat. Appl. *03 76 201.*
16. Knight, A. T. *(1990) PCT* Int. Pat. Appl. WO *90/14938.*
17. Bastioli, C., Lombi, R., Del Tredici, G. F. and Guanella, 1. *(1990)* Eur. Pat. Appl. *90/110069.*
18. Otey, F. H. and Doane, W. M. *(1984) Starch Chemistry and Technology,* (eds R. L. Whistler *et al.),* Academic Press, pp. *154-5, pp. 667-69.*
19. Otey, F. H. and Doane, W. M. *(1984) Starch Chemistry and Technology,* (eds R. L. Whistler *et al.),* Academic Press, pp. *397-403.*
20. Cael, J. J., Koening, J. L. and Blackwell, J. *(1975)* Biopolymers, ***14**, 1885-1903.*
21. Casu,B. and Reggiani,M. *(1966)Die Starke, 7, 218-228*
22. Henry F. Zobel *(1984) Starch Chemistry and Technology,* (eds R. L. Whistler *et al.),* Academic Press, pp. *287.*
23. Griffin, G. J. L. *(1991) PCT* Int. Pat. Appl. *W091/04286.*
24. Griffin, G. J. L. *(1978) IT 10,24,922.*
25. Griffin, G. J. L. (1977) US Pat. 4,02,13,88.
26. Griffin, G. J. L. (1978) UK Pat. 14,85,833.
27. Griffin, G. J. L. (1978) US Pat. 41,25,495.
28. Griffin, G. J. L. (1982) US Pat. 43,24,709.
29. Griffn, G. J. L. (1995) Eur. Pat. 363 383 B
30. Griffin, G. J. L. (1992) Int. Pat. Appl. WO 92/01741.
31. Griffin, G. J. L. (1985) *Prog. Biotechnol.,* **1,** *pp.* 201-10.
32. Warren, R. (1989) *Bridges,* 23-5.
33. Griffin, G. J. L. (1980) *Pure Appl. Chem.,* **52,** 399-407.
34. Griffin, J. L. (1975) *ACS Advances in Chemistry Ser., No.* 18424.
35. Evans, J. D. and Sikdar, S. K. (1990) *Chemtech,* **20**, 38-42.
36. Johnson, R. (1987) in *Proceedings of Symposium on Degradable Plastics,* Society of the Plastica Industry, Washington DC, June (0, pp. 6-13.

37. Hiltz, W. (1989) *Nat. Living,* 13.
38. Maddever, W. J. (1987) in *Proceedings of Symposium on Degradable Plastics,* Society of the Plastics Industry, Washington DC, June 10, pp. 41-4.
39. Scott, G. (1971) UK Pat. 1,356,107.
40. Gilead, D. and Scott, G. (1978) US Patent 4,519,161.
41. Sipinen, A. J., Jaeger, J. T., Rutherford, D. R. and Edblom, E. C. (1993) US Patent 5,216,043.
42. Hocking, P. J. (1992) J. M *S.-Rev. Macromol. Chem. Phys.,* **C32** (1*),* 35-54.
43. Andrady, A. L., Pegram, J. E. and Nakatsuka, S. (1993) J. *Environ. Polymer Degradation,* **1**(1), 31-43.
44. Andrady, A. L., Pegram, J. E. and Tropsha, Y. (1993) J. *Environ. Polymer Degradation,* **1** (3), 171-9.
45. Wilder, R. (1989) *Modern Plastic International,* September.
46. Narayan, R. and Lafayette, W. (1989) *Kunstoffe German Plastics,* **79**, 92-5.
47. Thomas, R. H. (1990) *Polymer News,* **14** (9), 271-3.
48. *High-Tech Materials Alert,* (1990) **7** (3), 4-6.
49. Chuo Chemical Co. (1992) J. P. Pat. Appl. 146953.
50. Aime, J. M. Mention, G. and Thouzeau, A. (1989) US Pat. 48,73,270.
51. Gallagher, F. G., Shin, H. and Tietz, R. F. (1993) US Patent.
51a.Conway, H. F. (1971) *Food Produet Development, pp.* 141-31.
52. Chiang, B. Y. and Johnson, A. J. (1977) *Cereal Chem.,* **54** (3), 429-35.
53. Conway, H. F., Lancester, E. B., Bookwalter, G. N. (1968) *Food Engineering,* 1021.
54. Mercier, C. (1971) *Feedstufs,* **4**, 33-9.
55. Lawton, B. T., Henderson, G. A. and Derlatka, E. J. (1972) *Canadian Journal of Chemical Eng.,* **50**, 168-72.
56. Anderson, R. A., Conway, H. F. and Pfeifer, V. F. (1969) *Griffin Cereal Science Today*, **14** (1), 4-7.
57. Derby, R. L, Miller, B. S. and Trimbo, H. B. (1975) *Cereal Chem.,* **52** (5), 702-13.
58. Harmann, D. V. and Harper, J. M. (1973) *Transactions of the ASAE,* 1175-8.
59. De La Gueriviere, J. F. (1976) Bull. *Anc. Eleves Ec. Fr. Meum.*, No. 276, 305-14.
60. Stevens, D. J. and Elton, G. A. H. (1971) *Die Starke,* **23** (1*),* 8-I 1.
61. Anderson, R. A., Conway, H. F. and Peplinski, A. J. (1970) *Die Starke,* **22** (4),130-5.
62. Mottern, H. H., Spadaro, J. J. and Gallo, A. S. (1969) *Food Technology,* **23,** 169-71.
63. Shetty, R. M., Lineback, D. R. and Seib, P. A. (1974) *Cereal Chem.,* **51,** 364-75.
64. Miller, B. S., Derby, R. I. and Trimbo, H. B. (1973) *Cereal Chem.,* **50,** 271-80.
65. Chiang, B. Y. and Johnson, A. J. (1977) *Cereal Chem.,* **54** (3), 436-43.
66. Anderson, R. A., Conway, H. F., Pfeifer, V. F. and Griffin, E. L. (1969) *Cereal Science Today,* **14,** 372-81.
67. Sterling, C: (1978) J. *Texture Stud.,* **9,** 225-55.
68. Mercier, C. and Feillet, P. (1975) *Cereal Chemistry,* **52** (3), 283-97.
69. Anderson, R. A., Conway, H. F., Pfeifer, V. F. and Griffin, (1969) *Cereal Science Today,* **14** (1), 4-7.
70. Donovan, J. W. (1979) *Biopolymers,* **18**, 263.
71. Colonna, P. and Mercier, C. (1985) *Phytochemistry*, **24** (8), 1667-74.
72. Sacchetto, J. P. and Stepto, R. F. T. (1990) US Patent 4,900,361.
73. Stepto, R. F. T. and Dobler, B. (1994) Eur. Pat. 0 326 517 B.
74. Lay, G., Rehm, J., Stepto, R. F. T. and Thoma, M. (1997) Eur. Pat. 0 327 505 B.
75. Sacchetto, J. P., Egil, M., Stepto, R. F. T. and Zeller, J. (1990) Eur. Pat. Appl. 0 391 853.

76. Sacchetto, J. P., Lentz, D. J. and Silbiger (1994) Eur. Pat. . 0 404723 B.
77. Silbiger, J., Sacchetto, J. P. and Lentz, D. J. (1990) Eur. Pat. Appl. 0404 728.
78. Sacchetto, J. P. and Rehm, J. (1991) Eur. Pat. Appl. 0 407 350.
79. Silbiger, J., Sacchetto, J. P. and Lentz, D. (1991) Eur. Pat. Appl. 0408501.
80. Sacchetto, J. P., Silbiger, J. and Lentz, D. J. (1991) Eur. Pat. Appl. 0 408502.
81. Silbiger, J., Lentz, D. and Sacchetto, J. P. (1994) Eur. Pat. 0 408 503 B.
82. Sacchetto, J. P., Silbiger, J. and Lentz, D. J. (1994) Eur. Pat. 0 409 781 B.
83. Lentz, D., Sacchetto, J. P. and Silbiger, J. (1994) Eur. Pat. 0 409 782 B.
84. Sacchetto, J. P., Silbiger, J. and Lentz, D. J. (1994) Eur. Pat. 0 409 783 B.
85. Lentz, D., Sacchetto, J. P. and Silbiger, J. (1994) Eur. Pat. 0 409 788 B.
86. Silbiger, J., Lentz, D. J. and Sacchetto, J. P. (1994) Eur. Pat. 0 409 783 B.
87. E.T.Cole, M. Egli (1990) Eur. Pat. Appl. 0 500 885
88. M. J. Izbicki , G. Loomis , A. Flammino (1993) Eur. Pat. Appl. 0 679, 172
89. F. Degli Innocenti , R. Lombi,C. Bastioli, M. Nicolini, M.Tosin (1998), Eur. Pat. Appl. 1 109 858
90. V. Bellotti, R.Lombi, C. Bastioli,I. Guanella,G. Del Tredici (1998) Eur. Pat. Appl. 965 615
91. Wittwer F. and Tomka, I. (1989) Eur. Pat. 0 I 18 240.
92. H.B. Wigman,W.W. Leathen, M.J. Brackenbeyer (1956) *Food Technol.* **10**,179-184
93. Bastioli, C., Bellotti, V. and Del Giudice, L. *et al.* (1993) US Pat. 52,62,458.
94. Bastioli, C., Bellotti, V., Del Giudice, L. and Lombi, R. (1990) Eur. Pat. Appl. 90/110070.
95. Bastioli, C., Bellotti, V., Del Giudice, L. and Lombi, R. (1990) Int. Pat..Appl. WO 90/EP 1 253.
96. Bastioli, C., Bellotti, V. and Del Tredici, G. F. (1990) Eur. Pat. Appl. WO 90/EP 1286.
97. Bastioli, C., Bellotti, V. and Montino, A. *et al.* (1991) Int. Pat. Appl. WO 91 /EP 1373.
97a.Wittwer, F. and Tomka, I. (1987) US Pat. 4,673,438.
98. Russell, C. R. and Hellies, L. (1957) US Pat. 27,88,546.
99. Van den Berg, C. (1981) *Vapour Sorption Equilibria and Other Water-Starch Interactions; a Physico-Chemical Approach,* Agricultural University Wageningen, The Netherland p. 49.
100. Tomka I., (1995) Eur. Pat. 0 397819 B
101. Tomka I. (1994) Eur. Pat. 0 539 544 B
102. Tomka I., Meissner J., Menard R. (1997) Eur. Pat. 0 537 657 B
103. Tomka I. (1998) Eur. Pat. 0 542 155 B
104. Tomka I. (2001) US 6,242 102
105. Tomka I. (1997) Eur. Pat. 0 596 437 B
106. Otey, F. H. (1979) US Pat. 41,33,784.
107. Otey, F. H. (1985) Eur. Pat. 1,32,299.
108. Otey, F. H. (1984) US Pat. 44,54,268.
109. Otey, F. H. (1982) US Pat. 43,37,181.
110. Otey, F. H., Westhoff, R. P. and Doane, W. M. (1987) *Ind. Eng. Chem. Res.,* **26,** 1659-63.
111. Otey, F. H., Westhoff, R. P. and Doane, W. M. (1980) *Ind. Eng. Chem. Prod. Res. Dev.,* **19**, 592-5.
112. Westhopp, R. P., Otey, F. H., Mehltretter, Ch. L. and Russell, C. R. (1974) *Ind. Eng. Chem. Prod. Res. Dev.,* **13**(2), 123-5.

113. Otey, F. H. (1976) *Polymer Plast. Technol. Eng.*, **7** (2) 221-34.
114. Otey, F. H., Westhofh R. P. and Russell, C. R. (1976) *Ind. Eng. Chem. Prod. Res. Dev.*, **15/2**, 1392.
115. Otey, F. H., Mark, A. M., Mehltretter and Russell, C. R. (1974) *Ind Eng. Chem. Prod. Res. Develop.* , **13/** 1 90-2.
116. Fanta, G. F. and Otey, F. H. (1989) US Pat. 48,39,450.
117. Swanson, C. L., Shogren, R. L., Fanta, G. F. and Imam, S. H. (1993) J. *Environ Polymer Soc.*, **1** (2), 155-66.
118. Shogren, R. L. (1992) *Carbohyd Polym.***19,** 83-90.
119. Fanta, G. F., Swanson, C. L. and Doane, W. M. (1988) *Polym. Prepr.* (A.C.S. Div. Polym. Chem.), **29** (2), 453-4.
120. Fanta, G. F., Swanson, C. L. and Doane, W. M. (1990) *Journal of Applied Polymer Science,* **40,** 811-21.
121. Fanta, G. F., Swanson, C. L. and Doane, W. M. (1992) *Carbohydr.Polym.*,**17**, 51-8.
122. Fanta, G. F. and Swanson, C. L., Shogren, R. L. (1992) J. *Appl Polym.Sci.*, **44,** 2037-42.
123. Shogren, R. L., Thompson, A. R. and Felker, F. C. *et al.* (1992) J. *Appl Polym. Sc.*, **44**, 1971-8.
124. Bastioli, C., Rallis, A., Cangiatosi, F. *et al.* (1991) *Proceeding of the Polyme Processing Society -European Regional Meeting,* Palermo, September **15**-18.
125. Bastioli, C. and Bellotti, V. (1990) Agrobiotecnologie e nuove produzioni chimiche da risorse rinnovabili, Accademia Nazionale dei Lincei, Rome, November 26-27.
126. Bastioli, C., Bellotti, V., Del Giudice, L. and Gilli, G. (1993) J. *Environ. Polym. Degradation,* **1**(3), 181-91.
127. Bastioli, C., Bellotti, V., Del Giudice, L. and Gilli, G. (1992) in *Biodegradable Polymers and Plastics,* (ed. M. Vert et al.), The Royal Society of Chemistry, pp. 101-10.
128. Bastioli, C., Bellotti, V. and Camia, M., Del Giudice, L and Rallis, A (1994) B*iodegradable Plastics and Polymers,* Eds Y Doi and K.Fukuda, Elsevier, p.200-213
129. Bastioli, C., Bellotti, V. and Del Tredici, G. F. *et al.* (1992) Italian Appl. T092 A000199.
130. Bastioli, C., Bellotti, V. and Rallis, A. (1994) *Rheologica Acta,* 33, 307-316.
131. Herschel, W. H. and Bulkley, R. (1926) *Proc. ASTM, 26,* 621.
132. Bastioli, C., Bellotti, V. and Montino, A. (1992) Int. Pat. Appl. 92/14782.
133. Roemesser, J. (1991) Presentation at Plastics Waste Management Workshop, American Chemical Society, Polymer Chemistry Division, New Orleans, LA, December.
134. Kaplan, D. L., Mayer, J. M. and Ball, D. (1992) *Proceedings Biodegradable Packaging Symposium,* Natick, MA, June.
135. Bastioli, C., Bellotti, V., Del Giudice, L. and Lombi, R. (1991) Int. Pat. Appl. WO 91 /02024.
136. Bastioli, C., Bellotti, V. and Montino, A. *et al.* (1992) Int. Pat. Appl. W092/01743.
137. Bastioli, C., Bellotti, V., Romano, G. C. and Tosin, M. (1992) Int. Pat. Appl. WO 92/02363.
138. Bastioli, C., Bellotti, V. and Del Giudice, L. (1991) Int. Pat. Appl. WO 91/02023.
139. Yoshida, Y. and Uemura, T. (1994) *Biodegradable Plastics and Polymer,* Eds. Y. Doi and K.Fukuda, Elsevier, p. 450,

140. Bastioli, C., Bellotti, V. and Del Tredici, G. F. et al. (1992) Int. Pat Appl. WO92/ 19680.
141. Bastioli C. *Polymer Degradation and Stability*, 59,(1998), 263-272
142. Bastioli C. *Starch based materials: properties and applications-* Renewable Bioproducts, Industrial outlets and Research for the 21st Century *(1997)* Wageningen, The Netherlands
143. Bastioli C.*Global Status for Biodegradable Polymers Production* –7th Annual Meeting of Bio/Environmentally Degradable Polymer Society (1998) Cambridge, Massachussets
144. Tomka, I. (1992) Int. Pat. Appl. W092/20740.
145. Buehler, F. S., Baron, V., Schmid, E. and Meir, P. (1992) Eur. Pat. Appl. 0541050 A2.
146. Schmid, E. and Buehler, F. S. (1992) Eur. Pat. Appl. 0522358.
147. Bastioli, C:, Bellotti, V. and Del Tredici, G. F. (1996) Int. Pat. 437 589
148. Tomka, I. (1992) Eur. Pat. Appl. 0542155.
149. Schroeter J. (1997) Eur. Pat. 0 551 125 B.
150. Maxwell, C. S. Tappi, (1970) 53(8), 1464-6.
151. Lacourse, N. L., Alberi, R. A. (1989) US Pat. 48,63,655.
152. Lacourse, N. L. (1990) Eur. Pat. Appl. 0375831.
153. Lacourse, N. L. (1996) Eur. Pat. 0 376 201 B.
154. Lacourse, N. L. and Altieri, P. A. (1991) US Pat. 50,35,930.
155. Lacourse, N. L. and Alberi, P. A. (1991) US Pat. 50,43,196.
156. Neumann, P. E. (1993) US Pat. 51,85,382.
157. Anfinsen, J. R. and Garrison, R. R. (1992) Int. Pat. Appl. WO 92/08759.
158. Bastioli, C. et al. (1995) Eur. Pat. Appl. 0 667 369
159. Bastioli, C. et al. (1995) Eur. Pat. 0 696 611
160. Bastioli, C. et al. (1995) Eur. Pat. Appl. 0 696 612
161. Bastioli,C. and Facco,S. (2001) Biodegradabile Plastics 2001 Conference,Frankfurt, Germany, November 26-27
162. Bastioli C. *Global Status of the production of biobased materials* –(2001) Actin Conference; Birmingham, UK-
163. Automotive News Europe, October 22, 2001 pa.15 „BMW, Ford adopt Corn Tire"
164. *Prufung der Kompostierbarkeit von Kunstoffen* (1998)– DIN V 54900-2
165. Requirements for packaging recoverable through composting and biodegradation. Test scheme and evaluation criteria f or the final acceptance of packaging – EN 13432
166. Compostabilita' dei Materiali Plastici – Requisiti e Metodi di Prova –UNI10785 – January 1999
167. *Life Cycle Assessment of Mater-Bi bags for the collection of compostable waste* (1998)– Composto; Switzerland,
168. *Life Cycle Assessment of Mater-Bi and EPS Loose-Fills* (2000)– Composto, Switzeland,
169. *Actual situation and prospects of EU industry using renewable raw materials* (2002) Edited by DG Enterprise /E.1

7

POLYMERS FROM RENEWABLE RESOURCES

E. CHIELLINI, F.CHIELLINI, P.CINELLI
Department of Chemistry and Industrial Chemistry
University of Pisa
Via Risorgimento 35, 56126 Pisa, Italy

1. Introduction

The current utilization of natural resources cannot be sustained forever. Most of the fuel utilized in our societies comes from fossil fuel, such as oil that, other than being subjected to price fluctuations, must eventually be depleted. Rising atmospheric carbon dioxide levels from combustion of fossil fuels are thought to be increasing global temperature that, in turn, may cause droughts, crop losses, storm damage, etc [1]. Fuel shortage and the waste accumulation in the environment are generating a worldwide interest in alternative resources and particularly for the use of renewable resources both as an energy source [2] and as raw materials for polymers and plastics [3]. There is increasing pressure for a wider utilization of biomass feed-stocks for specialty items. The total biomass produced on earth is estimated as approximately 170 billion tons, of which a very small portion, less than 4 %, is used. [4].

Concerns about climate change and the preservation of natural resources, represents a worldwide driving force to reduce the consumption of fossil fuel feedstock that is currently stimulating academic and industrial researchers as well as decision makers [5].

TABLE 1. Fossil and Renewable Resources [4]

Biomass	Consumption[a]	Reserves[b]
Renewable biomass[a]	6.0	170
Mineral oil	3.2	135
Natural gas	1,900 [d]	140,000 [d]
Coal	3.4	850

a) Billions metric tons/year if not otherwise stated
b) Billions metric tons if not otherwise stated
c) Through photosynthetic process
d) Cubic meters

In industrial production sustainability must be achieved, but keeping in mind that business will fail unless a minimum margin profit is guaranteed.

A normative strategy has been proposed for resource choice and recycling to

G. Scott (ed.), Degradable Polymers, 2nd Edition, 163-233.

meet the criteria of sustainability [6]. Also the use of biowaste as a resource for biobased productions has been proposed because of its high content of cellulose, hemicellulose and lignin [7]. The term "polymers from renewable resources" refers to natural products that are polymeric in character as grown or can be converted to polymeric materials by conventional or enzymatic synthetic procedures [8]. Thus under that heading one can include *natural polymers* used as direct feedstock for plastic production as well as *artificial polymers* as those obtained by chemical modification of preformed natural polymers or by polymerization of monomers deriving from renewables [9]. Traditional natural material are represented by cellulose, starch, leather, wool, silk, proteins, natural rubber, gums and vegetable oils binders.

The majority of uses now served by petroleum-derived plastics demand a long life span and nearly indestructibility. New bio-plastics should not be introduced as substitutes of the synthetic ones, but as more appropriate options where the degradation constitutes a plus in specific applications by defraying the cost inherent in the management of the disposal of post-consume items. Moreover the availability of raw materials from renewable resources should not interfere with food production [10]. Some products at present made from polyethylene could well be replaced by bio-plastics, while others could not. The same applies to polystyrene and polypropylene products [11]. Thus biodegradability is an advantageous property in those cases where it is implicitly demanded by the application that is where recycling is controlled by a fairly high management cost for disposal [12]. This means that marketing efforts must be focused on specific products, rather than on introducing a new material to the whole plastics industry. Plastic articles that are used once and then disposed of, are targeted as the primary market areas. Such application include packaging films, foams and bags, food service items (cups, plates, cutlery, containers of milk, water, soft drinks), personal care items, and agricultural mulch films are prime candidates for replacement with biodegradable polymers [13]. It is however taken for granted that infrastructures have to be available to allow for their bio-recycling as a final stage of their recovery.

Polymers from renewable resources include among the others: poly-saccharides, such as cellulose, starch, chitosan, lignin and proteins, like wool, silk and gelatin, oils, and microbial poly (ester)s, such as PHAs.

Polymers derived from renewable resources can be broadly classified according to the method of production. A first category encompasses polymers directly extracted/removed from natural materials, especially plants such as carbohydrates, aromatic plant products, polyisoprenes, and proteins.

Natural polymers, or biopolymers, are produced in nature by living organisms, and by plants through biosynthetic processes that involve carbon dioxide consumption [14]. Natural polymers are ultimately degraded and consumed in nature in a continuous recycling of resources. Arguments in favor of "natural" polymers are: biodegradability, renewability, recyclability, non-waste producing, neutrality on green house effect, and functionality. However, in some cases natural polymers such as rubber, lignin and humus present a slow rate of biodegradation that will not satisfy the rapid mineralization criteria currently advocated by standards committees for synthetic polymers [15]. However, because they are produced in nature there is no major concern about it, like some synthetic polymers that are notoriously recalcitrant to biodegradation.

A second category is constituted by polymers produced by chemical synthesis from renewable resources bio-derived monomers such as polylactate from starch or chemical modification of naturally occurring polymers, *artificial polymers*, such as cellulose esters and ether.

A last category may be identified in polymers produced by native microorganisms or genetically transformed bacteria. The best known example of this category is constituted by poly (hydroxyalkanoate)s (PHA), mainly poly (hydroxybutyrate) (PHB) and copolymers of hydroxy-butyrate and valerate produced by Monsanto and Metabolix under the trade name Biopol.

There are certain similarities in the structures of synthetic and natural polymers such as nylon and protein, synthetic and natural rubbers, but sometimes the breadth of function is far greater in natural polymers. For example spider silk has the strength of Kevlar combined with greater stretch [17]. Several polymers may also be produced by modification of natural polymers or polymerization of monomers from renewable resources.

In the present chapter we wish to focus on some polymeric materials from renewable resources, which should complement somehow the content of chapter 5, 6, 8, 9, 10, 13.

2. Natural Polymers

The most widespread natural polymers are polysaccharides such as cellulose and starch and chitin, but also lignin, proteins and others find several applications. Most polysaccharides are composed of five or six-membered rings, usually with two or three hydrolysis attached, respectively. Chemically, they are hemiacetals with ether linkages joining the monomeric units. Cellulose in particular is enjoying a worldwide consumption volume for paper and cardboard manufacturing comparablc to the overall synthetic polymeric materials that is above 205 Mtons [18].

2.1 CELLULOSE

Cellulose is the main component of higher plant cell walls. About 7.5 Gtons of cellulose grow and disappear each year, thus establishing it as the most abundant regenerated organic matter on earth [19]. In the secondary cell wall of plants, cellulose molecules are unbranched chains of up to 17,000 1,4 linked β-D-glucose residues (Figure 1) but shorter chains occur under other circumstances.

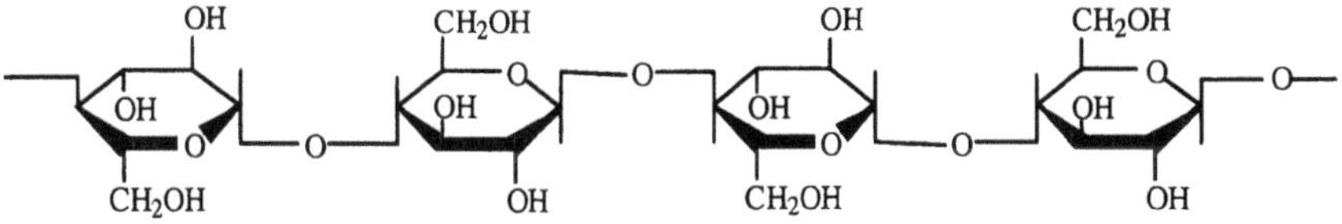

Figure 1. Schematic Representation of Cellulose Structure

Cellulose for industrial conversion comes from wood and scores of minor sources such as bagasse, the stalks of sugar cane after the extraction of the juice by press technology. Cellulose in wood, along with lignin, serves directly as fuel. After minimal processing of natural cellulosic materials, they are used as lumber and as textiles based on cotton, jute, ramie, flax (linen), and hemp. Cellulose is a relatively cheap raw material costing 0.5-1 €/Kg before derivatization [20]. After industrial treatment, with and without chemical derivatization, cellulose is made into diverse products including paper, membranes, explosives, textiles (rayon and cellulose acetate), and dietary fibers. For example cellulose acetate can be synthesized by the reaction of acetic anhydride with cotton linters or wood pulp, and cellulose esters from recycled paper and sugar cane bagasse have also been proposed [21]. Most celluloses have a high degree of polymerization; the intermediate glucose residues determine the chemical and physical properties and the weight of the terminal units can be practically ignored. The glycosidic bonds in cellulose are strong and this polymer is stable under a wide variety of reaction conditions. It is a generally insoluble, highly crystalline polymer. Industrially important chemical modifications of this polymer generally involve reaction with free hydroxyl groups in 2, 3, and 6 position [22]. These reactive sites undergo most of the reactions characteristic of alcohols. Etherification and esterification of individual hydroxyl groups, of the polysaccharide backbone, are of particular importance for cellulose. The chemical modification of cellulose from the melt or in solution facilitates its processing under conditions used for thermoplastic polymeric materials. Numerous derivatives are commercially available such as cellulose acetate, ethyl cellulose, hydroxy-ethyl cellulose, and hydroxypropyl cellulose. These cellulosic materials have been widely used for the fabrication of membranes and hollow fibers suited for immobilization of enzymes, (catalase, alcohol oxidase and glucose oxidase) [23,24] and in the practice of hemo-dialysis reverse osmosis, and chromatographic supports [23]. Hydroxyalkyl cellulose and carboxymethyl celluloses have found applications as matrices for drug delivery and as wound dressing [25].

Fatty acid esters of cellulose present interesting properties as water repellency, thermal stability and thermoplasticity. Conventional synthesis of such materials employs fatty-acid chlorides or anhydrides in organic solvents [26]. Solvent-free methods have also been developed using formic acid and octanoyl chloride as the gelatinizing and acylating agents respectively. Acylation reaction has often accomplished with the help of a co-reagent and the solvent exchange technique including pretreatment of cellulose by soaking into water followed by washing with ethanol, and finally with the fatty acid [27]. Recent technique proposes the preparation of a homogeneous mixture of cellulose, water, soap and fatty acid by emulsification. The esterification reaction is then performed after water removal by distillation [28]. Also long-chain fatty cellulose esters have been synthesized, up to stearate by reaction of stearic acid chloride with hydrolyzed cellulose in pyridine and 1,4-dioxane [29].

Esterification of cellulose can also be accomplished with lactones as acylating agents, by reacting low-substituted cellulose hydroxyalkyl ethers with lactone an ethylene- or propylene oxide spacer was inserted between the ester side chain and the main chain, in order to completely separate the esters units from the backbone and provide a sterically preferred link points for the lactone units [22]. These reactions were carried out in relatively dipolar aprotic media such as dimethysulfoxide, dimethylformalamide, dimethylacetamide or dioxane. While the introduction of ether groups imposes a serious problem to subsequent enzymatic and microbial degradation, esters retard the degradation without however preventing it [30]. Biodegradation of cellulose acetate and cellulose propionate have been established by Komarek et al. [31]

with a naturally derived mixed microbial culture derived from activated sludge. Microorganism were able to extensively degrade cellulose acetate with degree of substitution (DS) ranging from 1.85 to 2.57 over periods of 14-31 days. Cellulose acetate degradation was also evaluated *in vitro* and in a system in which cellulose diacetate films were suspended in a waste-water treatment system. The *in vitro* assay employed a stable enrichment culture initiated by inoculating a basal salt medium containing cellulose acetate with 5% (v/v) activated sludge. Cellulose diacetate with DS=1.7 was 80% degraded in 4-5 days while cellulose diacetate with DS =2.5 required 10-12 days [32]. One possible pathway for cellulose acetate biodegradation would involve attack by cellulase enzymes on the unsubstituted residues in the polymer backbone. Enzymatic cleavage of the acetyls by esterases, or simple chemical hydrolysis, would serve to expose additional unsubstituted residues, which could also be digested by the action of cellulase enzymes. The combined action of the esterase and cellulase enzymes would serve eventually to degrade completely cellulose acetate in the environment.

Cross-linking of the polymer chains imparts durable press properties to cellulosic textiles and dimensional stability to wood products. Formaldehyde or N-methylol derivatives are used for this purpose [8]. The reaction with the hydroxyl groups usually takes place under heterogeneous conditions because of the insoluble and crystalline nature of cellulose. Under such mild heterogeneous conditions, the hydroxyl groups are tightly engaged in the formation of stable hydrogen bonds, that inhibit them from reacting. Compared with soluble polysaccharides, therefore, the extent of such reactions is limited thus resulting in a fairly difficult and expensive process [20].

Cellulose is soluble only in unusual and complex systems [19]. When dissolved, cellulose molecules are still fairly extended, but exist as random coils with relatively large end-to-end distances. Commercially, dissolving pulps, which have lower molecular weights, are used along with strong alkali and derivatization. Cellulose subjected to high temperature and pressure during the steam explosion process can be dissolved in strong base. For film production cellulose is dissolved in an aggressive toxic mixture of sodium hydroxide and carbon disulphide ("Xanthation") and then recast into sulphuric acid to give cellophane films. This procedure that was largely applied in the past for the production of renegerated cellulose will probably be banned in the future because of its negative environmental impact.

A way to impart solubility and melt processabilty to cellulose, and other hydroxy polymers, has been identified in trimethylsilylation using different silylating agents. Recently Mormann and Demeter reported a method for cellulose silylation with hexamethyldisilizane in liquid ammonia [33]. Ammonia is known to activate cellulose by intercalation into the lattice by breaking up the inter- and intramolecular hydrogen bonds. In the process reported by Mormann and Demeter ammonia is the only byproducts generated from hexamethyldisilizane upon conversion of a hydroxy into a trimethylsiloxy group, and can be removed together with the ammonia used as reaction medium.

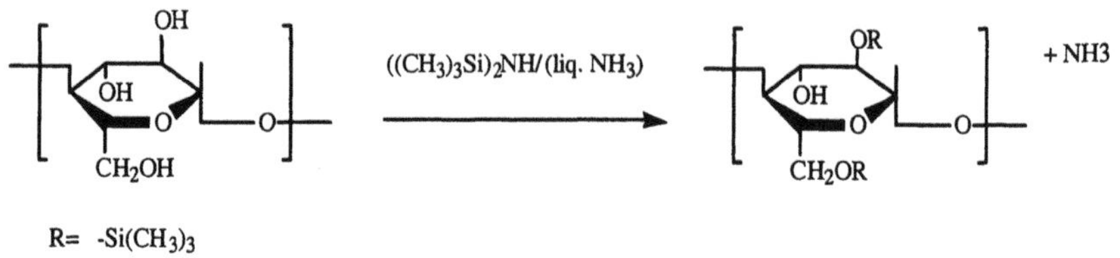

$((CH_3)_3Si)_2NH/(liq.\ NH_3)$ + NH3

R= $-Si(CH_3)_3$

Figure 2. Cellulose Silylation with Hexamethyldisilazane in Liquid Ammonia [33]

Saccharin was used as a catalyst at concentration of 0.5 mol % saccharin/mol of hydroxy groups. A ratio of trimethylsilyl groups to OH of 3.4 was found to be suitable for complete silylation. With a similar process Mormann and Spitzer have reported the silylation of OH-containing polymers, such as cellulose and poly(vinyl alcohol) by reactive extrusion [34]. A possible application of silylated cellulose is in the field of regeneration of cellulose after spinning or molding. Silylation of cellulose avoids the problems connected with the huge amounts of salt, wastewater and toxic reagents like carbon disulfide.

Aqueous salt solutions such as saturated zinc chloride or calcium thiocyanate can dissolve limited amounts of cellulose. Two non- aqueous solvents are ammonium thiocyanate in ammonia and lithium chloride in N,N-dimethylacetamide. Cellulose solutions up to about 15% can be made with these solvents. Blends of cellulose and poly(vinyl alcohol) have been prepared in N,N-dimethylacetamide-lithium chloride, and exhibited good miscibility due to their mutual ability to form intra-intermolecular hydrogen bonds between hydroxyl groups [35]. Miscible blends of cellulose and poly(vinylpyrrolidone) have been prepared by dissolution in dimethyl sulfoxide-paraformaldehyde and blending with poly(vinylpyrrolidone) dissolved in dimethyl sulfoxide [36,37]. Cellulose has also been blended with poly(ethylene glycol) in dimethylsulfoxide and paraformaldehyde [38].

Currently the application of heterogeneous processing conditions still prevails in industry, particularly for high volume polymers. Hence non-uniform distribution of the substituents groups on the cellulose matrix is a major concern, because they have profound effect on mechanical properties, physical and biological properties of modified products [39].

Thermoplastic cellulose derivatives such as esters can be used for exstrusion and moulding. Cellulose has been used as filler to reduce polymer cost as well as for reinforcement in composite materials [40,41]. Thus cellulose fibers are relatively cheap and light-weight compared to inorganic fillers. Native cellulose fibers are among the strongest and stiffest fibers available with a theoretical value for stiffness of a single crystal of more than 130 Gpa [42]. Drawbacks in the use of cellulose fibers are the lower processing temperature permissible and the high moisture absorption. This last problem can be minimized by chemical modification, such as acetylation, of some of the hydroxyl groups [43]. Cellulose fibers, when used in a hydrophobic polymeric matrix often present a poor adhesion at the fiber-matrix interface and tend to aggregate. Coupling or compatibilizing agents interact with both fibers and the matrix forming a link between the components. For cellulose fibers several coupling agents have been tested, as reported by Felix, such as chlorotriazines in cellulose fiber-polyesters composites, isocyanates and silanes in wood cellulose fibers and various thermoplastics [44,45]. Stearic acid has been used for fibers dispersion, maleated ethylene has been used as coupling agent in blends with polyethylene [46]. To improve fiber dispersion and the matrix-fiber adhesion several modification processes have been evaluated. Treatment of cellulose fibers prevents hydrogen bonds formation and makes the fiber surfaces and the matrix more similar. For example graft polymerization by attaching a suitable polymer segment on the surface with a similar solubility as the polymer matrix. A large number of graft copolymers of cellulose have been reported. Graft polymerization has been performed by radical polymerization, ionic polymerization and condensation, and ring opening polymerization [39]. In the grafting reaction several factors must be kept in consideration such as minimization of concurrent homopolymer formation, involvement of most of the cellulose molecules in the grafting process, control of the molecular weights and molecular weight distribution of the grafted side chains, reproducibility of the grafting yields [47].

In radical polymerization the molecular weight of the side chains grafts is difficult to control and may be very high and disperse. Moreover grafting of only a few high molecular weight chains occurs and considerable amounts of homopolymers are sometimes formed [48]. In spite of these problems many graft copolymers have been produced with this method and comprehensively reviewed by Hon [39].

Modification of lignocellulosic materials surface by copolymerization with vinyl monomers has been reported. The polymerization reaction is initiated at the surface of the fibers by incorporation of peroxides or oxidation-reduction agents, or by treatment with gamma radiation or cold plasma [49]. These reactions form free radicals on the fibers, which initiates the free chain reaction with the vinyl monomers. Different types of properties can be conferred to the fibers using different vinyl monomers, such as increased hydrophilicity with poly(vinyl alcohol), increased hydrophobicity with polystyrene or polyvinylacetate, increased reactivity with polyvinylamine etc.

Graft copolymers of polypropylene (PP) and maleic anhydride have shown to be very effective additives for wood cellulose/PP composites [50]. Thus cellulose fibers have been surface modified with polypropylene-maleic anhydride copolymer. The modified fibers have been compounded with polypropylene [51].

Cellulose fibers have been immersed in a solution of polypropylene-maleic anhydride copolymer in hot toluene (100 °C) for 5 min, (5% copolymer/fibers proportion) and then Sohxlet-extracted for 48 h in toluene. Contact angle measurements on treated cellulose fibers showed that the fibers had become totally hydrophobic. The reaction between cellulose and the copolymer has been reported as divided in a first step where the copolymer is converted into the more reactive anhydride form, and a second step where esterification of the cellulose fibers takes place, as shown in Figure 3.

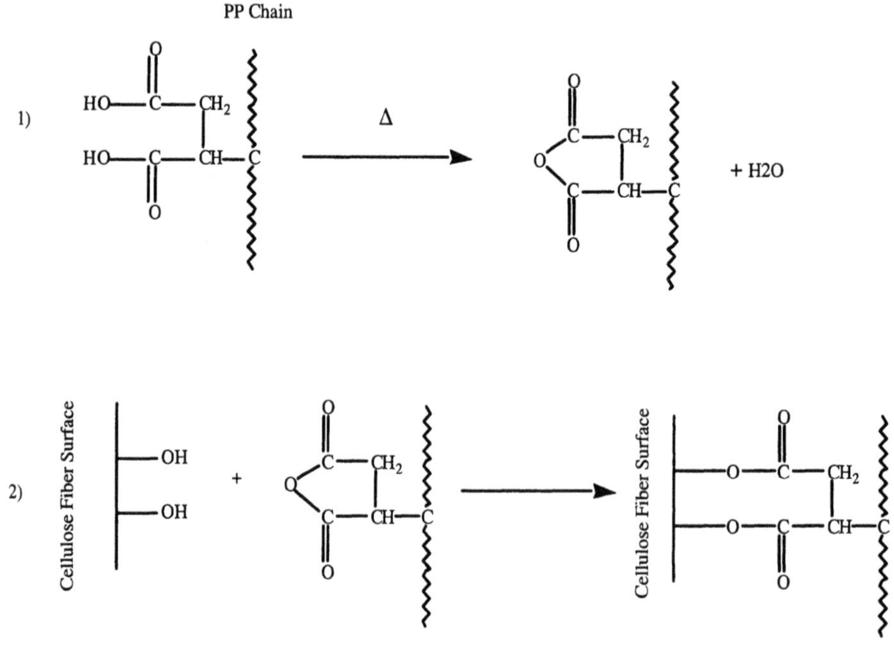

Figure 3. Reactions of Cellulose Fibers when Surface is Treated with Polyproylene-Maleic Anhydride Copolymer [51].

Tailor-made cellulose/polystyrene graft copolymers have been used as compatibilizers/interfacial agents in the preparation of cellulose-polystyrene alloys. Cellulose–polystyrene graft copolymers have also been prepared by reacting a anhydride terminated polystyrene directly with cellulose acetate. The reaction involves the free hydroxyl groups on cellulose acetate with the anhydride group on the polystyrene, resulting in the formation of an hemiester moiety. Using similar chemistry the preparation of cellulose acetate-polystyrene maleic anhydride was performed by direct reactive extrusion of cellulose acetate with a commercial random copolymer of polystyrene and maleic anhydride [52]. The graft copolymers function as emulsifying agents and provide for a stabilized, fine dispersion of the polystyrene phase in the continuous phase of the cellulosic matrix [52].

Cellulose finds extensive application as a membrane material in separation and enrichment technologies, in both its native and derivatized forms. However, it suffers from low stability and poor interactions in aqueous media despite its high hydrophilic surface area. Cross-linking, radiation grafting, and surfactant adsorption techniques aimed at achieving excellence in membrane workability has been applied [53]

Incorporation of maleic anhydride and styrene into cellulose produces a symmetrical amphiphilic graft copolymer. Poly(styrene-*alt*-acrylonitrile) has been grafted onto cellulose in the presence of zinc chloride [54]. Similarly poly(styrene-*ran*-maleic anhydride) has been grafted onto cellulose acetate, obtaining up to 70% graft yield [55]. Recently binary mixtures of styrene and maleic anhydride has been graft copolymerized onto cellulose extracted from Pinus needles by using a simultaneous irradiation method [56].

Activity for the production of industrial products of cellulose with specific properties by means of chemical modification is still receiving a worldwide attention as evidenced by papers in conferences, journals and numerous patents. In all cases there is a strong pressure to design processes with reduced environmental impact and fair Life Cycle Assessment balance with respect to other competitive materials.

2.2 STARCH

Starch is the major form of carbohydrate storage in green plants. It is the principal component of most seeds, tubers, and roots and is produced commercially from corn, wheat, rice, tapioca, potato, sago, and other sources. Starch price ranges in 0.5-1.5 €/Kg (20). Most commercial starch is produced from corn that is comparatively cheap and abundant throughout the world. Wheat, tapioca, and potato starch are produced on smaller scale and at higher prices.

Starch consists of six-membered ring glucose units (glucopyranose). The majority of starch molecules (up to 100% in waxy starches, 72% in normal maize starch, and 80% in potato starch) have highly branched structures, known as amylopectin. Amylopectin molecules are composed of chains of α-D-glucopyranosyl units joined by α-1-4 linkages. Branches are formed by joining these chains with α-1-6 linkages. The branch chains are present in double helical, crystalline structures. The average branch chain length is 20-30, glucosidic units, with an average degree of polymerization (DPn) ranging between 3 10^5-2.5 10^6 (Mn $5x10^7$-$2x10^8$ Dalton) [57]. The short branches in amylopectin are the source of crystallinity of starch [58]. Amylopectin structure is shown in Figure 4.

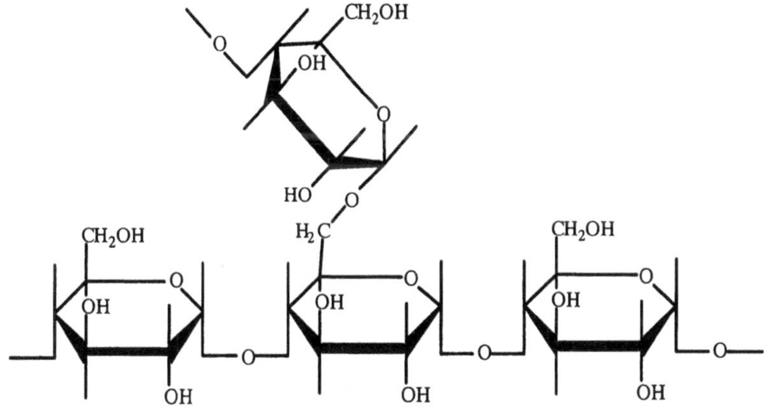

Figure 4. Schematic Representation of Amylopectin Structure

The other constituent of starch is known as amylose a primary linear molecule with very few branches; constituted by repeating units of 1-4-α-glucose. In certain high-amylose varieties, the amylose content can be as high as 50 or 70%. Its DPn is comprised between 200-220 10^3 depending on the source and method of attainment. Amylose can have several conformations, its structure is shown in Figure 5.

The ratio of amylose and amylopectin varies according to starch source. Chemical structures of starch, such as molecular sizes of amylose, branch-chain lengths of amylopectin, and the proportion of amylose to amylopectin, are found to affect the functional properties of starch.

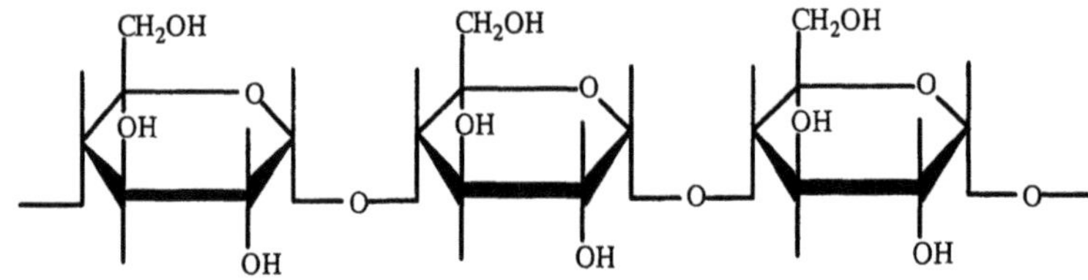

Figure 5. Schematic Representation of Amylose Structure

Amylopectin interacts more effectively with amylose and, thus, generates greater viscosity and gel strength. When the amylose content of the starch increases, the viscosity and gel strength of the starch paste also increase. Amylose alone and high-amylose maize starch, produce strong though films.

Starch is largely utilized, especially in the U.S. for ethanol production by a fermentation process, giving glucose as intermediate product. Starch finds wide usage in several non-food sectors such as sizing and coating of papers, adhesives, thickeners, and as an environmentally friendly additive in composite materials. Appropriate selection of chemical reagents allow for introduction of anionic or cationic charge into the starch molecules. The modification of starch and the properties and uses of such modified starches have been reviewed several times [11,59,60]. Several polyhydroxy compounds have been developed from starch for industrial applications that are less expensive than comparable products made from crude oil. Thus starch can be easily converted to glucose from which a variety of cyclic and acyclic polyols, aldehydes, ketones, acids, esters and ether can be obtained [61]. Recently a wood adhesive has been prepared by crosslinking cornstarch and polyvinyl alcohol with hexamethoxymethylmelamine in the presence of citric acid as catalyst [62].

Modified starch finds several applications in industry, such as cationic starches for paper treatment. Thus the dominant use for starch is in paper making applications [59]. Since starch is less crystalline than cellulose it is more susceptible to chemical modification and hence more vulnerable to degradation during modification. Research into starch modification is mostly focused on improving starch moisture resistance without losing the favorable factors such as easy degradation and relatively low price.

Starch esters represent an important class of starch derivates and have been recently reviewed by Tessler and Billmers [63]. Starch esters can be produced by an aqueous process, at low alkalinity, under controlled pH, and low temperature reactions usually reaching a fairly low degree of substitution (DS<0,2) [64]. Starch acetates with a DS of 2.4 or higher are not biodegradable, like cellulose, while intermediate DS acetate would be easily biodegradable [63,31,32]. Most commercially used starch derivates have a DS less than 0.2 [60]. Pure amylose starch is considered the most desirable precursor for starch-ester based thermoplastics since amylopectin has an adverse impact on mechanical and physical properties of these derivatives [64].

Starch can react with organic anhydride in water to yield starch esters, such as starch acetate that have been produced commercially by this process. Esters have also been prepared by aqueous reaction with vinyl esters, the byproduct acetaldehyde can be used to cross-link the starch by lowering the pH once esterification has been completed.

Mixed carbonic anhydrides, acyl guanidine, N-acylimidazoles, N-acyl-N'-methylimidazolium chloride and acyl phosphates are other agents that have been used to esterify starch in water [65-76]. Starch esters containing carboxylic acid group have been prepared by reaction with cyclic dibasic anhydrides [65].

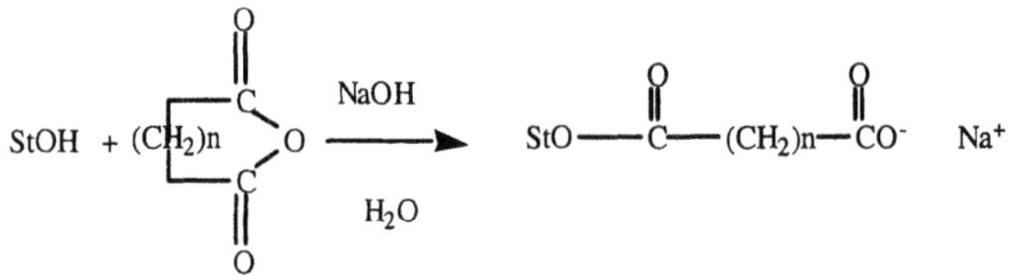

$$StOH + (CH_2)n\left<\begin{matrix}C{=}O \\ \quad O \\ C{=}O\end{matrix}\right. \xrightarrow[H_2O]{NaOH} StO{-}\overset{O}{\overset{\|}{C}}{-}(CH_2)n{-}\overset{O}{\overset{\|}{C}}O^- \quad Na^+$$

Figure 6. Starch Reaction with Cyclic Dibasic Anhydride [63]

When substituted succinic anhydrides, such as octenyl succinic anhydride, have been used for starch esterification the resulting products presented emulsion stabilizing properties. Alkyl sulfosuccinate hemiesters are formed by reaction of sulfonate alkyl succinic anhydrides with starch in water and when starch contain amylose and the alkyl group is a tetradecenyl or similar long-chain alkyl, the amylose tends to self aggregate and "hot-gel" properties are observed [77]. Non aqueous esterification of starch can make use of solvents such as pyridine, which serves the dual purpose of catalyst as well as solvent [26]. Starch esters in pyridine have been prepared with carboxylic acid anhydrides or acid chlorides, the latter being more effective for higher esters. Therefore solvents such as pyridine are costly and polluting. Moreover esterification under heterogeneous conditions, such as starch with p-toluensulfonyl chloride, may be accompanied by several side reactions [78,79]. Recently p-toluenesulfonyl starch sample were prepared by homogeneous reaction of starch with tosyl chloride in the presence of triethylamine dissolved in N,N-dimethyl acetamide in combination with LiCl [80].

More recently starch esters have been prepared in a solvent-free method by using formic acid and octanoyl chloride as the gelatinizing and acylating agent respectively [81]. Starch octenoate was thus produced with DS of 1.7 and an overall yield of 89%.

Starches have been oxidized with numerous oxidizing reagents mainly to reduce molecular weight and to increase solubility, for paper and food applications. Commercial oxidized starches are batch prepared utilizing room temperature conditions and low oxidant concentration (<3%) such as hypochlorite, permanganate, hydrogen peroxide, persulfate, periodate, and dichromate [60, 82-84]. Starch oxides have found a large application in adhesive production.

Reactive extrusion processing has been used as a technique for converting starch to chemicals, derivatives, and copolymers [85-90]. By this technique cationic derivatives, alkyl glucosides, carboxylates and oxidized starches have been produced [91]. Cationic starch has been prepared from native corn-starch and 3-chloro-2-hydroxypropyltrimethyl-ammonium chloride in aqueous sodium hydroxide using a twin-screw extruder as the reactor [92].

Starch succinates are polyanions and offer a number of very desirable properties such as high viscosity, low temperature viscosity stability, high thickening power, low gelatinization temperature, clarity of cooks, and good film forming properties. Therefore, succinate derivatives have been recommended as binders and thickening agents in foods, tablet disintegrants in pharmaceuticals, surface sizing agents and coating binders in paper [93]. Succinoylation of starch was earlier performed by via conventional batch processes but more recently has been investigated by reactive extrusion in the presence of succinic anhydride, water and a catalyst [93,94]. A reactive extrusion process of different starches with hydrogen peroxide and a ferrous-cupric catalyst has given water-soluble products of extremely high carboxyl and carbonyl content [91,95]. Starch has been reacted with synthetic polymer by graft polymerization. In this technique a free radical is initiated on the starch backbone and then allowed to react with polymerizable vinyl or acrylic monomers. Initiation can be chemically or radioactively induced. These systems have been reviewed by Fanta and Barely and Fanta and Doane [96,97]. Several of the starch graft polymers have been proposed as thickners for aqueous system, flocculants, clarification aids for wastewaters, retention aids in paper making, and many other uses. One of these materials has found commercial application under the trade name of Super Slurpers, because of its ability to rapidly absorb and retain many hundreds times its weight of water whithout dissolving or disintegrating. The polymer was made by graft polymerizing acrylonitrile onto gelatinized starch and subjecting the resulting starch-polyacrylonitrile graft copolymer to alkaline saponification to convert the nitrile groups to a mixture of carbamoyl and alkali metal carboxylate groups [98-100].

In its gelatinized form, starch is readily accessible to natural enzymes, amylases, and it is available from several renewable plant resources [101]. To achieve an effective degradability, blends or composite materials have been produced by processing of starch with biodegradable polymers such as: poly(vinyl alcohol), poly(lactic acid), poly(ε-caprolactone), poly(hydroxybutyrate-*co*-valerate), poly-esteramide, etc [102]. The costs of most biodegradable polymers is in the range of 4-8 €/Kg, whereas the cost of starch is almost one order of magnitude lower and the price for other agricultural surpluses or by-products may be even lower [103]. Thus low cost of starch makes it attractive to be blended with high cost biodegradable polymers [104-106] such as poly(hydroxybutyrate-*co*-valerate) (PHBV) [107-109] and their degradation has been investigated [110].

2.3 CHITIN AND CHITOSAN

Chitin is one of the most abundant natural polymer on the earth next to cellulose [111]. This important structural polysaccharide can be found in the cell walls of lower plants and fungi, insect cuticles and in the exoskeleton of arthropods and mollusks. An estimated billion tons of chitins are synthesized every year in nature [112] Chitin is a polymer of β-(1-4)-linked N-acetyl-D-glucosamine (NG) residues, and thus is similar to cellulose, but an acetamido group replaces the hydroxyl group at C-2 position of the glucosydic ring (Figure 7). For its inertness it gained less attention in respect to cellulose, remaining an almost unutilized resource.

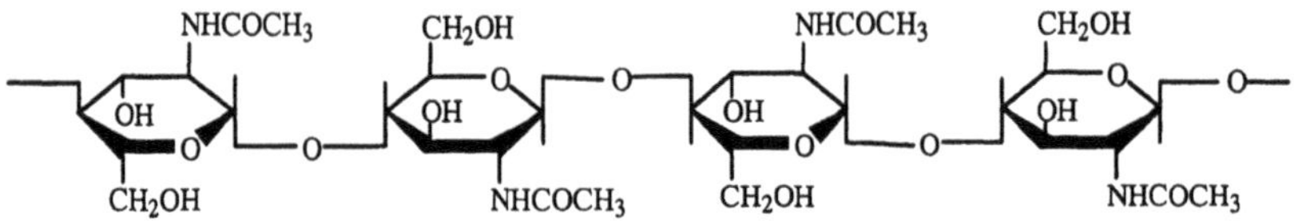

NHCOCH$_3$ CH$_2$OH NHCOCH$_3$ CH$_2$OH
OH O OH O OH O OH O O O O O
CH$_2$OH NHCOCH$_3$ CH$_2$OH NHCOCH$_3$

Figure 7. Schematic Representation of Chitin Structure

Chitin is insoluble in water, and is thus difficult to isolate without degradation. It is soluble in concentrated mineral acids. Three polymeric crystal structure of chitin are known, α-, β- and γ-structure. α–Chitin is the tightly compacted, most crystalline orthorombic, form where the chains are arranged in an anti-parallel fashion and can be obtained from the shell of crabs, lobsters and shrimps. β–Chitin has the monoclinic form where the chains are parallel, and is ususally obtained from the pen of loligo and squid. γ-Chitin is a mixture of α-, and β-chitin [111]. By treatment with 50% NaOH and subsequent washing in water α-, and β-chitin undergo polymorphic transformations. During swelling in NaOH the original lateral structure is destroyed and an alkali chitin complex is formed, in which the general orientation of the chitin chains remains parallel to the microfibrils axis. After washing in water the alkali chitin is converted to the α–chitin crystal structure [113]. In spite of the similarity in structure with cellulose, the chemical and physical properties of α–chitin are significantly distinct from those of cellulose. Derivates have been produced to overcome the solubility problem. α–Chitin is rather resistant to chemical modifications due to the peptide-like hydrogen bonds between chains, and harsh reaction conditions are required. Many kinds of chitin derivatives, including benzyl chitin, carboxymethyl chitin, etc have been prepared via alkali chitin which is a soda-chitin complex. β–Chitin can on the contrary be modified under milder conditions. Reactions such as acetylation, tosylation, tritylation, and acetolysis proceed much more easily with β–chitin than with α–chitin [114-116]. Hydroxypropyl chitin has been prepared by reaction of chitin and propylene oxide and structural and thermal properties have been investigated [117]. Chitin biodegradability is important in relation to its uses as a carrier for delayed release of drugs, pesticides and other applications. Biodeterioration by chitinoclastic bacteria in soils and oceans is very slow, thus chitinous compounds resist biodegradation under normal conditions. However, chitin can be degraded by chitinases associated with chitiobiases β-N-acetylhexos-aminidases, and lysozomes [118]

Chitosan is the fully or partially deacetylated product of chitin, is soluble in acetic acid and other organic solvents, thus it has more applications than chitin. Water resistant adhesive proteins used by marine animals (mussels) to adhere to wet or submerged surface has stimulated the studies to confer water-resistant adhesive properties to semidilute solutions of chitosan. High viscosities and water-resistant adhesive strength of chitosan semi-dilute solutions have been obtained by tyrosinase catalyzed reaction with 3,4-dihydroxyphenethylamine (Dopamine) or by reaction with glutaraldehyde [119]. The deacetylation process is rarely complete, and most commercial and laboratory products tend to be a copolymer of N-acetylglucosamine and N-glucosamine repeat units. The ratio of the two repeating units depends on the source and preparation conditions of chitosan, but in many cases the glucosamine units predominate.

Fully deacetylated chitosan has been treated with phthalic anhydride in N,N-dimethylformamide to give N-phthaloyl-chitosan that resulted soluble in polar organic solvents [120]. The C-6 hydroxyl groups of N-phtaloyl-chitosan were tritylated (triphenyl methylated) to give a derivative useful for substitution at C-3. Acetylation followed by detritylation gave another derivative easily convertable to the trimethylsilylated derivative (Figure 8).

All these derivatives resulted soluble in common polar organic solvents, and the trimethylsilylated derivative was also soluble in dichloromethane and dichloroethane that are conveniently used for glycosilation. Artificial branched polysaccharide can be prepared by introduction of sugar branches into chitin. The trimethylsilylated derivative has been reacted with an orthoester of D-mannose in dichloromethane in the presence of trimethylsilyltrifluoromethanesulfonate as a catalyst resulting in the formation of a branched product. Deprotection with hydrazine yielded chitosan with α–mannoside branches [120].

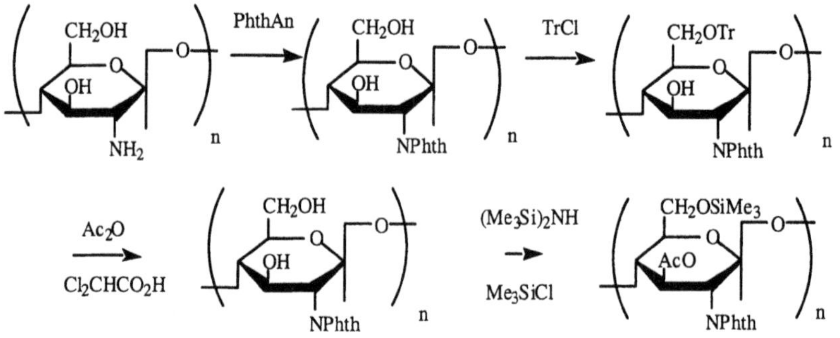

Figure 8. Schematic Synthesis of Chitosan Trimethylsilylated Derirative [120]

Most commercial polysaccharides, such as cellulose, dextran, pectin and starch are either neutral or acidic, but chitosan is a basic polysaccharide [112]. In neutral or basic pH conditions, chitosan displays free amino groups and is insoluble in water, while in acid conditions becomes water soluble due to the protonation of amino functions, and its solubility depends on the distribution of free amino and *N*-aceyl groups [111,121-123].

Chitosan is a linear polyelectrolyte at acidic pH and it has a high charge density [124]. Indeed the pH and ionic strength have an important impact on the intrinsic viscosity of polyelectrolytes. Chitosan in acid aqueous solutions has been surface reacted with polyanion aqueous solutions such as heparin, sodium alginate or polyacrylic acid to give polyelectrolyte complexes [125,126]. Chitosan has also been proposed for waste water treatment, since by electrostatic interactions of chitosan NH_3^+ groups with COO^- and SO_3^- groups in proteins or polyanions such as alginates, carrageenan, and pectin, chitosan can form polymeric complexes. Suspended proteins found in waste effluents, fermentation media and other industrial streams can be recovered as chitosan-protein or chitosan-polyanion-protein complexes. Chitosan in its free amino group form can function on the contrary as a ligand-exchanger to recover amino acids [113]. Chitosan in the copper and amino copper forms has been used for the production of ligand-exchangers that have shown to be particularly effective for the recovery and separation of aspartic acid, glutamic acid, tryptophan and cysteine. Thus higly porous ion exchangers and adsorbent developed as quaternary ammonium crosslinked chitosan, resulted able to adsorb proteins. Chitosan is a strong positively charged molecule and acts as a chelating agent for various heavy metals, that makes it interesting for application in waste water treatment also because chitosan can adsorb many toxic substances such as pesticides and commercial dyes from soil and water streams [127,128].

Blending is also an important process for developing industrial applications of polymeric materials, thus chitin and chitosan have been blended with polyvinyl alcohol and poly(ethylene oxide) [129-132]. Extruded plastic films have been prepared containing 20% corn starch, 20% chitin, 25% low density polyethylene, 25% etylene-co-acrylic acid and 10% urea as a plasticizer [133].

2.3.1 Biomedical Applications of Chitosan

Chitin and chitosan have a very low toxicity, infact the LD_{50} of chitosan in laboratory mice is 16g/Kg body weight, which is very close to that of salt or sugar. Chitosan is safe in rats up to 10% in the diet [134]. Thanks to its biocompatibility and biodegradability chitosan and its derivatives have found a wide range of applications in the biomedical field as supported from commercialization of many chitin and chitosan based products for clinical use (Table 2).

Artificial kidney membranes prepared from modified and albumin-blended chitosan membranes were found to be potential candidates for dialysis applications [135]. These membranes displayed a superior permeability properties for smaller molecules compared to the standard cellulose membranes, and interestingly showed maximum reduction in platelet adhesion in comparison to other membranes [136,137].

TABLE 2. Chitin and Chitosan Based Commercial Biomedical Products [112]

Product	Application	Manufacturer
Evalson R	Personal care	Chito-Bios, Ancona Italy
Depolymerized Chitosan	Hair care	Wella Inc., Germany
Noodels containing chitosan	Dietary foods	NihonKayaku Inc., Japan
Chitin liquid (CM-chitin)	Skin care	Ichimarn-FarukosuInc., Japan
Nonwoven chitin fabric	Burn therapy	Yunichika Inc., Japan
Chitin fiber	Biodegradable suture	Yunichika Inc., Japan
Chitosan-collagen composite	Artificial skin	Katakurachikkarin Inc., Japan

Chitosan and its derivatives have aroused much interest also as wound healing accelerators and have been investigated by many researchers for a long time [138,139]. There is considerable biochemical evidence linking N-acetylglucosamine (NAG) with the metabolism of hexamines which are assumed to originate and cross-link wound collagen. The physiological importance of glucosamine has been stressed by various authors; it takes part in the detoxification function of liver and kidneys and possesses anti-inflammatory, hepatoprotective, antireactive, antihypoxic activities [140,141]. The repeating monomer subunit of chitin derivatives, is a major component of dermal tissue and its presence is essential for repair of scar tissue. In fact, glycoproteins, which contain large amounts of NAG are one of the predominant proteins isolated in the early stage of would healing. It is possible that chitin derivatives, being easily degraded by lysozymes naturally present in wound fluid, could promote wound healing by functioning as a controlled delivery source for NAG to the healing wound [142]. Moreover it has already been shown, that during tissue regeneration and reorganization processes, chitosan and its derivatives support blood coagulation, prevent abnormal fibroblastic reactivities and act as bactericide and wound-healing accelerators [143,144].

A series of biofibers based on chitin and chitosan has been prepared by wet spinning and post modification of chitosan fibers as recently reviewed by Hirano [145]. These biofibers based on chitin, chitosan, chitin-cellulose, chitosan-cellulose-silk fibroin, chitosan-tropocollagen, and chitin-cellulo-silk fibroin have been proposed for applications including antithrombogenic, antimicrobial, hemostatic, and wound healing [145]. N-hexanoyl, and N-octanoyl chitosan are antithrombogenic and emocompatible,

N-acylchitosan filaments have been used for surgical suture production [145].
Another important aspect of health care is the controlled release of drugs. The potential of chitosan in the design of carriers for controlled release of drug delivery is enormous in view of its useful properties like degradability, non toxicity and biocompatibility. It is important to remember that the degraded products of chitosan do not introduce any disturbance in the body, thus it can be a suitable matrix available in different forms for a sustained release of various drug formulations. Nevertheless up to know the study of most chitosan-based drug delivery systems is limited to *in vitro* tests and the systems that showed promising results may be considered for commercialization after their *in vivo* evaluation [111].

2.4 LIGNIN

Lignin (from the Latin lignum, wood) is, after cellulose, the principal constituent of wood. Lignin is closely associate with cellulose, and bound to plant polysaccharides, especially hemicelluloses. Lignin in wood appears to be insoluble in water unless modified by physical or chemical treatment, which seems to change its state of polymerization, or hydrolyze its bonding to other wood constituents. Free lignins are insoluble even in strong mineral acids and hydrocarbons. Their insolubility in 72% sulfuric acid is the basis for a quantitative test to differentiate them from cellulose and other carbohydrates [146,147]. Commercial lignins are soluble in aqueous alkaline solutions and some are also soluble in several oxygenated organic compounds and in amines. Lignin is a complex three-dimensional biopolymer composed of three different phenylpropane units crosslinked by at least 10 types of C-C and C-O-C bonds. The distribution of the different links can differ in relation to the biological constraints of the environment in which conditions is plant growing thus leading to a large variability in the polymer architecture. The three main constitutive lignin units are p-hydroxyphenyl (H), guaiacyl (G), and syringic (S) aldehydes (Figure 9).

It was assumed that coniferous lignin contains about 14 % p-hydroxyphenyl units, 6 % syringic units, and 80 % guaiacyl units. The polymer is believed to have an average molecular weight of over 10,000. Exact determination of the ratios of the units in deciduous lignin is complicated by apparent variations in lignins isolated from different tree species. Indications are that the average molecular weight for deciduous lignin probably does not exceed 5,000.

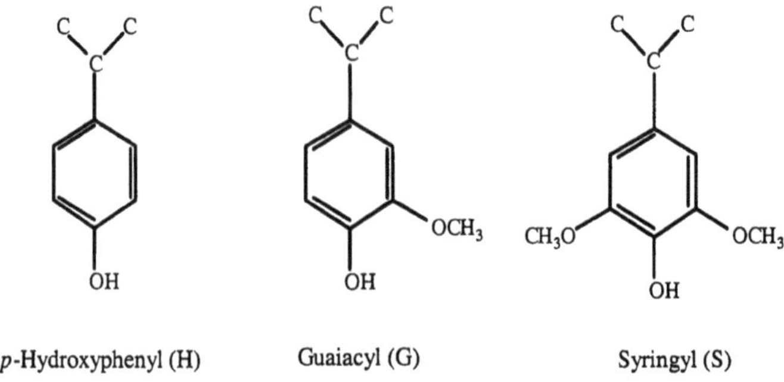

Figure 9. Representation of Typical Subunits of Lignin

Coniferous lignin contains 2% benzyl groups of the phenolic type and 6% of the alkanoic type (see Fig. 10). About 0.24 oxygen atoms in lignin are present as benzyl alcohols or benzyl aryl ethers. These groups are involved in the delignification reactions in chemical pulping operations.

Phenylpropane units in lignin are linked by aliphatic and aromatic carbon bonds and ether bonds. Wood lignin has a definite structure that cannot be represented by a single formula. A plausible schematic structure for lignin is shown in Figure 10

Very little of the millions of tons of lignins potentially available from wood wastes every year is utilized except for fuel. The heat of combustion of lignin is about 29.5 MJ/kg and thus it comprises nearly 40% of the fuel value of softwood, whereas it constitutes only some 28% of its dry mass. The waste liquors of the wood pulp industry are currently the principal source of the small amount of lignins that is being used. From this source lignins are available as such, as sodium ligninates, and as lignin sulfonates in a wide range of purities, in the form of brown fluids or powders.

Lignin can be extracted from sugar cane bagasse by the acetosolv process with 93% acetic acid, 0.2% HCl, 6.8% water, (w/w) at 109 °C after 3 h pulping treatment. The black liquor is concentrated under reduced pressure and lignin is precipitated in water at 70 °C and purified by dissolution in acetone [148].

Liquefaction of lignocellulosics was earlier performed by several hours treatment at 300-400 °C in aqueous or organic solvents. Subsequently liquefaction in organic solvents at temperatures of 240-270 °C without catalyst or at temperatures around 80-150 °C with acidic catalysts have been developed with very high yield (90-95%) [149]. Liquefation of wood in the presence of phenol and its applications to thermosetting materials have been also proposed. Liquefaction has been realized by using sulfuric acid [150], phosphoric [151,152], or oxalic acid as catalysts [153]. The resulting liquefied woods appear as novolac like resins ands can be converted in adhesives, moldings, fibers and carbon fibers.

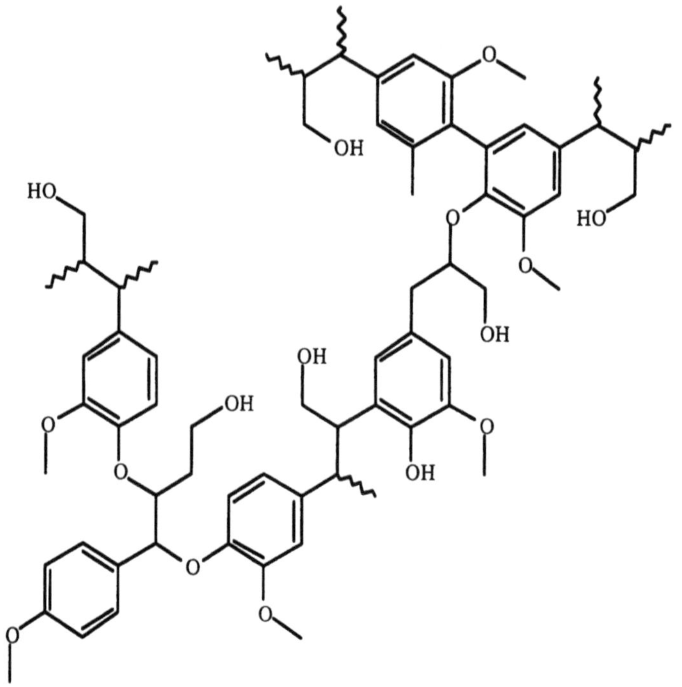

Figure 10. Schematic Representation of Proposed Structure of Lignin.

Many uses have been proposed for lignin and some have been developed commercially. Low cost uses of lignin include surface active agents to prevent the formation of aggregates of finely divided insoluble particles in suspension, binders, emulsifying agents, sequestering agents, fuel source, concrete admixtures, oil well drilling muds, and others as reviewed by Northey [154].

Other uses involve conversion to vanillin, to tanning materials, to dispersing agents, and to reinforcing fibers for rubber [147]. An increase in lignin utilization value might be achieved by copolymerization of lignin with synthetic polymers [155]. For example a method has been developed for free radically grafting styrene onto lignin [156-159].

Plant polyphenols such as lignins and tannins have also been investigated as substitutes of synthetic phenolic adhesives since the largest volume application of phenolic resins is in plywood adhesives [160-162]. Other applications have been investigated and envisaged in the production of molded items. Resols have been prepared with the partial replacement of phenol by organosolv sugar cane bagasse lignin and processed in a mold. In this study phenol was partially substituted by lignin [10, 20, 40 100% (w/w)] with a 1/1 phenol/formaldehyde molar ratio and 4% (total weight) of NaOH as a catalyst [148]. The cure reaction was performed in a mold initially submitted to 100bar pressure at 50 °C for 20 min and then released and kept at 120 °C for 60 min and at 150 °C for 40 min. In this procedure lignin reacted with phenol moiety of the polymer chain, acting first as a chain extender rather than as a filler. Molded resins containing up to 40% lignin presented modulus retention at elevated temperatures (220 °C) and appeared suitable for preparation of phenolic molded resins [148].

The chemical and macromolecular properties of lignin as well as its supra molecular organization within the cell wall make it particularly recalcitrant to biological degradation. Among the various organisms involved in plant and wood decay in nature, the white rot fungi are apart, as having evolved a unique but complex system to oxidatively degrade lignin and mineralize it [163,164]. The key step in this process is the formation of radical species on lignin monomers. The radical-generating systems of white rot fungi may involve low molecular weight metal chelators, such as siderophores and glycopeptides, enzymes, such as peroxidases and laccase, and abiotic high valence manganese complexes formed by interaction between accumulated MnO_2 and fungal metabolites, such as oxalates [165-167]. Either phenoxyl, cation and/or benzyl radicals generated by these oxidants on simple lignin model compounds finally result in various carbon-carbon and alkyl-aryl ether bond cleavage as well as specific oxidation products [166].

Peroxidases and laccases are key enzymes in the lignin biodegradation process. They oxidize phenolic and non-phenolic lignin model compounds into their phenoxyl radicals and cation radicals respectively. Further non enzymatic evolution then leads to the cleavage of various C-C and ether bonds [166]. Enzyme-based reactions that involve the formation of reactive free radicals might be expected to be relatively indiscriminate and non-selective of their substrate [168,169]. It was found that polystyrene could be degraded by white rot fungi when their traditional substrate, lignin, was present [169]. It is reported that, white rot *Basidiomycetes* were able to biodegrade styrene or methacrylate graft co-polymers of lignin containing different proportions of lignin and polystyrene or poly(methyl methacrylate). Lignin/styrene copolymerization products contained 10.3, 32.3, 50.4 wt% lignin, and methyl methacrylate copolymerization products contained 11 to 18 wt% of lignin. The styrene polymer samples were incubated with white rot *Pleurotus ostreatus*, *Phanerochaete chrysosporium*, *Trametes versicolor*, while the methyl methacrylate polymer samples were incubated with white rot *Pleurotus ostreatus*, *Trametes versicolor* and *Phlebia radiata*. Bioconversion and degradation of lignin-styrene graft copolymer was verified by weight loss, quantitative ultraviolet spectrophotometric analysis and scanning electron microscopy. The most efficient degradation of lignin and polystyrene constituents of the copolymer by white rot fungi was observed for the plastic with the highest lignin content. The capacity of the white rot fungi to induce weight loss was in the order: *Phanerochaete chrysosporium*, *Trametes versicolor*, *Pleurotus ostreatus*. Sample based on lignin and polymethyl methacrylate also showed mass losses after 60 days of incubation higher than the corresponding control or homopolymer. In this case *Pleurotus ostreatus* was the more effective for reducing the mass of the polymer sheet [169].

Commercial poly(vinylalcohol) samples (PVA) of average molecular mass (Mw) 120,000 g/mol, were subjected to biodegradation with enzymatic extracts of the *Phanerochaete chrysosporium fungus*, in which the lignin peroxidase activity was detected. The results of differential refractive index and ultraviolet absorption(UV), Gel Permeation Chromatography (GPC), Fourier Transform Infrared Spectroscopy (FTIR) and Gas Chromatography coupled to a Mass Detector (GC-MS) showed that in 15 days, under the conditions of this study, the average molar mass of the polymer decreases by 79.3 %. Benzaldehyde was detected as the main degradation product [170,171]. Degradation tests on composite materials, prepared by casting of aqueous suspensions [172], based on PVA of similar molecular weight (100,000-146,000) and 95% hydrolysis degree (Airvol425) in blends with sugar cane bagasse or orange peel outlined a positive effect on the degradation by the presence of the lignocellulosic fillers [173]. Recent soil burial test experiments on composite material based on potato starch, PVA and corn and wheat fibers (by-products from ethanol production) indicate higher mineralization rate and final extent, for composite containing the lignocellulosic fillers in comparison with composites only based on starch and PVA. Thus lignin is increasing in importance as filler in biodegradable materials as a possible promoter of synthetic polymers degradation.

Lignin co-products of pulping treatment have attracted great attention for their utilization in engineering plastic design because of lignin's tetrafunctional branch points. Typical technical lignins are kraft lignin and lignin sulfonate. Kraft lignin is utilized in preparation of three-dimensional polymers such as polyurethanes or phenol resins [174,175]. Thus polyester networks have been synthetized by condensation of both aliphatic and phenolic OH groups from different types of lignin with difucntional reagents [176].

Lignins surface and lignin solubility characteristics can be modified by treating with monofunctional isocyanates, thus the attachment of relatively short side arms (hairs) possessing hydrophilic properties, e.g. oligomeric ethylene oxide chains, did indeed provide a means of modifying quite drastically those properties, even when only about one half of the available OH groups were involved [176]. These materials showed a marked solubility in water and displayed glass transition temperatures which decreased as the extent of grafting increased and went from about 60 °C for pristine lignin to -60 °C with 80% substitution 176].

Sulfur free lignin, derived from the methanol based organosolve pulping of spruce wood as well as from beech wood prehydrolysis, has been proposed as a filler for polypropylene films [177].

As reported by Ghosh et al. [178] a certain degree of compatibility has been found between modified cellulose and modified lignin. Blend studies on melt-extruded and solvent cast films of hydroxypropyl cellulose and lignin revealed the presence of secondary interactions between the two components. Polymer blends of hydroxypropyl cellulose and organosolv lignin have been prepared by mixing in solution of both pyridine and dioxane, and casting as films. Mixing of two components in the melt followed by extrusion [179] gave partially miscible blends displaying a mesomorphic type behavior. The lignin component was found to reinforce the amorphous cellulose derivative matrix, that forms an oriented crystal mesophase structure, and this behaves as a nanocomposite structure [179]. Further studies were reported on interactions of cellulose and lignin by modifying either or both polymers [180]. Many cellulose derivatives, such as ethyl cellulose and cellulose acetate-butyrate over a wide range of compositions revealed lower compatibility with lignin thus leading to the formation of heterogeneous, two-phase morphology as shown by differential scanning calorimetry (DSC) and scanning electron microscopy (SEM) [180]. Further investigations with the intent of understanding the interaction between mesomorphic cellulose derivatives and lignin revealed higher degrees of phase mixing for continuous fibers prepared from anisotropic solutions of cellulose acetate-butyrate and lignin in dimethylacetamide [181]. More recently commercially available cellulose acetate-butyrate has been blended in melt and solution with lignin esters having different ester substituents (acetate butyrate, hexanoate, and laurate). All lignin esters formed phase-separated blends with cellulose acetate-butyrate with domain size depending on processing conditions and the interaction between phases depending upon blend components. Cellulose acetate-butyrate /lignin acetate, and cellulose acetate butyrate/lignin butyrate revealed the strongest interactions with domain sizes on the 15-30 nm scale as probed by DMTA and DSC techniques [178].

Lignin has been also modified by enzymatic reactions as recently reviewed by Huttermann et al. [182]. The obvious candidates for use in technical processes involving lignin are the enzymes involved in lignin degradation from white rot fungi [183]. Lignin peroxidase oxidizes phenolic and also non-phenolic lignin subunits by abstracting one electron and generating radicals that are further decomposed through a non-enzymatic process [184]. Manganese peroxidase acts basically in the same way as lignin peroxidase except that this enzyme uses the Mn(II)-Mn(III) system as redox mediator. Laccase, is widely distributed and is reported to have quite different physiological functions [185]. Laccase has a lower redox potential than the two previous enzymes, and needs a free phenolic hydroxyl group for its action. In the presence of certain organic redox systems it is able to oxidize substrates such as lignin from pulp that would not otherwise react with this enzyme. Laccase is also known to be able to polymerize phenols [186]. In the review of Hutterman et al. three different approaches have been reported in detail. In the first one *in situ* polymerization of lignin for the production of particle boards has been investigated. Adhesive cure was based on the oxidative polymerization of lignin using phenol oxidases (laccase) as radical donors. This lignin-based bio-adhesive have been applied under conventional pressing conditions. By this process 80% of the petrochemical binders in the wood-composite industry, could be replaced by analogous materials from renewable resources. The second approach was focused on enzymatic copolymerization of lignin and alkenes. In the presence of organic hydroperoxides, laccase catalyzes the reaction between lignin and olefins. Detailed studies on the reaction between lignin and acrylate monomers showed that chemo-enzymatic copolymerization offers the possibility to produce structurally defined lignin-acrylate copolymers. The system permits the control of the molecular weights of the products in a way that has not been possible with chemical catalysts. This is a novel attempt to enzymically induce grafting of polymeric side chains onto the lignin backbone, and it enables the utilization of lignin as part of new wave of engineering materials. In the last approach enzymatic activation of the middle-lamella lignin in wood fibers was performed for the production of wood composites. The incubation of wood fibers with a phenol-oxidizing enzyme resulted in oxidative activation of the lignin crust on the fiber surface. The fibers were bound together in a way that come close to naturally grown wood. This process will, for the first time, yield wood composites that are produced solely from naturally grown products without any addition of resins based on fossil resources. The commercial availability of laccase by Novo Nordisk (Denmark) should promote the development of industrial production for the above quoted materials.

2.5 PROTEINS

Proteins are built up of α-aminoacids differing in the presence of reactive groups such as amino, carboxyl, amido, hydroxyl, thiol and imidazole groups. The structure of proteins is an extended chain of aminoacid residues joined by amide linkages which are readily degraded by enzymes, particularly proteases.

Residues from 20 different natural α-aminoacids can be present. Except for glycine, the aminoacids are optically active, but only the L-isomers occur in nature. Examples of the most common structural proteins are listed in Table 3.

TABLE 3. Common Structural Proteins [8].

Protein	Morphology	Occurrence
Keratin	Three –helix wound around each other in a superhelix	Hair, skin, fur, wool, horn
Fibroin	Antiparallel, pleated sheets	Silk, spider webs
Elastin	Covalently cross-linked elastomers	Arterial walls, ligaments
Actin & Myosin	Myofibrils	Striated muscles
Collagen	Triple helix	Skin, tendon, cartilage, bone, teeth

Proteins play an extremely important role in all living cells for which they perform a large variety of functions, and govern most of the physical and chemical activities necessary to life [14]. In addition to the natural function of proteins in living organism, and foodstuff the properties of several proteins have enabled them for a wide variety of applications such as fermentation processes, pharmaceuticals etc. Even if protein plastics based on casein and soy protein have been known since 1930s and 1940s, research on the technical applications of industrial proteins has been limited so far. To some extent this can be explained as due by the higher price of proteins when compared to some other biopolymers, especially starch. Most protein applications are specialty products based on industrial proteins. Examples of these bulk applications are coatings, encapsulant in the pharmaceutical and food industries, adhesives, surfactants and plastics [103]. Proteins have also considerable potential for the production of slowly degrading packaging or mulching films [187].

Examples of industrial proteins are plant proteins, such as wheat and corn gluten, soy, pea and potato proteins, and animal proteins, such as casein, whey, keratin, collagen, and gelatin.

Alpha keratin fibers occur in hairs, wool, quills, and together with fibroin fibers such as silks and spiders webs are all highly extensible fibrous protein while collagen is a relatively inextensible fibrous protein. Because of their commercial applications and the relatively complex structure at the molecular and near molecular level, the interpretation of the physical properties of α-keratin fibers has been object of recent studies [188].

Gelatins are high molecular weight polypeptides derived from collagen, the primary protein component of animal connective tissues, such as bones, skin, and tendons [189,190]. Modern technological applications of gelatin depend on its high solubility in hot water, polyampholyte character, availability in a wide range of viscosity, and thermally reversible gel formation. Among proteins, the last-mentioned property is unique, suggesting that it depends upon aspects of the collagen primary structure that survives the denaturation and degradation of the conversion processes.

The collagen molecule consists of three helical peptide chains (alpha chains) held in close, parallel association. Collagen is transformed into gelatin by denaturation and physical and chemical degradation. The commercial interest in collagen primarily arises from its use as the basic material for the manufacturing of leather products [191]. The relative inextensibility and stiffness of the individual collagen fibers, which form the leather composite, and which control the mechanical properties of the leather itself. Denaturation, particularly thermal denaturation, is central to any conversion process. It is described most readily in terms of a special soluble form of collagen, which is not a commercial source of gelatin. Thermal denaturation occurs near 40 °C for solutions of most mammalian collagens and is marked by an abrupt change in intrinsic viscosity, specific optical rotation, optical rotatory dispersion, sedimentation, and light scattering. The properties above this temperature are no longer those of the regular, highly extended triple helix (collagen fold), but of flexible random coils. This is the conformational disordered state called gelatin. Commercial gelatins are not derived from soluble collagen but from the more abundant bone and skin collagen fibrils, which are stabilized by intra-chain Schiff's base and aldol group formation. Manufacturing processes employ lyotropic and hydrolytic treatments with aqueous solutions of acid (type-A gelatin) or base (type-B gelatin), which swell the collagen and ultimately solubilise it.

Gelatin has a high glycine content (33 mol %) in addition to the presence of two unusual amino acids, hydroxyproline (10 mol %) and hydroxylysine (0.5 mol %) [192]. The glycine residues spaced throughout the chain, create a profusion of spaced "hinges" around which relatively unrestricted rotation can occur. The presence of chain-chain bonding induces restrictions to rotation around the skeletal bonds. The addition of components such as water or glycerol, which are able to separate effectively the chains from each other, eliminates these constraints [193]. The diluent is supposed to occupy the space between collagen molecules forming hydrogen bonding regularly placed along the molecular length. Thereby linking one molecule to another while also promoting the arrangement of the molecules in parallel does occur [192].

The molecules of commercial gelatin are heterogeneous with respect to amino acid composition, except for glycine that is the main non-chiral aminoacid constituent of gelatin. Glucosidic impurities, like hexoses, glucose and galactose, that are covalently bonded to hydroxylysine by glycosidic linkages and are present at 0.5 % by weight. The yellow color and turbidity of commercial gelatins have been attributed to mucoprotein impurities, type-A gelatin being usually superior to type-B in this respect. Gelatin is soluble in a number of organic solvents, including acetic acid, trifluoroethanol, formamide, ethylene glycol, glycerol, and dimethyl sulfoxide.

Gelatin is currently used in a variety of applications comprising manufacturing of pharmaceutical products, x-ray and photographic films development and food processing [103,194]. As a biomaterial, gelatin displays several advantages, since it is a natural polymer that has not shown antigenity, it is completely resorbable in vivo and its physicochemical properties can be suitably modulated [195]. The only serious problem, being nowdays represented by viral contamination in animal derived proteinaceous materials, may constitute a limiting factor for an expansion in this field of application.

The cost of industrial proteins ranges from 0.5-5€/kg for plant proteins to more than 5 €/kg for animal proteins. The amount of protein is available for technical applications, for example wheat gluten is produced at 400,000 tons/year worldwide. Proteins have specific properties which compensate the price disadvantage: good processability, both in the melt and in solution, even at contents higher than 30%, good film forming properties and good mechanical properties of the films, adhesion to various substrates, high resistance towards UV-radiation, oils and organic solvents, surface active properties [103].

TABLE 4. Technical Applications of Proteins [103]

Protein	Application
Casein	Adhesives, paper coatings, leather finishes
Gelatin	Pharmaceutical capsules, photographic emulsions, adhesives, encapsulation, edible films
Soy proteins	Paper coatings, plywood adhesives, plastics
Corn zein	Printing inks, grease proof paper, floor coatings
Keratin	Textile, cosmetics
Wheat gluten	Adhesives, cosmetics, detergents, edible films

Proteins are versatile materials, which combine many characteristics relevant for technical applications. The protein properties depend on both the source (amino acid composition) and the modifications that have been performed to improve specific features. The interest in non-food use of protein is increasing over the last two decades because of their renewability and biodegradability. Proteins have thus been used for the fabrication of materials such as edible films, films and coating, adhesives, thermoplastics and surfactants [196,197,198].

As reported by Arvanitoyannis et al. the interest in edible films is bound to their numerous applications such as coating for chocolate nuts, fruits and vegetables. and occasionally wax [199,200]. Several publications have been reported dealing with edible film based on protein with particular reference to restriction of moisture loss, gas permeability, control of microbial activity, preservation of structural integrity of the product and gradual release of enbodied flavours and antioxidants in foods [201-205]. For example edible films and coating have been prepared by wheat gluten, and wheat and corn protein (206), sodium caseinate, starches, sugars or glycerol [199]. Thus protein films are effective as gas barrier (O_2 and CO_2), but their water vapour transmission rates are high [207]. Collagen and its derivate gelatin have also been used for sausage casings and as gelling agents because their abundant supply (208). Collagen films extruded in the form of tubular sausage casing may be viewed as a convenient edible packaging material [209]. Edible wrappings based on gelatin with farinaceous constituents have been marketed [210] and films made from gelatin, soluble starch and polyols have been reported [198,199]. Many of the major food-related and other industrial uses of gelatin are based on the structure/property relationship of gelatin for which water is an excellent plasticizer of the predominant amorphous regions [211].

Proteins valorization as thermoplastic or thermosetting materials has been performed by chemical modifications [212]. Improved functional properties can be obtained by esterification of protein carboxyl and amide groups by fatty alcohol that lead to a protein-derivative with improved functional properties which would result into a lower water sensitivity. For example sunflower and wheat gluten have been esterified by octanol in the presence of an acid catalyst. Esterified protein result less soluble at basic pH compared to native proteins, indicating a hydrophobation effect [213].

Some applications of proteins demand a high water resistance. One way to increase the water resistance is to cross-link the protein chains. Dialdehyde crosslinking specifically reacts on free amine functions, while not specific crosslinkers as a formaldehyde resin, reacts both with carboxyl, amine and hydroxyl groups [214]. By crosslinking with formaldehyde, the water sensitivity of protein plastics can be dramatically reduced, particularly at high relative humidity levels [215]. Glutaraldehyde has been shown to be very reactive toward the N-terminal amino groups of peptides as well as the α-amino groups of the aminoacids [216]. Formaldehyde, glyoxal, or glutaraldehyde resulted effective in enhancing the maximum puncture force of films made from cotton seed proteins [217]. Soy isolate has been treated with formaldehyde and glyoxal at 1.0, 2.5 and 5.0% (w/w) and with adipic and acetic anhydrides to produce materials suited for the fabrication of plastics items by compression molding [216].

Wheat gluten has been crosslinked using water soluble 1-ethyl-3(3-dimethyl amino propyl) carbiimide chlorohydrate together with N-hydroxysuccinimide [218]. Crosslinking of the protein occurred through activation of the carboxylic acids groups, followed by reaction with the free amino groups of the same or of an other protein chain. N-hydroxysuccinimide was added as a catalyst to enhance cross-linking efficiency. Thus its action is supposed to reduce the possibility of hydrolysis of activated species and to suppress side reactions. The degree of crosslinking was dependent on the reaction time, the molar ratio of added reagents and the pH of the mixture. Increasing pH decreased the number of modified amino groups in wheat gluten. Modification at pH 5-7 yielded very low soluble polymers since at that pH the protein has a zero net charge thus protein-protein interactions are favored creating aggregates. In these conditions reagents can only reach the protein surface promoting intermolecular crosslinking between carboxylic acid and amino groups of different protein chains.

Crosslinking of proteins has also been performed by enzymatic reaction. Peroxidase and transglutaminase have been used for this purpose. A treatment by horseradish peroxidase in the preparation of films of thermally denaturated soy proteins reduced the elongation at elastic limit and increased the tensile strength [219]. Similarly behavior was recorded for films based on casein or whey protein when treated with transglutaminase [220,221]. Transglutaminase catalyses the self polymer-ization of proteins through ε-(γ-glutamyl)-lysine bonds but can also introduce crosslinks between external primary amines and glutamine residues of proteins. Despite the low content of lysine in gluten, some intermolecular ε-(γ-glutamyl)-lysine bonds were formed as well. Thus in the case of gluten in doughy state transglutaminase increased the proportion of high weight polymers molecules and strengthened the network rheological properties [222].

Films have been produced by cross-linking gluten or deaminated gluten using transglutaminase and adding various diamines, such as putrescine, cadaverine, diaminohexane and diaminooctane, in the film-forming solution in order to modulate the length. The diamine are able to react at their own extremities acting as a spacer between gluten proteins. Loss of solubility was related to the formation of high molecular weight polymers [223].

Hardening of gelatin with low molecular weight aldehydes is well documented in the literature [224-226]. Crosslinking is predominantly due to Schiff's base formation by condensation of the formyl group and the ε-amino groups present in lysine and hydroxylysine residues (Figure 11). In DSC analysis of hot cast films crosslinked with formaldehyde Fraga observed a broader temperature range of gelatin glass transition attributed at the crosslinking of gelatin chains through the α-amino acid present in the soft blocks with a consequent increase of rigidity in these blocks. The second glass transition was associated to gelatin rigid blocks composed of sequences mainly made up of the imino acids proline and hydroxyproline including glycine at every third position [227].

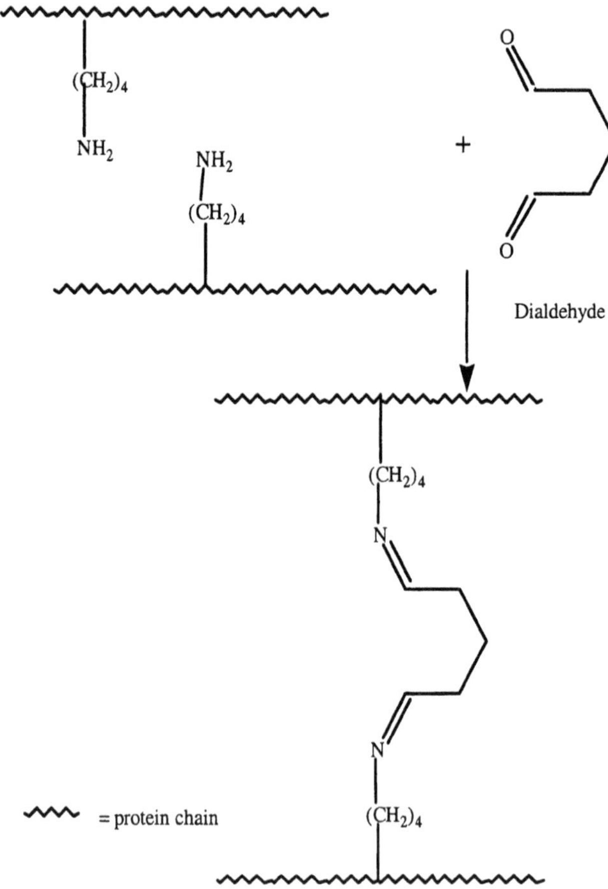

Figure 11. Schematic Representation of Reaction Between the ε-Amino Group of the Lysine Residue of a Protein Chain and a Dialdehyde

Gel formation obtained by cooling gelatin aqueous solutions is accompanied by some characteristic changes, which have been ascribed to a partial regain of the collagen triple-helix structure. Over the years much work has been devoted to the study of gel formation by biopolymers such as gelatin [228-234]. Hydrogels are materials which, when placed in excess water, are able to swell and retain large volumes of water in its swollen three-dimensional structure without dissolution. Many materials of this type are considered to be biocompatible and a wide range of biomedical applications has been described. Among them are contact lenses, artificial tendons, matrices for tissue engineering, and drug delivery systems. Nevertheless, there is still a need to develop nontoxic biodegradable hydrogels for specific biomedical applications, e.g., wound treatment [235].

Gelatin gels have a relatively low melting point, they are not stable at body temperature. Therefore, it is imperative to stabilize these gels by establishing chemical cross-links between the protein chains. A variety of hardening procedures are described in the literature [236] Chemical crosslinking of gelatin gel typically utilizes bifunctional reagents such as glutaraldehyde (Figure 11) [237,238] and diisocyanates [239], as well as carbodiimides [240], polyepoxy compounds [241], and acyl azides [242]. Glutaraldehyde is by far the most widely used agent, due to its high efficiency to stabilize collagen-based biomaterials and despite local cytotoxicity [243] and calcification of long-term implants [244].

Hydrogels have been prepared by reacting gelatin with partial periodate oxidized dextran as well [245,246]. Chemical aging of the gelatin-dextran dialdehyde hydrogel occurred, influencing the release patterns as a function of storage time. Furthermore, immobilization of incorporated peptides or drugs was observed [247-249].

A recent investigation reports the preparation of hydrogels based on methacrylamide-modified gelatin. The crosslinking reaction has been performed in two steps [250]. Gelatin is first derivatized by reaction with methacrylic anhydride. The gelatin methacrylamide can then be cross-linked in a subsequent step. The water-soluble gelatin obtained after derivatization, can be cross-linked by a number of suitable polymerization processes, such as redox, thermal, and UV treatment; -irradiation; or e-beam curing [250]. The rheological properties of the gelatin-based hydrogels were controlled by the degree of substitution, polymer concentration, initiator concentration, and UV irradiation conditions.

Gelatin scraps generated in different manufacturing processes comprising pharmaceutical, cosmetics, tannery, and food segments, may often constitute a concern for the environment connected to their disposal [251]. They strongly swell in water and have high carbon and nitrogen content that may lead to high oxygen demand once they reach the sewage drainage system, waste water treatment plants and eventually fresh water streams. A labor intensive and expensive treatment is required for a correct management in their disposal. Value added solutions as aimed at defraying their disposal costs are then sought. Moreover proteic waste materials have attracted attention because of their intrinsic agronomic values bound to the fairly high nitrogen content (12-15%).

Among the pharmaceutical uses, gelatin capsules are probably the best known. Hard capsules are made by a dipping process that uses high gel-strength gelatins, while soft capsules are made by a continuous rotary die method from plasticised (water-glycerol) gelatin of low-to-medium gel strength. The soft capsules are formed filled with pastes or liquids, and sealed in one operation. Scraps of animal gelatin, generated in pharmaceutical industry have been considered for the production of liquid mulches and films from aqueous suspensions [252,253]. To improve consistency of the films and regulate time of degradation in the soil, gelatin scraps were blended with fillers from renewable resources such as sugar cane bagasse and with synthetic biodegradable polymers such as poly(vinyl alcohol) at different grades of hydrolysis (PVA) [254]. When cast films based on gelatin scraps were applied on the soil the introduction of sugar cane bagasse in the formulations increased time of permanence conferring higher cohesiveness to the samples and lowering the mineralization rate of the composites in soil burial tests [252,254]. The presence of a substantial amount of water and glycerol in gelatin scraps allowed for a partial compatibility between gelatin and PVA, particularly in blends containing an amount of gelatin up to 20% by weight. Thus addition of up to 20% of gelatin scraps to PVA increased the elongation at break of cast films, and reduced tensile strength and Young's modulus [255]. This behavior was attributed to the additional plasticization of PVA by the glycerol present in gelatin scraps. In accordance to this plasticizer effect played by the glycerol, tensile strength and Young's modulus decreased. However, for blends with higher gelatin scraps content these three mechanical properties dropped down to the values typical of films based only on gelatin scraps. Thus gelatin/PVA blends with compositions up to 40 wt-% of gelatin appeared to be more flexible than those with higher amounts of waste gelatin whose films became harder and started to turn opaque.

Glutaraldehyde was used as crosslinking agent to improve water resistance and regulate degradation rate in soil [256]. When increasing the glutaraldehyde concentration from 1 to 2.5 wt-%, crosslinked gelatin showed a lowering in the Young's modulus and at the same time a significant increase on elongation at break [257]. Similar behavior was found on water sensitivity and mineralization rates [252]. This behavior of gelatin has been attributed to the disruption of the helical structure allowing the chains to assume a random coil conformation with glutaraldehyde acting as a crosslinker among the chains [239].

Crosslinking of gelatin with glutaraldehyde gives rise to formation of short aliphatic segments between gelatin chains. When a warm solution of gelatin is cooled not only chemical crosslinking will be present but physical ones too as attributable to recovered collagen triple-helix structure [192]. Consequently, the crosslinked gelatin at ambient temperature will be the sum of these two processes. By increasing of the glutaraldehyde concentration, the probability of triple-helix recovering will be reduced and the structure of the crosslinked gelatin will be like that of the random coil polymers. Thus the balance of the chemical and physical crosslinking plays an important role on the ultimate properties of gelatin based materials.

In the leather industry the tanning process generates a huge amount of protein waste, thus it has been estimated that one ton of wet salted hides yields 200 kg of leather and over 600 Kg of solid waste [258]. Approximately 600,000 metric tons of chromium-containing solid waste, chrome shavings, are generated worldwide. The problem associated with this mass of collected waste are compounded by the nature of the leaching medium. Atmospheric pollution in some part of Europe has resulted in high levels of acid in the rain and the low pH helps to mobilize the chromium content of the waste. Tannery wastes have been used for fertilizer or composite boards but nowadays the tannery pays for transportation and reprocess of these wastes for disposal [259]. Historically, shavings, trimmings and splits from the chrome tanning of hides and skins have been disposed of in landfills, but tighter local restrictions and shortage in landfills have caused the tanning industries to look for different alternatives. Combustion has the disadvantage of the formation of carcinogenic calcium chromate in the ashes. Pyrolysis is the subject of intense research but not yet applied. In Figure 12 are summarized the various options for the disposal of tanned waste.

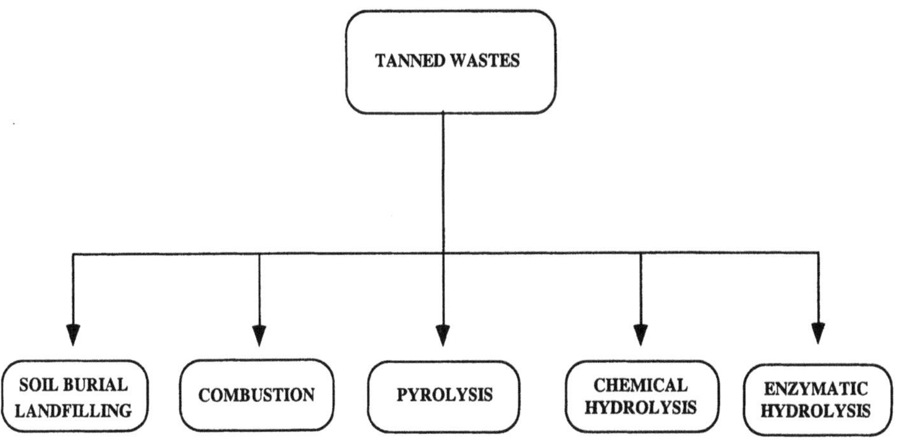

Figure 12. Primary Methods Used in the Processing or Disposal of Tannery Wastes [260]

Several methods have been developed to treat these wastes such as chemical hydrolysis that takes place either in alkaline or acid medium, at high pressure and temperature [261,262]. The final product is a solution containing both protein and precipitated chromic or chromous salts. The main disadvantage is the relatively high content of chromium in the recovered protein (300-500 ppm).

A recent technology has been developed to isolate protein products (gelatin and collagen hydrolysate) from chrome shavings by using alkaline protease under mild conditions. The chromium-containing solid waste is first treated with alkali agents for six hours at 70-72 °C, to extract a high molecular weight gellable collagen protein. The sludge that remains is further treated with enzymes to recover a lower molecular weight protein hydrolizate and a recyclabe chromium product. The pH at which the reaction takes place (8.3 to 10.5) prevents the chromium from going into solution, thus averting the poisoning of the enzyme by chromium and enabling the recovery of chromium as $Cr(OH)_3$ by filtration [259]. In Figure 13 is reported a block scheme relevant to enzymatic treatment of proteinaceous tannery waste.

The recovered protein fractions, practically chromium-free, can be used in a wide range of formulations for the production of adhesives, cosmetics, films, animal feed and fertilizers. The isolated residue containing chromium and organic matter (chrome cake) has the potential to be recycled into the tanning process by treatment with sulfuric acid [263].

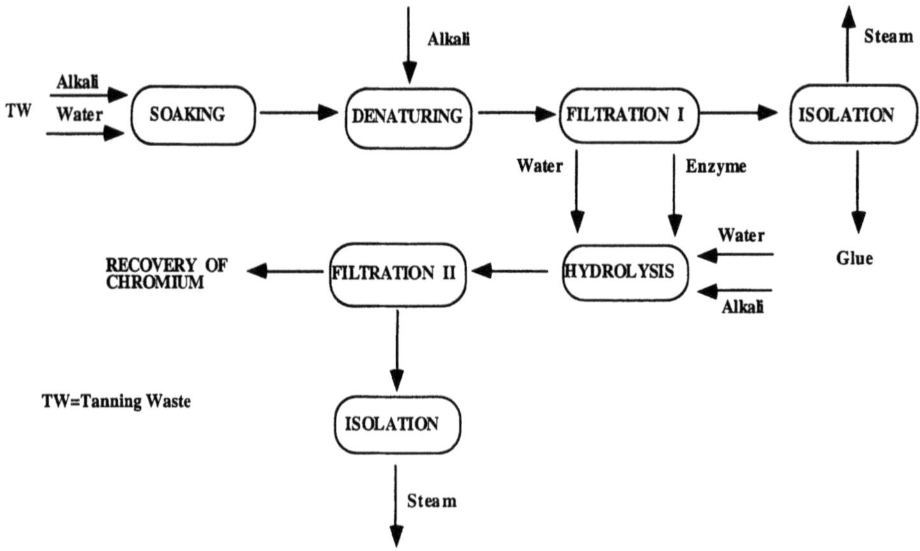

Figure 13. Schematic Representation of the Enzymatic Hydrolysis of Proteinaceous Tannery Watse [260]

2.6 MISCELLANEOUS NATURAL POLYMERS

Numerous other polymeric materials from natural resource have been applied as raw material or after appropriate modification. Lignocellulosics have found extensive attention in composite production as light-weight, cheap, degradable fillers and their modification by chemical reaction has been extensively studied. Lignocellulose mainly consists of cellulose, hemicellulose, and lignin. Sugar cane bagasse has been crosslinked with epichlorohydrine in the presence of NH_4OH or imidazole. With a 2:1 epichlorohydrine-to-imidazole molar ratio were made strong anion exchangers with dye-binding capacities that made them suitable for removing anionic dyes from waste water of textile industry. Epichlorohydrine is also commonly used crosslinking agent that effectively stabilizes agricultural residues for the preparation of weakly acidic cation exchangers [265].

Graft polymerization has been known as an useful way to improve the properties of many natural polymers with the goal of extending their applications. Many initiation systems can be used for the modification of lignocellulosic materials, including bamboo, unbleached pulp, bagasse, jute fibers. Some examples are represented by Ce^{4+} (266), xanthate-Fe^{2+}-H_2O_2 (267), Fe^{2+}-H_2O_2 (267), $NaHSO_3^-$ soda lime glass [268], $KMnO_4$ [269].

2.6.1 *Cardanol and Derivatives*

Other polymeric materials from agricultural resource have recently received strong attention such as cashew nut shell liquid, a byproduct of the cashew industry, which constitutes a source of unsaturated hydrocarbon phenols. It is available as a reddish brown viscous fluid in the soft honeycomb structure of the shell of the cashewnut, a plantation product obtained from the cashew tree, Anacardium oxidentale [270]. Since cashew nut shell liquid is nearly one third of the total nut weight, much of the cashew nut shell liquid is formed as a by-product of mechanical processes for the edible use of the cashew kernel. Recent investigations, carried on particularly in India, have revealed that the constituents of cashew nut shell liquid possess special structural features which can be chemically transformed into specialty and high value polymeric products [271].

Simple chemical modifications such as those amenable to introduce phosphorus and/or bromine gave a series of multi-purpose resins and products for use as flame retardants, adhesives, matrix resins for brake linings and composites, etc.

Cardanol (Figure 14), the main component obtained by thermal treatment of cashew nut shell liquid, is a phenol derivative mainly having as *meta* substituent a C15 unsaturated hydrocarbon chain with mostly 1–3 double bond [272].

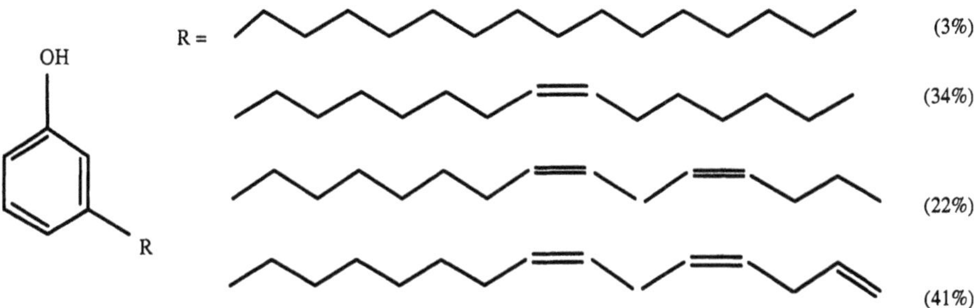

Figure 14. Schematic representation of Cardanol Structure [272]

Only a small part of cardanol obtained in the production of the cashew kernel is used in the industrial field, though it has various potential industrial uses such as resins, friction lining materials, and surface coatings [273]. Therefore, development of new applications for cardanol is strongly desired. Phenolic resins from cardanol and formaldehyde are industrially produced as the prepolymer of coating materials with high gloss surface mainly for indoor use. However, resins containing formaldehyde have much concern about the toxic nature of formaldehyde in their manufacturing and use [274]. By oxidative polymerization of cardanol using iron-*N,N*-9-ethylene bis (salicylideneamine) (Fe-salen) as catalyst the polymer, reported in Figure 15, can be produced. The curing reaction is regarded as a new formaldehyde-free coating system.

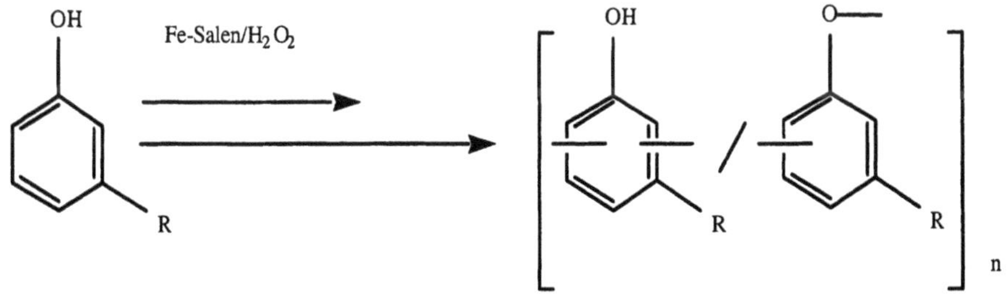

Figure 15. Schematic Representation of Oxidative Polymerization of Cardanol by Using Fe-salen as Catalyst. [272]

Enzymatic syntheses of polyaromatics using oxidoreductase as the catalyst have been extensively developed over the past decade. Peroxidases induced an oxidative polymerization of various phenol derivatives [275-279], yielding a class of polyphenols showing high thermal stability. This enzymatic process is expected to be an alternative way of preparing phenol polymers without use of toxic formaldehyde. The resulting polymer was cured by cobalt naphthenate or thermal treatment to give a brilliant film with high gloss surface. Polymerization of cardanol has been carried out using hydrogen peroxide as oxidizing agent in organic solvents. Fe(II)-salen was oxidized by hydrogen peroxide to form Fe(III)-salen that acts as a catalyst for the polymerization. In the case of the polymerization of cardanol, the considerable increase in polymer yield and molecular weight was achieved by addition of pyridine [280] into the reaction medium.

A novel polyether was synthesized by cationic polymerization of glycidyl 3-pentadecenyl phenyl ether, synthesized from cardanol, in the presence of a latent thermal initiator, N-(benzyl-)N,N-di-metil anilinium hexafluoroantimonate [281].

Novel polyurethanes were prepared by solvent polycondensation reaction of 1,6-hexane -diisocyanate with 4-[(4-hydroxy-2-pentadecenylphenyl)azo]phenol, synthesized from cardanol, and 1,4-butanediol [281]. A novel terpolyester was synthesized by solvent polycondensation of tere-phthaloyl chloride with 4-[(4-hydroxy-2-pentadecenyl phenyl) diazenyl]phenol (synthesized from cardanol) with 1,4-butanediol [282].

Glycosylation of cardanol with penta-O-acetyl-β-D-glucopyranose followed by deprotection afforded a glycolipid mixture that self-assembled into nanofibers in water and acted as gelation agents. The helical morphology of the fibers could be controlled by altering the degree of side-chain unsaturation [283]

Cardanyl acrylate, a derivate of cardanol, has been used for the preparation of a number of resins by condensing cardanyl acrylate with furfural and selective organic compdounds in the presence of acid as catalyst [284]). Cardanol has also been reacted with acetic anhydride and orthophosphoric acid to become acetylated cardanol and phosphorylated cardanol respectively [285].

2.6.2 *Plant Oils and Animal Fats*

Plant oils and animal fats play an important part in renewable resources because of their sufficient availability and their versatile applications. The triglyceride oils constitute the main portion of animal fats, butter, vegetable oils and margarine as well as oils from oil-bearing seeds [8]. When liquid at room temperature triglycerides are called oils, when solid they are fats, thus usually a few degrees change in temperature will cause the change from solid to liquid, or liquid to solid. Commercially important oils are produced by seeds of soybean, corn (maize), cotton, sunflowers, flax, rapeseed, castor beans, tung, palms, peanuts, olive, coconut, almonds and canola. Flax (linseed) and tung oils are example of "drying oils" that form a tough, elastic film on drying and are used in paints, varnishes, and enamels [286]. Soybeans, corn etc are examples of "semidrying oils" mainly used for cooking and in foods. Castor and rapeseed opils are examples of "nondrying oils", which remain greasy. Some triglyceride oils such as tallow are saturated and cannot be polymerized easily, but most oils contain at least one double bond per acid residue. About 80% of the worldwide production of oils are plant oils, soybean oil and palm oil being the most important ones. European oilcrops (rapeseed, sunflower, linseed) contain more than 90%. of unsaturated fatty acids. According to plant species, these oils have main constituents (oleic acid, linoleic acid, linolenic acid) of approximately 60% [4]. Figure 16 shows a schematic representation of a polyunsaturated triglyceride.

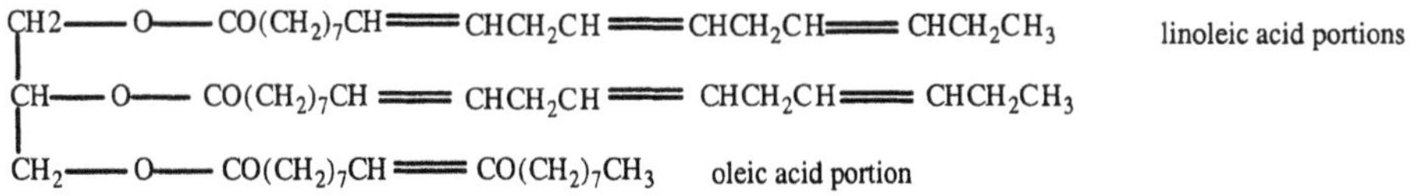

$$
\begin{array}{l}
CH2-O-CO(CH_2)_7CH{=}CHCH_2CH{=}CHCH_2CH{=}CHCH_2CH_3 \quad \text{linoleic acid portions} \\
CH-O-CO(CH_2)_7CH{=}CHCH_2CH{=}CHCH_2CH{=}CHCH_2CH_3 \\
CH_2-O-CO(CH_2)_7CH{=}CO(CH_2)_7CH_3 \quad \text{oleic acid portion}
\end{array}
$$

Figure 16. Schematic Representation of Linseed Oil Structure [8]

Whereas the largest share of approximately 100 million t/a of fats and oils are used for human foodstuff 15% are available for oleo-chemistry (soaps, detergents, cosmetics, biodiesel, lubricants and polymer additives).

Plant oils that are mainly used in industrial non-food applications often have unusal chemical compositions. Thus oils such as castor oil contain reactive groups in addition to unsaturation. Castor oil contains ricinoleic acid, which is identical to oleic acid except that carbon 12 (counting the –C=0 carbon as carbon 1) has a hydroxyl group in place of the hydrogen atom.

This change increases the viscosity of ricinoleic acid in comparison with oleic acid, and enhances castor oil value for industrial applications such as lubricating oils and greases, coatings, sealants, plasticizers for nitrocellulose. Moreover the presence of the –OH group allows polymer formation. The hydroxyl group participates in the formation of polyesters and polyurethane [8].

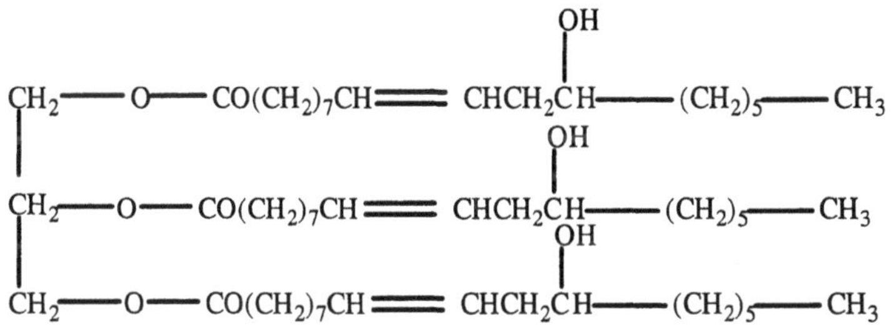

$$
\begin{array}{l}
CH_2-O-CO(CH_2)_7CH{=}CHCH_2CH(OH)-(CH_2)_5-CH_3 \\
CH_2-O-CO(CH_2)_7CH{=}CHCH_2CH(OH)-(CH_2)_5-CH_3 \\
CH_2-O-CO(CH_2)_7CH{=}CHCH_2CH(OH)-(CH_2)_5-CH_3
\end{array}
$$

Figure 17. Schematic Representation of Castor Oil Structure [8]

Unsaturated fatty compounds such as oleic acid, petroselinic acid, erucic acid, ricinoleic acid, linoleic and linolenic acid, and10-undecenoic acid and also the respective esters, alcohols and native oils are alkenes, and contain an electron-rich C—C double bond that can be functionalized in many different ways by reactions with electrophilic reagents [287].

Metzger et al. have been reporting on numerous C,C-bond forming additions, for example to oleic acid, giving a great variety of interesting branched and long-chain fatty compounds with potentially new and interesting properties [287]. Epoxidation is a method to convert the triglycerides to a more chemically reactive form capable of polymerization. Epoxidized vegetable oils have been applied as crosslinking agents in environment-friendly solvent-free powder coatings. A variety of vegetable oil-based epoxidized crosslinking agents is available by both chemical and enzymatic routes with high chemical purity levels. Functionalized vegetable oils could serve as building blocks for the preparation of biodegradable hydrophobic plastics, polyhydroxy alkanoates exhibiting good drying properties. 4-Acetylphenyl methacrylate containing a polymerizable vinyl group was prepared by reacting 4-hydroxyacetophenone with methacryloyl chloride in the presence of triethylamine in methyl ethyl keton. A series of interpenetrating polymer networks based on polyurethane of castor oil and isophorone diisocyanate condensed with cardanyl acrylate/cardanyl methacrylate using ethylene glycol dimethacrylate as a cross-linker and benzoyl peroxide as an initiator were synthesized and characterized [288]. The polyurethanes obtained from castor oil, toluene-2-4-diisocyanate, and hexamethylene diisocyanate, with a varying NCO/OH ratio, were reacted with the new vinyl monomer 4-acetyl phenyl methacrylate to give interpenetrating polymer networks by using ethylene glycol dimethacrylate as cross-linker and benzoyl peroxide as initiator [289,290].

Warwel et al. applied catalytic methods of olefin chemistry to achieve polymer building blocks as well as polymers like functionalized polyolefins, polyesters, polyethers, polyamides as well as sugar-based surfactants [4]. The fundamental approach was the polymer synthesis based on unsaturated fatty acid methyl esters, which are available by industrially applied transesterification of fats and oils with methanol. In Figure 18 is reported a schematic representation of the potential of plant oil components in the preparation of different polymeric materials.

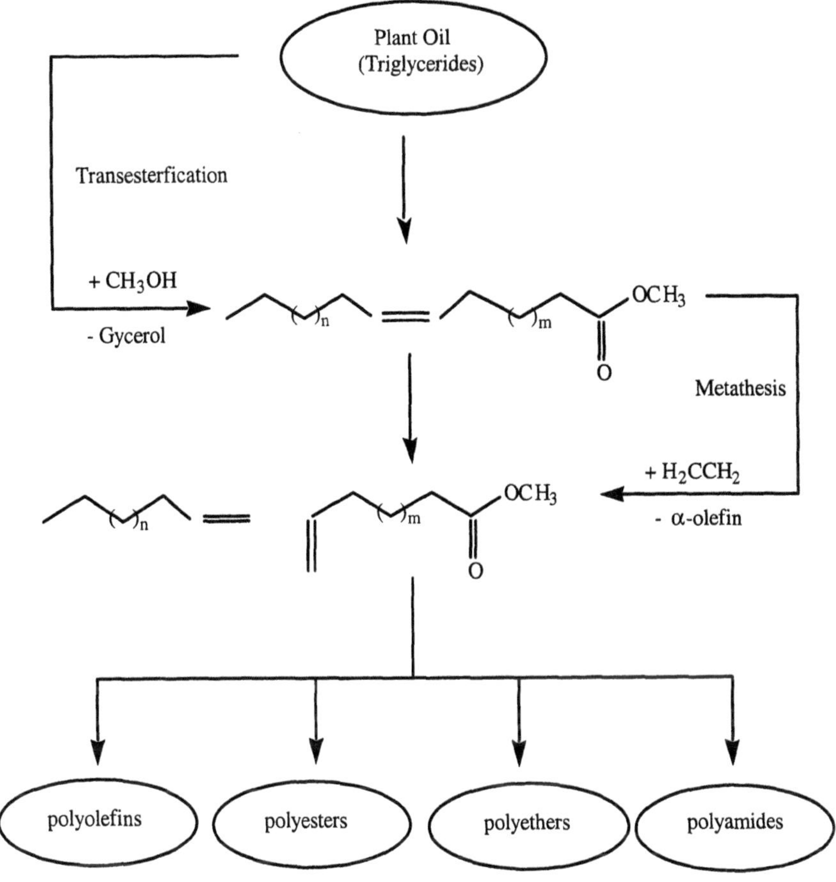

Figure 18. Schematic Representation of the Production of Polymeric Materials from Plants Oils [4]

First, unsaturated fatty acid methyl esters obtained from plant oils were converted to terminally unsaturated esters and α-olefins by metathesis with ethylene using heterogeneous rhenium or homogeneous ruthenium catalysts. These esters were directly copolymerized with ethylene by an insertion-type palladium-catalyzed polymerization to functionalized polyolefins. Polyesters were synthesized by metathesis dimerization of ω-unsaturated esters and subsequent polycondensation of the produced internally unsaturated dicarboxylic esters or by acid transesterification with petrochemical diols and additional acyclic diene metathesis polymerization. Fatty acid methyl esters ω-epoxidized, obtained by a new method of chemo-enzymatic epoxidation, were converted into polyethers displaying a comb-like structure by using aluminoxanes as catalyst. The same epoxy functionalized derivatives can be converted into sugar surfactants by nucleophilic ring-opening with amino carbohydrates.

Also animal oils have been used for polymers production, for example native or conjugated fish oil has been used for production of thermosetting polymers by cationic polymerization with divinylbenzene, norbornadiene or dicyclopentadiene comonomers, initiated by boron trifluoride diethyl etherate [291]. The produced polymers ranged from rubbers to hard plastics.

In the last decade, industry has been trying to formulate biodegradable lubricants with technical characteristics superior to those based on mineral oil (petroleum) [292]. Volumes of lubricants, especially engine oils and hydraulic fluids, are relatively large and most of them are based on mineral oils. Compared to the lubricants based on crude oil, vegetable-based lubricants do display a higher propensity to biodegradation [293-295] but appear inferior in many other technical characteristics and hence performances [296]. Lubricants based on vegetable oils still comprise a narrow market segment; however, they are finding their way in relatively short lasting and light duty applications such as chainsaw bar lubricants, drilling muds and oils, straight metalworking fluids, food industry lubricants, open gear oils, biodegradable grease, hydraulic fluids, marine oils and outboard engine lubricants, oils for water and underground pumps, rail flange lubricants, shock absorber lubricants, tractor oils, agricultural equipment, elevator oils, mould release oils, two stroke engine lubrificants and other [296]. However since the early development of lubricants for special applications (e.g. turbojet engine oils) it was recognized, that fatty acid polyol esters have comparable or even better technical properties than mineral oil. Subsequently, innumerable synthetic esters have been synthesized by systematic variation of the fatty acid and the alcohol components [297]. Whereas the alcohol moieties of the synthetic esters are usually of petrochemical origin, the fatty acids are almost exclusively based on renewable resources. The physico-chemical properties of oleochemical esters can cover the broad spectrum of technical requirements for the development of high-performance industrial oils and lubricants. These imply excellent lubricating properties, good heat stability, high viscosity index, low volatility and superior shear stability [298]. While mineral oil is a finite resource, whose availability may be highly subjected to political considerations, synthetic esters as mainly derived from renewable resources (animal fats and plant oils) are less affected by political shortcomings. Moreover LCA studies recently performed on the above systems [298] did provide positive vision and expectations for an ever increasing acceptance of oil and fats from renewable resources. In Figure 19 are reported the various steps and parameters utilized in the LCA relevant to the production consumption and disposal of industrial consumer products.

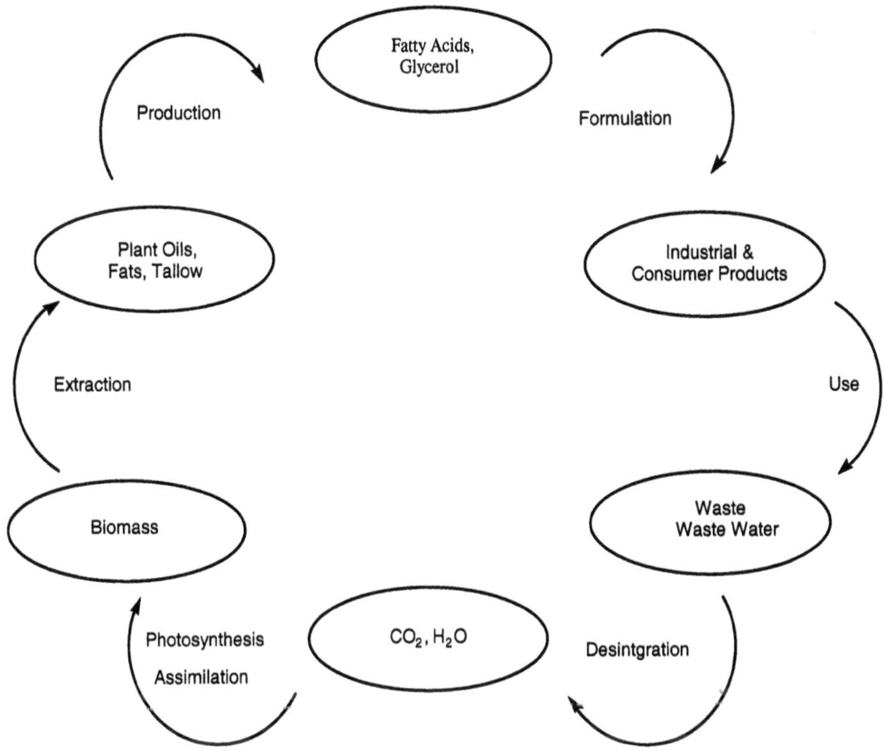

Figure 19. Schematic Representation of LCA Steps Sequence for the Use of Natural Fat and Oil Resources as Raw Materials in Productive Processes for Industrial and Consumer Foods. [298]

The resulting oleochemicals are then used as raw materials for the formulation of industrial (technical) or consumer products. After their use, the products are discharged, the majority via wastewater. A small fraction ends up in landfills, and an even smaller fraction of waste is energetically used (burned). Regardless of the pathway of discharge, at the end of their lifespan the organic chemicals are disintegrated into carbon dioxide and water, either by microorganisms present in wastewater treatment plants or in landfills, or by incineration. In the case of oleochemicals, the carbon dioxide liberated equals the amount of carbon dioxide that was originally taken up by the plants from the atmosphere thus resulting neutral to green house effect increase. Fossil-based products by contrast, whatsoever they are disposed of, lead to an increase of atmospheric carbon dioxide, thus contributing to global warming.

3. Polyesters

Aliphatic polyesters which have excellent mechanical properties and biodegradability and are well suited to disposable applications, are deserving a particular attention in the area of environmentally degradable polymeric materials. A major drawback is nowadays represented by a relatively high cost. In an effort to reduce their cost, blending with low cost natural polymers has been pursued [299]. Modified polyester and powdered cellulose, sodium alginate and chitosan in lyophilized form have been used as fillers. The samples were prepared in the form of films of different thickness and contained various amounts of natural components [300]. Polyhydroxyalkanoates (PHA)s are extensively reported in chapter 9 and 10, thus in this chapter attention will be focused more on poly (lactic acid) (PLA).

3.1 POLY(LACTIC ACID) (PLA)

Poly(lactic acid) is a biodegradable aliphatic polyester attainable by polycondensation of lactic acid, a monomeric precursor that can be obtained from renewable resources. Lactic acid is a chiral molecule available in the L and D stereoisomer forms (Figure 20). In the fermentation process sugar feedstocks, such as dextrose (glucose), are obtained either directly from sources such as sugar beets or sugar cane, or through the conversion of starch from corn, corn steep liquor, potato peels, wheat, rice and other starch source. Lactic acid in the L-form is actually produced by fermentation from nearly any renewable resource Yields of lactic acid are greater than 90%, and in the batch production lactic acid can be produced at the rate of 2 g per liter per hour [286]. L-lactic acid occurs in the metabolism of all animals and microorganism, and thus is an absolutely non-toxic degradation product of polylactides, as proved by the successful application of polylactides as resorbable medical suture over a period of three decades [301]. Lactic acid from crude oil is usually obtained in racemic form, and equimolar mixture of the L and D enantiomers. The ability to produce the L-isomer in high chemical and optical purity has important ramification in the chemistry and ultimate process/property relationship achievable in the polymers produced from lactic acid.

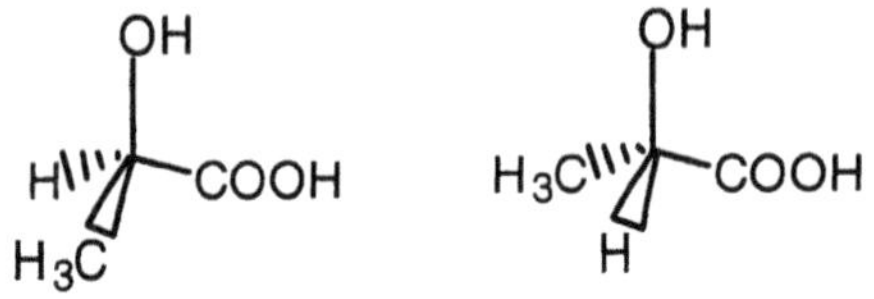

D-lactic acid (R-enantiomer) L-lactic acid (S-enantiomer)

Figure 20. Representation of Lactic Acid Stereoisomers

Poly(lactic acid) has an old history as reviewed by Lunt [302]. Already in the early 30's Carothers was investigating the production of an aliphatic polyester from lactic acid, even if this product had low molecular weight and poor mechanical properties. In 1954 DuPont patented [303] a higher molecular weight product. Further studies were limited by the material susceptibility to hydrolytic degradation. In 1972 copolymers of lactic acid and glycolic acid were developed by Ethicon and proposed as high strength fibers for medical resorbable sutures. These products slowly hydrolyze within the body to the constituent acids that are afterward susceptible to be metabolized. In 1980s advances in the bacterial production of D-glucose, obtained from corn, permitted to drastically reduce production costs of lactic acid [302].

Poly(lactic acid) or polylactide can be synthesized by two different pathways implying a single step or two step pathway respectively. The abbreviation PLA is commonly used for both poly (lactic acid) and polylactide that from a structural point of view identify the same macromolecular compounds.

Polycondensation reaction produces low molecular weight poly(lactic acid). Moreover the water produced during the polymerization process has to be removed and usually high temperatures and long reaction time are required. The process developed by Mitsui Toatsu [304] starts with aqueous lactic acid solution that is purified and concentrated. The direct condensation and cyclization processes are carried on at elevated temperature with addition of a transesterification catalyst, using azeotropic distillation to remove the water of condensation. High molecular weight PLA is produced and no reaction intermediates are isolated. This process is relatively simple and economic but molecular weight, molecular weight distribution and the end groups are fairly difficult to control.

In contrast to the more traditional polycondensation, lactide ring opening polymerization provides a direct and easy access to the corresponding high molecular weight polylactide. Thus lactic acid can be transformed into lactides, by means of a combined process of oligomerization and cyclization. The racemic mixture of L and D lactic acid generate three different types of lactides two optically active forms and a mesoform.

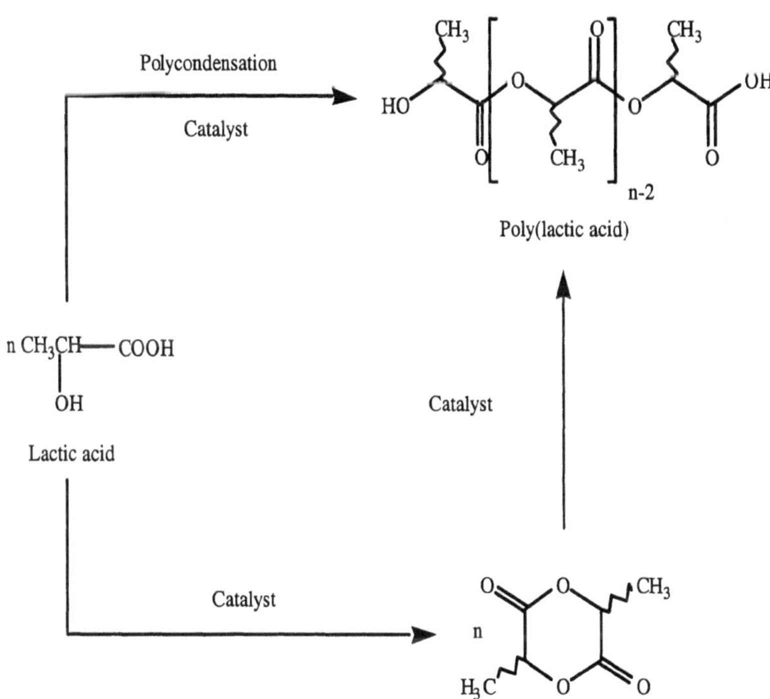

Figure 21. Schematic Representation of PLA Synthesis[302]

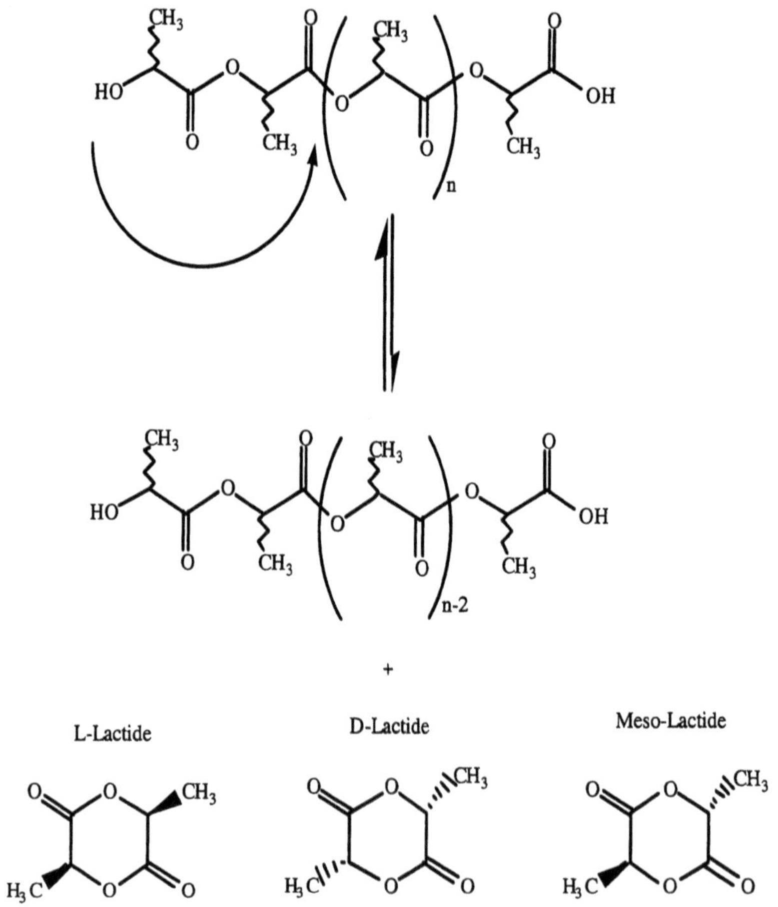

L-Lactide D-Lactide Meso-Lactide

Figure 22. Schematic Representation of the Polymerization of Lactide [302]

In the polymerization of lactides to polylactide the crude lactides have to be purified with regard to residual water (<50 ppm) and free acidity (<0.1% meq/kg). Depending on the type of lactides used, for the ring opening polymerization, polymers with different properties can be generated. For this process the necessary step of dimerization of lactic acid has the disadvantage of increasing production costs.

The technology to produce PLA economically on a commercial scale has been intensively developed in the last two decades. For the conversion of lactic acid to high-molecular PLA Cargill-Dow uses a solvent free process and a novel distillation process to produce a range of polymers [305]. In this process lactic acid is converted in to a low molecular weight poly (lactic acid) that is then converted in the cyclic dimer, lactide, by controlled depolymerization.

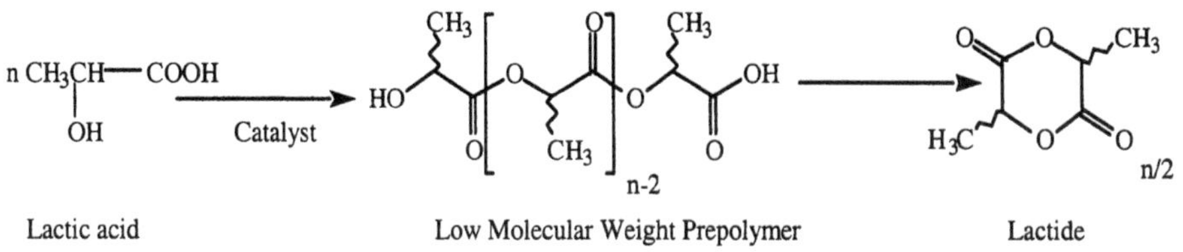

Figure 23. Schematic Representation of the Polymerization of Lactide [302]

The lactide is maintained in liquid form and purified by distillation. PLA with controlled molecular weight is then produced by catalytic ring opening of the lactide intermediate.

The ring opening polymerization of lactides is catalyzed by organo-metallic compounds, which have been first classified into anionic or cationic initiators. Polymerization is usually assumed to proceed through a coordination-insertion mechanism. A great deal of investigation has been carried out during the years on the polymerization mechanism [306-311]. Lewis acid type catalyst such as metals, metal halogenides, oxides, aryls and carboxylates are of special interest for this type of reaction. A commonly used catalyst is tin(II)octoate [312]. The control of ring-opening polymerization with these catalysts has some problems. Lewis acid catalysts are not chemically bonded to the growing chain and can therefore activate more than one growing chain so that the monomer-to-catalyst molar ratio does not properly control the degree of polymerization. Moreover intermolecular and intramolecular transesterification reactions may perturb the chain propagation, broaden the molecular weight distribution and yield cyclic oligomers [312]. Protic compounds such as water, alcohols, thiols and amines are able also to initiate the lactide polymerization.

Ring opening polymerization is promoted as well by other metal alkoxides especially Mg, Sn, Ti, Zr, Zn, and Al-alkoxides. With these compounds in a first step a complex between the initiator and the monomer is formed, followed by a rearrangement of the covalent bonds. The monomer is included in between the metal-oxygen bond of the cyclic monomer, so that the metal is incorporated with an alkoxide bond into the growing chain (coordinate type mechanism) [312].

In 1996 Jacobsen reported the preparation of PLA continuously by reactive extrusion [313] requiring that the bulk polymerization was close to completeness within a very short time (< 10 min.) and that the polylactide stability was high enough at the processing temperature. In this type of process $Sn(Oct)_2$ was used as a catalyst for lactic acid polymerization, even specific studies had evidenced adverse effects on PLA molecular weight and properties. This was due to backbiting and intermolecular transesterification reactions, which occurred during lactic acid polymerization, and also further melt processing [314,315]. The addition of an equimolar amount of a Lewis base, particularly triphenylphosphine to $Sn(Oct)_2$, significantly enhanced the lactide polymerisation rate in bulk. This kinetic effect was accounted for by coordination of the Lewis base onto the metal atom of the initiator, making the insertion of the monomer into the metal alkoxide bond of the initiator easier [309,316]. The Sn-OR bond was formed in situ by the reaction of the alcohol and the tin(II)dicarboxylate. Thus the ring opening polymerization proceeded via the same coordination-insertion mechanism involving the selective O-acyl cleavage of the cyclic ester monomer, as already known from alkoxide initiators [318, 319]. The addition of one equivalent of triphenylphosphine to $Sn(Oct)_2$, allowed an acceptable balance to be achieved between propagation and depolymerization rates, so that the polymerization was fast enough to be performed through a continuous single-stage process in an extruder. Reactive extrusion polymerization using laboratory scale closely inter-meshing co-rotating extruders have been achieved and the influence of various processing parameters on the resulting polymeric properties has been reported.

3.1.1 Applications of Polylactic Acid (PLA)

PLA applications include wide variety of biomedical products, food packaging for meat and soft drinks, films for agro-industry and non-woven materials in hygienic products. Poly(lactide) (PLA) may play a major role in future markets for biodegradable polymers from renewable resources. An optically pure poly(L-lactide), or poly(D-lactide), is a crystalline, hard and rather brittle material melting in the temperature range 175-185 °C, depending on molecular weight, and size of the crystallites [301]. A poly(lactide) from racemic monomer has a random stereosequence, thus is amorphous and transparent with a glass transition temperature of 50-60 °C , depending on molecular weight, and can be used for the production of glues and films. These thermoplastic polymers are amenable to different procedures including extrusion, injection molding, blow molding, or fiber spinning processes, into various products [302,321].

PLA has potential as a future packaging material [322] because of its good mechanical properties, comparable to standard packaging polymers such as polystyrene with an elasticity modulus of 3500 MPa, a maximum tensile strength of 50 MPa, and an elongation at break of 4% [321].

After use PLA based items can be recycled by re-melting and processing a second time, incinerated or landfilled though it is mainly intended for disposal by composting or soil degradation. Taking into account the production of lactic acid by fermentation of renewable resources, PLA could provide a closed natural loop, being produced from plants and crops, polymerized and processed into a packaging product and degraded after use into soil and humus, which is the basic necessity for growth of new plants and crops (Figure 24).

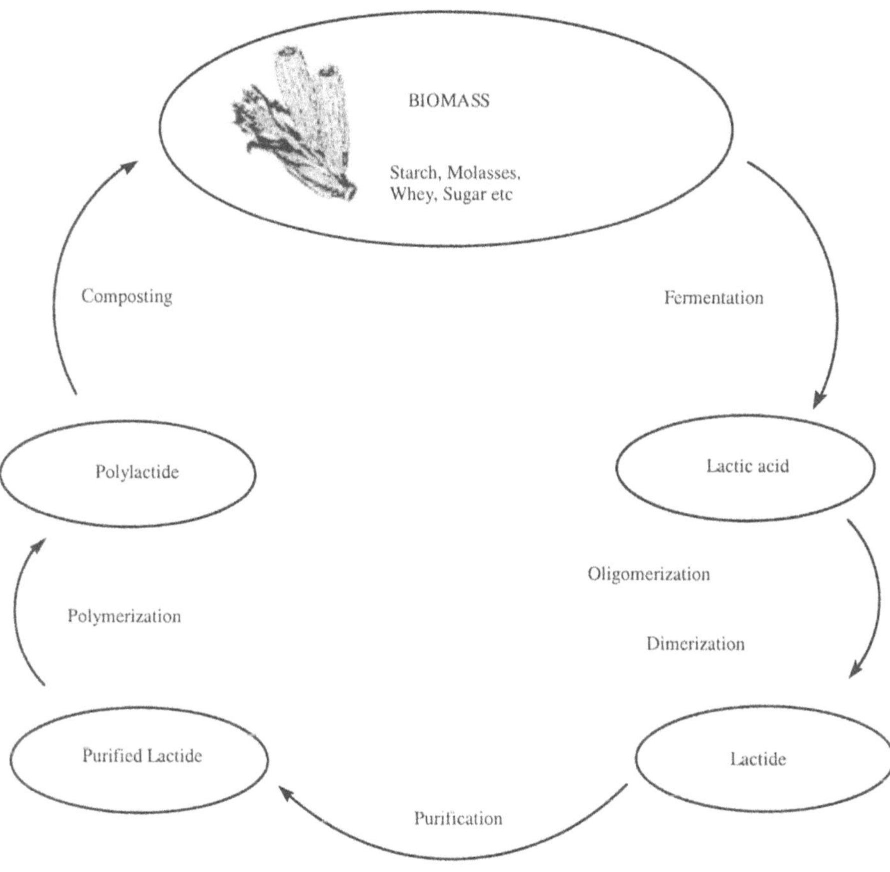

Figure 24. Schematic Representation of the PLA Production in a Close Loop System [312]

Variation of the properties of PLA can be achieved by copolymerization with various lactones or by physical blending with other polymers. The most widely used comonomers comprise glycolide and ε-caprolactone. The incorporation of glycolide enhances the rate of hydrolytic degradation and is usually desired, when the copolylactides are designed for pharmaceutical applications in controlled drug release practice. Incorporation of ε-caprolactone reduces the rate of hydrolytic degradation and lowers the glass transition temperature thereby enhancing the flexibility of films and fibers [301]. Block copolymers of L-lactide are normally crystalline while random copolyesters are usually amorphous and are of interest as transparent films or as matrix for drug delivery systems.

Homopolymers of lactide and lactones and relevant copolymers can be also obtained by polymerization in the chambers of twin-screw extruders thus reducing production costs and allowing for a more facile tuning of the ultimate properties of the obtained polymeric materials [323].

3.1.2 Biomedical Applications of Poly (α-hydroxy acid)s

Hydroxy acids derived from natural resources such as lactic and glycolic acid have been employed to synthesize a wide variety of useful biodegradable polymers for large number and type of biomedical product applications. As an example, bioresorbable surgical sutures made form poly(α-hydroxy acid)s have been in clinical use since 1970.

The ester bond of the poly(α-hydroxy acids) is cleaved by hydrolysis which leads to a decrease of the polymer molecular weight but does not affect the mass of a possible implant [324]. The initial degradation of the polymers occurs until the molecular weight is in the range of 5000 Dalton, at which point cellular degradation takes over. The phenomena of polymer resorption involves the action of inflammatory cells such macrophages and/or lymphocytes; although such late stage inflammatory response may affect negatively some of the healing events. These polymers have been successfully employed as matrices for cell transplantation and tissue regeneration [325,326].

An important feature of such polymers is the degradation rate that can be controlled by tuning the initial molecular weight, the exposed surface area, the crystallinity and, in the case of copolymers, the ratio of the hydroxy acid monomers. Drawbacks are their modest range of mechanical properties and a correspondingly modest range of processing conditions. Nevertheless, they can generally be processed into tubes, films and matrices by using standard processing techniques as molding, extrusion, solvent and spin casting. Moreover, ordered fibers, meshes, and open-cell foams have been formed to fulfill the surface area and cellular requirements of a variety of tissue engineering constructs [327-329]. Poly (α-hydroxy acid)s in combination with other materials such as polyethylene glycol were shown to give better results in terms of reduction of inflammatory response to degradation products and to modification of cellular response [330].

Poly (glycolic acid) (PGA), poly (lactic acid) (PLA) and their copolymers are the most widely investigated and used synthetic degradable polymers for biomedical applications [331]. They are discussed further in Chapters 5 and 11 of this book.

PGA is highly crystalline and has a high melting point and low solubility in organic solvents and was used in the development of the first totally synthetic absorbable suture [332]. Due to its hydrophilic nature it was shown that surgical sutures made of PGA tend to lose their mechanical strength rapidly over a period of 2 to 4 weeks postimplantation [333]. An intensive investigation aimed at the improvement of the material properties of PGA was undertaken by many researchers, focusing on the preparation of copolymers of PGA with the more hydrophobic poly (lactic acid) (PLA). Sutures composed of copolymers of glycolic and lactic acid are currently marketed under the trade names of Vicryl and Polyglactin 910 [334,335].

The presence of an extra methyl group renders lactic acid more hydrophobic. The hydrophobicity of PLA limits the water uptake of films to 2% [334] and decreases the backbone hydrolysis rate in respect to the one of the PGA homopolymer [333]. Moreover PLA is more soluble in organic solvents than PGA.

It is important to note that no linear relationship exists between the ratio of glycolic acid to lactic acid and the physicomechanical properties of their relevant copolymers. The highly crystallinity of PGA is rapidly lost in PGA-PLA copolymers. These morphological changes lead to an increase in the hydration and hydrolysis rate and copolymers tend to degrade more rapidly than the homopolymers of PGA or PLA [334,336].

As already mentioned the two different stereoisomeric forms of lactic acid (D and L) give rise to four morphologically distinct polymers. Generally speaking L-PLA is more frequently employed than D-PLA since the hydrolysis of L-PLA yields the naturally occurring stereoisomer of lactic acid, L(+) lactic acid.

The differences in the crystallinity of D,L-PLA and L-PLA have important practical implications. Since DL-PLA is an amorphous polymer it is usually employed for applications such as drug delivery where it is important to have a homogeneous dispersion of the active species within a monophasic matrix [337-339]. The semicrystalline L-PLA instead is preferred in applications where high mechanical strength and toughness are required such as sutures and orthopedic devices [340-343].
Copolymers of PLA and PGA are being intensively used for a large number of drug delivery applications as well [343,344].

3.1.3 *Other Applications of Poly(lactic acid)*

Plastic items can be fabricated from PLA as films, fibers and shaped containers including among the others grocery bags, pens, television cabinets and toys. So far, researchers haven't found anything made from conventional plastic resins that can't be made from PLA. PLA is known to loose all mechanical properties when composted for five weeks. The horticulture industry is one of several possible target areas for marketing PLA. The horticulture industry uses millions of pounds of polystyrene and other plastics each year to produce small disposable bedding plant containers, as well as pots for producing and merchandising plants [345]. PLA is very promising for use as mulch films or as clips for vineyards [346].

As a packaging material PLA is readily extruded, gives enhanced properties when oriented, and can be converted into high clarity end-products offering specific attributes such as stiffness, twist retention and sealability.

Several companies in Europe, Japan and USA, are involved in PLA production for commodity applications, such as Galactic Laboratories (Belgium), Hycail (The Netherlands), Dai-Nippon Chemical (Japan), Shimadzu (Japan), Toyobo (Japan), Mitsui Chemicals (Japan) and Cargill Dow Polymers (USA). These last two companies are the most active in PLA production and marketing [346]. PLA developed by Cargill Dow as "NatureWorks" has been proposed for common consumer items such as clothing, cups, food containers, food wrappers, as well as home and office furnishings [347]. NatureWorks has been proposed in a broad range of film end-uses including: high clarity/high stiffness films as an alternative to cellophane in uses from confectionery twist wrap to premium wrapping for flowers, toiletries and prestige gifts. Bags for compost and garden refuse as well as agricultural mulch films to replace paper (when wet strength is required) or blended polymer alternatives. A wide variety of multi-layer films for packaging uses such as flavored cereals, coffee packs and pet food. Window films for envelopes, cartons and other packages. Lamination films including end-uses where cellulose acetate can be replaced. Low temperature heat seal layers and/or flavor and aroma barriers in co-extruded structures where its combination of properties allows layer simplification or replacement of nylons. Shrink sleeve films and high modulus label films. Non-fogging films for fresh produce packaging [347].

PLA is readily converted into a variety of fiber forms, using conventional melt-spinning processes. Mono component and bicomponent, continuous (flat and textured) and staple fibers of various types are easily produced. Fibers from PLA demonstrate excellent resilience, outstanding crimp retention and improved wicking compared with natural fibers.

TABLE 5. Items Based on PLA or different Segmens of Application [347]

End-Use Example	Benefits
Dairy	Viable alternative with additional disposal options
Candy Wrap	Dead-fold, clarity, gloss, "natural packaging
Fast Food Cups	Hot and cold performance, "natural packaging"
Fresh Food Containers	Flavor and aroma barrier, "natural packaging"
Food Containers	Packaging waste reduction, flavor, and aroma barrier
Paper Ice Cream Holders	Structural integrity, "natural packaging"
Flavored Cereal Packaging	Easy heat seal and flavor retention
Coffee Packs	Flavor and aroma retention
Snack Foods	Flavor, crisp feel
Milk and Fresh Yogurt	Flavorable barrier properties,"natural packaging"
Compost bags and biodegradable packing	Completely biodegradable in compost plants

Microdenier fibers are readily produced. Fabrics produced from PLA are being utilized for their silky feel, drape, durability and moisture-management properties. Independent testing have confirmed that PLA fibers have other key advantages over PET fibers, including: low flammability and smoke generation, excellent UV stability, high luster, lower density, controllable bonding temperatures (achieved by control of the optical composition.

PLA polymers have been proposed also as water-based emulsions. Emulsion applications may include: paper and board coatings and pigments, paints, binders for non-woven fabrics, binders for building products, and adhesives. Lactic acid derivatives, to be used as chemical intermediates in products such as solvents, hot-melt adhesives, coatings, surfactants, acrylic esters, and agricultural intermediates [348].

Applications of PLA are expected to increase so that the production cost of PLA is likely to decrease in the near future as production increases due to economies of scale. A world-scale production facility for Cargill Dow's resin, NatureWorks™ PLA, has just been constructed in Nebraska (USA). At full capacity the facility is expected to produce 140,000 metric tons of PLA annually with appropriate price reduction. Recently Cargill Dow and Mitsui Chemicals have announced collaboration for business. The two companies have signed a collaboration agreement to accelerate worldwide market development for PLA. Mitsui Toatsu is a large chemical company that currently manufactures two types of biodegradable polymers for medical applications: polyglycolic acid and glycolic-lactic acid copolymers. Mitsui Toatsu has patented a novel process to polymerize lactic acid through direct condensation polymerization to obtain high molecular weight poly(lactic acid) [349]. Mitsui produces its own Lacea-brand PLA resin in a pilot plant. Dow found that Mitsui's technology for materials, processing, and applications can be applied to NatureWorks PLA [350,351].

TABLE 6. End Uses of PLA Polymeric Materials [347]

End-Use Example	Benefits
PLA Based Apparel	
Performance active wear, fashion active wear	Excellent hand and drape moisture management and wicking
Casual blends	Easy care (resists wrinkling)
Wool blends	Natural-based fiber
Carpet tile	Sustainability
Industrial wall panels	Resiliency
Upholstery, interior furnishings, and automotive	Reduced flammability, smoke generation and toxicity
Outdoor Fabrics, Awnings	Good soiling and UV resistance and stain removal lcompared with Nylon 6 and 6.6
PLA Based Non Woven	
Agriculturals and geotextiles	Degradability
Wipes	Enhanced wicking
Diapers/ADL	Low linting
Binder fiber	Resilience , controllable thermal bonding, enhanced wet strength, natural-based fiber

4. Genomics and Renewable Resources

Recent advances in molecular biology may provide tools for a continued sustainable development in combination with investment in new infrastructure [352]. Sources of renewable raw material over a short time-period (annually) are prerequisites for longer-term sustainability, and in this scenario it is likely that a gradual transition from finite resources to renewable will occur thanks to the important knowledge that come from biological science [353]. The employment of structural and functional genomics could provide gene discovery methods, allowing for the understanding of the genetic basis for the regulation of particular natural molecular structures. Moreover, the possibility of engineering metabolic pathways as well as obtaining large-scale production of discovered genes may be achieved by means of biotechnology tools. Understanding from a genomic point of view the biosynthetic pathway in different species allows for the isolation and modification of potentially valuable candidates which could be transformed into higher-yelding production systems.

For their ability to capture primary solar energy and convert it into carbon-based molecular structures, plants may be considered a large source of renewable resources. Crop production systems in forestry and agriculture have been successful in supporting human progress for centuries. New processing systems have been developed to offset the major limitations inherent in plants, to take out some of the undesirable materials (trypsin inhibitors, alkaloid toxins, mycotoxins) and to add extra nutrients that are missing. In terms of plant use as industrial raw materials the cost and energy requirement for such processing has typically not been competitive with raw inputs from hydrocarbons. Recently the possibility to obtain transgenic crops in combination with the demonstration that plant constituents can be altered in beneficial ways, has opened up the possibility that valuable molecular structures could be produced *in planta* [354].

Nowadays crop production gives some essential material and some material that is not used very effectively. For example maize and soybean are major crops in the US and produce a large amount of biomaterial per unit area. From the synergic activity of plant breeding and transgenic techniques it will be possible to more efficiently develop traits of value in the major crops. Recent study carried out by the US National Academy of Science indicates that biotechnology could play a major role in developing new approaches to renewable resources [355].

Besides the major genomic projects such as the structural sequencing of the human genome [356]. increasing attention is being devoted to genomics research that include crop plant projects, new methods for more rapid sequencing and the emergence of a large area called functional genomics [357]. Improvement of breeding and detection of production may be achieved thanks to the structural information arising from crop genomics projects. New techniques such as nucleotide micro-arrays may provide important knowledge on gene expression, regulation and function. A recent example is the improved understanding of the regulation mechanisms for gene expression in the plant flavonoid pathway [358]. Genomics is beginning to be applied to the improvement of the nutritional and nutraceutical features of food plants [359]. Conventional breeding with recurrent selections is certanly limited in terms of modifying metabolism in specific beneficial ways, thus the application of functional genomics shows very high potential.

For example, functional genomics may help to improve genetic alteration of the enzymes and organisms in the bio-refineries. The largest current processing of a renewable resource in bio-refineries is the wet-milling of maize. The most well-known product is the 1.3 billion gallons of fuel ethanol produced from 14 million metric tons of maize grain. However the fermentation process allows for the conversion of oxygenated carbohydrate molecules into several low molecular weight chemicals such as lactic and succinic acid, lysine and alcohols. Several of the organism used in the fermentation process have been genetically altered via traditional microbial genetics. Genomics could allow further progress through genetic-based improvements in the mechanisms for direct fermentation of multisubstrates, including C_5 pentose sugars [360]. Moreover, another important contribution from genomics may allow for the manipulation of raw input materials.

Several other materials are also produced in maize and a fairly high amount of solar energy is trapped in unused lignocellulosic stover. Ongoing studies are focused on the development of enzymatic biocatalyst and/or thermochemical processes to utilize lignocellulosic biomass, with improved conversion into energy or into molecular building blocks such as levulinic acid [361].

Anaerobic fermentation of maize carbohydrate can also be modified to produce high yields of the 3-carbon molecule lactic acid [362]. Esterification with ethanol produces ethyl lactate which has good solvent properties and can be used as a chemical intermediate. Conversion of lactate to the lactide dimer produces the intermediate for polylactate resin manufacture.

A novel polyester G3T from DuPont may also be obtain by a renewable production of 1-3 propanediol via fermentation of glucose/starch. Improving the process through identifying genes for appropriate metabolic enzymes is being attempted [362].

The expansion of renewable resources applications as well as the developing of economically attractive and efficient bio-systems may be obtained thanks to the development of functional and structural genomics techniques [353]. For example spider silk proteins which have become topical in the past few years due to improved understanding of the function of the different proteins and the underlying genes [363]. These remarkable proteins are formed at room temperature and extruded from silk glands in fluid form that self-assembles into polymer filaments. Seven different types of spider silks have been identified, each with different and remarkable properties [364]. For example flagelliform silk is an elastic-type polymer with a capability ot extend by over 200%. The *flag* gene was found to contain repetitive exon and intron regions, and to code for a protein with glycine-rich tandem arrays.

In the marine environment as well there are many species with different metabolic pathways that produce very useful molecular substance such glues, solvents, enzymes able to operate under extreme conditions [365]. Structural and functional genomics in the area of natural materials will allow expanded searches for homologous genes in other species. Genomics libraries may allow improved gene selection and optimization for expression within particular production systems as bacteria, plant or animal, allowing for the *de novo* design of polymers with desirable functional characteristics.

The high potential of structural and functional genomics appears to be a key point for the development of future renewable resources. Nevertheless, genomics alone cannot be sufficient to achieve the desired changes. Genomics combined with advances in metabolic engineering, improved production and harvesting systems and investment in sophisticated bio-refineries will make a significant difference in the sustainable production of renewable resources and hence in all the valuable functional and structural materials that can be derived therefrom.

5. Conclusions

There is a perception in many countries that greater utilization of products from plants, animal and microbes from land and sea can and should play a more significant role in meeting society's needs for a broad range of industrial and consumer products. As the chemical industries come under federal and state regulations on discharged air, water and solid wastes, much attention has been directed to improve the effectiveness and efficiency of existing technologies, and to the production of derivatives form renewable resources which are environmentally acceptable in their full life cycle. Some of these materials, particularly from excess production, as byproducts or as agricultural or food production could be cheap and could provide product that are reusable and recyclable [366].

Several industries are focusing their research activity on materials from renewable resources. The conversion of agricultural products and byproducts from industrial manufacture, polymers, biotechnology, fuel technology, and other applications have also been the object of increasing interest (367-374). Many major companies such as DuPont, Dow, BASF, Monsanto, Mitsui, Mitsubishi have pursued biotechnology [347,375-378]. Some medium-sized companies are nowadays discovering unique formulations and fabrication processes to produce organic materials which, serve as hydrocolloid aqueous delivery systems for variably-viscous industrial, agricultural, nutritional and household solutions. The materials are swellable, water-soluble polymers derived from natural fibers, sugars, gums and starches. These polymers, at low concentrations, quickly hydrate and disperse in water, producing gels or highly viscous solutions. A revisiting of the technologies that were applied for the production and manufacturing of industrial and consumer products from renewable resources such as cellulose, lignin, starch, is at present ongoing and seeking for more environmentally friendly and locally viable applications.

These natural materials, formulated from abundant, renewable natural carbohydrates resources, are thought to replace acrylics and sodium aspartate in a variety of cleaning products such as laundry detergents, dishwashing liquids, household and industrial cleansers, textile and mining processing liquids, as well as personal care items such as shampoos, cleansers and cosmetics. Agricultural applications include delivery or nutrients, herbicides, pesticides and microbes to infertile soil. Midwest Grain's Research and Development efforts with natural polymers from wheat starch and wheat proteins are ongoing. New resins have been developed that address some shortcomings of other starch and protein products [378]. Blends have been developed using mixtures of starch, proteins, and aliphatic polyesters that display a range of properties typical for low-density polyethylene to polystyrene. These blends can be processed into articles using conventional plastics processing technology such as injection molding, blow molding, film blowing, sheeting, calendaring, and other conventional polymer operations [366].

The worldwide increasing demand of polymeric materials and plastics will certainly offer room to options for tailor-made materials obtainable from renewable resources. These have the potential to reduce the burden associated with plastic waste generation and their associated greenhouse gases.

References

1. Shogren, R.L. and Bagley, E.B. (1999) Natural Polymer as Advanced Materials: Some Research Needs and Directions, in S.H. Imam, R.V. Greene, B.R. Zaidi, (eds.) *Biopolymers - Utilizing Nature's, Advanced Materials*; ACS Symp. Ser. **723**, ACS, Washington DC, pp. 2-11.
2. El Bassam, N. (2001) Renewable Energy for Rural Communities, *Renewable Energy*, **24**, 401-408
3. Rowell, R.M., Sanadi, A.R., Caulfield D.F., and Jacobson R.E., (1997) Utilization of Natural Fibers in Plastic Composites: Problems and Opportunites, in A.L. Leao, F.X. Carvalho, E.Frollini, (eds.) *Lignocelluloisc-Plastic Composites,* Sao Paulo, Brazil, pp.23-52.
4. Warwel, S., Brüse, F., Demes, C., Kunz, M. and Klaas M.R. (2001) Polymers and Surfactants on the Basis of Renewable Resources, *Chemosphere*, **43**, 39-48.
5. Ayong Le Kama, A.D. (2001) Sustainable Growth, Renewable Resources and Pollution, *Journal of Economic Dynamics & Control*, **25**, 1911–1918
6. Reijnders, L. (2000) A Normative Strategy for Sustainable Resource Choice and Recycling, *Resou.r Conserv. Recycl.,* **28**, 121-133.
7. Van Wyk, J.P.H. (2001) Biotechnology and the Utilization of Biowaste as a Resource for Bioproduct Development, *Trends in Biotechnology* **19**, 172-177
8. Sperling L.H. and Carraher C.E. (1988) Polymers from Renewable Resources in H. F. Mark, N. M. Bikales, C.G. Overberger, and G. Menges (eds.) *Encyclopedia of Polymer Science and Engineering*, Wiley, New York, **Vol.12**, pp.658-690
9. Chemicals and Materials from Renewable Resources (2001), Proceedings of a Symposium Held at the 218th National Meeting of the ACS in New Orleans, Louisiana, 22-26 August 1999, ACS *Symp. Ser.* **784**, J.J. Bozell (ed.) ACS,

Washington D.C., USA, 226 pp.

10. Scott, G. (2000) Green Polymers, *Polym. Degrad. Stab.* **68**, 1–7.
11. Doane, W. M. (1994) Biodegradable Plastics. *J. Polym. Mater.* **11**, 229-237.
12. Heyde, M. (1998) Ecological Consideration on the Use Production of Biosynthetic and Synthetic Biodegradable Polymers, *Polym. Degrad. Stab.* **59**, 3–6.
13. Mayer, J.M. and Kaplan D.L. (1994) Biodegradable Materials: Balancing Degradability and Performance, *TRIP / Reviews* **2**, 227-235.
14. MacGregor, E. A., and Greenwood, C .T. (1980) *Polymers in Nature*, John Wiley & Sons, New York, USA, 391 pp.
15. Scott, G. and Wiles, D.M. (2001) Programmed-Life Plastics from Polyolefins: A New Look at Sustainability, *Biomacromolecules* **2**, 615-622.
16. Feil H. (1998) Sense and Non-Sense About Polymers *Macromol.Symp.* **127**, 7-11.
17. O'Brien, J., Fahnestock, S.R., Termonia, Y. and Garnder, K.C.H. (1998) Nylon from Nature: Syntethic Analogs to Spider Silk, *Adv. Mater* **10**, 1185 - 1195.
18. Van Os, G. (2001) The European Plastics Industry - A Sunset Industry?, *Polymers in Europe Quo Vadis? EPF-Special Issue*, 19-22.
19. French, A.D., Bertoniere, N.R., Battista, O.A., Cuculo, J.A., and Gray, D.G. (1993) Cellulose, Chapter in: *Kirk-Othmer Encyclopedia of Chemical Technology*, 4th Ed. **Vol. 5**, John Wiley & Sons, New York, pp. 476-496.
20. Petersen, K., Nielsen, P.V., Bertelsen, G., Lawther, M., Olsen, M.B., Nilsson, N.H. and Mortensen, G. (1999) Potential of Biobased Materials for Food Packaging, *Trends Food Sci. Technol.* **10**, 52-68.
21. Elion, G.R. (1993) Synthesis of Plastic from Recycled Paper and Sugar Cane U.S. Patent 5,244,945.
22. Simon, J., Müller, H.P., Koch, R., and Müller V. (1998). Thermoplastic and Biodegradable Polymers of Cellulose, *Polym. Degrad. Stab.* **59**, 107-115
23. Lagoa, R., Murtinho, D., and Gil, M. H. (1999) Membranes of Cellulose Derivatives as Supports for Immobilization of Enzymes, *Biopolymers - Utilizing Nature's, Advanced, Materials*; Acs *Symp. Ser.* **723**, S. H. Imam, R.V. Greene, B. R. Zaidi, (eds.) ACS, Washington DC, pp.228 - 234.
24. Hoenich, N., A., and Stamp, S. (2000) Clinical Investigation of the Role of Membrane Structure on Blood Contact and Solute Transport Characteristics of a Cellulose Membrane, *Biomaterials,* **21**, 317-324.
25. Vert, M. (2001) Biopolymers and Artificial Biopolymers in Biomedical Applications, an Overview, In *Biorelated Polymers, Sustainable Polymer Science and Technology*, in E. Chiellini, H. Gil, G. Braunegg, J. Buchert, P. Gatenholm, and M. Van der Zee (eds.), Kluwer Academic/Plenum Publishers, pp. 63-79.
26. Malm, C., Mench, J., Kendall, D., and Hiatt, G. (1951) Aliphatic Acid Esters of Cellulose. Preparation by Acid Chloride-Pyridine Procedure, *Ind. Eng. Chem.* **43**, 684-688.
27. Vaca-Garcia, C., and Borredon, M.E. (1999) Solvent-Free Fatty Acylation of Cellulose and Lignocellulosic Wastes II. Reactions with Fatty Acids., *Bioresource Technol.* **70**, 135-142.
28. Girardeau, S., Aburto, J., Vaca-Garcia, C., Alric, I., Borredon, E., and Gaset, A. (2001) An Original Method of Esterification of Cellulose and Starch, In

Biorelated Polymers, Sustainable Polymer Science and Technology, E. Chiellini, H. Gil, G. Braunegg, J. Buchert, P. Gatenholm, and M. Van der Zee (eds.), Kluwer Academic/Plenum Publishers, pp.53-59.

29. Wang, P., Tao B.Y. (1999) Characterization of Plasticized and Mixed Long-Chain Fatty Cellulose Esters, *Biopolymers - Utilizing Nature's, Advanced, Materials; ACS Symp. Ser.* **723**, S. H. Imam, R. V. Greene, B. R. Zaidi, (eds.) ACS, Washington DC, pp. 77 - 87.
30. Just E.K., and Majewicz T.G., Cellulose Ethers (1987), In *Encycorklopedia of Polymer Science and Engineering*, 2nd ed. J.I.Kroschwitz (ed.) John Wiley & Sons; New Y, **Vol.3**. pp. 226-269
31. Komarek, R.J., Gardner, R.M. and Buchanan, C.M. (1993) Biodegradation of Radiolabeled Cellulose Acetate and Cellulose Propionate *J. Appl. Polym. Sci.* **50**, 1739-1746
32. Buchanan, C. M., Gardner, C. M., and Komarek, R. J. (1993) Aerobic Biodegradation of Cellulose Acetate *J. Appl. Polym. Sci.* **47**, 1709-1719
33. Mormann, W., and Demeter, J. (1999) Silylation of Cellulose with Hexamethyldisilizane in Liquid Ammonia-First Examples of Compltely Trimethylsilylated Cellulose, *Macromolecules*, **32**, 1706-1710
34. Mormann, W. and Spitzer, D. (2001) Silylation of OH-Polymers by Reactive Extrusion, *Macromol.Symp.* **176**, 279-287.
35. Nishio, Y., Hratani, T., Takahashi, T., and Manley, R.S. (1989) Cellulose/Poly(vinyl alcohol) Blends: An Estimation of Thermodynamic Polymer-Polymer Interaction by Melting point Depression Analysis, *Macromolecules.* **22**, 2547-2549.
36. Masson, J.F. and Manley, R.St.J. (1991) Cellulose/Poly(4-vinylpyridine) Blends, *Macromolecules* **24**, 5914-5921.
37. Masson, .J.F. and Manley, R.St.J (1991) Miscible Blends of Cellulose and Poly(vinylpyrrolidone), *Macromolecules* **24**, 6670-6679.
38 Nishioka, N., Hamabe, S., Murakami, T. and Kitagawa, T. (1998) Thermal Decomposition Behavior of Miscible Cellulose / Synthetic Polymer Blends, *J. Appl. Polym. Sci.* **69**, 2133-2137.
39 Hon D.N.S. (1992) New Development in Cellulosic Derivatives and Copolymers, in *Emerging Technologies for Materials and Chemicals from Biomass* 1992, R.M. Rowell, T.P. Scultz, R. Narayan (eds.) *ACS Symp. Ser.* **476**, pp.176-196.
40 Gatenholm, P., Kubat, J. and Mathiasson, A. (1992) Biodegradable Natural Composites.1. Processing and Properties *J. Appl. Polym. Sci.* **45**, 1667-1677.
41 Collier, J.R., Lu, M., Fahrurrozi, M., and Collier B.J. (1996) Cellulosic Reinforcement in Reactive Composite System, *J.Appl.Polym.Sci.*, **61**, 1423-1430
42 Gatenholm, P. (1997) Interfacial Adhesion and Dispersion in Biobased Compoasites. Molecular Interacions Between cellulose and Other Polymers, in *Lignocellulosic-Plastic Composites*, A. Leao, F.X.Carvalho, E. Frollini (eds.) Sao Paulo, Brasil, 53-59.
43 Rowell, R.M., Tillman, A.M., Simonson, R. (1986) A Simplified Procedure for the Acetylation of Hardwood and Softwood Flakes for Flakeboard Production *J.Wood Chem.Tech*, **6**, 427

44 Maldas, D., Kokta, B. V., and Daneault, C. (1989) Influence of Couplng Agents and Treatments. on the Mechanical Properties of Cellulose Fiber-Polystyrene Composites. *J. Appl. Polym. Sci.*, **37**, 751—775.

45 Raj, R. G., Kokta, B.V., Maldas, D., and Daneault, C. (1989) Use of Wood Fibers in Thermoplastics. VII. The Effect of Coupling Agents in Polyethylene-Wood Fiber Composites. *J. Appl. Polym. Sci.*, **37**, 1089—1103.

46 Raj, R.G. and Kokta, B.V. (1992) Mechanical Properties of Surface-Modified Cellulose Fibers-Thermoplastic Composites in *Emerging Technologies for Materials and Chemicals from Biomass*, R.M. Rowell, T.P. Scultz, R. Narayan (eds.) *ACS Symp. Ser.* **476**, pp.76-87

47. Stannet, V. (1982) Some Challenges in Grafting to Cellulose and Cellulose Derivatives in Graft Copolymerization of Lignocellulosic Materials, Hon, D.N.S. (ed.), *ACS Symp. Series*, **187**, pp.1-20.

48. Narayan, R. (1988) Synthesis of Controlled Cellulose – Synthetic Polymer Graft Copolymer Structures in *Cellulose Wood: Chemistry and Technology*; C. Schuerch (ed), John Wiley & Sons: New York, p.945.

49. Young, R.A. (1997) Utilization of Natural Fibers: Characterization, Modification, and Applications in *Lignocellulosic-Plastic Composites*, A. Leao, F.X.Carvalho, E. Frollini (eds.), Sao Paulo, Brasil, pp.1-22.

50. Takase, S., and Shiraishi, N. (1989) Studies on Composites from Wood and Polypropilenes. II. *J. Appl. Polym. Sci.*, **37**, 645—659.

51. Felix, J. M., and Gatenholm, P. (1991) The Nature of Adhesion in Composites of Modified Cellulose Fibers and Polypropylene. *J. Appl. Polym. Sci.*, **42**, 609—620.

52. Narayan, R. (1992) Compatibilization of Lignocellulosics with Plastics, in *Emerging Technologies for Materials and Chemicals from Biomass,* R.M. Rowell, T.P. Scultz, R. Narayan (eds.) *ACS Symp. Ser.* **476**, pp. 57-75.

53. Byun, H.S., Burford, R.P. and Fane, A.G. (1994) Sulfonation of Cross-linked Asymmetric Membrane Based on Polystirene and Divinyl Benzene *J. Appl. Polym. Sci.* **52**, 825-835

54. Gaylord, N. G. and Anand, L.C. (1971). Alternating Copolymer Graft Copolymers. III. Cellulose Graft Copolymers. I. Grafting of Alternating Styrene-Acrylonitrile Copolymers onto Cellulose in Presence of Zink Chloride. *J. Polym. Sci.: Part B: Polym. Phys.*, **9**, 617—621.

55. Nie, L. and Narayan, R. (1993). Upper Limit on Grafting Conversion and Phase Homogeneity in Grafting Reaction Products of Cellulose Acetate/Poly[Styrene-*ran*-(Maleic Anhydride). *Polym. Mater. Sci. Eng.*, **69**, 309—311.

56. Chauhan, G.S., Mahajan, S. and Guleria, L.K. (2000) Polymers from Renewable Resources: Sorption of Cu+2 Ions by Cellulose Graft Copolymers, *Desalination* **130**, 85–88.

57. BeMiller, J.N. (1994) Carbohydrates, in: *Kirk-Othmer Encyclopedia of Chemical Technology,* 4th ed., **Vol. 4**, John Wilet & Sons, New York, pp. 911-948

58. Zdrahala, R. J. (1997) Thermoplastic Starch Revisited, *Macromol. Symp.*, **123**, 113-121

59. Doane, W.M. (1992) Emerging Polymeric Materials Based on Starch, in *Emerging Technologies for Materials and Chemicals from Biomass*, R.M. Rowell, T.P. Scultz, R. Narayan (eds.) *ACS Symp. Ser.* **476**, pp.197-230.

60. Wurzburg, O. B. (1986) Modified Starches: Preparation and Uses, in O. B. Wurzberg (ed.), CRC, Inc.; Boca Raton, FL, pp. 23-28.
61. Otey, F.H., and Doane W.M. (1984) *Starch Chemistry and Technology*, 2nd ed., R. L. Whistler, J. M. Bemiller, and E. F. Paschall (eds) Academic Press, New York, pp. 389-416.
62. Imam, S.H., Mao, L., Chen, L., and Greene R. V. (1999) Wood Adhesive from Crosslinked Poly(vinylalcohol) and Partially Gelatinized Starch: Preparation and Propreties, *Starch/Stärke* **51**, 225-229
63. Tessler, M.M., and Billmers, R.L. (1996) Preparation of Starch Esters, *J.Env.Polym.Degr*. **4**, 85-89
64. Sagar, A. D., and Merril, E.W. (1995) Properties of Fatty-Acid Esters of Starch, *J. Appl. Polym. Sci.* **58**, 1647-1656.
65. Caldwell, C.G. (1949) Starch Ester Derivatives and Method of Making the Same U.S. Patent 2,461,139.
66. Smith, C.E. and Tuschoff, J.V. (1960) Acylation of Hydroxy Compounds with Vinyl Esters, U.S. Patent 2,928,828.
67. Smith, C.E. and Tuschoff, J.V. (1962) Acylation of Starch, U.S. Patent 3,022,289
68. Tessler, M.M. and Rutenberg, M.W. (1972) Process for Reacting Ungelatinized Starch with a Mixed Carbonic -Carboxylic Anhydride of a Polycarboxylic Acid, U.S. Patent 3,699,095
69. Tessler, M.M. and Rutenberg, M.W. (1973) Preparation of Starch Esters, U.S. Patent 3,720,662.
70. Tessler, M.M. and Rutenberg, M.W. (1973) Preparation of Starch Esters, U.S. Patent 3,728,332
71. Tessler, M.M. (1973) Preparation of Starch esters, U.S. Patent 3,720,663
72. Tessler, M.M. (1977) Preparation of Starch Esters, U.S. Patent 4,020,272
73. Tessler, M.M. (1975) Preparation of Starch Esters, U.S. Patent 3,928,321
74. Jarowenko, W. and Wurzburg, O.B. (1966) Novel Starch Ester Derivatives, U.S. Patent 3,284,442
75. Tessler, M.M. (1974) Starch Phosphate Esters, U.S. Patent 3,838,149
76. Tessler, M.M. (1978) Method for Preparing Starch Sulfate Esters, U.S. Patent 4,093,798
77. Tessler, MM., Wurbzurg O.B. and Dirscherl T.A. (1983) Alkyl and Alkenyl Sulfosuccinate Starch Half Esters, a Method for the Preparation Thereofore, U.S. Patent 4,387,221
78. Clode, D.M., and Horton, D. (1971) Preparation and Characterization of the 6-Aldehyde Derivatives of Amylose and Whole Starch, *Carbohydr. Res.* **17**, 365-373.
79. Horton, D., and Meshreki, M.H. (1975) Syntheses of 2,3-Unsaturated Polysaccharides from Amylose and Xylan., *Carbohydr. Res.* **40**, 345-352.
80. Heinze, U., Haack, V. and Heinze, T. (2001) New Highly Functionalised Starch Derivatives, In: *Biorelated Polymers-Sustainable Polymer Science and Technology*, E. Chiellini, H. Gil, G. Braunegg, J. Buchert, P. Gatenholm, and M. Van der Zee (eds.), Kluwer Academic/Plenum Publishers, pp. 20 -218.
81. Aburto, J., Alric, I., and Borredon, E. (1999) Preparation of Long-Chain Esters of Starch Using Fatty Acid Chlorides in the Absence of an Organic Solvent., *Starch/Stärke*, **51**, 132 - 135.

82. Rutenberg, M.W., and Solarek, D. (1984) *Starch: Chemistry and Technology*; R. L. Whister, J. N. BeMiller and E. F. Paschall, (eds.), Academic Press: New York, NY pp.315 - 323.
83. Kruger L.H. (1989) Degradation of Granular Starch U.S. Patent 4,838,944
84. Ewing, F.G. (1970) Starch Process U.S. Patent 3,539,366
85. Marsman, J.H., Pieters, R.T., Janssen, L.P.B.M., and Beenackers, A.A.C.M. (1990) Determination of Degree of Substitation of Extruded Benzylated Starch by H-NMR and UV Spectrometry, *Starch/Starke* **42,** 191 - 196.
86. Meuser, F., Gimmler, N., and Oeding, J. (1990) System Analytical Consideration of Derivatization of Starch with a Cooking Extruder as Reactor, *Starch/Starke* **42**, 330—336.
87. Chinnaswamy, R., and Hanna, M.A. (1991) Extrusion - Grafting Starch onto Vinyling Polymers, *Starch/Starke* **43**, 396—402.
88. Carr, M.E., Kim, S., Yoon, K.J., and Stanley, K.D. (1992) Graft - Polymerization of Cationic Methacrylate and Acrylonitrile Monomers onto Starch, *Cereal Chem.* **69**, 70—75.
89. Chang, Y.H., and Lii, C.Y. (1992) Preparation of Starch Phosphates by Extrusion, *J. Food Sci.* **57**, 203—205.
90. Esan, M., Brummer,T.M., and Meuser, F. (1996) Chemical and Processing Aspects of the Production of Cationic Starch Using Extrusion Cooking, *Starch/Starke* **48**, 131—136.
91. Wing, R. E. and Willett J. L. (1999) Thermochemical Processes for Derivatization of Starches with Different Amylose Content in *Biopolymers - Utilizing Nature's, Advanced, Materials*; Acs Symp. Ser. **723**, in S. H. Imam, R.V. Greene, and B.R. Zaidi, (eds.) ACS, Washington DC, pp.55-64.
92. Carr, M.E. (1994) Preparation of Cationic Starch Containing Quaternary Ammonium Substituents by Reactive Twin-Screw Extrusion Processing, *J.Appl.Polym.Sci*, **54**, 1855-1861.
93. Wang, L., Shogren, R.L. and Willett, J.L. (1997) Preparation of Starch Succinates by Reactive Extrusion, *Starch / Stärke* **49**, 116-120.
94. Tomasik P., Wang, Y.J., and Jane, J.L. (1995) Facile Route to Anionic Starches-Succinylation, Maleination, and Phatalation of Corn Starch on Extrusion, *Starch/Starke*, **47**, 96-99
95. Wing, R. E., and Willett, J.L. (1997) Water Soluble Oxidized Starches by Peroxide Reactive Extrusion, *Ind. Crop. Prod.* **7**, 45-52.
96. Fanta, G.F. and Bagley, E.B. (1977) Starch, Graft Copolymers, in: *Encyclopedia of Polymer Science and Technology*, Suppl. **Vol. 2**, Wiley-Interscience, New York, pp. 665-699.
97. Fanta, G. F., and Doane W. M. (1986) Grafted Starches, Modified Starches: Properties and Uses, Wurzburg, O. B. (ed.), CRC Press. Boca Raton, pp. 149-178.
98. Weaver, M.O., Bagley, E.B., Fanta, G.F. and Doane, W.M. (1976) Method of Reducing Water Content of Emulsions, Suspensions, and Dispersions with Highly Absorbent Starch-Containing Polymeric Compositions, U.S. Patent 3,935,099.
99. Weaver, M.O., Bagley, E.B., Fanta, G.F. and Doane, W.M. (1976) Highly-Absorbent Starch-Containing Polymeric Compositions, U.S. Patent 3,997,484.
100. Skinner, E. L., and Elmquist, L.F. (1979) Film Forming SGP, U.S. Patent

4,156,664.
101. Swanson, C.L., Shogren, R.L., Fanta, G.F. and Imam, S.H. (1993) Starch-Plastic Materials- Preparation, Physical Properties, and Biodegradability (A review of Recent USDA Research), *J. Environ. Polym. Degrad.* **1,** 155-166.
102. Ratto, J.A., Stenhouse, P.J., Auerbach, M., Mitchell, J. and Farrell, R. (1999) Processing Performance and Biodegradability of a Thermoplastic Aliphatic Polyester/Starch System, *Polymer*, **40**, 6777-6788.
103. de Graaf L.A. and Kolster P. (1998) Industrial Proteins as a Green Alternative for "Petro" Polymers: Potentials and Limitatioms, *Macromol. Symp.* **127**: 51-58
104. Biresaw, G. and Carriere C.J. (2000) Interfacial Properties of Starch/Biodegradable Esters Blends, *Polym. Prepr*, **41**, 64-65
105. Averous, L., Moro, L., Dole, P., and Fringant, C. (2000) Properties of Thermoplastic Blends: Starch-Polycapolactone, *Polymer*, **41**, 4157-4167.
106. Avella, M., Errico, M.E., Laurienzo, P., Martuscelli E., Raimo, M., and Rimedio, R. (2000) Preparation and Characterization of Compatibilised Polycaprolactone/Starch Composites, *Polymer*, **41**, 3875-3881
107. Ramsay, B.A, Langlade, .Carreau, P.J., Ramsay, J.A. (1993) Biodegradability amd Mechanical Properties of poly(beta hydroxy butirate-co-beta-hydroxyvalerate) starch blends, *Appl .Environm. Microbiol*., **59**, 1242-1246.
108. Shogren, R.L. (1995) Poly(ethylene oxide)-Coated Granular Starch-Poly(hydroxybutyrate-co-hydroxyvalrate) Composite Materials *J.Environm.Polym.Deg*., **3**, 75-80.
109. Kotnis, M.A., O'Brien, G.S and Willett, J.L. (1995) Processing and Mechanical Properties of Biodegradable Poly(hydroxybutyrate-co-valerate)-Starch Compositions, *J .Environm. Polym Deg*. **3**, 97-105
110. Imam, S.H., Gordon, S.H., Shogren, R.L. and Greene, R.V. (1995) Biodegradation of Starch-Poly(b-hydroxybuterate-co-valerate) Composites in Municipal Activated Sludge, *J. Environ. Polym. Deg*. **3,** 205-213.
111. Muzzarelli, R. (1977) Chitin, *Chitin*, Pergamon Press, New York.
112. Singh, D.K., and Ray, A.R. (2000) Biomedical Application of Chitin, Chitosan, and Their Derivatives, *J.M.S. Rev. Macromol. Chem. Phys*., **C40**, 69 – 83.
113. Li, J., Revol, J.F. and Marchessault, R.H. (1999) Alkali Induced Polymorphic Changes of Chitin, *Biopolymers - Utilizing Nature's, Advanced, Materials; Acs Symp. Ser*. **723**, S.H. Imam, R.V. Greene, B.R. Zaidi, (eds.) ACS, Washington DC, pp.88 - 96.
114 Kurita, K., Tomita, K., Tada, T., Niscimura, S., and Scimoda, K. (1993) Squid Chitin as a Potential Alternative Chitin Source-Deacetylation Behaviour and Characteristics Properties. *J. Polym. Sci., Part A: Polym. Chem*. **31**, 485-491.
115. Kurita, K., Tomita, K., Ishi,i S., Nishimura, S., and Shimoda, K. (1993) Beta-Chitin as a Convenient Starting Material for Acetolysis for Efficient Preparation of N-Acetyl Chitooligosaccharides, *J. Polym. Sci., Part A: Polym. Chem*. **31**, 2393-2395.
116. Kurita, K., Ishii, S., Tomita, K., Nishimura, S., and Shimoda, K. (1993) Reactivity Charactestics of Squid Beta-Chitin as Compared with Those of Shrimp Chitin-High Potentials of Squid Chitin as a Starting Material for Facile Chemical Modifications *J. Polym. Sci., Part A: Polym. Chem.* **32**, 1027-1032.
117. Kim, S.S., Kim, S.J. , Moon, Y.D. and Lee, Y.M. (1994) Thermal Characteristics of Chitin and Hydroxypropyl Chitin, *Polymer* **35**, 3212 - 3216.

118. Muzzarelli, R. (1985). Chitin, in: *Encyclopedia of Polymer Science and Engineering*, John Wiley, New York, **Vol.3**, pp.430-441.
119. Yamada, K., Chen, T., Kumar, G., Vesnovsky, O., Topoleski, L.D.T and Payne, G.F. (2000) Chitosan Based Water-Resistant Adhesive. Analogy to Mussel Glue, *Biomacromolecules* **1**, 252-258.
120. Kurita, K. (1998) Chemistry and Application of Chitin and Chitosan, *Polym. Degrad. Stab.* **59**, 117 - 120.
121. Wu, A.C.M., and Bough, W.A. (1978) Proceedings of 1ts International Conference on Chitin/Chitosan, in R. Muzzarelli and E.R. Pariser (eds.), MIT-SG, Cambridge p.88
122. Sannan, T., Kurita K. and Iwakura Y. (1976) Studies on Chitin, 2. Effect of Deacetylation on Solubility. *Makromol.Chem.*, **177**, 3589-6000
123. Brugnerotto, J., Desbrières, J., Heux, L., Mazeau, K. and Rinaudo, M. (2001) Overview on Structural Characterization of Chitosan Molecules in relation with the Behaviour in Solution, *Macromol.Symp.* **168**, 1-20.
124. Rinaudo, M., and Domard, A. (1989) Solution Properties of Chitosan, in *Chitin and Chitosan*, G. Sjak-Braek, T. Anthonsen, and P. Sandford (eds.), Elsevier, New York, p.77
125. Dutkiewicz, J., and Tuora, M. (1992) Advances in Chitin and Chitosan, in *Advances in Chitin and Chitosan*, C. J. Brine, P. A. Sanford, and J. P. Zikakis, (eds.), Elsevier, New York, p.496.
126. Peniche, C., and Argüelles-Monal, W. (2001) Chitosan Based Polyelectrolyte Complexes, *Macromol.Symp.*, **168**, 103-116.
127 Torres, J. A., Dewitt-Mireles, C., and Savant, V. (1999) Two Food Applications of Biopolymers: Edible Coatings Controlling Microbial Surface Spoilage and Chitosan Use to Recover Proteins from Aqueous Processing Wastes, In *Biopolymers - Utilizing Nature's, Advanced, Materials; ACS Symp. Ser.* **723,** S.H. Imam, R.V. Greene, B.R. Zaidi, (eds.) ACS, Washington DC, pp. 248-282.
128. McKay, G., Blair, H.S. and Gardner J.R. (1982) Adsorption of Dyes on Chitin. I. Equilibrium Studies *J.Appl.Polym.Sci.* **27**, 3043-3057.
129. Lee, Y.M., Kim, S.H., and Kim, S.J. (1996) Preparation and Characteristics of β-Chitin and Poly(vinyl alcohol) Blend, *Polymer*, **37**, 5897 - 5905.
130. Mima, S., Miya, M., and Yoshikawa, S. (1983) Highly Deacetylated Chitosan and Its Properties *J. Appl. Polym. Sci.* **28**, 1909-1917.
131. Kim, J.H., Lee, Y.M., and Kim, K.Y. (1992) Properties and Swelling Characteristics of Crosslinked Poly(vinyl alcohol) Chitosan Blend Membrane *J. Appl. Polym. Sci.* **45**, 1711-1717.
132. Mucha, M., Piekielna, J., Wieczorek, A. (1999) Characterisation and Morphology of Biodegradable Chitosan / Synthetic Polymer Blends, *Macromol Symp.*, **144**, 391-412.
133. Niño, K.A., Imam, S.H., Gordon, S. H., and Wong, L.J.G. (1999) Extruded Plastics Containing Starch and Chitin: Physical Properties and Evaluation of Biodegradability, in *Biopolymers - Utilizing Nature's, Advanced, Materials; Acs Symp. Ser.* **723**, S.H. Imam, R.V. Greene, B.R. Zaidi, (eds.) ACS, Washington DC, pp. 198 - 203.
134. Kurita, K. (1986) Chitin in Nature and Technology, in: *Chitin in Nature and Technology*, R. Muzzarelli, C. Jeuniaux, G.W. Gooday, (eds.), Plenum Press, New York, p.287.

135. Reinhart, C.T. and Peppas, N.A. (1984) Solute Diffusion in Swollen Membranes. Part II. Influence of Crosslinking on Diffusive Properties *J.Membr.Sci.* **18**, 227-239.

136. Chandy, T., and Sharma, C. P. (1990) Chitosan - as a Biomaterial, *Artif. Cells Artif. Org*. **18**, 1-24.

137. Chandy, T. and Sharma, C.P. (1992) Prostaglandin-E1-Immobilized Poly(vinyl alcohol)-Blended
Chitosan Membranes - Blood Compatibility and Permeability Properties, *J. Appl. Polym. Sci.* **44**, 2145-2156.

138. Muzzarelli, R., Baldassarre, V., Conti, F., Ferrara, P., Biagini, G., Gazzanelli, G. and Vasi, V. (1988) Biological Activity of Chitosan: Ultrastructural Study, *Biomaterials*, **9**, 247-252.

139. Muzzarelli R. (1994) In Vivo Biochemical Significance of Chitin-Based Medical Items, in *Polymeric Biomater*ials, D.S.Dekker (ed.), New York, pp.179-197.

140. Zupanets, I.A., Drogovoz, S.M., Yakovleva, L.V., Pavlii, A.I., and Bykova, O.V. (1990) Physiological Importance of Glucosamine, *Fiziol.Zh.*, **36**, 115-120

141. Setnikar, I., Greda, R., Pacini, M. A., and Revel, L. (1991) Antireactive Properties of Glicosamine Sulfate, *Arzneim.-Forsch.* **41**, 157-161.

142. Carlozzi M., and Iezzoni, D.G. (1966) Facilitating Healing of Body Surface Wounds by Intravenous Administration of n-Acetyl Glucosamine, Glucosamine, or Pharmaceutically Acceptable Acid Salts of Glucosamine, U.S. Pat. 3,232,836

143. Okamoto, Y., Shibazaki, K., Minami, S., Matsuhashi, A., Tanioka, S. and Shigemasa, Y. (1995) Evaluation of Chitin and Chitosan on Open Wound-Healing in Dogs, *J.Veterinary Medical Science*, **57**, 851-854.

144. Shigemasa Y., and Minami, S. (1996) Application of Chitin and Chitosan for Biomaterials, in *Biotechnology and Genetic Engineering Reviews*, **Vol.13**, 383-420.

145. Hirano, S. (2001) Fibers Based on Chitin and Chitosan, *Macromol.Symp.*, **168**, 21-30

146. Goheen, D. H., and Hoyt, C. H. (1981) Lignin, in: *Kirk-Othmer Encyclopedia of Polymer Science and Technology*, 3th Ed., **Vol. 14**,. John Wilet & Sons, New York pp. 294-312.

147. Pollak, A. (1952) Lignin, in: *Kirk-Othmer Encyclopedia of Chemical Technology*, 1st Ed., **Vol. 8**, John Wilet & Sons, New York, pp. 327-338.

148. Piccolo, R. S. J., Santos, F., and Frollini, E. (1997) Sugar Cane Bagasse Lignin in Resol - Type Resin: Alternative Application for Lignin-Phenol-Formaldehyde Resins, *J.M.S. Pure Appl. Chem.* **A34**, 153-164.

149. Shirashi, N. (1992) Liquefaction of Lignocellulosics in Organic Solvents and its Applications, in *Emerging Technologies for Materials and Chemicals from Biomass*, R.M. Rowell, T.P.Scultz, R. Narayan (eds.), ACS Symp. Ser. **476**, pp.136-145.

150. Shiraishi, N., and Kishi, N. (1986) Wood-Phenol Adhesives Prepared from Carboxymethylated Wood. I. *J. Appl. Polym. Sci.*, **32**, 3189-3209.

151. Lin, L., Yoshioka, M., Yao, Y. and Shiraiahi N.. (1994) Liquefaction of Wood in the Presence of Phenol Using Phosphoric Acid as a Catalyst and the Flow Properties of the Liquefied Wood, *J. Appl. Polym. Sci.* **52**, 1629-1636

152. Lin, L., Yoshioka, M., Yao, Y. and Shiraishi, N. (1995) Physical Properties of

Moldings from Liquefied Wood Resins, *J. Appl. Polym. Sci.* **55**, 1563 - 1571.

153. Alma, M.H., Yoshioka, M., Yao, Y., and Shiraishi, N. (1996) Phenolation of Wood Using Oxalic Acid as a Catalyst: Effect of Temperature and Hydrochloric Acid Addition, *J. Appl. Polym. Sci.* **61**, 675-683.
154. Northey, R.A. (1992) Low Cost Uses of Lignin, in *Emerging Technologies for Materials and Chemicals from Biomass, ACS Symp.* Ser. **476**, R.M. Rowell, T.P. Scultz, R. Narayan (eds.) ACS, Washington, D.C., pp.146-175.
155. Glasser, W.G., Riasl, T.G., Kelly, S.L., and Ward, T.C. (1989) Engineered Lignin-Containing Material with Multiphase Morphology, *TAPPI Proc. Wood Pulping.Chem.*, pp.35-38.
156. Meister, J.J. and Chen, M.J. (1991) Graft 1-Phenylethylene Copolymers of Lignin. I. Synthesis and Proof of Copolymerization, *Macromolecules*, **24**, 6843-6848.
157. Meister, J.J. (1991) Soluble or Crosslinked Graft Copolymers of Lignin, Acrylamide and Hydroxymethacrylate, U.S. Patent 5,037,931
158. Meister, J. J. and Li, C.T. (1992) Synthesis and Properties of Several Cationic Graft Copolymers of Lignin, *Macromolecules* **25**, 611 - 616
159. Narayan R., Stacy N., Ratcliff M., and Chum H.L. (1989) Engineering Lignopolystirene Materials of Controlled Strucutres, in *Lignin:Properties and Materials, ACS Symp.Ser.* **397**, W.G.Glasser, S.Sarkanen (eds.) ACS, Washington, D.C., pp.475-485.
160. Matuana, L. M., Riedl, B. and Barry, O. (1993). Kinetic Characterization by Differential Enthalpy Analysis of Lignosulfonate Based Phenol-Formaldehyde Resins. *Eur. Polym. J.*, **29**, 483-490.
161. Ash, T., Wu, C.F., Creamer, A.W., and Lora J.H. (1992) Improved Lignin-Based Wood Adhesives, PCT Int. Appl. Wo 92/18557.
162. Peng, W., Riedl, B. and Barry, A.O. (1993) Study on the Kinetics of Lignin Methylolation. *J. Appl. Polym. Sci.*, **48**, 1757-1763.
163. Hall, R. L. (1980) Enzymatic Transformation of Lignin, *Enzyme Microbiol. Technol.* **2**, 170-176.
164. Kirk, T. K., and Farrell, R. L. (1987) Enzymatic Combustion: The Microbial Degradation of Lignin, *Annu. Rev. Microbiol.* **41**, 465-505.
165. Lequart C., Kurek, B., Debeire P., Monites, B. (1998) MnO_2 and Oxalate: An Abiotic Route for the Oxidation of Aromatic Components in Wheat Straw *J.Agr.Fodd.Chem.*, **46**, 3868-3874
166. Kurek, B., Martinez-Inigo, M.J., Artaud I., Hames, B.R., Lequart, C. and Monties, B. (1998)
Structural Features of Lignin Determining its Biodegradation by Oxidative Enzymes and Related Systems, *Polym.Deg.Stab.*, **59**, 359-364.
167. Martinez Inigo, M.J. and Kurek, B. (1997) Oxidative Degradation of Alkali Wheat Straw Lignin by Fungal Lignin Peroxidase, Manganese Peroxidase and Laccase: A Comparative Study *Holzforschung*, **51**, 543-548.
168. Chapman, G. M. (1994) Status of Technology and Applications of Degradable Products in *Polymers from Agricultural Coproducts*, ACS Symp. Ser. **575**, M. L. Fishman, R. B. Friedman and S. J. Huang (eds.), ACS, Washington,D.C., pp. 29 -47.
169. Milstein, O., Gersonde, R., Huttermann, A., Chen, M.J. and Meister, J.J. (1996) Fungal Biodegradation of Lignin Graft Copolymers from Ethene Monomers,

J.M.S. Pure Appl. Chem. **A33**, 685 - 702.

170. Lopez, B.L., Mejia A.I., and Sierra, L. (1999) Biodegradability of Poly(vinyl alcohol) *Polym.Eng.Sci.,* **39**, 1346-1352.
171. Mejia, A., Lopez, B.L., Mulet, A., (1999) Biodegradation of Poly (vinylalcohol) with Enzymatic Extracts of Phanerochaete Chrysosporium, *Mac.Symp.* **148**, 131-147.
172. Chiellini, E., Cinelli P., Imam S.H., and Mao L. (2001) Composite Films Based on Biorelated Agro-Industrial Waste and Poly(vinyl alcohol). Preparation and Mechancial Propertie Characterization, *Biomacromolecules*, **2**, 1029-1037.
173. Chiellini, E., Cinelli P., Imam S.H., and Mao L. (2001) Composite Films Based on Poly(vinyl alcohol) and Lignocellulosic Fibres: Preparation and Characterization, In: *Biorelated Polymers-Sustainable Polymer Science and Technology*, E. Chiellini, H. Gil, G. Braunegg, J. Buchert, P. Gatenholm, and M. Van der Zee (eds.), Kluwer Academic/Plenum Publishers, pp. 87-100.
174. Saraf, V. P., Glasser, W. G., Wilkes, G. L., and McGrath, J. E. (1985). Engineering Plastics from Lignin. VI. Structure-Property Relationships of PEG-Containing Polyurethane Networks, *J. Appl. Polym. Sci.*, **30**, 2207-2224.
175. Yoshida, H., Mörck, R., Kringstadt, K.R. and Hatakeyama, H. (1987) Kraft Lignin in Polyurethanes I. Mechanical Properties of Polyurethanes from a Kraft Lignin-Polyether Triol-Polymeric MDI System, *J. Appl. Polym. Sci.* **34**, 1187-1198.
176. Gandini, A., Belgacem, N.M. (1998) Recent Advances in the Elaboration of Polymeric Materials Derived from Biomass Components, *Polym. Int.* **47**, 267-276.
177. Kosikova, B., Demianova, V., and Kacuracova M. (1993) Sulfur-Free Lignins as Composites of Polypropylene Films, *J. Appl. Polym. Sci.* **47**, 1065-1073.
178. Ghosh, I., Jain, R.K., and Glasser, W.G. (1999) Multiphase Materials with Lignin. XV. Blends of Cellulose Acetate Butyrate with Lignin Esters, *J.Appl.Polym.Sci.* **74**, 448-457.
179. Rials, T.G., and Glasser, W.G. (1989) Multiphase Materials with Lignin. IV, *J. Appl. Polym. Sci.* **37**, 2399 - 2415.
180. Rials, T. G., and Glasser, W.G. (1990) Multiphase Materials with Lignin.5. Effect of Lignin Strucutre on Hydroxypropylcellulose Blend Morphology, *Polymer.*, **31**, 1333-1338.
181. Dave, V., and Glasser, W.G. (1997) Cellulose-Based Fibres from Liquid Crystalline Solution.5. Processing and Morphology of CAB Blends with Lignin, *Polymer*, **38**, 2121-2126.
182. Hüttermann, A., Mai, C., and Kharazipour A. (2001) Modification of Lignin for the Production of New Compounded Materials, *Appl. Microbiol. Biotechnol.* **55**, 387 - 394.
183. Mester, T., and Tien, M. (2000) Oxidation Mechanism of Ligninolytic Enzymes Involved in the Degradation of Environmental Pollutants, *Int.Biodeter.Biodegr.*, **46,** 51-59.
184. Tien, M., and Kirk T.K. (1983) Lignin Degrading Enzyme from Phanaerochaete Chrysosporium, *Science*, **221**, 661-663.
185. Hutterman A., and Haars A. (1987) Biochemical Control of Forest Pathogen Inside the Tree, In: I. Chet (ed) *Innovative Approaches toPlant Disease Control.* John Wiley & Sons, New York, pp.275-296.

186. Kruus K., Niku-Paavola, M.L. and Viikari L.(2001) Laccase-a Useful Enzyme for Modification of Biopolymers, In: *Biorelated Polymers-Sustainable Polymer Science and Technology*, E. Chiellini, H. Gil, G. Braunegg, J. Buchert, P. Gatenholm, and M. Van der Zee (eds.), Kluwer Academic/Plenum Publishers, pp.255-261

187. Chiellini E., Cinelli P., Dantone S.and Ilieva V.I. (2002) Environmentally Degradable Polymeric Materials in Agriculture Applications-An Overview, *Polymery*, **47**, 538-544.

188. Feughelman, M. (2002) Natural Protein Fibers, *J. Appl. Polym. Sci.* **83**, 489-507.

189. Ward, A. G., and Courts, A. (1977) *The Science and Technology of Gelatin*; Academic Press: New York.

190. Veis, A. (1964) *The Macromolecular Chemistry of Gelatin*; Academic Press: New York.

191. Feughelman, M. (2002) Natural Protein Fibers. *J. Appl. Polym. Sci.*, **83**, 489-507.

192. Rose P.I. (1987) Gelatin in *Encyclopedia of Polymer Science and Engineering*, H.F. Mark, N.M. Bikales, C.G. Overberger, and G. Menges (eds.) John Wiley, New York, **Vol.7**, pp.488-513.

193. Yannas, I.V., and Tobolsky A.V. (1968) Stress Relaxation of Anhydrous Gelatin Rubbers, *J. Appl. Polym. Sci.*, **12**, 1-8

194. Yannas I.V. (1972) Collagen and Gelatin in the Solid State, *J. Macomol. Sci. Macromol. Chem.,* **C7**, 49-104

195. Panduranga R.K. (1995) Recent Developments of Collagen-Based Materials for Medical Applications and Drug Delivery Systems. *J. Biomater. Sci. Polym. Ed.*, **7**, 623-645.

196. Cuq B., Gontard N., and Guilbert S.I. (1998) Proteins as Agricultural Polymers for Packaging Production. *Ceral.Chem*, **75**, 1-9.

197. Kester, J. J., and Fennema, O. R. (1986) Edible Films and Coatings : a Review., *Food Technol.* **40**, 47 - 59.

198. De Graaf L.A., Kolster P., and Vereijken, J.M. (1998), Plants Protein from European Crops, Food and non-food applications; Springer Verlag Berlin, Heidelberg, New York, pp.335-339.

199. Arvanitoyannis, I., Psomiadou, E. and Nakayama A. (1996). Edible Films Made from Sodium Caseinate, Starches, Sugars or Glycerol. Part 1. *Carbohydr. Polym.*, **31**, 179-192.

200. Arvanitoyannis, I., Psomiadou, E., Nakayami, A., Aiba, S., and Yamamoto, N. (1997) Edible Films Made from Gelatin, Soluble Starch and Polyols, Part 3., *Food Chem.* **60**, 593-604.

201. Garcia-Rodenas, C.L., Cuq, J.L., and Aymard, C. (1994) Comparison of *in Vitro* Proteolysis of Casein and Gluten as Edible Films or as Untreated Proteins, *Food.Chem.*, **51**, 275-280.

202. Gennadios, A., Weller, C.L., and Testin, R.F. (1993) Modification of Physical and Barrier Properties of Edible Wheat-Gluten Based Films, *Cereal.Chem.*, **70**, 426-429

203. Kester, J.J. and Fennema, O. (1986) Edible Films and Coatings: a Review. *Food.technol.*, **00**, 47-59.

204. McHugh, T.H., Aujard, J.F., and Krochta, J.M. (1994) Plasticized Whey Protein Edible Films: Water Vapor Permeability Properties, *J.Food.Sci.*, **59**, 394-398.

205. Tolstoguzow, V.B. (1994) Some Physiochemical Aspects of Protein Processing in Foods. In *Gums and Stabilizers for the Food Industry*, G.O.Philips, P.A. Williams and D.J. Wedlock (eds.), **Vol.7**, IRL Press, Oxford, pp.115-154.

206. Gennadios, A., and Weller, C. L. (1990) Edible Films and Coatings from Wheat and Corn Proteins., *Food Technol.* **44**, 63 - 69.

207. Baldwin, E. A., Nisperos-Carriedo, M. O., and Baker, R. A. (1995) Use of Edible Coatings to Preserve Quality of Lightly (and Slightly) Processed Products., *Crit. Rev. Food Sci. & Nutr.* **35**, 509-524.

208. Jonston-Banks, F. A. (1990) Gelatin, in *Food Gels*, P. Harris, (ed.), Elsevier Appl. Sci., London, pp. 233-289.

209. Hood, L.L. (1987) Collagen in Sausage Casings., *Adv. Meat. Res.* **4**, 109-129.

210. Torres, J. A. (1994) Edible Films and Coatings from Proteins., Protein Functionality in Food Systems, in N. S. Hettiarachchy and G. R. Ziegler (eds.), *IFT Basic Sympos. Ser.,* Marcel Dekker Inc., New York, pp. 467-507.

211. Slade, L., and Levine, H. (1987). Polymer Chemical Properties of Gelatin In *Advances in Meat Research. Collagen as Food*, A.M. Pearson, T.R. Dutson and A.Q.J. Bailey (eds.), Van Nostrand Reinhold Co, New York, **Vol.4**, pp. 251-266.

212. Lens, J.P., Mulder, W.J., and Kolster, P. (1999) Modification of Wheat Gluten for Non-food Applications, *Cereal Foods World*, **44**, 5-9.

213. Ayhllon-Meixuero F., Tropini V., and Silvestre F. (2001) Fatty Esterification of Plant Proteins, in: *Biorelated Polymers-Sustainable Polymer Science and Technology*, E. Chiellini, H. Gil, G. Braunegg, J. Buchert, P. Gatenholm, and M. Van der Zee (eds.), Kluwer Academic/Plenum Publishers, pp.231-236

214. Brother, G.H., and McKinney, L.L. (1938). Protein Plastics from Soyben Products. Action of Hardening or Tanning Agents on Protein Material. *Ind. Eng. Chem.*, **30**, 1236-1240.

215. Fraenkel-Conrat, H., and Olcott, H.S. (1948) The Reaction of Formaldehyde with Proteins. V. Cross-Linking between Amino and Primary Amide or Guanidyl Groups., *J. Am. Chem. Soc.*, **70**, 2673—2684.

216. Paetau, I., Chen, C.Z., and Jane, J. (1994) Biodegradable Plastic Made from Soybean Products. II. Effects of Cross-Linking and Cellulose Incorporation on MechanicalProperties and Water Absorbtion, *J. Environ. Polym. Degrad.* **2**, 211-217.

217. Marquié, C., Aymard, C., Cuq, J. L., and Guilbert, S. (1995) Biodegradable Packaging Made from Cottonseed Flour: Formation and Improvement by Chemical Treatments with Gossypol, Formaldehyde and Glutaraldehyde. *J. Agric. Food Chem.* **43**, 2762—2767.

218. Tropini, V., Lens, J.P., Mulder, W. J., and Silvestre, F. (2001) Chemical Modification of Wheat Gluten, in *Biorelated Polymers, Sustainable Polymer Science and Technology*, E. Chiellini, H. Gil, G. Braunegg, J. Buchert, P. Gatenholm, and M. Van der Zee (eds.), Kluwer Academic/Plenum Publishers, pp. 237 - 242.

219. Stuchell, Y. M., and Krochta, J. M. (1994) Enzymatic Treatments and Thermal Effects on Edible Soy Protein Films. *J. Food Sci.* **59**, 1332 - 1337.

220. Yildirim, M., and Hettiarachchy, N. S. (1997) Biopolymers Produced by Cross-Linking Soybean 11S Globulin with Whey Proteins Using Transglutaminase., *J. Food Sci.* **62**, 270 - 275.

221. Motoki, M., Aso, H., Seguro, K., and Nio, N. (1987) a s1 Casein Film Prepared

Using Transglutaminase., *Agric. Biol. Chem.* **51**, 993—996.

222. Larré, C., Deshaye, G., Lefebre, J., and Popineau, Y. (1998) Hydrated Gluten Modified by a Transglutaminase., *Nahrung*. **42**, 155—157.

223. Larré C., Desserme, C., Barbot, J., Mangavel, C., and Guéguen J. (2001). Enzymatic Crosslinking Enhance Film Properties of Deamidated Gluten, In *Biorelated Polymers, Sustainable Polymer Science and Technology*, E. Chiellini, H. Gil, G. Braunegg, J. Buchert, P. Gatenholm, and M. Van der Zee (eds.), Kluwer Academic/Plenum Publishers, pp. 243—253.

224. Akin H., and Harisci N., (1995) Preparation and Characterization of Crosslinked Gelatin Microspheres, *J. Appl. Polym. Sci.*, **58**: 95-100

225. Chatterji P. R. (1989) Gelatin with Hydrophilic/Hydrophobic Grafts and Glutaraldehyde Crosslinks, *J. Appl. Polym. Sci.*, **37**, 2203-2212

226. Digenis G.A., Gold T.B., and Shah V.P. (1994) Cross-linking of Gelatin Capsules and its Relevance to their in Vitro-in Vivo Performance, *J. Pharm. Sci.*, **83**: 915-921

227. Fraga A.N., and Williams R.J.J. (1985) Thermal Properties of Gelatin Films, *Polymer*, **26**: 113-118

228. Clark, A.H., and Ross-Murphy, S.B. (1987) Structural and Mechanical Properties of Biopolymer Gels. *Adv. Polym. Sci*, **83**, 57-192.

229. Ross-Murphy, S. B. (1997) Structure and Rheology of Gelatin Gels. *Imag. Sci. J.* **45**, 205-209.

230. Todd, A. (1961) Rigidity Factor of Gelatin Gels. *Nature*, **191**, 567-569.

231. Te Nijenhuis, K. (1981) Investigation into the Ageing Process in Gels of Gelatin/Water Systems by the Measurement of their Dynamic Moduli: Part I. *Colloid Polym. Sci.*, **259**, 522-530.

232. Te Nijenhuis, K. (1981) Investigation into the Ageing Process in Gels of Gelatin/Water Systems by the Measurement of their Dynamic Moduli: Part II. *Colloid Polym.Sci.* **259**, 1017-1026

233. Ross-Murphy, S. B. (1992) Structure and Rheology of Gelatin Gels: Recent Progress. *Polymer*, **33**, 2622-2627.

234. Te Nijenhuis, K. (1997) Thermoreversible Networks: Viscoelastic Properties and Structure of Gels. *Adv. Polym. Sci.* **130**, 160-194.

235. Stanton, J. S., Salik, V., Bentley, G., and Dawnes, S. (1995) The Growth of Chondrocytes Using Gelfoam as a Biodegradable Scaffold. *J. Mater. Sci. Mater. Med.*, **6**, 739-744.

236. Rault, I., Herbage, F.D., Abdul-Malak, N., and Huc, A. (1996) Evaluation of Different Chemical Methods for Cross-linking Collagen Gel, Films and Sponges. *J. Mater. Sci. Mater. Med.*, **7**, 215-221.

237. Jayakrishnan, A., and Jameela, S.R. (1996) Glutaraldehyde as a Fixative in Bioprostheses and Drug Delivery Matrices, *Biomaterials*, **17**, 471-484.

238. Nieuwenhuis, P., and Feijen, J. (1995) Glutaraldehyde as a Cross-linking Agent for Collagen-Based Biomaterials. *J. Mater. Sci. Mater. Med.*, **6**, 460-472.

239. Olde Damink, L.H, Dijkstra, P.J., Van Luyn, M.J. van Wachem, P.B., Nieuwenhuis,P. and Feijen, J. (1995) Cross-linking of Dermal Sheep Collagen Using Hexamethylene Diisocyanate *J. Mater. Sci. Mater. Med.*, **6**, 429-434

240. Ofner, C.M., and Bubnis, W.A. (. 1996) Chemical and Swelling Evaluations of Amino Group Cross-linking in Gelatin and Modified Gelatin Matrices. *Pharm. Res.*, **13**, 1821-1827.

241. Sung, H.W., Hsu, H.L., Shih, C.C. and Lin, D.S. (1996) Cross-linking Characteristics of Biological Tissues Fixed with Monofunctional or Multifunctional Epoxy Compounds. *Biomaterials,* **17**, 1405-1410.

242. Petite, H., Rault, I., Huc, A., Mesnache, P. and Herbage, D.J. (1990) Use of the Acyl Azide Method for Cross-Linking Collagen-Rich Tissue such as Pericardium *Biomed. Mater. Res.*, **24**, 179-188.

243. Speer, D. P., Chvapil, M., Eskelson, C.D., Ulreich, J. (1980) Biological Effect of Residual Glutaraldehyde in Glutaraldehyde-Tanned Collagen Biomaterials. *J. Biomed. Mater. Res.*, **14**, 753-764.

244. Khor, E., Wee, A., Loke, W.K. and Tan, B.L. (1996) Dimethyl Sulfoxide as an Anticalcification Agent for Glutaraldehyde-Fixed Biological Tissue. *J. Mater. Sci. Mater. Med.*, **7**, 691-693.

245. Schacht, E., Nobels, M., Vansteenkiste, S., Demeester, J., Franssen, J. and Lemahieu, A. (1993) Some Aspects of the Cross-Linking of Gelatin by Dextran Dialdhydes. *Polym. Gels Networks*, **1**, 213-224.

246. Bogdanov, B., Schacht, E. and Van Den Bulcke, A. (1997) Thermal and Rheological Properties of Gelatin-Dextran Hydrogels, *J.Therm.Anal.*, **49**, 847-856

247. Schacht, E., Bogdanov, B., Van Den Bulcke, A., and De Rooze, N. (1997) Hydrogels Prepared by Cross-Linking of Gelatin with Dextran Dialdehyde. *React. Funct. Polym.*, **33**, 109-116.

248. Draye, J.P., Delaey, B., Van de Voorde, A., Van Den Bulcke, A., De Reu, B., Schacht, E. (1998) In Vitro and in Vivo Biocompatibility of Dextran Dialdehyde Cross-Linked Gelatin Hydrogel Films. *Biomaterials*, **19**, 1677-1687.

249. Draye, J.P., Delaey, B., Van de Voorde, A., Van Den Bulcke, A., Bogdanov, B., and Schacht, E. (1998) In Vitro Release Characteristics of Bioactive Molecules from Dextran Dialdehyde Cross-linked Gelatin Hydrogel Films, *Biomaterials*, **19**, 99-107.

250. Van Den Bulcke I., Bogdanov B., De Rooze N., Schacht E.H., Cornelissen M. and Berghmans H., (2000) Structural and Rheological Properties of Methacrylamide Modified Gelatin Hydrogels, *Bioacromolecules,* **1**, 31-38.

251. Environmental Bio-Process (M) Sdn. Bhd., GEL-OUT-Gelatin Eradication System. http://www.malaysia-web.com/cyberdir/environ/gelout.htm

252. Kenawy E. R., Cinelli P., Corti A., Miertus S. and Chiellini E. (1999) Biodegradable Composite Films Based on Waste Gelatin, *Macromol. Symp.*, **144**, 351-364

253. Cinelli P., 1999, *Formulation and Characterization of Envrionmentally Compatible Polymeric materials for Agriculture Applications*. PhD Thesis, University of Pisa.

254. Chiellini E., Cinelli P., Corti A., Kenawy E.R., Grillo F. E., Solaro R. (2000) Environmentally Sound Blends and Composites Based on Water-Soluble Polymer Matrices, *Macromol. Symp*. **152**: 83-94.

255. Chiellini E., Cinelli P., Grillo Fernandes E., Kenawy E.R. and Lazzeri A. (2001) Composite Materials Based on Gelatin and Fillers from Renewable Resources: Thermal and Mechanical Properties, in *Biorelated Polymers, Sustainable Polymer Science and Technology*, E. Chiellini, H. Gil, G. Braunegg, J. Buchert, P. Gatenholm, and M. Van der Zee (eds.), Kluwer Academic/Plenum Publishers, pp. 101-112.

256. Chiellini E., Cinelli P., Corti A., and Kenawy E.L. (2001) Composite Films Based on Waste Gelatin: Themal-Mechanical Properties and Biodegradation Testing, *Polym.Degrad Stab.* **73**, 549-555.

257. Chiellini E., Cinelli P., Grillo Fernandes E., Kenawy E.R. and Lazzeri (2001) Gelatin-Based Blends and Composites. Morphological and Thermal Mechanical Characterization, *Biomacromolecules*, **2**, 806-811.

258. Cabeza L.F., Taylor M.M., DiMaio G.L., Brown E.M., Mermer W.N., Carrio R., Celma P.J. and Cot J. (1998) Processing of Leather Waste: Pilot Scale Studies on Chrome Shavings. Isolation of Potentially Valuable Protein Products and Chromium, *Waste Management* **18**, 211-218.

259. Taylor M.M., Diefendorf E.J., Thompson C.J., Brown E.M, and Marmer W.N. (1994) Extraction and Characterization of "Chrome Free" Protein from Chromium-Containing Collagenous Waste Generated in the Leather Industry, In: *Polymer from Agricultural Coproducts*, M.L.Fishman, R.B. Friedman, S.J.Huang (eds.) *ACS Symp.Ser.* **575**, pp.171-187.

260. Kolomaznik, K, Kupec, J.and Taylor, M. (1997) *CSCE/ASCE Environmental Engineering Conference,* Edmonton, Alberta, Canada 22-26, July 1997.

261. Holloway, D. F. (1978) Process for Recovery and Separation of Nutritious Protein Hydrolysate and Chromium from Chrome Leather Scrap. U.S. Patent 4,100,154.

262. Guardini, G. (1984). Process for Recovering Proteins and Chromium from Chrome-Tanning Waste. U.S. Patent 4,483,829.

263. Taylor, M. M., Diefendorf, E. J., Na, G. C., and Marmer, W. N. (1992) Enzymatic Processing of Materials Containing Chromium and Protein. U.S. Patent 5,094,946.

264. Laszlo, J.A., and Dintzisi, F.R. (1994) Crop Residues as Ion-Exchange Materials-Treatment of Soybean Hull and Sugar-Beet Fiber (Pulp) with Epichlorohydrin to Improve Cation-Exchange Capacity and Physical Stability, *J. Appl. Polym. Sci.* **52**, 531-538.

265. Kubota, H., and Ogiwara, Y. (1969) Effect of Lignin in Graft Copolymerization of Methyl Methacrylate on Cellulose by Ceric Ion, *J. Appl. Polym. Sci.*, **13**, 1569-1575.

266. Hornof, V., Kokta, B.V., and Valade, J.L. (1975) The Xanthate Method of Grafting. III. Effect of Lignin Content on the Graftability of Wood Pulp, *J. Appl. Polym. Sci.*, **19**, 1573-1584.

267. Huang, Y.F., Zhao, B.A., He, S.J., and Gao, J. (1992) Graft-Copolymerization of Methyl-Methacrylate on Stone Ground Wood Using the H_2O_2-Fe^{2+} Method, *J. Appl. Polym. Sci.* **45**, 71-77

268. Nagaty, A., Mustafa, A.B., and Mansour, O.F. (1979) Lignocellulose-Polymer Composite, I. *J. Appl. Polym. Sci.*, **23**, 3263-3269.

269. Zheng, G.Z., Zhao, B.A., He, S.J., and Gao, J. (1995) Initiation of Graft Copolymerization by Direct Oxidation of Lignocellulose with KMnO4 and its Mechanism, *J.M.S. Pure Appl. Chem.* 1281-1292.

270. Bhunia, H.P., Nando, G.B., Chaki, T.K., Basak, A., Lenka, S., and Nayak. P.L. (1999) Synthesis and Characterization of Polyners from Cashewnut Shell Liquid (CNSL), a Renewable Resource II. Synthesis of Polyurethanes, *Eur. Polym. J.*, **35**, 1381-1391.

271. Pillai, C.K.S. (2000) Polymeric Materials from Renewable Resources: High Value Polymers from Cashewnut Shell Liquid, *Pop. Plast. Packag.*, (Spec.issue), 79-84, 86-90.

272. Ikeda R., Tanaka, H., Uyama, H, and Kobayashi, S. (2000) A New Crosslinkable Polyphenol from a Renewable Resource, *Macromol. Rapid Commun.* **21**, 496-499.

273. Mahanwar P.A., and Kale D.D. (1996) Effect of Cashew Nut Shell Liquid (CNSL) on Properties of Phenolic Resins *J.Appl.Polym.Sci.*, **61**, 2107-2111.

274. Kopf P.W. (1986) Phenolic Resins, In *Encyclopedia of Polymer Science and Engineering*, 2nd Ed., Wiley, New York, **Vol.11**, p.45-95.

275. Akkara, J.A., Senecal, K.J., and Kaplan, D.L. (1991). Synthesis and Characterization of Polymers Produced by Horseradish-Peroxidase in Dioxane, *J. Polym. Sci.: A: Polym. Chem.*, **29**, 1561—1574.

276. Uyama, H., Kurioka H., Kaneko, I., and Kobayashi, S. (1994) Synthesis of a New Family of Phenol Resin by Enzymatic Oxidative Polymerization, *Chem. Lett.*, **3**, 423—426.

277. Wang, P., Martin, B.D., Parida, S., Rethwisch, D.G., and Dordick, J.S. (1995) Multienzymic Synthesis of Poly(hydroquinone) for Use as a Redox Polymer, *J. Am. Chem. Soc.*, **117**, 12885—12886.

278. Ayyagari, M., Akkara, J. A., and Kaplan, D. L. (1996). Enzyme-Mediated Polymerization Reactions: Peroxidase-Catalyzed Polyphenol Synthesis, *Acta Polym.*, **47**, 193—203.

279. Uyama, H., Lohavisavapanich, C., Ikeda, R., Kobayashi, S. (1998) Chemoselective Polymerization of a Phenol Derivative Having a Methacryl Group by Peroxidase Catalyst, *Macromol.*, **31**, 554—556.

280. Tonami, H., Uyama, H., Kobayashi, S., Higashimura, H., and Oguchi, T. (1999) Oxidative Polymerization of 2,6-disubstituted Phenols Catalyzed by Iron-Salen Complex. *J. Macrom. Sci.-Pure Appl. Chem.*, **A36**, 719—730.

281. Bhunia, H.P., Nando, G.B., Basak, A., Lenka, S., and Nayak, P.L. (1999) Synthesis and Caracterization of Polymers from Cashewnut Shell Liquid (CNSL), a Renewable Resource III. Synthesis of a Polyether, *Eur. Polym. J.* **35**, 1713-1722.

282. Bhunia, H. P., Basak, A., Chaki, T. K., and Nando, G. B. (2000) Synthesis and Characterization of Polymers from Cashewnut Shell Liquid: a Renewable Resource V. Synthesis of Copolyester, *Eur. Polym. J.* **36**, 1157-1165

283. John, G., Masuda, M., Okada. Y., Yase, K., and Toshimi, S. (2001) Nanotube Formation from Renewable Resources via Coiled Nanofibers, *Adv. Mater.* **13**, 715-718

284. Guru, B.N., Das, T.K.and Lenka, S. (1999) Polymers from Renewable Resources. XXVII. Studies on Synthesis, Characterization, and Thermal Properties of Resins Derived from Cardanyl Acrylate-Furfural-Organic Compounds, *Polym.-Plast. Technol. Eng.*, **38**, 179-187

285. Das, D., Nayak, P.L., and Lenka, S. (1998) Polymers from Renewable Resources. XXV. Interpenetrating Polymer Networks Derived from Castor Oil-Isophorone Diisocyanate-Cardanyl Acrylate/Cardanyl Methacrylate: Thermal and XRD Studies, *Polym.-Plast. Technol. Eng.*, **37**, 419-426.

286. Stevens E.S. (2002) Biopolymers, in *Green Plastics*, Princenton University

Press, Princenton, New Jersey, UK, pp.83-103.

287. Metzger, J. O. (2001) Organic Reactions without Organic Solvents and Oils and Fats as Renewable Raw Materials for the Chemical Industry, *Chemosphere*, **43**, 83—87.

288. Biermann, U., Metzger, J.O., (1999) Friedel-Crafts Alkylation of Alkenes: Ethylaluminum Sesquichloride Induced Alkylations with Alkyl Chloroformates, *Angew. Chem. Int. Ed.* **38**, 3675-3677.

289. Das, T.K., Das, D., Guru, B.N., Das, K.N., and Lenka, S. (1998) Polymers from Renewable Resources. XXVIII. Synthesis, Characterization, and Thermal Studies of Semi-Interpenetrating Polymer Networks Derived from Castor-Oil-Based Polyurethanes and Cardanol Derivatives, *Polym.-Plast. Technol. Eng.*, **37**, 427-435

290. Das, S.K., and Lenka, S. (1999) Polymers from Renewable Resources. XXIX. Synthesis and Characterization Of Interpenetrating Networks Derived From Castor-Oil-Based Polyurethane-4-Acetyl Phenyl Methacrylate, *Polym.-Plast. Technol. Eng.*, **38**, 149-157.

291. Li, F., Marks, D.W., Larock, R.C., and Otaigbe J.U. (2000) Fish Oil Thermosetting Polymers: Synthesis, Structure, Properties and Their Relationships, *Polymer*, **41**, 7925-7939

292. Padavich, R.A. and Honary, L. (1995) A Market Research and Analysis Report on Vegetable-Based Industrial Lubricants, *Soc. Automotive Eng. Tech. Pap.* 952077.

293. Battersby, N.S., Pack, S.E., and Watkinson, R.J. (1992) A Correlation Between the Biodegradability of Oil Products In the CEC L-33-T-82 and Modified Sturm Tests, *Chemosphere*, **24**, 1989–2000.

294. Battersby, N.S., Ciccognani, D., Evans, M.R., King, D., Painter, H.A., Peterson, D.R and Starkey, M. (1999) An Inherent Biodegradability Test for Oil Products: Description and Results of an International Ring Test, *Chemosphere*, **38**, 3219-3235.

295. Randles, S.J., and Wright, M. (1992) Environmentally Considerate Ester Lubricants for Automotive and Engineering Industries, *J. Syn. Lub.* **9**, 145–161.

296. Erhan S.Z., and Asadauskas S. (2000) Lubricant Basestock from Vegetable Oils *Ind.Crop.Prod.* **11**, 277-282

297. Willing, A. (1999) Oleochemical Esters Environmentally Compatible Raw Materials For Oils and Lubricants From Renewable Resources, *Fett/Lipid*, **101**, 192-198.

298. Willing, A. (2001) Lubricants Based on Renewable Resources- an Environmentally Alternative to Mineral Oil Products, *Chemosphere* **43**, 89-98.

299. Lim, S.W., Jung, I.K., Lee, K.H., and Jin, B.S. (1999) Structure and Properties of Biodegradable Gluten/Aliphatic Polyester Blends, *Eur.Polym.J,*. **35** , 1875-1881.

300. Ratajska M., and Boryniec, S. (1999) Biodegradation of Some Natural Polymers in Blends with Polyolefines, *Polym. Adv. Tech.*, **10**, 625-633.

301. Kricheldorf, H. R. (2001) Syntheses and Application of Polylactides, *Chemosphere*, **43**, 49-54.

302. Lunt, J. (1998) Large-Scale Production, Properties and Commercial Applications of Polylactic Acid Polymers, *Polym. Degrad. Stab.*, **59**, 145-152.

303. Lowe C.E. (1954) Preparation of High Molecular Weight Polyhydroxy acetic

ester, U.S. Pat.ent 2,668,162.

304. Enomoto, K., Ajioka, M., and Yamaguchi, A. (1994) Polyhydroxy carboxylic Acid and Preparation Process Thereofore, U.S. Patent 5,310,865

305. Kolstad, J.J., Hall, E., Iwen M.L., Benson, R.D., Borchardt, R.L., and Gruber P.R. (1992). Continuous Process for Manufacture of Lactide Polymers with Controlled Optical Purity, U.S. Patent 5,142,023.

306. Spassky, N., Simic, V., Hubert-Pfalzgraf, G., and Montaudo, M. S. (1999). Synthesis of Aliphatic Polyesters by Controlled Ring-Opening Polymerization of Cyclic Esters. Characterization, Properties, Transesterification Reactions, *Macrol.Symp.*, **144**, 257-268

307. Kricheldorf, H. R., Kreiser-Saunders, I., and Damrau, D. O. (1999). Resorbable Initiators for Polymerizations of Lactones, *Macrol.Symp.*, **144**, 269—276.

308. Kim, S.H., and Kim, Y.H. (1999) Direct Condensation Polymerization of Lactic Acid, *Macromol. Symp.*, **144**, 277—287.

309. Degée, P., Dubois, PH., Jérôme, R., Jacobsen, S., and Fritzh, G. (1999). New Catalysis for Fast Bulk Ring-Opening Polymerization of Lactide Monomers, *Macromol. Symp.*, **144,** 289—302.

310. Ovchinnikova, T., Zhuravleva, I., Bush, L., and Il'in, N. (1999). Kinetics and Mechanism of Glycolide and Ethylenoxalate Copolymerization. Characteristics of the Copolymers Formed and Mechanism if the Biodegradation, *Macromol. Symp.,* **144**, 303—311.

311 Ph. Dubois and Ph. Degée (eds.), (2000) Advances in Ring Opening (Methathesis) Polymerization, *Macromol.Symp*. **153**,1-342 pp.

312. Jacobsen, S, Degée, Ph., Fritz, H.G., Dubois, Ph. and Jérome R. (1999) Polylactide (PLA)-A New Way of Production, *Polym.Eng.Sci.*, **39**, 1311-1319.

313. Jérôme, R., Degée, Ph., Dubois, Ph., Jacobsen, S., and Fritzh, G. (1998Aliphatischer Polyester und/oder Copolyester und Verfahren zu seiner Herstellung, DE19628472.

314. Gogolewski, S., Jovanovic, M., Perren, S. M., Dillon, J.G., and Hughes, M. K. (1993). The Effect of Melt-Processing on the Degradation of Selected Polyhydroxyacids - Polylactides, Polyhydroxybutyrate, and Polyhydroxybutyrate-co-Valerates, *Polym. Degrad. Stab.*, **40**, 313—322.

315. Södergard, A., and Näsman, J. H. (1994) Stabilization of Poly(L-Lactide) in the Melt, *Polym. Degrad. Stab.*, **46**, 25—30.

316. Degée, Ph., Dubois, Ph., Jacobsen, S., Fritz, H.G., and Jérôme, R. (1999) Beneficial Effect of Triphenylphosphine on the Bulk Polymerization of L,L-Lactide Promoted by 2-Ethylhexanoic Acid Tin (II) Salt, *J. Polym. Sci.: A: Polym. Chem.*, **37**, 2413—2420.

317. Kricheldorf, H. R., Berl, M., and Scharnagl, N. (1988). Poly(lactones). 9. Polymerization Mechanism of Metal Alkoxide Initiated Polymerizations of Lactide and Various Lactones, *Macromolecules*, **21**, 286—293.

318. Löfgren, A., Albertsson, A. C., Dubois, P., and Jérôme, R. (1995) Recent Advances in Ring-Opening Polymerization of Lactones and Related-Compounds, *J. Macromol. Sci.-Rev. Macromol. Chem. Phys.*, **C35**, 379—418.

319. Jacobsen, S., Fritz, H.G., Degée, Ph., Dubois, Ph., and Jérôme, R. (2000) Continuous Reactive Extrusion Polymerisation of L-Lactide - An Engineering View, *Macromol. Symp.*, **153**, 261-273.

320. Jacobsen, S., Fritz, H.G., Degée, Ph., Dubois, Ph., and Jérôme, R. (2000) New

Developments on the Ring Opening Polymerisation of Polylactide, *Ind. Crops Prod.*, **11**, 265-275.

321. Jacobsen, S., and Fritz, Hans-Gerhard (1999) Plasticizing Polylactide-The Effect of Different Plasticizers on the Mechanical Properties, *Polym. Eng. Sci.*, **39**, 1303-1310.
322. Narayan, R. (1992) Biomass (Renewable) Resources for Production of Materials, Chemicals, and Fuels/ A. Paradigm Shift, in: *Emerging Technologies for Materials and Chemicals from Biomass, ACS Symp. Ser.* **476**, in R.M. Rowell, T.P. Schultz, and R. Narayan (eds.), ACS, Washington, DC, pp.1-10.
323. Wang, L., Ma, R., Gross, R.A., and McCarthy S.P. (1998) Reactive Compatibilization of Biodegradable Blends of Poly(lactic acid) and Poly(ε-caprolactone), *Polym.Deg.Stab.*, **59**, 161-168.
324. Vert, M., and LI, S.M. (1992) Bioresorbability and Biocompatibility of Aliphatic Polyesters, *J. Mater. Sci. Mater. Med.*, **3**, 432-446
325. Freed, L.E., Grande D.A., Lingbin, Z., Emmanual, J., Marquis, J.C., and Langer, R. (1994) Joint Resurfacing Using Allograft Chondrocytes and Synthetic Biodegradable Polymer Scaffolds, *J. Biomed. Mater. Res*, **28**, 891-899.
326. Freed, L.E., Vunjak-Novakovic, G., Biron, RJ., Eagles, DB., Lesnoy, DC., Barlow, SK., and Langer, R. (1994) Biodegradable Polymeric Scaffolds for Tissue Engineering, *BioTechnology*, **12**, 689-693
327. Helmus, M.N., and Hubbell, J.A. (1993) Materials Selection, *Cardiov. Pathol.* **2**, 53S-71S.
328. Hubbell, J.A. (1995) Biomaterials in Tissue Engineering, *BioTechnology*, **13**, 565-576
329. Cima, L.G., Ingber, D.E., Vacanti, J.P., and Langer, R. (1991) Hepatocyte Culture on Biodegradable Polymeric Substrates, *Biotech. Bioeng*, **38**, 145-158
330. Sawhney, A.S., Pathak, C.P., Hubbell, J.A., (1993) Bioeridible Hydrogels Based on Photopolymerized Poly(ethyleneglycol)-co-poly(a-hydroxy acid) Diacrylate Macromers, *Macromolecules*, **26**, 581-587.
331. Shalaby, S.W., and Johnson, R.A. (1994) Synthetic Absorbable Polyesters, *Biomedical Polymers*, 2-34.
332. Frazza, E.J., Schmitt E.E. (1971) A New Sorbable Suture, *J.Biomed.Mater.Res. Symp.*, **1**, 43-58
333. Reed, A. M., and Gilding, D. K. (1981). Biodegradable Polymers for Use in Surgery -Poly(glycolic)/Poly(lactic acid) Homo and Copolymers: 2. In vitro Degradation, *Polymer*, **22**, 494-498.
334. Gilding, D.K., and Reed, A.M. (1979) Biodegradable Polymers for Use in Surgery-Polyglycolic/Poly(lactic acid) Homo-and Copolymers: 1, *Polymer*, **20**, 1459-1464.
335. Wasserman, D. and Versfeit, C.C. (1975) Use of Stannous Octoate Catalyst in the Manufacture of L(-) Lactide-Glycolide Copolymer Sutures, U.S. Patent 3, 839, 297.
336. Miller, R.A., Brady, J.M., Curtright, D.E., (1978) Degradation Rates of Oral Resorbable Implants: Rate Modification with Changes in PLA/PGA Copolymers Ratios, *J. Biomed. Mater. Res.*, **11**, 711-719
337. Allemann, E., Doelker, E. and Gurny, R. (1993) New Approach for the Preparation of Nanoparticles By an Emulsification-Diffusion Method, *Eur. J. Pharm. Biopharm.*, **41**, 14-18

338. Allemann, E., Doelker, E. and Gurny, R. (1993) Drug-Loaded Poly(Lactic Acid) Nanoparticles Produced by a Reversible Salting Out Process-Purification of an Injectable Dosage Form, *Eur. J. Pharm. Biopharm,* **39**, 13-18.

339. Fishbein, I., Chorny, M., Rabinovich, L., Banai, S., Gati, I. and Golomb, G. (2000) Nanoparticulate Delivery System of a Tyrphostin For the Treatment of Restenosis, *J. Contr. Rel.*, **65**, 221-229.

340. Leenslag, J.W., Pennings, A.J., Bos, Rund R.M., Rozema, F.R., and Boering. G. (1987) Resorbable Materials of Poly(L-lactide). VI. Plates and Screws for Internal Fracture Fixation, *Biomaterials*, **8**, 70—73.

341. Vainionpää, S., Kilpikari, J., Laiho, J., Helevirta, J., Rokkanen, P., and Törmälä, P. (1987) Strength and Strength Retention in Vitro, of Absorbable, Self-Reinforced Polyglycolide (PGA) Rods for Fracture Fixation, *Biomaterials*, **8**, 45—48.

342. Hay, D. L., von Fraunhofer, J. A., Chegini, N., and Masterson, B. J. (1988) Locking Mechanism Strength of Absorbable Ligating Devices, *J. Biomed. Mater. Res.*, **22**, 179—190.

343. Lewis, DH., *in Biodegradable Polymers as Drug Delivery Systems.* (1990) Chasin M., Langer R. (eds.) Marcel Dekker, New York, NY, pp. 1-41

344. Vila, A., Sanchez, A., Tobio, M., Calvo, P., and Alonso, MJ. (2002) Design of Biodegradable Particles from Protein, *J. Contr. Rel.*, **78**, 15-24.

345. http://agproducts.unl.edu/pla.htm

346. Bogaert, J.C. and Coszach, Ph. (2000) Poly(lactic acid): A Potential Solution to Plastic Waste Dilemma, *Macromol.Symp.*, **153**, 287-303.

347. http://www.cargilldow.com

348. http://www.cdpoly.com/emerge.asp

349. http://itri.loyola.edu/biopoly/mitsui.htm

350. http://www.mitsuichemicals.com

351. http://www.plasticstechnology.com/articles/200201bib1.html

352. McLaren, J. S. (2001). The Vision for Renewable Resources, in *Chemicals and Materials from Renewable Resources*, *ACS Symp. Ser.* **784**, in J.J. Bozell (ed.), ACS, Washington, DC, pp.24—36.

353. McLaren, J.S. (2000). Future Renewable Resource Needs: Will Genomics Help?, *J. Chem. Technol. Biotechnol.*, **75**, 927—932.

354. McLaren, J.S. (1998). The Success of Transgenic Crops in The USA, *Pesticide Outlook*, **Dec.**, 36—41.

355. National Academy of Sciences (1999) Biobased Industrial Products: Priorities for Research and Commercialization. National Academy Press, Washington, DC.

356. Clark, M.S. (1999). Comparative Genomics: the Key to Understanding the Human Genome, *Project. Bioessays*, **21**, 121—130.

357. National Science and Technology Council (1999) National Plant Genome Initiative. National Plant Genome Initiative: Progress Report, Office of Science and Technology Policy, Washington, DC.

358. Bruce, W., Folkerts, O., Garnaat, C., Crasta, O., Roth, B., and Bowen, B. (2000). Expression Profiling of the Maize Flavonoid Pathwey Genes Controlled by Estradiol-Inducible Transcription Factors CRC and P, *The Plant Cell*, **12**, 65—79.

359. Della Penna, D. (1999) Nutritional Genomics: Manipulating Plant

Micronutrients to Improve Human Health. *Science*, **285,** 375—379.

360. Gulati, M., Kohlmann, K., Ladisch, M. R., Hespell, R., and Bothast, R.J. (1996). Assessment of Ethanol Production Options for Corn products, *Bioresource Technol.*, **58**, 253—264.
361. Bozell, J.J., and Landucci, R. (1993) Alternative Feedstocks Program: Technical and Ecomomic Assessment, Thermal/Chimical and Bioprocessing Components., U.S. Department of Energy, Washington, DC.
362. Wilke, D. (1999). Chemicals from Biotechnology: Molecular plant Genetics will Challenge the Chemicaland Fermentation Industry, *Appl.Microbiol. Biotechnol.*, **52**, 135—145.
363. Tirrell, D. (1996) Putting a New Spin on Spider Silk, *Science*, **271**, 39—40.
364. Hayashi, C.Y., and Lewis, R.V. (2000). Molecular Architecture and Evolution of a Modular Spider Silk Protein Gene, *Science*, **287**, 1477—1479.
365. Smith, B.L. (1998). Studying Shells: a Grouwth Industry. *Chemistry & Industry*, **16**, 649—653.
366. http://www.bedps.org/july2000.html
367. Chandra, R., and Rustigi, R. (1998) Biodegradable Polymers, *Prog.Polym.Sci.*, **23**, 1273-1335.
368. Agricultural Material as Renewable Resources, Nonfood and Industrial Applications, Fuller G., McKeon T.A. Bills D.D.Eds., *ACS Symp Ser.*, 1999, 280 pp.
369. Kalia, V.C., and Raizada, N., and Sonakya, V. (2000) Bioplastics, *J. Sci. Ind. Res.*, **59**, 433-445 .
370. Mishra, D. P. and Mahanwar, P. A (2000) Advances in Bioplastic Materials, *Pop.Plast. Packag.*, **45**, 68-76.
371. Albertsson, A.C. (2000) Biodegradation of Polymers in Historical Perspective Versus Modern Polymer Chemistry, *Environ.Sci. Pollut. Control. Ser.*, **21**, 421-439.
372. Hokens, D.; Mohanty, A. K.; Misra, M.; Drzal, L. T. (2001) Environment-Friendly "Green" Biodegradable Composites from Natural Fiber andCellulosic Plastic, *Polym. Prepr*, **42**, 71-72.
373. Sun, X.S. (2001) Novel Materials from Agroproteins: Current and Potential Applications of Soy Protein Polymers, *ACS Symp. Ser.*, **786** (*Biopolymers from Polysaccharides and Agroproteins*) 132-148.
374. Gerngross, T.U., Slater, S.C. (2000) How Green Are Green Plastics?, *Scientific American*, **Aug.**, 37-41.
375. http://www.dupont.com, http://www.plastics.dupont.com
376. http://www.basf.com
377. http://www.monsanto.com

378 http://www.midwestgrain.com

8

SUSTAINABLE POLY(HYDROXYALKANOATE) (PHA) PRODUCTION

G. BRAUNEGG
Institute of Biotechnology
Graz University of Technology
Petersgasse 12
A-8010 Graz
Austria

1. Historical review

Polyhydroxyalkanoates (PHAs) are homo- or heteropolyesters synthesized and intracellularly stored by numerous prokaryotes. They can be produced in large quantities from renewable resources by means of well known fermentation processes and the imposition of particular culture conditions, and a number of physical or chemical methods are known to extract them from the producing biomass. Production processes such as batch, semi-batch and continuous fermentation are all known to work. PHAs have properties similar to those of some polyolefins. This, combined with the fact that they are fully and rapidly biodegraded under the appropriate conditions, has generated a high interest in them as substitutes to petroleum-based polymers in many applications [1].

PHAs are stored in the form of granules by bacteria. The observation of the granules as refractile bodies in bacterial cells under the microscope goes back at least to Beijerinck in 1888 [reported in 2]. The first determination of the composition of a PHA had to wait until 1927 and the work of Lemoigne [3]. In a soil bacillus resembling *Bacillus megaterium*, the anaerobic degradation of an unknown material led to the excretion of 3-hydroxybutyric acid. Lemoigne identified this material as the homopolyester of the hydroxyacid, 3-hydroxybutyrate, or poly-3-hydroxybutyrate (PHB), and described it tentatively as a reserve material. During the following thirty years, interest in PHB was scant and nearly restricted to the description of detection and cell-content estimation methods and to culture conditions that lead to its synthesis and degradation inside *Bacillus* cells [cited in 4]. A convincing proposal for a functional role for PHB first came from Macrae and Wilkinson in 1958, who observed that *B. megaterium* stored the homopolymer especially swiftly when the glucose-to-nitrogen source ratio of the medium was high [5], and that subsequent degradation by *Bacillus*

G. Scott (ed.), Degradable Polymers, 2nd Edition, 235-293.

cereus and *B. megaterium* occurred rapidly in the absence of an exogenous carbon and energy source [4]. The authors concluded that PHB was a carbon- and energy-reserve material that slowed down cell autolysis and death, and correctly speculated on the involvement of acetate and coenzyme A complexes in the pathway of PHB synthesis [4].

These papers mark the time when interest in PHB began to increase dramatically. The following fifteen years saw intense research on the subject and featured important developments in knowledge about the polymer's widespread occurence in micro-organisms other than *Bacillus*, including the bacterial genera *Pseudomonas* [6,7], *Azotobacter* [8], *Hydrogenomonas* [9], *Chromatium* [10], a cyanobacterium [11], and many others [reviewed in 12]; PHB physical and chemical properties, including molecular mass [13], in particular that following different methods of extraction [14], and melting point [10, 15], continuing Lemoigne's original observations [3]; PHB crystalline structure [14, 16, 17]; granule morphology and properties [18-23]; new methods of PHB detection or quantification [24-35]; PHB metabolism and its regulation [12, 27-33], including investigations in continuous culture [17, 31, 34, 35]; enzymology of PHB biosynthesis [12, 20, 28, 29, 35, 36, 37]; PHB intra- and extracellular degradation, and its enzymology [2, 6, 20, 35–41]; and physiological function of PHB, including confirmation of its survival value for some of the micro-organisms that store it [9, 12, 27, 42]. Of particular importance was the recognition of the relationship between PHB biosynthesis and extracellular environment. Conditions that favor PHB accumulation were advanced as those which lead to high NAD(P)H, high acetyl-CoA and low free coenzyme A intracellular concentrations [12]. These conditions varied according to the micro-organism, but all involved a growth-limiting factor, such as a deficiency in nitrogen [9], potassium or sulfur [17], or oxygen [31]. Table 1 lists up the genera of procaryots known for accumulation of PHAs.

The field of PHAs was thus well developed by the end of 1973, but interest in the biopolymers remained directed almost solely at their physiological significance as microbiological substances. Little consideration was given to their possible use by humans as commodity materials. Their potential usefulness in this area had already been recognized in the first half of the 1960's, as patents related to PHB production by fermentation [43]; extraction from the producing biomass [44]; plastisation with additives [45]; and use unextracted as a polymer mixed with other cell material [46] and pure for absorbable prosthetic devices [47] were filed. However, petroleum-based plastics could be manufactured with more ease and at much lower costs, and there was no easily identifiable reason to think that fossil fuels would not remain low priced in the foreseeable future. This gave little incentive to push for the development of a natural plastics industry, especially in the absence of a strong public concern for environmental issues.

The 35th conference of the Organization of Petroleum Exporting Countries (OPEC) in Vienna in September-October 1973 started a chain of events that would change the situation over the following decade. The members of the OPEC decided on an immediate rise in oil prices by 70 percent [48]. This increase was followed by another one at the Tehran conference in December of the same year, of 130 percent this time, so that the price of a barrel (159 liters) of crude oil, which was US $3.00 in

September 1973, stood at US $11.70 at the beginning of 1974. Other increases were imposed in 1975, 1977, 1979 and 1980, which brought the price of oil to US $30.00 a barrel by 1980. The influence of the OPEC on the world oil market started to wane in the early 1980's - and no-one these days talks of an 'oil crisis', the average price of one benchmark crude oil in 1996 having been about US $20.29 a barrel, approximately the same price, after correcting for inflation, as that before the oil crisis of 1973 [49]. But OPEC policies during the seventies led to fears that the price of fossil-fuels could no longer be regarded as reliable. This cast doubts on the future of the petroleum-based-polymer industry and set the stage for the search for alternate types of plastic materials.

Table 1: PHA Accumulating Genera of Prokaryotic Microorganisms

Acinetobacter	Escherichia [a]	Paracoccus
Actinomycetes	Ferrobacillus	Pedomicrobium
Alcaligenes [a,b]	Gamphospheria	Photobacterium
Aphanothece [a]	Haemophilus	Pseudomonas [a,b]
Aquaspirillum	Halobacterium [a]	Rhizobium [a,b]
Asticcaulus	Hyphomicrobium	Rhodobacter
Azomonas	Lamprocystis	Rhodococcus [b]
Azospirillum	Lampropedia	Rhodopseudomonas
Azotobacter[a]	Leptothrix	Rhodospirillum [b]
Bacillus [a,b]	Methanomonas	Sphaerotilus [a]
Beggiatoa	Methylobacterium [b]	Spirillum
Beijerinckia [b]	Methylocystis	Spirulina
Beneckea	Methylomonas	Stella
Caryophanon	Methylovibrio	Streptomyces
Caulobacter	Micrococcus	Syntrophomonas
Chlorofrexeus	Microcoleus	Thiobacillus
Chlorogloea	Microcystis	Thiocystis
Chromatium	Moraxella	Thiodictyon
Chromobacterium	Mycoplana [a]	Thiopedia
Clostridium	Nitrobacter	Thiosphaera
Corynebacterium [b]	Nitrococcus	Vibrio
Derxia [b]	Nocardia [a,b]	Xanthobacter
Ectothiorhodospira	Oceanospirillum	Zoogloea [a]
a Detailed knowledge about growth and production kinetics available		
b Accumulation of Copolyesters known		

In 1976, Imperial Chemical Industries (ICI) of England started investigating whether PHB could be profitably produced by bacterial fermentation from photosynthesis-derived carbohydrate feedstocks [50]. Not only could PHB be synthesized from renewable resources, but some of its properties resembled those of polypropylene [51]. Other PHB features, like biodegradability and biocompatibility, piezoelectric properties, and the possibility of using it as a source of optically active molecules [52], were recognized early on as additional assets and kept ICI's interest in

PHB alive after the oil crisis had begun to pass. The bending piezoelectric effect was observed in oriented films of PHB. The coefficient between the electrical polarization and the stress gradient was found to be in the order of 10(-18) Cm/N, which was similar to the value reported for bone. Anisotropy in the value of the coefficient was also observed [53].

Pure PHB is brittle and has a low extension to break [54]. This lack of flexibility limits its range of applications, and if PHB were the only existing polyhydroxyalkanoate, it is dubious that a large market niche could be found for PHAs. However, Wallen and Davis reported in 1972 the isolation from activated sludge of a polyester with physical and chemical properties not identical (but similar) to those of PHB [54]. Analysis later revealed the presence of 3-hydroxyvaleric acid (3HV) and 3-hydroxybutyric acid units as major components, and 3-hydroxyhexanoic acid, and possibly 3-hydroxyheptanoic-acid units as minor components of the new compound [55]. This was the first report of a heteropolymeric PHA. There would be more such findings from the scientific community, though not before the middle of the 1980's [56-58]. But the potential significance of the existence of PHAs other than pure PHB was recognized virtually right away, so that when ICI filed patents in the early 1980's for the production by various processes [59-61], extraction from producing cells [62-67], and blending with other organic polymers [68], of PHB, the company also claimed a process for the production by fermentation of bacterial copolyesters of 3HB and a range of other monomers, including 3HV units, from a variety of substrates, including carbohydrates such as glucose, and organic acids such as propionic acid [69].

Interest in copolymers, in particular in copolymers of 3HB and 3HV (i.e., PHBVs), stemmed from the fact that they have melting points much lower, and are less crystalline, more ductile, easier to mold and tougher, than pure PHB [70], and are thus better candidates for commodity materials. Variation in their 3HV content leads to a range of properties spanning a wide variety of thermomechanical properties.

Today, a range of PHBVs with 0 to 24 mol% 3HV produced with the bacterium *Ralstonia eutropha* is marketed under the trademark BIOPOL® by Monsanto. Monsanto acquired the BIOPOL® business in April 1996 from Zeneca Bio Products, England, a company that resulted from the partial splitting of ICI in June 1993. Production capacity is of the order of 800 tons per annum [71]. BIOPOL® is also produced and sold in the U.S., under the tradename PHBV [70], and sold in Japan [72]. At about the same time that ICI was developing its production of PHB, work done by Lafferty and Braunegg [73, 74] showed the advantages of using a strain of *Alcaligenes latus* for such an endeavor. The authors found that this fast-growing bacterium could synthesize the homopolymer at substantial rates during normal growth, which is not the case with *R. eutropha*. This eventually led to the commercial production of PHB by Chemie Linz in Austria in the late 1980's and early 1990's [75].

2. The need for affordable, biodegradable composite materials

The rationale for the development and marketing of biodegradable plastics has been presented here as the necessity to replace oil-based polymers in view of the problems their after-use persistence pose to our society. But the fact is that biodegradables have not yet replaced conventional plastics in a significant way. Annual production of BIOPOL® is but a tiny fraction of that of synthetic polymers. The principal reason for this has been hinted at above: petroleum-based plastics can still be manufactured much more cost effectively. The cost of polyolefins like polyethylene and polypropylene is less than US \$1 kg^{-1} [reported in 76]. BIOPOL®, whose price has been drastically reduced since its production began [70], still sells at about seventeen times the price of synthetic plastics [77]. This impediment to the marketability of PHAs can only partly be alleviated by the willingness of the public to pay more for products that are considered environmentally friendly.

It is actually not realistic to expect PHAs to entirely replace oil-based polymers. At the present, specialty applications are being targeted for the BIOPOL® group of materials, where its biodegradability and biocompatibility are particularly desirable [70]. As Monsanto's production capacity increases, the wholesale cost of PHAs will be driven down, and new market niches will open. Other efforts to increase the cost effectiveness of PHA production, including the development of novel fermentation strategies, must in any case be pursued.

As the solid-waste problem confronts us with increasing urgency, as the polluting combustion of fossil fuels is more and more frowned upon, and as new applications for biodegradable polymers will undoubtedly be found, the prospects for polyhydroxyalkanoates look bright. The future will tell us how big a role they will play in the survival of our modern societies. Whatever this role is, it will be humbling to remember that these polymers were invented long before our coming by bacteria, the Earth's old survivors, and that we should accept the help of such tiny beings to contribute to the comfort of future generations.

In the next section, a review of the field of PHAs is presented. Since detailed general surveys of polyhydroxyalkanoates have been published in the past [e.g., 12, 78-80], emphasis is given on the significant developments of the 1990's. The majority of PHAs are aliphatic polyesters of carbon, oxygen and hydrogen. Their general formula is shown in Figure 1. The composition of the side chain or atom R and value of x determine together the identity of a monomer unit. For poly(3-hydroxyalkanoates) (or poly(3-hydroxyalkanoates)), the most common PHAs, x is equal to 1. Pure PHB is composed of monomers with a methyl group as side chain, and the P(3HV) units of PHBVs contain an ethyl group on carbon number 3. A large number of PHAs other than PHB and PHBVs are now known.

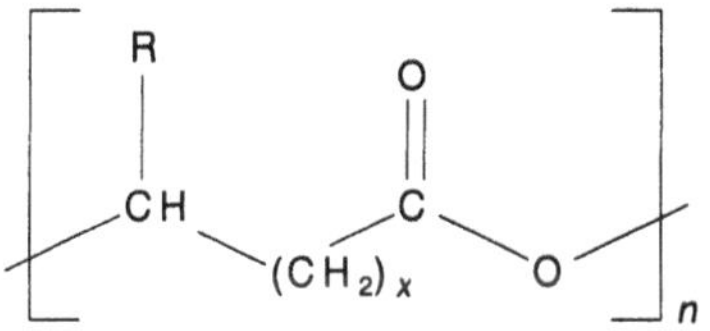

$$\left[-\underset{\underset{R}{|}}{CH}-(CH_2)_x-\overset{\overset{O}{\|}}{C}-O- \right]_n$$

Figure 1: General formula of PHAs

PHAs are widely distributed in the natural world. In addition to its and other PHAs' occurrence in numerous genera of eubacteria [reviewed in 81 and 82], low-molecular-mass PHB has been found, as a short-chain oligomer ($n = 100$ to 200)

complexed with other large molecules, in the cytoplasmic membrane of enterobacteria like *Escherichia coli* and in eukaryotes ranging from plants to humans [83, 84].

3. Novel PHAs from Ralstonia eutropha and Alcaligenes latus

In addition to PHB and PHBVs [66], *R. eutropha* has been shown to produce P(3HB-*co*-4HB) [85, 86], P(3HB-*co*-3HV-*co*-5HV) [87], and P(3HB-*co*-4HB-*co*-3HV) [88] polymers in the past. *A. latus* has also already been reported to synthesize PHB [89] and PHBV [90, 91].

Nakamura et al. [92] reported the production of a pure P(4HB) homopolymer by *R. eutropha*. From mixtures of 4-hydroxybutyric acid and citrate supplemented with ammonium sulfate during polymer accumulation, a range of copolyesters containing 70 to 100 mol% 4HB was obtained, although the presence of the nitrogen source markedly decreased total-polymer production. The P(4HB) homopolymer was accumulated at less than 2% of the cell dry mass (CDM). Trials without ammonium sulfate yielded copolyesters of 24 to 47 mol% 4HB at an average of 21% of CDM. A series of copolyesters of 3-hydroxybutyrate (3HB) and 3-hydroxyvalerate (3HV) was produced by *Alcaligenes eutrophus* H16 from an amino acid, threonine. The 3HV content of these polyesters ranged from less than 0.1% to 30% [93]. With a recombinant, PHA-leaky (defective in genes presumably regulating PHB mobilization) mutant of *R. eutropha* in nitrogen limitation, Steinbüchel and colleagues [94] were able to produce P(4HB) homopolymer at 27.2% of CDM, and at almost 30% in other mutants.

Valentin et al. [95] were able to obtain a P(3HB-*co*-3HV-*co*-4HV) terpolyester with up to 8.8 mol% 4HV from 4-hydroxyvaleric acid or 4-valerolactone as sole carbon sources in batch, fed-batch or two-stage fed-batch cultures of various strains of *R. eutropha*.

A poly(3-hydroxybutyrate-*co*-3-hydroxypropionate) copolyester has been produced by *R. eutropha* in a nitrogen-free medium containing 3-hydroxypropionic (3HP) acid, 1, 5-pentanediol or 1, 7-heptanediol [96]. The polymer contained up to 7 mol% 3HP with 3-HP acid as sole carbon source, and the cells stored a 3-mol% 3HP polymer at 42% of their dry mass. Valentin and co-workers [95] also obtained P(3HB-*co*-2-mol% 3HP) from *R. eutropha* grown on 3-HP acid. Hiramitsu et al. [97] found that the 3HP fraction could be increased up to 26 mol% when mixtures of sucrose and 3-HP acid were used with *A. latus*. In these experiments, the bacterium accumulated the copolymer from 29 to 49% of its dry mass.

The same strain of *A. latus* has also been reported to produce P(3HB-*co*-4HB). Hiramitsu et al. [98] obtained the copolyester from mixtures of sucrose and γ-butyrolactone in one-stage cultures, where growth and polymer accumulation take place simultaneously. The final copolyester content of the cells varied between 17 and 47%, and up to 45 mol% 4HV units were incorporated in it. γ-Butyrolactone was converted into the corresponding 4HB units with a 60% yield, which is substantially higher than that observed with *R. eutropha*.

4. Novel PHAs from the *Pseudomonas* genus

Pseudomonads are undoubtedly the most versatile accumulators of PHAs. The syntheses by *P. oleovorans* of four- to twelve-carbon monomers (R = CH_3 to $(CH_2)_8CH_3$) from n-alkanes, n-alkanoates and n-alcohols, unsaturated monomers from n-alkenes, and of branched-side-chain units from branched substrates, have been reviewed [78], along with PHA accumulation from n-alkanoic acids by other pseudomonads. Huisman et al. [99] advanced that the capacity to accumulate a wide range of PHAs with very little or no 3HB units were a distinguishing trait of fluorescent pseudomonads. The composition of PHA monomers synthesized by pseudomonads is related to that of their substrates, with most units containing 2 carbon atoms less than the carbon source.

More recent investigations by Huijberts et al. [100] with *P. putida* revealed the synthesis by this micro-organism growing on glucose of PHAs composed of seven different monomers, including units of 3-hydroxydecanoate (3HD; the major constituent), 3-hydroxyhexanoate (3HHx), 3-hydroxyoctanoate (3HO), and saturated and mono-unsaturated monomers of 12 and 14 carbon atoms.

Other unsaturated, medium-side-chain (msc) PHAs from pseudomonads have been lately reported. Lee and colleagues [97] used *Pseudomonas* sp. A33 and other related organisms isolated by Schirmer et al. [98] to produce various copolyesters. *Pseudomonas* sp. A33 in the presence of 1, 3-butanediol stored a PHA of 3HB units and nine other different constituents, including the saturated, 16-carbon 3-hydroxyhexadecanoate (3HHD; 0.2 mol%), and three unsaturated: 3-hydroxydodecenoate (3HDDE; 21.0 mol%), 3-hydroxytetradecenoate (3HTDE; 3.4 mol%), and 3-hydroxyhexadecenoate (3HHDE; 1.4 mol%). The authors used several techniques to demonstrate that this PHA was a real copolymer and not a blend of polymers, but did not determine whether it had a random distribution of monomers or consisted of block structures.

Pseudomonas sp. 61-3, isolated by Doi's group from soil was found to produce a polyester consisting of 3-hydroxyalkanoic acids of even carbon numbers C4, C6, C8, C10 and C12 when sodium gluconate was fed as the sole carbon source [103]. The polyester produced was fractionated with boiling acetone. The acetone-insoluble fraction (28 wt%) of the polyester was a poly-3-hydroxybutyrate homopolymer, while the acetone-soluble fraction (72 wt%) was composed of seven different 3-hydroxyalkanoate (3HA) units ranging from C4 to C12: 40 mol% 3-hydroxybutyrate (3HB), 5 mol% 3-hydroxyhexanoate (3HH), 20 mol% 3-hydroxyoctanoate (3HO), 24 mol% 3-hydroxydecanoate (3HD), 1 mol% 3-hydroxy-5-cis-decenoate (3H5D), 4 mol% 3-hydroxydodecanoate (3HDD) and 6 mol% 3-hydroxy-5-cis- dodecenoate (3H5DD). The copolymer was characterized by nuclear magnetic resonance (NMR) spectroscopy, gel permeation chromatography and differential scanning calorimetry. The acetone-soluble fraction of this amorphous copolymer was shown to have a random sequence distribution of the seven 3HA units of C4 to C12 by analysis of the 150 MHz 13C-NMR spectrum, and this was the first example of microbial synthesis of a random copolyester consisting of 3HB and medium-chain-length 3HA units.

Unsaturated, msc PHAs produced by *P. oleovorans* from n-octane and 1-octene were crosslinked with electron-beam irradiation by de Koning et al. [104]. The resulting material had the properties of a true rubber yet retained its biodegradability.
Poly(3-hydroxyalkanoates) with phenyl units as part of the functional group have been produced by *P. oleovorans*. Kim et al. [105] fed the organism with mixtures of 5-phenylvaleric acid and either n-nonanoic acid or n-octanoic acid, to obtain two different polymers, one of 3-hydroxyalkanoate units corresponding to the fed alkanoate, the other of 3-hydroxy-5-phenylvalerate (3H5PV). 3H5PV made up to 40.6 mol% of the total polymer, which reached 31.6% in mass of the CDM. The biomass yield was low however.

When they supplied 11-cyanoundecanoic acid and n-nonanoic acid as carbon sources for polymer accumulation by *P. oleovorans*, Lenz and co-workers [106] obtained a PHA composed of up to 32 mol% of cyano-containing monomers, most likely of 9-cyano-3-hydroxynonanoate and 7-cyano-3-hydroxyheptanoate. Biomass yield and cyano-unit content increased, but total polymer production decreased, as the proportion of the cyano substrate in the feed mixture increased. In the same series of investigations, a cyano-containing PHA was also produced from a mixture of 11-cyanoundecanoic acid and n-octanoate.

PHAs with halogenated functional groups can be synthesized by *P. oleovorans*. In addition to the chlorinated and fluorinated polymers reported [107-108], poly(3-hydroxyalkanoate) copolymers containing brominated repeating units have been produced. Kim et al. [110] grew *P. oleovorans* on mixtures of nonanoic or octanoic acid and 6-bromohexanoic acid, 8-bromo-octanoic acid or 11-bromoundecanoic acid. The molar percentage of brominated units in the polymer reached 37.5 when an equimolar mixture of nonanoic acid and 11-bromoundecanoic acid was used, and PHA and biomass yields increased with increasing length of brominated substrate. All PHAs obtained were random copolymers. Recently, Bear et al. were able to produce a copolyester containing up to 37 % terminal epoxy groups in the side chains, when *P. oleovorans* was fed with a mixture of 10-epoxyundecanoic acid and sodium octanoate [111].

5. Novel PHAs from other microorganisms

The role, synthesis and depolymerization of PHB in the purple bacterium *Rhodospirillum rubrum* have been extensively investigated in the past [7, 22, 40]. Now, *R. rubrum* has recently been used by Ulmer et al. [112] to synthesize unusual polymers containing 3HB, 3HV, and 3-hydroxy-4-pentenoate (3H4PE) repeating units from both 4-pentenoic acid and pentanoic acid. Long growth times and low cell and polymer yields were observed.

Poly(3HB-*co*-3H4PE) was also produced by new soil isolates from unrelated carbon sources [113]. Two strains of the recently redesignated *Burkholderia cepacia* (formerly *Pseudomonas cepacia*) accumulated the copolymer from either gluconate or sucrose. In one case, a 6.9-mol%-3H4PE polymer made up 70% of the CDM. Starting from tiglic acid with or without gluconic acid as a second carbon source, *B. cepacia* formed a terpolyester containing 3-HB, 3-HV, and 2-methyl-3-hydroxybutyric acid [114], the same polymer was formed as well by *R. eutropha*.

Copolyesters of 3-hydroxybutyrate (3HB) and 3-hydroxyvalerate (3HV) were produced by *Burkholderia cepacia* D1 at 30 °C in nitrogen-free culture media containing n-butyric acid and/or n-valeric acid [115]. When n-valeric acid was used as the sole carbon source, the 3HV fraction in copolyester increased from 36 to 90 mol% as the concentration of n-valeric acid in the culture medium was increased from 1 to 20 g/l. The addition of n-butyric acid to the culture solution resulted in a decrease in the 3HV fraction in copolyester. The copolymers biosynthesized by this method were mixtures of random copolymers having a wide variety of composition of the 3HV component. The melting points of the fractionated copolymers show a concave curve with the minimum at the 3HV content of approximately 40 mol%. The alpha-parameter of lattice indices of the PHB crystal for the fractionated copolymers largely increased as the 3HV composition increased. Biodegradability of the copolymer increased with the lower content of 3HV composition and/or the lower crystallinity.

A poly(3HV) homopolymer has been synthesized by three strains of *Chromobacterium violaceum* fed on sodium valerate [116]. Steinbüchel et al. showed that one strain in particular (DSM 30191) could accumulate 45 to 65% of its dry mass as a low-molecular-mass P(3HV). Other PHA-producing substrates yielded with this strain PHB only.

Poly(3HB-*co*-3HP), produced by *R. eutropha* and *A. latus* as noted above, has also been synthesized by a soil isolate of *Methylobacterium* [117]. In nitrogen-limited cultures containing methanol as the main carbon source (0.5% (v/v)), the addition of 3-hydroxypropionate led to accumulation of the copolyester. The 3HP fraction in the polymer could exceed 10 mol%, but total copolymer content in the cell was always below 12%.

Most polyhydroxyalkanoates (PHAs) reported to date fall into one of two broad classes: either hydroxybutyrate-hydroxyvalerate copolymers (typified by the PHA produced by *Alcaligenes eutrophus*), or hydroxyoctanoate-rich heteropolymers (typified by the PHA produced by *Pseudomonas oleovorans*). Few reports of copolymers rich in hydroxybutyrate (HB), but containing a minor proportion of a co-monomer with a higher carbon number than valerate, have appeared. Caballero et al. [118] report on the biosynthesis and characterization of HB-rich polymers containing 2-4 mol% of hydroxycaproate (HC) units, as well as a terpolymer containing HC and hydroxyoctanoate (HO) units. These polymers were produced in good yields by Comomonas testosteroni, *Bacillus cereus* and an unidentified third organism when grown on caproate or octanoate. The minor co-monomers were found to be rejected from the PHB crystallites by X-ray analysis and by quantitative analysis of the melting point depression. The greatly reduced melting point, coupled with the retention of a high degree of crystallinity, could make these materials attractive as melt-processible thermoplastics.

The copolyester Poly(3HB-co-3HV) was synthesized from the combined carbon sources of glucose and sodium propionate by a filamentation-defective mutant of *Sphaerotilus natans*, which is a typical filamentous bacterium often found in activated sludge. The 3-hydroxyvalerate content in the produced polymer increased with increasing concentrations of propionate. Cell growth and polyester synthesis were observed even when 0.6% sodium propionate was added to the medium, and the 3-hydroxyvalerate content in the polymer produced was about 60 mol%. The monomer composition of the copolymer was also varied by aeration conditions, time of propionate feeding, and cultivation time. The strain flocculated in accordance with cell growth, allowing rapid and convenient separation of the biomass from the culture fluid [119].

6. Intracellular aspect of PHA granules

The number of PHB granules in the cytoplasm of *R. eutropha* has been observed to remain constant at eight to twelve during cultivation in nitrogen limitation [120]. For *B. megaterium* McCool et al. reported inclusion body proliferation by budding [121]. Accommodation of additional polymer was done through increases in the diameter of the granules, gradually forcing the cells to change their shape from cylindrical to spherical. The cells invariably stopped increasing their PHB content at around 80% of the CDM in polymer, and as PHA synthase activity and intracellular monomer concentration remained high at this point, the authors concluded that physical constraints at the level of cell geometry were the limiting factor for polymer accumulation. At the same time, the molecular mass (MM) of the polymer decreased gradually during a fermentation. In phosphate-limited culture, the MM of PHB in *R. eutropha* went down from 2×10^6 to 6×10^5 Da as the polymer content of the cells increased [120]. Similar results were obtained under different culture conditions.

Below figure 2 shows distinct granules of PHB in a cell of *Alcaligenes latus DSM 1122*, growing under ammonium limiting conditions.

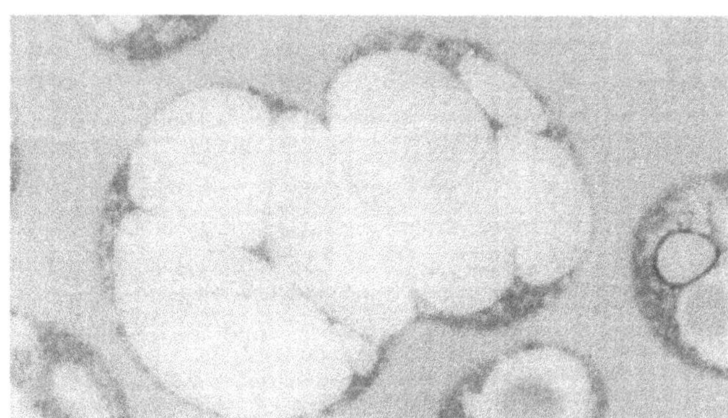

Figure 2: Alcaligenes latus DSM 1122, growth under ammonium limiting conditions: A Cell in the Late Accumulation Phase showing distinct granules of PHB. The PHB content is about 93% of cell dry weight. Magnification: 6000 x.

Studies with numerous organisms have shown that *in vivo* PHB granules have diameters of 0.2 to 0.7 μm and are surrounded by a membrane coat composed of lipid and protein about 2 nm thick [18, 21, 122-124]. The polymer chains generally form helices, and each granule probably contains a minimum of 1000 molecules [78]. X-ray-diffraction analysis of solid PHB led initially to the belief that the core of the inclusion bodies was crystalline [13, 14], but more recent ^{13}C-NMR spectroscopy of whole cells of *Methylobacterium* and *R. eutropha* by Barnard and Sanders showed that the bulk of the PHB homopolymer and PHBV copolymer is in fact in a very labile state, well above its glass-transition point [125, 126]. In a recombinant strain of *E. coli* harboring the PHA genes of *R. eutropha*, however, Hahn et al. [127] have deduced that the accumulated PHB was in a quasicrystalline form, possibly due to hydrogen bonding to other molecules or cations.

The presence of a plastifier has been proposed to explain the amorphous state of PHB granules *in vivo*. Barnard and Sanders [125] believed water to be the most likely plasticizer, and Mas et al. used density measurements to estimate that PHB granules of *R. eutropha* contain 40% of water [128]. This contention has been challenged by de Koning and Lemstra [129], who also argue that the presence of a highly effective plasticizer need not be supposed, and that crystallization kinetics suffice to explain the amorphous state.

This claim has been in turn recently contested by Lauzier and Marchessault [130], who proposed a model where the granules contain hydrogen-bound water that prevents crystallization and possibly plays a role in the orientation of the monomer chains during polymer biosynthesis. According to these authors, water molecules share hydrogen bonds with the carbonyl groups of the polyester backbone to form a so-called 'pseudo-network' that keeps some segments of the molecule almost straight, instead of the near helical conformation, thus preventing entanglement as the chains increase in length during biosynthesis. In this model, 5 to 10% water would be sufficient to account for the pseudo-network.

The activities of PHA synthase and PHA depolymerase are closely related to the membrane protein layer of the granules [23, 36, 40, 131-132]. PHB extraction methods that damage or destroy the membrane lead to loss of synthase activity and increased susceptibility to depolymerization.

7. Polyhydroxyalkanoate metabolism in *R. eutropha* and *A. latus*

The pathways and enzymology of PHA synthesis and degradation have been studied in many organisms. PHB (actually poly-[R-3-hydroxybutyrate]) synthesis and degradation were shown to be the complementary parts of a cycle in *R. eutropha* and *Azotobacter beijerinckii* more than twenty years ago [29, 32]. Since then, synthesis of PHAs other than PHB has been partly or completely elucidated in *R. eutropha*. In contrast, much less is known about *A. latus*.

7.1.METABOLISM DURING BALANCED GROWTH

R. eutropha and *A. latus* catabolize carbohydrates via the Entner-Doudoroff pathway to pyruvate, which can then be converted through dehydrogenation to acetyl-CoA. During reproductive growth, acetyl-CoA enters the tricarboxylic acid (TCA) cycle with the release of CoASH and is terminally oxidized to CO_2 generating energy, in the form of ATP (actually GTP) and reducing equivalents (NADH, NADPH, and $FADH_2$), and biosynthetic precursors (2-oxoglutarate, oxaloacetate) [133]. Direct amination or transamination of the oxaloacetate leads to the synthesis of amino acids, which are incorporated into the polypeptide chains of nascent proteins. Oxidation of the TCA-produced pyridine nucleotides in the respiratory chain generates additional phosphate-dependent ATP, which can support the endergonic requirements of protein biosynthesis. The rate of admission of acetyl-CoA into the TCA cycle is thus contingent on the availability of sources of nitrogen, phosphorus and other elements, as well as on the oxidative potential of the environment.

Synthesis of PHB from acetyl-CoA condensation (see below) during cell reproduction is never completely non-existent - and can be substantial in *A. latus* - reflecting the availability of acetyl-CoA to be used for purposes other than oxidation even in these conditions.

7.2. TRIGGERING MECHANISM FOR INCREASED POLYMER ACCUMULATION

The rate of PHB synthesis can however increase significantly when cells encounter growth-limiting conditions other than a limitation in the carbon source. In *R. eutropha*, deficiencies in nitrogen, phosphorus, oxygen [9, 34, 135], magnesium, or sulfate [136] are known to work. Cessation of protein synthesis leads to high concentrations of NADH and NADPH. These in turn inhibit citrate synthase and isocitrate dehydrogenase, resulting in a slowdown of the TCA cycle and the channeling of acetyl-CoA towards PHB biosynthesis [12]. The potential role of citrate synthase in the regulation of PHB production via its ability to control carbon flux into the tricarboxylic acid cycle is discussed by Henderson and Jones [137]. This can result in massive accumulation of the polymer. Metabolic flux analysis for PHB synthesis from various carbon sources have shown that the maximum PHB yield may be limited by the available NADPH [138]. In recombinant *E. coli*, the level of NADPH and/or the NADPH/NADP ratio seem to be the most critical factor regulating the activity of acetoacetyl-CoA reductase and, subsequently, PHB synthesis [139].

When carbon sources other than strictly acetyl-CoA precursors (e.g. valeric acid) are also present under these conditions, they can be incorporated into the polymer chain, leading to monomers other than 3HB units.

7.3.PATHWAYS OF PHA SYNTHESIS

The pathways of PHA synthesis in *R. eutropha* from various substrates in nitrogen-free conditions are shown in Figure 3.

PHB is produced from acetyl-CoA by the sequential action of three enzymes, 3-ketothiolase, acetoacetyl-CoA reductase and PHA synthase (Pathway I) [29]. 3-ketothiolase reversibly combines two acetyl-CoAs into acetoacetyl-CoA and is competitively inhibited by high concentrations of CoASH, which is released when acetyl-CoA enters the TCA cycle [140]. NADPH-dependent acetoacetyl-CoA reductase reduces its substrate to R-3-hydroxybutyryl-CoA, and this is incorporated by PHA synthase into the polymer chain. 3HB units can also be synthesized form butyric acid directly via acetoacetyl-CoA without its prior degradation to acetyl-CoA (Pathway II) [140]. In this sequence of reactions featuring β-oxidation of the substrate, both NADH-linked and NADPH-linked acetoacetyl-CoA reductases effect the epimerization of S-3-hydroxybutyryl-CoA to the R isomer, and 3-ketothiolase is not involved.

When propionic acid is present in the medium, 3-ketothiolase condenses one propionyl-CoA with one acetyl-CoA to form 3-ketovaleryl-CoA, which is polymerized after reduction to 3-hydroxyvalerate monomers by PHA synthase (Pathway III) [141]. Acetyl-CoA can be provided by a second substrate, but elimination of the carbonyl carbon of propionyl-CoA always takes place to some extent, so that both 3HB and 3HV units are synthesized when propionic acid is the sole carbon source. Still, the 3HV fraction in the copolymer rises with increasing ratio of propionic acid to acetyl-CoA-generating substrate [141]. With propionic acid alone as substrate, up to 40 mol% 3HV units are incorporated in the copolyester.

Valeric acid can also serve as a precursor for 3-hydroxyvalerate units (Pathway IV) [140]. Similarly to 3HB synthesis from butyric acid, this does not involve the catabolism of the acid to a shorter alkyl-CoA, but rather its direct incorporation into the polymer via valeryl-CoA and its β-oxidation to 3-hydroxyvaleryl-CoA. Use of valeric acid leads thus to higher 3HV contents in the polymer in *R. eutropha* than propionic acid, as decarboxylation of propionyl-CoA and loss of units with an odd number of carbons is reduced - but not totally eliminated, as the intermediary S-3-hydroxyvaleryl-CoA can be degraded to propionyl-CoA and acetyl-CoA [142]. When valeric acid is the sole carbon source, *R. eutropha* synthesizes a copolyester with 90 mol% 3HV monomers.

A. latus was also shown to incorporate more 3HV units from valeric acid into the copolymer than from the same amount of propionic acid [91]. This may be an indication of at least very similar pathways of PHBV in *A. latus* and in *R. eutropha*, although the authors did not comment on this.

Figure 3 also shows the pathway of 4-hydroxybutyrate synthesis from 4-hydroxybutyric acid (Pathway V) [81, 143]. 4-Hydroxybutyryl-CoA is first formed, and part of it is polymerized by PHA synthase. A portion of the hydroxyacyl-CoA is however dehydrated to the corresponding enoyl-CoA, which enters the pathway of 3HB synthesis via R-3-hydroxybutyryl-CoA. According to Valentin et al. [136] it is more likely that formation of 3-hydroxybutyryl-CoA occurs via succinate semialdehyde, succinate, pyruvate, and acetyl-CoA from 4-hydroxybutyrate. A copolymer of 3HB and 4HB is thus usually produced from 4HB acid. Presence of butyric acid in the medium somewhat inhibits the conversion of 4-hydroxybutyryl-CoA into R-3-hydroxybutyryl-CoA. The apparently complete inhibition of 3HB synthesis from 4HB acid by citrate and ammonium sulfate observed by Nakamura et al. [92] has to date not been explained. Substrates which can be converted to 4-hydroxybutyric acid, such as γ-butyrolactone, 1, 4-butanediol, 1, 6-hexanediol, and 4-chlorobutyric acid, also lead to the formation of 4HB monomers [143].

Nakamura et al. [96] have proposed a pathway for the synthesis of 3-hydroxypropionate in *R. eutropha* (Pathway VI). When 3HP acid is present as the sole carbon source in nitrogen-free medium, its metabolization into 3-hydroxypropionyl-

CoA is largely followed by a decarboxylation to acetyl-CoA and 3HB synthesis. But a portion of the 3-hydroxypropionyl-CoA is also directly polymerized into 3HP, presumably under the action of PHA synthase, producing a random copolyester of 3HB and 3HP units. When 1, 5-pentanediol or 1, 7-heptanediol are used as carbon sources, formation of 3-hydroxypropionyl-CoA probably results from the β-oxidation of the longer acyl-CoAs derived from the substrates. The authors noted that since *R. eutropha* can grow on 3-HP acid, conversion to acetyl-CoA is probably the favored step, and only low contents of 3HP (7 mol%) in the polyester are observed with this micro-organism when 3HP acid is the sole carbon source.

In contrast, *A. latus* DSM 1124 cannot grow on 3HP acid. This may explain the P(3HB-*co*-26-mol% 3HP) obtained by Hiramitsu and Doi [97] even when sucrose accompanied the carboxylic acid in the medium. The author proposed that decarboxylation of 3-hydroxypropionyl-CoA to acetyl-CoA is impossible in *A. latus*, therefore leading to its polymerization being favored. Presumably, synthesis of 4-hydroxyvalerate and 5-hydroxyvalerate repeating units from the corresponding substrates occurs in *R. eutropha* in a fashion similar to that of P(3HP) from 3HP acid.

Synthesis of PHBV from 2-hydroxyoctanoic (2HO) acid or 12-hydroxystearic acid (18 carbon atoms) by *Alcaligenes* AK 201 has been tentatively explained by Akiyama and Doi [145] (Pathway VIII). When 2HO acid is used as the sole carbon source, β-oxidation of the compound would yield CO_2 and n-heptanoate, which could be further incorporated as acetyl-CoA and valeryl-CoA in the copolyester through pathways I and IV in Fig. 2. 12-Hydroxystearate would first be degraded to 2HO by a five successive β-oxidative cleavages, producing acetyl-CoAs in the process. The authors could however not explain the large differences observed in the 3HV content of the copolymers obtained from the two substrates, that from 2HO acid being much higher (47 mol%) than that from 12-hydroxystearic acid (1.8 mol%).

Nocardia corallina accumulates a copolyester of 3HB and 3HV containing more than 60 mol% 3HV from most carbon sources, the majority of 3HV units are synthesized via the methylmalonyl-CoA pathway [146].

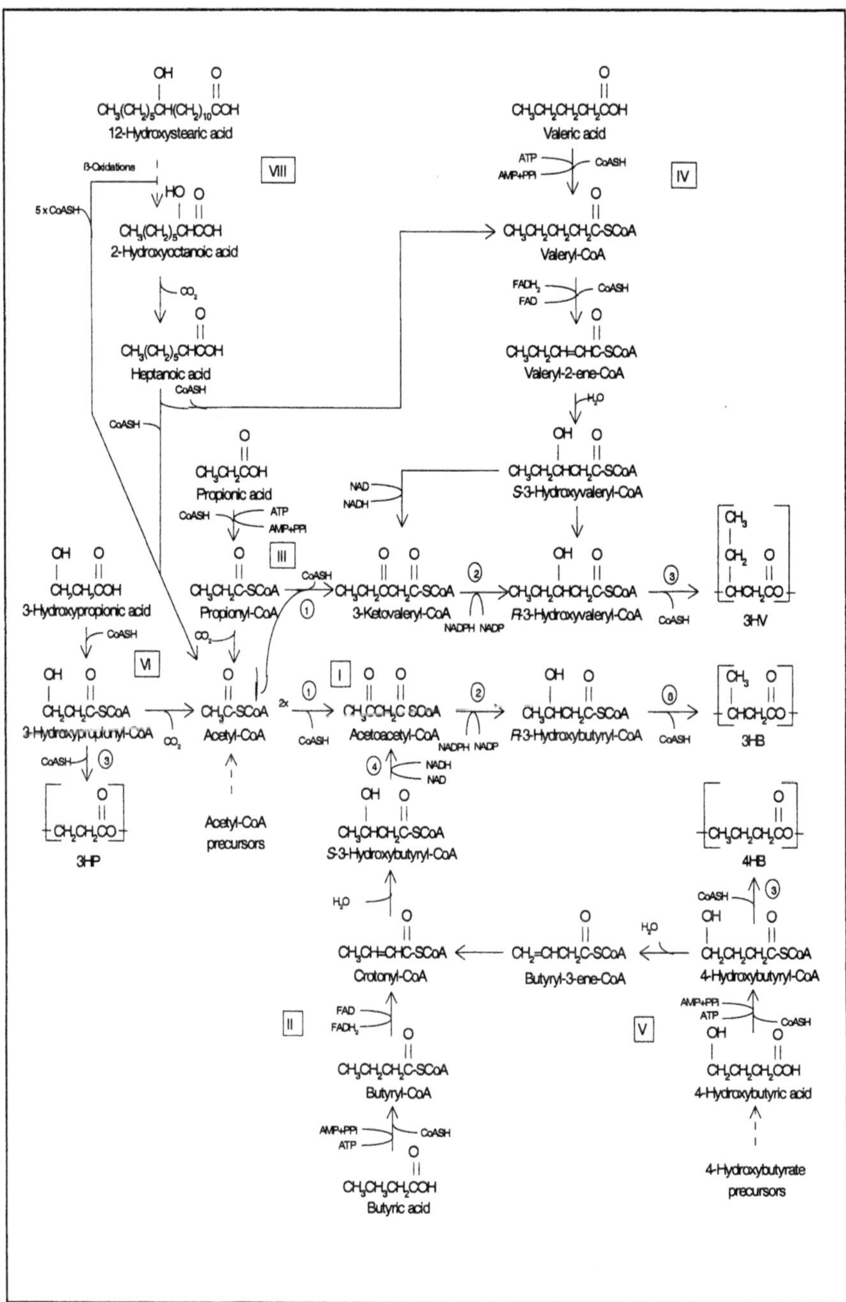

Figure 3: Pathways of PHA synthesis in R. eutropha, except for Pathway VII (Alcaligenes AK 201) [29, 81, 96, 138, 141, 142, 143, 145]. Enzymes: ①, 3-ketothiolase; ②, NADPH-dependent acetoacetyl-CoA reductase; ③, PHA synthase; ④, NADH-dependent acetoacetyl-CoA reductase. See text for details.

7.4.ENZYMOLOGY OF PHA SYNTHESIS IN *R. EUTROPHA*

7.4.1 3-Ketothiolase

Haywood et al. purified 3-ketothiolase (acetyl-CoA acetyltransferase) from a glucose-utilizing strain of R. eutropha [147]. They found it to consist of two distinct constitutive isoenzymes, 3-ketothiolase A and B, each with its own substrate specificity. 3-Ketothiolase A is active with only with four- or five-carbon 3-ketoacyl-CoAs, and must solely be responsible for PHA synthesis in R. eutropha, as this micro-organism does not accumulate PHAs with hydroxyacid repeating units of more than five carbon atoms. 3-Ketothiolase B has a broader specificity (four- to ten-carbon 3-ketoacyl-CoAs), and Haywood and colleagues have speculated that it may have a function other than PHA accumulation.

The authors observed that the condensation reaction effected by 3-ketothiolase was strongly inhibited by free coenzyme A, as mentioned above. This accounts for the low PHB levels in R. eutropha during unrestricted growth on acetyl-CoA-generating substrates, when high amounts of CoASH are released by TCA cycle. When Doi et al. grew R. eutropha on butyric acid in the presence of high amounts of ammonium sulfate (3 to 10 g L^{-1}) [131], PHB levels in the cells still reached about 25 to 30% of the CDM during balanced growth. Butyric-acid incorporation into the polymer does not involve 3-ketothiolase (Fig. 2). The same experiments with glucose as carbon source showed total inhibition of PHB synthesis above 4 g ammonium sulfate L^{-1}. The involvement or not of 3-ketothiolase can thus be a key factor in PHB-metabolism regulation in R. eutropha. The use of pentanoic acid as 3HV precursor also limits the role of 3-ketothiolase. !

7.4.2 NADPH- or NADH-dependent acetoacetyl-CoA reductases

Two acetoacyl-CoA reductases were found by Haywood et al. in *R. eutropha* [cited in 78], each with a distinct substrate and coenzyme (NADH or NADPH) specificity. NADPH-dependent acetoacetyl-CoA reductase mediates the reversible reduction of four- to six-carbon 3-ketoacyl-CoAs to the R isomers only of hydroxyacyl-CoAs. NADH-linked acetoacetyl-CoA reductase effects the oxidation of both R and S substrates to acetoacetyl-CoA, but produces only S-3-hydroxybutyryl-CoA in the reverse reduction reaction. Doi and co-workers [156] have advanced that the reduction of 3-ketoacyl-CoA by the NADPH-dependent reductase may be the rate-determining step in PHA synthesis.

7.4.3 PHA synthase

The PHA synthase of *R. eutropha* is capable of polymerizing 3-hydroxy-, 4-hydroxy-, and 5-hydroxyalkanoates from R isomers of four- to five-carbon hydroxyacyl-CoAs [131, 95], although its activity is markedly higher with four-carbon substrates [131]. If Nakamura et al. [96] (see above) are right in their speculation on the pathway of 3HP synthesis, the enzyme must also effect the polymerization of 3-carbon compounds. It has been isolated in two forms, a soluble one predominating during unrestricted growth of the cells, and a granule-associated one when culture conditions favored PHB accumulation [131]. It is possible that PHA synthase is inhibited by the cellular concentration of CoASH, albeit at higher levels than 3-ketothiolase. Doi et al. have estimated the number of PHA synthase molecules in *R. eutropha* to be about 18000 [148]. The amount of polyhydroxyalkanoate synthase in a host organism plays a key role in controlling the molecular weight and the polydispersity of the polymer [149]. Thiol groups were suspected to be the active sites of the enzyme [20, 120], Wodzinska et al. have shown acylation of cystein 319 and subsequent activation of PHA synthase by dimerization [150].

7.5.GENES FOR PHA SYNTHESIS

A review of the recent body of knowledge on the structure and function of PHA-biosynthetic genes in a wide range of micro-organisms has been published by Steinbüchel and colleagues [151]. The three genes for PHA biosynthesis in *R. eutropha* have been characterized and cloned in *E. coli* [152-155]. The resulting recombinant strains were able to accumulate large amounts of the polymer. The hypothesis by Schubert et al. [154] that the genes for 3-ketothiolase, acetyl-CoA reductase and PHA synthase are clustered has been confirmed, they form a single operon (combination of promoter, operator and structural genes) [156], and Peoples and Sinskey [153] have showed that the three enzymes are coded in the order synthase-thiolase-reductase. They could not determine whether one or more promoters were involved, but Schubert et al. identified the promoter for the PHA synthase gene and noted that it is probably that of the whole operon [156]. PHA synthesis is subject to transcriptional control [155], as is usual for metabolic pathways affected by environmental conditions.

Genser and co-workers [157] have isolated, sequenced, and expressed in *E. coli* the three PHA genes of *A. latus*. Sequencing revealed high respective homologies (71 to 80%) to the *R. eutropha* genes and the same orientation and organization, suggesting a single-operon arrangement here also. A functional promoter, of different structure and possibly more active than that from *R. eutropha*, was located upstream of the PHA synthase gene. Based on the nature of the homologies, the authors concluded that *R. eutropha* and *A. latus* inherited their PHA genes by horizontal transfer from a common ancestor.

7.6.INTRACELLULAR PHA DEGRADATION

The intracellular degradation of PHB has been studied in detail in a number of organisms [12]. PHB hydrolysis is effected by the sequential action of PHA depolymerase, R-3-hydroxybutyrate dehydrogenase and acetoacetyl-CoA synthetase, to form R-3-hydroxybutyric acid and acetyl-CoA [81]. In *R. eutropha*, the sole product of PHB hydrolysis is R-3-hydroxybutyric acid, but a mixture of dimers and monomers of the acid can be obtained in other organisms [36]. In nitrogen-free cultures, Doi et al. [148] have studied the kinetics of PHB accumulation and degradation in *R. eutropha* during carbon excess and limitation, respectively. They found that the rate of polymer hydrolysis (degradation) was about 10 times lower than that of its synthesis, and proposed a reaction scheme for the depolymerization process.

Studies with *B. megaterium* showed that, unlike PHA synthase, PHA depolymerase is in a soluble form, not bound to the granule [20]. There may however be a protein component of the granule outer-membrane that inhibits depolymerase activity [75]. This would explain the increase in polymer hydrolysis upon damage on granule structure by solvent action.

7.7.CYCLIC NATURE OF PHA METABOLISM

Whether polymer synthesis and degradation simultaneously take place in *R. eutropha* has been investigated by two groups of researchers. Haywood et al. [131] studied PHB turnover in *R. eutropha* growing in continuous fermentation at a high carbon-to-nitrogen ratio (10 g glucose L^{-1}, 1.23 g ammonium sulfate L^{-1} [147]). Steady-state was reached with a radioactive-glucose (20 μCi [U-^{14}C]-glucose L^{-1}) feed medium, after which the culture was supplied with an unlabelled-glucose mixture. Other steady-state conditions were not affected. Radioactivity in the whole culture and in extracted PHB was measured at intervals following the change in medium. The observed rate of radioactivity decrease in the extracted PHB was equal to that of the whole culture (i.e., to the dilution rate). The authors concluded that during growth, no significant turnover of PHB takes place, which implies a low or absent PHA depolymerase activity in these conditions.

On the contrary, PHA metabolism in *R. eutropha* was however shown to be a cyclic process during PHB and PHBV synthesis in nitrogen-free medium. In shake-flask experiments, Doi et al. [158] submitted *R. eutropha* cells containing PHB from butyric acid to PHBV-accumulating conditions (pentanoic acid as the sole carbon source). By analysis of the change in polymer composition with differential-scanning-calorimetry melting curves, they suggest that the PHB homopolymer was gradually replaced by a copolymer at a relatively constant 3HV fraction (50±5 mol%). Almost complete replacement of the homopolymer was achieved after 96 h. In the reverse experiment, PHBV was similarly replaced by PHB. These investigations showed that polymer synthesis is concomitant with its degradation in *R. eutropha* under nitrogen starvation, resulting in cyclic PHA metabolism.

8. Physiological role of PHAs

The role of PHB in eubacteria has been well reviewed by Anderson and Dawes [78] and ranges from a carbon and energy reserve material, including an energy source for sporulation in *Bacillus* [33], to oxidizable substrate for respiratory protection in azotobacteria [32]. The physiological function of other PHAs is assumed to be similar to that of PHB.

PHA synthesis has been advanced as an energy-producing mechanism in anaerobic syntrophic (H_2-producing, acetogenic) bacteria [159]. Extracellular PHBV degradation concurrent with sulfate reduction in anoxic lake-sediment samples was seen by Urmeneta et al. [160] as an indication that PHAs could serve as carbon and electron sources.

The complexed PHB that has been found in eukaryotes [83, 84] is in too minute amounts to serve as reserve material. Its biological function has still not been totally explained.

8.1. FUNCTION OF THE SYNTHETIC PROCESS DURING OXYGEN LIMITATION

Quite apart from the role stored PHAs can play in the survival of micro-organisms in the absence of an external carbon and energy source, the process of their synthesis has been demonstrated to be itself of importance in conditions of low oxidative capacity of the environment, for example when the availability of oxygen is restricted.

Senior et al. [31] observed that in dinitrogen-fixing *A. beijerinckii* supplied with air, PHB accumulation was not promoted in ammonium-free basal medium. In dinitrogen-limited chemostat cultures, the PHB content of the cells remained below 1.5% at all dilution rates investigated. But in the same series of experiments, oxygen limitation led to increased PHB accumulation and cell yield in both nitrogen-limited and nitrogen-sufficient cultures. The authors proposed that, under a restricted availability of oxygen, biosynthesis of PHB in *A. beijerinckii* replaces respiration for the re-oxidation of NAD(P)H. The pathways of PHA synthesis in *A. beijerinckii* are almost identical to those in *R. eutropha*, but the acetoacetyl-CoA reductase of the former micro-organism utilizes either NADH or NADPH as an electron donor for the synthesis of R-3-hydroxybutyryl-CoA [35]. Therefore, under this scenario, *A. beijerinckii* would use PHB not only as a store of carbon and energy, but also as an electron sink. Further chemostat studies by Jackson and Dawes [161] confirmed that in *A. beijerinckii* fixing nitrogen, NADH oxidase activity dropped upon oxygen limitation, but that intracellular NADH/NAD ratios, which immediately increased when oxygen became limiting, rapidly readjusted as PHB accumulation started. NADH oxidase activity rose swiftly again when oxygen limitation was relaxed. A somewhat different but similar behavior was observed in ammonium-grown cultures [162], where relaxation of oxygen limitation followed by a gradual increase of the dissolved-oxygen tension was accompanied by an initial decrease in PHB content, then an increase, and finally a decline to negligible values. This increase in polymer synthesis at a certain range of dissolved-oxygen concentrations was attributed to the lower requirements of growth on ammonium in reducing equivalents and ATP than those of the nitrogenase system. In other words, PHB accumulation in *A. beijerinckii* does not only replace respiration, but can also work in concert with it as electron-sink.

Because, in contrast to *A. beijerinckii*, *R. eutropha* does not fix molecular nitrogen, polymer synthesis can be favored in the latter bacterium by N limitation. Following the original work that demonstrated this in *R. eutropha* DSM 428 grown autotrophically (and that showed that the introduction of a nitrogen source into a culture of high polymer content leads to a re-utilization of PHB for protein synthesis) [18], Schlegel's group investigated the role of oxygen limitation on PHA storage by *R. eutropha*. In chemostat under autotrophic conditions [34, 163], the micro-organism accumulated PHB from CO_2 and H_2 to up to 25% of its dry mass when the oxygen was limited, which is less than a third of the concentrations that can be reached with nitrogen limitation [164]. When molecular hydrogen was the limited nutrient, the homopolymer concentration in the cells was extremely low (<1%), an effect Schuster et al. attributed, at least partly, to intracellular deficiencies in reduced pyridine nucleotides (NAD(P)H) and the ensuing shut-off of PHB synthesis [34]. In any case, biopolymer synthesis upon O_2 limitation in *R. eutropha*, in addition to shunting excess carbon away from the slowed down TCA cycle, most probably also serves the same function as in *A. beijerinckii*: substituting the pyridine nucleotide-reducing step of 3-ketoacyl-CoA oxidation for respiration. But Steinbüchel and Schlegel [165] showed that out of seven different nutrient limitations (oxygen, ammonium, sulfate, magnesium, phosphate, potassium and iron), the one of oxygen yielded the slowest rate of PHB synthesis from *R. eutropha* growing heterotrophically on gluconate.

When the dissolved-oxygen concentration was kept between 1 and 4% of air saturation during the polymer accumulation phase with *R. eutropha*, the yield of (3HB) monomer from glucose was not affected, but the propionate-to-3hydroxyvalerate (3HV) monomer yield was two to three times (0.48 to 0.73 mol of 3HV per mol of propionate consumed) that observed in a control experiment (0.25 mol mol^{-1}), where the accumulation-phase dissolved-oxygen concentration was 50 to 70% of air saturation. The effect of a low dissolved-oxygen concentration is probably attributable to a reduction of the oxygen-requiring decarbonylation of propionyl-CoA to acetyl-CoA [166].

A. latus typically synthesizes PHB to up to 70% of its CDM during unrestricted growth [167], in contrast to 10 to 15% for *R. eutropha*. Until now, no satisfactory explanation has been advanced for this constitutive accumulation. But very recent work by Genser et al. [157], who were inspired by Manchak and Page's studies of *Azotobacter vinelandii* UWD [168], have shown no NADH oxidase activity in *A. latus*. In the same series of investigations, a recombinant strain of *E. coli* harboring the PHA genes of *A. latus* or *R. eutropha* produced less than 3% polymer during balanced growth. Since NADH oxidase in *E. coli* was highly active, the authors have proposed that the constitutive polymer synthesis in *A. latus* is a response to this organism's incapacity to regenerate NAD by the respiratory chain. NADH oxidase activity in *R. eutropha* was lower than in *E. coli*, a possible explanation for the synthesis of small amounts of PHB during balanced growth. Therefore, here also, PHA synthesis would serve as an alternative oxidative process for NADH.

9. Enzymology of extracellular PHA degradation

Native (intact) PHA granules are not degraded by extracellular PHA depolymerases, indicating a structural difference between them and the intracellular depolymerases [169].

Although a number of micro-organisms have long been known to be capable of extracellular PHA degradation [12], until recently only the extracellular PHB-depolymerase systems of *Pseudomonas lemoignei* and the activated-sludge isolate *Alcaligenes faecalis* T1 had been purified and characterized, and the PHB depolymerase gene of the latter cloned and sequenced [39, 170–172]. Recently, a new, marine species of *A. faecalis* was isolated and its extracellular depolymerase was characterized, and the respective gene cloned and sequenced [173, 174]. The enzyme's main products of PHB degradation were dimeric and trimeric esters of 3-hydroxybutyrate, which are also the products of the depolymerases of *A. faecalis* T1 and *P. lemoignei*. Jendrossek et al. [175], however, isolated a bacterium identified as *Comamonas* sp. whose extracellular depolymerase degraded PHB to monomeric 3-hydroxybutyrate, indicating a mechanism of hydrolysis not reported so far.

The depolymerase system and genes of *P. lemoignei* have been further investigated by Jendrossek's group [169, 176, 177]. The authors identified, cloned and sequenced five PHA depolymerase genes from this organism, including one for the degradation of pure P(3HV) in addition to that of PHB and PHBVs. An extracellular P(3HO) depolymerase from a new *P. fluorescens* isolate was also characterized by Jendrossek and colleagues [102], and its gene was cloned in *E. coli*, sequenced and characterized [178]. The enzyme's main product was the dimeric ester of 3HO.

Ramsay et al. [179] developed a method for the specific isolation of micro-organisms producing extracellular depolymerases for medium-side-chain PHAs. A sterile, colloidal suspension of a copolyester of 3HB, 3HHx, 3HO and 3HD monomers was used as enrichment medium, and six bacterial species capable of growing in it were isolated. Not surprisingly, all were identified as pseudomonads or members of related genera.

10. Detection and analysis of PHAs

PHAs can be detected and their content in cells determined by a number of methods [12, 78]. The most common detection technique *in vivo* is the fluorescent staining of granules [180]. Many methods for the analysis of cell content, structure and composition of PHB and other PHAs have been reported, including gas chromatography (GC) after solvent extraction and hydrolytic esterification of the polymer [181, 182], pyrolysis under nitrogen of extracted PHAs followed by GC-mass spectrometry [183], and a variety of ^{1}H- and ^{13}C-NMR techniques [184, 185]. PHB can also be determined by IR-spectrometry after extraction in chloroform [186]. Molecular-mass determinations are now typically performed by gel-permeation chromatography, which has mostly replaced the earlier calculations from the intrinsic viscosity of PHB solutions [see 12]. Glass transition and melting temperatures of solid-state PHAs are estimated by differential scanning calorimetry [78].

Zeiser [reported in 187] modified the conditions of the gas-chromatographic detection of poly(3-hydroxyalkanoates) developed in 1978 [181] to adapt the method for the determination of P(3HB-*co*-4HB). Measurement of PHB in live cells of *R. eutropha* by flow cytometry and spectrofluorometry after Nile-Red staining of granules has been recently investigated [188]. The authors noted that the rapidity of the methods could be useful for process control during cultivation.

Four narrow lines are observed in the high-resolution cross-polarization magic-angle spinning 13C NMR spectra of intact, lyophilized samples of *Pseudomonas sp.* LBr, in addition to the broader lines normally associated with bacterial cellular material. These narrow lines arise from poly-3-hydroxybutyrate. The cellular carbon contained in this storage material can be measured quantitatively and nondestructively from the solid-state NMR spectra. The authors found that cells starved for phosphorus store up to 50% of their total carbon as poly-3-hydroxybutyrate. When such cells are used to inoculate medium containing a source of phosphorus, all of the poly-3-hydroxybutyrate is metabolized by the time the culture has reached midlogarithmic growth phase [189]. Direct pyrolysis mass spectrometry (DPMS) was applied to poly(p-xylylene) (PPX) and poly(hydroxybutyrate) (PHB) samples. The validity of a mathematical model for pyrolytic behavior was tested using PPX and then applied to PHB. Simple cleavage at an ester bond is the most probable process in the latter. Examination of a random copolymer of PHB with 20% hydroxyvalerate shows that the latter behaves pyrolytically in a similar manner to PHB itself. DPMS can be used to detect very low levels of a 2nd monomer in a polymer sample [190].

11. Some physical properties of PHAs

11.1 SOLID-STATE CONFORMATION

Solid-state PHB is a compact right-handed helix with a two-fold screw axis (i.e., two monomer units complete one turn of the helix) and a fiber repeat of 0.596 nm [196]. The forces underlying this conformation are mainly van-der-Waals interactions between the carbonyl oxygens and the methyl groups. The stereoregularity of PHB makes it a highly crystalline material. It is optically active, with the chiral carbon always in the R absolute configuration in biologically produced PHB. Its melting point is around 177 °C [70], close to that of polypropylene, with which it has other similar properties, although the biopolymer is stiffer and more brittle [51, 197].

PHBV chains also have crystalline conformations [196]. The properties of copolymers of 3HB and 3HV vary with their content in 3HV. The melting temperature of PHBV is minimum (ca. 80 °C) at a 3HV molar fraction of about 30% (pseudo-eutectic point). Below this 3HV content, the PHB lattice is the sole crystalline phase, while above 30 mol% 3HV, PHB units are embedded in a P(3HV) crystalline matrix. The distribution of the two monomers is statistically random [198]. Their overall lower crystallinity and glass-transition temperatures confer on PHBVs enhanced mechanical properties, such as toughness and softness, that make them more interesting thermoplastics than pure PHB [197]. Random copolymers of 3HB and 4HB also display lower crystallinity and glass-transition points than PHB [85], resulting in mechanical behaviors close to those of elastic rubbers when the 4HB content exceeds 40 mol% [cited in 78]. Similarly, the crystallinity of P(3HB-*co*-3HP) copolymers decreases with increasing fraction of 3HP units [96].

Medium chain length polyhydroxyalkanoates, MCL-PHAs, produced by bacteria as inclusion bodies or granules were analyzed in situ by differential scanning calorimetry (DSC) without isolation from the cells. The kinetic DSC study of PHA granules, which contained mostly 3-hydroxyoctanoate units (PHO), in *Pseudomonas putida* BM01 cells showed that the polymer within the granules existed in an amorphous state, but it crystallized after dehydration of the cells under freeze-drying condition (below –50°C) followed by annealing at ambient temperature. In this manner, PHO within the cells readily crystallized to the maximum degree of crystallinity within 24 h at room temperature, which was much faster than for the same polymer isolated by solvent extraction. This observation suggests that the polymer within the cellular granules may be well organized. The DSC endothermic melting peak areas for the room-temperature annealed polymers within the cells were directly proportional to the amount of polymer in the cell, and the results from this type of quantitative analysis were essentially identical to those obtained by gas chromatographic and gravimetric analysis of the polymers. X-Ray diffraction analysis of the polymer in the freeze-dried, whole cells and of the isolated, fully crystallized polymer showed that the two types of PHO samples had similar crystal structures, but the polymer in the granules exhibited better side-chain packing and higher crystallinity [199].

11.2 VISCOELASTIC RELAXATIONS AND THERMAL PROPERTIES OF PHAS

3-Hydroxybutyrate-3-hydroxyvalerate (3HB-3HV) as well as 3-hydroxybutyrate-4-hydroxybutyrate (3HB-4HB) copolyesters have been investigated by differential scanning calorimetry, thermogravimetric analysis and dynamic mechanical spectroscopy, over a wide range of compositions (0-95 mol% 3HV; 0-82 mol% 4HB). Both series of isolated copolyesters are partially crystalline at all compositions. Quenched samples show a glass transition that decreases linearly with increasing co-monomer molar fraction, more markedly when the co-monomer is 4HB. Above Tg, all copolyesters, rich in 3HB units, show a cold crystallization phenomenon followed by melting, while at the other end crystallization on heating is observed only in 3HB-3HV copolymers. The viscoelastic spectrum, strongly affected by thermal history, shows two relaxation regions: the glass transition, whose location depends on copolymer type and composition, and a secondary dispersion region at low temperatures (-130/-80 °C). The latter results from a water-related relaxation analogous to that of PHB and, in 3HB-4HB copolymers, from another overlapping absorption peak centered at -130 °C, attributed to local motion of the methylene groups in the linear 4HB units [200].

11.3 SOLUBILITY PARAMETERS

The three dimensional solubility parameters defined by Hansen are based on dispersion forces between structural units, interaction between polar groups and hydrogen bonding. For polar polymers such as poly(3-hydroxyalkanoates), P(3HA), this approach was used to obtain the three coordinates of a solubility parameter in terms of: a dispersion part, a polar part and a hydrogen bonding part. Thirty-eight different solvents for poly-3-hydroxybutyrate, PHB, which are mentioned in the literature are examined by this method and the theoretical predictions are compared with the experimental reports. Another overall comparison between PHA polymers provides their Hansen and Hildebrand parameters for side chain lengths up to C13. In this series a linear progression in calculated solubility parameters with side chain length was found. An appendix provides information and data on calculation of the solubility parameters. While the solubility information is limited and only covers homopolymers, it should help to highlight some of the contradictions regarding PHB solubility. This semi-empirical approach is only valid for amorphous polymers hence crystallinity effects, which are important with PHB, as well as molecular weight effects still require analysis [201].

11.4. MOLECULAR MASS AND MOLECULAR-MASS DISTRIBUTION OF EXTRACTED PHAS

The molecular-mass distribution of a polymer is a measure of the distribution of its individual molecules' molecular mass (MM) around the average molecular mass; a narrow distribution around a high average is usually desired.

In addition to being a function of the producing organism and the strategy of production (duration of fermentation, growth rate, carbon-source concentration, etc.) [78], the average MM of PHAs is affected by the method of extraction. Values have up to recently typically ranged between 2 x 10^5 to 2 x 10^6 Da [78]. Zeneca considered MMs of about 600 000 Da acceptable for the thermoplastic applications of its BIOPOL® [70]. The degree of polymerization of some of the newly discovered PHAs, for example the P(3HV) homopolymer with a MM reaching as low as 60 000 Da produced by *C. violaceum* in Steinbüchel's laboratories [116], may have to be improved before their use as thermoplastics can be considered.

The mild NH_3 treatment used by Page and Cornish [195] to extract PHB from *A. vinelandii* yielded a product with a high MM of 1.7 x 10^6 to 2.0 x 10^6 Da and low dispersity. More recently, Page's group [202] used the same strain of *A. vinelandii* growing on beet molasses to produce a 4-million-Da PHB, quite possibly the highest MM PHA reported so far. The authors studied the effects of the substrate on the degree of polymerization.

Other investigations on the influence of culture conditions on PHA molecular mass include recent work by Anderson et al. [203, 204], who found that the MM of PHB produced by *R. eutropha* was generally unaffected by the choice or concentration of the carbon source. Koyama and Doi [205] could show, however, that *R. eutropha* growing on fructose and pentanoic acid in a chemostat yielded PHBVs with MMs that increased along with the dilution rate of the cultures. Asenjo and co-workers [206] studied the effect of magnesium and phosphate concentration on the MM and MM distribution of PHB from *R. eutropha* in different culture conditions. They found that in order to avoid a lowering of polymer MMs and a widening of the MM distribution in fed-batch or continuous, nitrogen-deficient cultures, excess in glucose should be limited, and Mg and phosphate concentrations kept above minimal levels. The authors also speculated that the oft-reported, time-dependent decrease in MM of PHB during fermentations of *R. eutropha* and other organisms [120, 207, 208] may be caused by insufficient control of these parameters near the end of cultures.

12 Biodegradability of PHAs

PHB and PHBVs are degraded in both aerobic and anaerobic environments by the action of extracellular enzymes from microbial populations [70]. Doi et al. [143] have further pursued their early studies on the hydrolytic and enzymatic degradation of films of PHB, PHBVs and P(3HB-*co*-4HB)s in various environments, studies that found that the presence of 4HB units enhances the rates of both types of erosion. Nakamura et al. [92] exposed P(3HB-*co*-4HB) films to extracellular PHA depolymerase isolated from *Alcaligenes faecalis*. Enzymatic degradation as measured by weight loss was accelerated by 4HB contents up to 28 mol%, but depolymerization was inhibited at 4HB fractions above 85 mol% of the copolyester. In another set of similar experiments [209], the critical 4HB fraction was 13 mol%, at which point the rate of degradation was about 10 times faster than that of the homopolymer PHB. Doi and colleagues [210] have speculated that this acceleration could be attributed to the decreased crystallinity of 4HB copolymers relative to PHB and PHBVs, offering the degradative enzymes better access to the polymer chains. Nishida and Tokiwa [211] confirmed that crystallinity depressed the microbial degradability of PHB. A P(3HB-*co*-4-mol% 3HP) copolyester was found to enzymatically degrade faster than PHB [96].

Nishida and Tokiwa's [211] observations suggested two different methods of microbial attack on PHB: a preferential degradation of amorphous regions of the polymer by extracellular depolymerase, followed by colonization by bacteria of the film surface and subsequent localized degradation.

Although thc conditions in conventional municipal landfills are reputed to be unfavorable to biodegradation [212], PHBV was observed to lose weight in a simulated landfill environment, albeit at slower rates than those estimated in anaerobic sewage conditions and for PHB in certain types of soils [70]. Mergaert and co-workers [213] investigated the decomposition of PHB, P(3HB-*co*-10-mol% 3HV) and P(3HB-*co*-20-mol% 3HV) in household compost heaps. After 150 days, a substantial mass loss was observed for the P(3HB-*co*-20-mol% 3HV) only, but the authors noted that degradation rates depend strongly on the microbial population involved, the substrate specificity of the extracellular enzymes and the temperature, as has been mentioned elsewhere [211]. Gilmore et al. [214] observed important (50 to 100%) mass losses from PHBV samples (unspecified 3HV content) exposed for 120 days to municipal activated sludge, and showed that the degradation resulted entirely from biological activity. From laboratory experiments, Briese et al. [215] reported that PHBV degradation by aerobic sewage sludge was strongly pH-dependent, with pH's between 7 and 8.5 giving maximal rates.

Biodegradation experiments of polymer-films in marine and soil environments indicated that, in general and depending on the environment, biodegradation rates for unblended polymers were: polyhydroxybutyrate-*co*-valerate > cellophane > chitosan > polycaprolactone. Results from blends are more difficult to interpret since different biodegradation rates of the component polymers and leaching of plasticizers and additives can impact the data [216].

13. Blends of PHAs with other materials

Wheat starch granules and poly-(3-hydroxybutyrate-co-3-hydroxyvalerate) [P(HB-co-HV), (19.1 mol% HV)] were blended at 160 °C. Increasing the starch content from 0 to 50% (wt/wt) decreased the tensile strength of P(HB-co-HV) from 18 MPa to 8 MPa and diminished flexibility as Young's modulus increased from 1, 525 MPa to 2, 498 MPa, but overall mechanical properties of the polymer remained in a useful range. A mixed microbial culture required more than 20 days to degrade 150-microns-thick samples of 100% P(HB-co-HV), whereas samples containing 50% (wt/wt) starch disappeared in fewer than 8 days. Starch granules degraded before P(HB-co-HV) did. Aerobic degradation proceeded more rapidly than anaerobic degradation [217].

More recent work in the same field has been performed by Imam et al. [218]. Accordingly, degradation of the bioplastic, as determined by weight loss and deterioration of tensile properties, correlated with the amount of starch present (100% starch >50% starch > 30% starch > 100% PHBV). Incorporation of PEO into blends slightly retarded the rate of degradation. The rate of loss of starch from the 100% starch samples was about 2%/day, while the rate of loss of PHBV from the 100% PHBV samples was about 0.1%/day. Biphasic weight loss was observed for the starch-PHBV blends at all of the stations. A predictive mathematical model for loss of individual polymers from a 30% starch-70% PHBV formulation was developed and experimentally validated. The model showed that PHBV degradation was delayed 50 days until more than 80% of the starch was consumed and predicted that starch and PHBV in the blend had half-lives of 19 and 158 days, respectively. Consistent with the relatively low microbial populations, bioplastic degradation in deeper water exhibited an initial lag period, after which degradation rates comparable to the degradation rates at the other stations were observed. Presumably, significant biodegradation occurred only after colonization of the plastic, a parameter that was dependent on the resident microbial populations. Therefore, it can be reasonably inferred that extended degradation lags would occur in open ocean water where microbes are sparse.

The structure and mechanical properties of melt-pressed sheets of bacterially produced poly-3-hydroxybutyrate (PHB) and 3-hydroxybutyrate-3-hydroxyvalerate copolymer (PHBV) filled with various amounts of particulate maize starch granules were investigated by Koller and Owen [219]. The experimental methods included stress-strain measurements, differential thermal analysis to detect heats of fusion and melting temperatures, SEM, and optical microscopy to obtain information on crysallization behavior and spherulite growth. The reinforcing effect of destructured starch was superior to that of native starch granules, presumably due to surface roughness, irregularity of shape, and its smaller particle size. Brittle PHB became even more brittle with the addition of starch, while PHBV was more suitable for the filler since it was far less brittle. Starch apparently increased the degree of crystallinity of PHBV without affecting nucleation behavior.

Blends of PHBV and polyalcohols have been prepared by solvent casting. The polyalcohols used were castor oil (CO) and poly-propylene glycol (PPG400 and PPG1000). Thermal behaviour, crystallinity, morphology and dynamic mechanical properties of systems with various PPG1000 compositions have been studied. Crystallinity was determined by means of Fourier transform infrared spectroscopy

(FTIR) and X-ray diffraction. Final morphology was studied by means of scanning electron microscopy (SEM). Dynamical mechanical analysis showed two glass transition temperatures for the blends, corresponding to separate phases of PHBV and polyalcohols. Blend immiscibility was found, and polyalcohol addition enhances the crystallinity of PHBV. However, the storage modulus value decreases, upon the addition of amorphous compound [220].

Atactic poly(epichlorohydrin) (PECH) rubber was blended in a wide range of ratios with poly[D(-)3-hydroxybutyrate] (PHB) isolated from *Alcaligenes eutrophus* cultures [221]. The crystallization and thermal behavior of PECH/PHB blends were analyzed by DSC and microscopy. All blends showed a single glass transition temperature, whose value was dependent on composition, in good agreement with the theoretical values calculated by the Fox equation. The influence of blend composition on the overall crystallization rate and on the spherulite growth rate suggested that the two components formed a miscible blend in the amorphous phase. The equilibrium melting temperature decreased with blending; the negative value of the interaction parameter of the PECH/PHB system suggested that the two components could form a miscible mixture which was thermodynamically stable above the equilibrium melting temperature The reorganization phenomenon of PECH crystals after first melting was also observed.

The biodegradability of solvent-cast films of poly(R-3-hydroxybutyrate) (PHB) blended with the melt-compatible component atactic polyepichlorohydrin (PECH) was investigated by Sadocco et al. [222]. A bacterium which produced extracellular enzymes that degraded PHB even when blended with PECH was isolated, and tentatively designated as *Aureobacterium saperdae*. The growth rate of *A. saperdae* decreased with increasing PECH content in the blend, at 60 wt.-% PECH in the blend growth was completely inhibited. The decrease in the bacterial growth rate could be due to the dilution. of PHA molecules on the blend film surface caused by the presence of PECH molecules. At the stationary phase of bacterial growth the percentage of weight loss of blend films decreased with increasing PECH fraction, which was probably due to the lower accessibility of PHB when blended with PECH. During the bacterial growth only PHB was metabolized, whereas neither degradation nor abiotic release of PECH was detected for blend films.

Gatenholm and Mathiasson report on the synergistic effects observed during the processing of cellulose with bacteria-produced PHB. Cellulose fibers processed with PHB-melt exhibited defibrillation as shown by SEM. Microscopic studies of extended. fibers demonstrated that a dramatic fiber-size reduction occurred during processing. The size reduction was related to the degree of processing, which, as the authors believe, depended on the amt. of crotonic acid produced during thermal degradation of PHB. Size exclusion chromatography studies of PHB samples thermally treated under well-defined conditions showed a decrease in molecular weight as a function of treatment time. The samples processed with and without cellulose were analyzed for their molecular weight, and the processing of cellulose was found to contribute to a greater amount of chain scission, plausibly caused by local overheating as a result of shear forces developed during processing [223].

Cyclic oligo(3-hydroxybutyrate), oligo(3-HB), was synthesized and purified by Brandl et al [224], resulting in oligolides that contained three to seven (R)-3-hydroxybutyrate units (triolides up to heptolides). In addition, linear 3-HB octamers obtained as either tertiary-butyl or methyl esters were substituted with different end groups at the hydroxy end. The hydroxy terminus was replaced by either a benzyloxy, trifluoroacetoxy, crotonyloxy (S)-3-hydroxybutyryloxy, or fluorenylmethylcarbonyloxy (FMOC) group. P(3-HB) hairpin loops occurred on the surface of certain regions of the polymer, especially of lamellar crystallites. Cyclic 3-HB oligomers provide a model system for these loops. It is assumed that they provide attachment points for the depolymerizing enzymes. All of the (R)-oligolides tested were degraded except the (R)-triolide. Triolides were not degraded, suggesting that enzymatic attack was prevented presumably by steric hindrance on the rigid ring system. Unsubstituted linear octamers were degraded. Biodegradation was prevented when the hydroxy terminus was protected by the FMOC group, but was not dependent on a free hydroxy terminal group; all other protecting groups did not prevent degradation. Substitution of the carboxy end of a methyl or tertiary butyl ester group did not influence biodegradation.

The compatibility of a crystalline/crystalline polymer blend, poly(vinyl alcohol) (PVA)/poly-3-hydroxybutyrate (PHB), was studied by high-resolution solid-state ^{13}C-NMR and ^{1}H NMR spectroscopy [225]. The ^{1}H spin relaxation time (T1) measurements indicate that both the compatibility and the domain size depend on the composition of the blend. The PVA/PHB blend is compatible only when the blend contains a large amount of PVA. The domain sizes of the compatible blend are less than 200 Å. In the pulse saturation transfer (PST) MAS NMR spectra, the carbonyl carbon resonance from PHB showed a down-field shift, which indicates that the compatibility of the PVA/PHB blends is due to the hydrogen-bonding interaction in the amorphous phase. The DSC measurement showed that the compatible blends adopt low crystallinity for both PHB and PVA. The crystallization in these blends is likely to be disturbed by the hydrogen-bonding interaction in the amorphous phase. The compatibility of PVA/PHB is also affected by the tacticity of PVA. The compatible composition range of the syndiotactic-rich PVA/PHB blend is wider than that of the atactic-PVA/PHB blend. The capacity to form the hydrogen bond depends on the tacticity of PVA.

The melting and crystallization behavior of PHB and poly(ethylene succinate) blends has been studied by differential scanning calorimetry and optical microscopy [226]. The results indicate that PHB and PES are miscible in the melt. Consequently the blend exhibits a depression of the melting temperature of both PHB and PES. In addition, a depression of the equilibrium melting temperature of PHB is observed. The Flory-Huggins interaction parameter, obtained from melting point depression data, is composition dependent, and its value is always negative. Isothermal crystallization in the miscible blend system PES/PHB is examined by polarized optical microscope. The presence of the PES component gives a wide variety of morphologies. The spherulites exhibit a banded structure and the band spacing decreases with increase PES content.

14. Strategies of PHA production

Since the early fermentation descriptions by Baptist [43, 44] and others [59-61, 69], an important amount of research has looked into the optimization of PHA production processes. While not directly concerning the matter of process improvement, much literature on PHAs has potential impacts on existing production technologies: the use of novel substrates, the utilization of new organisms, and the better understanding of known ones (and the role PHAs play in their lives). Other research efforts have specifically addressed the issue of productivity. Examples of these are reports on obtaining better substrate-to-product yields and production rates through improved control of conventional systems, and on the development of innovative fermentation techniques.

The description of present or past processes for the commercial production of PHAs have been the subject of recent reviews [75, 135, 227, 228]. Current industrial PHBV production utilizes a glucose-assimilating strain of *R. eutropha* in fed-batch regime with phosphate limitation as the triggering factor for biopolymer accumulation and propionic acid as 3HV precursor [227]. For convenience, this *R. eutropha*/carbohydrate/ propionate combination in fed-batch mode will be referred to here as a conventional strategy for production of PHBVs.

This section will review recent reports of attempts to produce PHA. Emphasis will be put on processes using *R. eutropha* or *A. latus*, and the survey will be limited to literature on PHA production in bioreactor only. While flask experiments are the necessary first stages in the development of new production strategies (e.g., the use of inexpensive substrates and sources of growth factors, as recently reported in [229] and [230], respectively) and therefore of high interest, the potential of a novel system can be truly assessed only once it has been 'scaled up' to fermentor operation, in which culture conditions at least approach those of industrial fermentations. The other area of research with a possible future impact on large-scale production of PHAs, the use of transgenic plants, will be briefly discussed.

14.1. INVESTIGATIONS AND VARIATIONS OF THE CONVENTIONAL STRATEGY

PHA concentrations of greater than 80 g l^{-1} with productivities of greater than 2 g PHA l^{-1} h^{-1} can be routinely obtained by fed-batch cultivation of several bacteria. Metabolic engineering approaches have been used to expand the spectrum of utilizable substrates and to improve PHA production. These advances will lower the price of PHA from the current market price of ca. US$ 16 kg^{-1}, and it is predicted that PHA will become a leading biodegradable plastic material in the near future [231].

Kim et al. [105, 232] have used on-line glucose control to obtain high-cell-density cultures of *R. eutropha* with high concentrations of PHB and PHBV. Following the observation that growth of *R. eutropha* was maximized at glucose concentrations between 10 and 20 g L^{-1}, the authors kept the sugar content of a 2.5-L culture within these limits during both growth and PHB-accumulation phases [105]. Close monitoring of the glucose concentration in the mineral-salts medium was achieved by either exit-gas analysis by mass spectrometry and stoichiometric deduction of glucose content from CO_2-evolution rate, or with automatic glucose assay of filtered broth samples. Nitrogen limitation was used as the inducing factor for polyester storage, and the effect of biomass concentration at the onset of accumulation on productivity was investigated. The dissolved-oxygen concentration (DOC) was kept above 20% of air saturation (AS) throughout. At its most productive, the culture produced in 50 h 164 g CDM L^{-1} containing 121 g PHB L^{-1} (76%). Overall polymer productivity was thus 2.42 g L^{-1} h^{-1}. In another, slightly less productive fermentation of the same type, the accumulation-phase yield of PHB from glucose varied between 0.30 and 0.23 g g^{-1}.

In similar experiments [232], these authors added propionic acid to the glucose solution to produce PHBV during the accumulation phase. The effect of the ratio of propionic acid to glucose in the solution on the final concentration, 3HV content, and productivity of copolymer was studied. The DOC was kept at 20% of AS at all times. Very high CDM and polymer concentrations were achieved here also. As the propionic acid-to-glucose feed ratio was increased from 0.17 to 0.52 mol mol^{-1}, the copolymer 3HV fraction went up from 4.3 to 14.3 mol%, but its productivity decreased from 2.55 to 1.64 g L^{-1} h^{-1}. Similarly, the yield of 3HV from propionic acid decreased from 0.33 to 0.28 mol mol^{-1}. These experiments by Kim and colleagues have yielded the highest copolymer productivities reported so far for PHA-producing fermentations. The higher homopolymer concentrations (136 g L^{-1}) obtained earlier by Suzuki et al. [233] with a methylotroph were reached only after 175 h.

Doi's group [234] has appropriately brought some of their previous work one step further by showing that PHBV synthesis from butyric and pentanoic acids, the mechanism of which was elucidated at the end of the last decade [81, 140, 143], can be exploited for the production of substantial amounts of the copolymer in a fermentor. In fed-batch cultures at high carbon-to-nitrogen ratio (40 mol mol^{-1}, but never a total nitrogen exhaustion), the presence of the two carbon sources during the growth and accumulation phases produced after 30 h up to 13.5 g P(3HB-*co*-27-mol% 3HV) L^{-1} representing 72% of cell dry mass (CDM) with high yields. Decreasing the C/N ratio led to a gradual inhibition of polymer synthesis simultaneous to an increase in its 3HV fraction. Similar results were obtained with fructose and pentanoic acid as carbon sources, but the 3HV fraction of the polymer was unaffected by the C/N ratio when fructose and pentanoic acid were used.

One possible significance of this work, not mentioned by its authors, is the effect of small quantities of nitrogen source during the accumulation phase. PHA production processes usually feature the total exhaustion of an element essential for growth. In this case residual biomass (all non-polymer cell material) stays constant during polymer storage. In the aforementioned cultures of Koyama and Doi, the supply of small amounts of nitrogen during the accumulation period supported a slight but constant growth of the cells. This may have played a role in the high amounts of polymer produced. These results were confirmed by Aragao et al. maintaining a controlled residual growth capacity by feeding suboptimal amounts of NH_4OH during PHA accumulation phase. Improved volumetric PHA productivities were obtained and the proportion of 3HV incorporated was approximately twofold higher in these cultures [235]. Bitar and Underhill [236] reported that limited feeding of ammonia during PHB accumulation from glucose by *R. eutropha* led to higher polymer production rates but no increases in the number of cells. Possibly, accommodation of enlarging polymer granules is realized by biosynthesis of additional cytoplasmic membrane and cell wall, thereby retarding the structural stresses that were proposed by Ballard et al. [120] as a stopping factor for polymer accumulation.

Large amounts of PHBV from glucose and valerate were also obtained in a nitrogen-free, fed-batch fermentation of *R. eutropha* by Lee et al. [237]. More than 90 g P(3HB-*co*-20-mol% 3HV) L^{-1} (75% of CDM) were produced in 50 h. The specific 3HV formation rate with valerate was higher than that obtained with propionate in the equivalent experiment, showing again the superior convertibility of valerate to 3-hydroxyvalerate units.

The ability of *Alcaligenes eutrophus* to grow and produce polyhydroxyalkanoates (PHA) on plant oils was evaluated by Fukui and Doi [238]. When olive oil, corn oil, or palm oil was fed as a sole carbon source, the wild-type strain of *A. eutrophus* grew well and accumulated poly-3-hydroxybutyrate homopolymer up to approximately 80% (w/w) of the cell dry weight during its stationary growth phase. In addition, a recombinant strain of *A. eutrophus* PHB-4 (a PHA-negative mutant), harboring a PHA synthase gene from *Aeromonas caviae*, was revealed to produce a random copolyester of 3-hydroxybutyrate and 3-hydroxyhexanoate from these plant oils with a high cellular content (approximately 80% w/w). The mole fraction of 3-hydroxyhexanoate units was 4-5 mol% whatever the structure of the triglycerides fed. The polyesters produced by the *A. eutrophus* strains from olive oil were 200-400 kDa (the number-average molecular mass). The results demonstrate that renewable and inexpensive plant oils are excellent carbon sources for efficient production of PHA using *A. eutrophus* strains.

In 1985 Braunegg and Bogensberger [167] have shown for the first time, that PHA production can also occur in association to the growth phase of microorganisms and not only in an accumulation phase. Dry biomass of *A. latus* DSM 1123 when grown on sucrose as a sole carbon source showed a PHB content between 58 % and 70% without any special growth limitation applied. Culture conditions for the optimum growth and biosynthesis of PHB in *A. latus* DSM 1123 were investigated by Cho et al [239]. Optimum carbon and nitrogen sources and their concentrations for growth were detected, and batch and fed-batch fermentations were performed in a 2.5 L jar type aerobic fermentor with various pH control solutions. Sucrose and $(NH_4)_2SO_4$ were the most effective carbon and nitrogen sources for the growth of *A. latus*. The optimum C/N ratio varied with the concentrations of carbon and nitrogen sources. The maximum specific growth rate was obtained at the sucrose concentration of 30 g/L and C/N ratio of 30. The specific growth rate increased more than two times and lag time was reduced when yeast extract and polypeptone were added. PHB was synthesized in the logarithmic growth phase. By using NH_4OH and NaOH solutions in the first and second stage as pH control solutions, significant increases in the specific growth rate, biomass and PHB concentrations were observed. Under optimal conditions, the maximal biomass and PHB accumulation yield ($Y_{P/X}$) attained after 40 h were 17.6 g/L and 46%, respectively.

Ten-liter-scale fed-batch fermentations (fermentations with addition of different substrates during the experiment) were used by Braunegg and co-workers [187] for the production of PHBV with *A. latus* growing on glucose and propionate, and P(3HB-*co*-4HB) with *R. eutropha* on glucose and γ-butyrolactone. In the first case, 6.6 g (72% of the CDM) of a copolymer containing 28 mol% 3HV were obtained per liter in only 33.75 h. With *R. eutropha*, 7.2 g P(3HB-*co*-7.9-mol% 4HB) L^{-1} was produced in 83 h.

A two-stage fed-batch method employing two different micro-organisms growing on two substrates in complex medium was reported by Tanaka et al. [240] to produce

PHB. In the first stage, the pentose xylose was converted by a strain of *Lactococcus lactis* to a mixture of lactic and acetic acids. After removal of the cells by (presumably aseptic) centrifugation, *R. eutropha* was used to inoculate the supernatant in the same 1-L fermentor. No nutrient deficiency was present to favor polymer synthesis, but the cells accumulated PHB to up to 55% of their CDM during growth on lactate. In 24 h, 4.7 g homopolymer L^{-1} were produced. With such a strategy for PHA production, the extent to which the advantages of the use of an inexpensive substrate like xylose are offset by the additional procedures needed to separate the two micro-organisms must be carefully calculated.

Enzymatically hydrolyzed potato processing waste has been studied as a possible source of a fermentable substrate for the production of PHB by *R. eutropha*. The results indicated that potato starch waste could be converted with high yield to a concentrated glucose solution. The most economical process used barley malt as a source of amylase enzyme with an optimal ratio of 10:90 g g^{-1} of potato waste. A conversion efficiency of 96% of the theoretical value was obtained with a final glucose concentration of 208 g L^{-1}. After dilution and addition of mineral salts the hydrolysate was converted by a batch culture to 5.0 g L^{-1} of PHB, comprising 77% of the cell dry weight [241].

Ishizaki and colleagues [242, 243] have perfected the operation of autotrophic cultures of *R. eutropha* for the production of high quantities of PHB. They addressed the two major difficulties of this strategy, namely poor utilization of gases and danger of explosion, by using a gas-recycling system and keeping the oxygen concentration in the gas feed below the lower explosion limit (approx. 7%), respectively. In a 2-L fermentor fed with recycled mixtures of H_2, O_2 and CO_2 in varying proportions, limited-dissolved-oxygen conditions led to the accumulation of 61.9 g PHB L^{-1} (68% of CDM) in 40 h, or an overall polymer productivity of 1.55 g L^{-1} h^{-1}. The authors noted however that the very high oxygen-transfer requirements of the fermentor, owing to the low O_2 content of the gas feed, would be problematic for the scaling-up of this system necessary for commercial exploitation.

One-liter and 75-L cyclone bioreactors have recently been used for the production of PHB [244]. Sheppard et al. utilized nitrogen-free cultures of *R. eutropha* on glucose and mineral salts to investigate the synthesis of the homopolymer in these stirrer-less vessels. The system could support up to 5 (75-L unit) to 8 g biomass L^{-1} (1-L unit) at DOCs above 80% of AS, suggesting possible higher cell concentrations with the appropriate media. The end homopolymer content of the cells varied between 80% and 96% in mass, which are unusually high values, and which led the authors to suspect important variations in the oxygen availability in the pumped recirculation loop, where the DOC was not monitored. The glucose-to-PHB yields were lower in the large-scale unit than in the smaller one. According to the authors, the potential advantage of cyclone reactors, compared to conventional stirred-tank vessels, are the lower fabrication costs owing to a simpler design.

An interesting approach to PHA production using a mixed culture is shown by Katoh et al. [245]. The mixed culture system was considered to be effective when sugars such as glucose are converted to lactate by *Lactobacillus delbrueckii* and the lactate is converted to poly-3-hydroxybutyrate (PHB) by *Alcaligenes eutrophus* in one fermentor. For the modeling of the effect of NH_3 concentration on the cell growth of A. eutrophus and PHB production rates, metabolic flux distributions were computed at two culture phases of cell growth and PHB production periods. It was found that the NADPH, generated through isocitrate dehydrogenate in the TCA cycle, was predominantly utilized for the reaction from α-ketoglutarate to glutamate when NH_3 was abundant, while it tended to be utilized for the PHB production through acetoacetyl CoA reductase as NH_3 concentration decreased. This phenomenon was reflected in the development of a mathematical model. In the mixed culture experiments, two phases were observed, namely the lactate production phase due to *L. delbrueckii* and the lactate consumption phase due *to A. eutrophus.* The lactate concentration could be estimated on-line by the amount of NaOH solution and HCl solution supplied to keep the culture pH at constant level. Several mixed culture experiments were conducted to see the dynamics of the system. Finally, a mathematical model which can describe the dynamic behavior of the present mixed culture was developed and the model parameters were tuned for fitting the experimental data. The model may be used for several purposes such as control, optimization, and understanding process dynamics.

Recombinant micro-organisms have also been investigated for large-scale production of PHAs. *E. coli* harboring the PHA synthesis genes of *R. eutropha* was used by Kim et al. [246] to produce PHB. In a 2.5-L, fed-batch pH-stat, 88.8 g PHB L^{-1} (76% of the CDM) were obtained after 42 h in a glucose-tryptone-yeast extract medium. Wang and Lee produced PHB by a fed-batch culture of filamentation-suppressed recombinant *E. coli.* A PHB concentration of 149 g L^{-1}, with a productivity of 3.4 g of PHB L^{-1} h^{-1}, could be obtained in pH-stat fed-batch culture in a defined medium. Insufficient oxygen supply at a DOC of 1 to 3% of air saturation during active PHB synthesis phase did not negatively affect PHB production. By growing cells to a concentration of 110 g L^{-1} and then controlling the DOC in the range of 1 to 3% of air saturation, a PHB concentration of 157 g L^{-1} and PHB productivity of 3.2 g of PHB L^{-1} h^{-1} were obtained. For scale-up studies, the fed-batch culture was carried out in a 50-liter stirred tank fermentor, in which the DOC decreased to zero when cell concentration reached 50 g L^{-1}. However, a relatively high PHB concentration of 101 g L^{-1} and PHB productivity of 2.8 g of PHB L^{-1} h^{-1} could still be obtained [247]. Poly(3HB-*co*-3HV) can be produced with recombinant *E. coli* strains efficiently from propionate. P(3HV) synthesis from valerate was only obtained upon induction with acetate and/or oleate [248].

The effects of rheological change by addition of sodium carboxymethyl cellulose (CMC) to culture medium in an air-lift-type fermentor on autotrophic production of poly-[D(-)3-hydroxybutyric acid] [PHB] by two-stage culture of *Alcaligenes eutrophus* was investigated by Taga et al [249]. Addition of 0.05% CMC increased PHB production rate during the PHB accumulation phase to twice that of the control culture. It was thought that addition of a small amount of CMC was beneficial for production of PHB employing the air-lift fermentor under safe autotrophic culture conditions in which oxygen concentration was maintained below 6.9% (v/v). The volumetric mass transfer coefficient (K_La) observed in the presence of CMC is shown to be correlated with the PHB production rate obtained.

Ramsay et al. [91] were the first to investigate PHB and PHBV production in one- and two-stage continuous cultures. In a one-stage chemostat, *R. eutropha* DSM 545 accumulated 33% of its dry mass as PHB when fed with a nitrogen-limited medium of glucose and mineral salts. PHBV was produced in similar experiments with *A. latus* when propionic or valeric acid was added to the feed mixture containing sucrose as main carbon source. In single-stage chemostat, feed propionic-acid concentrations of up to 5 g L^{-1} yielded a copolymer with a 3HV molar fraction reaching 20% without affecting the polymer productivity obtained with sucrose only. Substitution of the three-carbon acid with valeric acid led to higher 3HV contents in the copolymer. At high concentrations of propionic acid in the feed (8.5 g L^{-1}), assimilation of sucrose was inhibited. In this case, transfer of the reactor's effluent into a second chemostat led to complete consumption of the sugar and obtention of P(3HB-*co*-11-mol% 3HV) representing 58% in mass of the CDM. Unfortunately, the copolymer productivities and yields from substrates for these experiments were not reported. The authors pointed out, that the constitutively high PHA-synthesis rate of *A. latus* makes this micro-organism better suited for polymer production in single-stage continuous culture.

Koyama and Doi [205] also investigated PHBV production in chemostat by *R. eutropha* growing on fructose and pentanoic acid. By varying the dilution rate and ammonium sulfate concentration of the feed, they obtained a maximum productivity of 0.31 g of a 41-mol%-3HV copolymer L^{-1} h^{-1} (42% of CDM). Large amounts of unused fructose were detected in the culture broth.

Incomplete utilization of substrates, resulting from high carbon-to-nitrogen ratios in feeds, is often encountered in single-stage continuous PHA-producing processes. Unless the extra carbon can easily and cheaply be recycled back into the process, such losses entail a lower production profitability. The use of a second stage downstream from the first, as shown by Ramsay et al. [88], can advantageously increase the time of exposure for the organisms to conditions favorable for polymer accumulation, leading to higher yields and productivities. Attempts to develop continuous processes for a profitable production of PHAs will most probably be successful only when multi-stage arrangements are considered. Ramsay et al. [250] have argued that in the first stage, 50 to 60 g high-protein biomass L^{-1} would have to be produced. Braunegg et al. [187] have presented theoretical evidence that the use of a plug-flow tubular reactor for the second stage allows a maximal productivity for a number of organism/substrate systems, including *R. eutropha* and *A. latus* synthesizing PHAs from carbohydrates.

14.2. THE USE OF ALCOHOLS AS SUBSTRATES

Park et al. [251, 252] used a mutant strain of *R. eutropha* capable of using alcohols as a carbon source for the production of PHB and PHBV in fed-batch fermentors. With phosphate limitation as inducing factor, ethanol was used for the production in 7 L of 46.6 g PHB L^{-1} (74% of the CDM) in 50 h [251]. When 1-propanol was added to the medium, up to 15.1 mol% in 3HV units were incorporated to the polymer, and when propanol was the sole carbon source, the cells accumulated about 85% of their CDM in P(3HB-*co*-35.2-mol% 3HV). Both alcohols were completely consumed. In computer-controlled fermentations [252], switches between the alcohols or mixtures thereof led to improved copolymer yield from the substrates and production rates.

PHBVs were also produced, albeit in low concentrations, from alcohols in 3 L, fed-batch cultures of the methylotroph *Paracoccus denitrificans* [253]. With methanol and the five-carbon n-amyl alcohol as carbon sources periodically added during nitrogen limitation, the bacterium accumulated, in a total of about 140 h, 2.3 g P(3HB-*co*-60-mol% 3HV) L^{-1} (26% of the CDM). The authors noted that n-amyl alcohol incorporation into the copolymer was more efficient than that of methanol, which was mostly oxidized to carbon dioxide.

Several alcohols were examined as substrates for the PHA synthesis by *P. denitrificans*. From ethanol only PHB was formed, when n-pentanol was used the PHV was synthesized, whereas PHBV accumulated during growth on n-propanol. When alcohols were automatically fed as growth substrates, ethanol, n-propanol, and n-pentanol gave higher polyester content. Although PHB was synthesized from methanol or n-butanol, its content was very low. Under nitrogen-deficient conditions, PHA content in cells increased, especially with ethanol, n-propanol, and n-pentanol [254].

Bourque and co-workers [255] used a new soil isolate of *Methylobacterium extorquens* in preliminary experiments to produce PHB from methanol and PHBV from methanol and valerate in 2-L, fed-batch fermentations. For PHB production, the addition rate of methanol was computer-controlled in an attempt to keep the alcohol's concentration at an optimal level, which the authors previously determined to be 1.7 g L^{-1} for the new isolate. This only partially worked, and in spite of no growth-limiting factor, the CDM reached, after a long 160 h, 9 g L^{-1}, 30 to 33% of which as the homopolymer. The isolate was also able to accumulate a P(3HB-*co*-20-mol% 3HV) copolymer. The authors pointed out the advantages of using methanol as a carbon source: besides being a non-food substrate, it is relatively inexpensive, easily manipulable, completely miscible with water and of low viscosity.

PHB was produced in fermentor by Daniel et al. [256] using the facultative methylotroph *Pseudomonas* 135. In 40 h, the organism stored 6.1 g PHB L^{-1} (37% of its CDM) from methanol in ammonium-deficient conditions. PHB production triggered from deficiencies in magnesium and phosphate were also investigated.

14.3 THE USE OF PSEUDOMONADS

Ramsay's group has made extensive investigations of PHA production in fermentor by members of the genus *Pseudomonas*. *P. pseudoflava* was grown in batch fermentation on glucose, xylose or arabinose, and accumulated PHB from these, and was grown in chemostat on the hydrolysate from the hemicellulosic fraction of poplar-wood (mostly xylose as carbon source) [257]. Although Ramsay and colleagues did not specifically mention this, the use of this inexpensive hydrolysates to support the growth of *P. pseudoflava* in chemostat could quite possibly fulfill the first-stage requirements of a two-stage continuous process. *P. oleovorans* was used in chemostat to produce medium-side-chain PHAs from sodium octanoate [258].

PHA synthesis cultures by *P. putida* KT2442 growing on long-chain fatty acids in continuous culture was studied by Huiberts and Eggink [259]. The effects of growth rate on biomass and polymer concentration were determined, the highest volumetric productivity was 0.13 g PHA L^{-1} h^{-1} at a specific growth rate of 0.1 h^{-1}. The molecular mass of the polymer remained constant at all growth rates but changes in the monomeric composition of the P(3HHx-*co*-3HO-*co*-3HD-*co*-3HDD-*co*-3HDDE-*co*-3HTDE) synthesized were observed. Optimal PHA formation was observed at a C/N ratio of 20 mol mol^{-1}. In order to optimize PHA production *P. putida* KT2442 was cultivated to high cell densities (30 g L^{-1}, 23% PHA) in oxygen-limited continuous cultures. This corresponds to a volumetric productivity of 0.69 g L^{-1} h^{-1} [260]. Due to the fact that some of the substrates, such as octanoic acid, alkenoic acids, and halogenated derivatives, are toxic to *Pseudomonas putida* KT2442, when present in excess, efficient production of mcl-PHAs therefore requires control of the carbon source concentration in the supernatant. Therefore a closed-loop control system based on on-line gas chromatography was developed in order to maintain continuously fed substrates at desired levels. In combination with the graphical programming environment LABVIEW, a flexible process control system was set up that allows users to perform supervisory process control and permits remote access to the fermentation system over the internet. Single-substrate supernatant concentration in a high-cell-density fed-batch fermentation process was controlled by a proportional (P) controller (P = 50%) acting on the substrate pump feed rate. Sodium octanoate concentrations oscillated around the setpoint of 10 mM and could be maintained between 0 and 25 mM at substrate uptake rates as high as 90 mmol L^{-1} h^{-1}. Under co-feeding conditions Na-10-undecenoate and Na-octanoate could be individually controlled at 2.5mM and 9mM, respectively, by applying a proportional integral (PI) controller for each substrate. The resulting copolymer contained 43.5 mol% unsaturated monomers and reflected the ratio of 10-undecenoate in the feed. It was suggested that both substrates were consumed at similar rates. These results showed that this control system was suitable for avoiding substrate toxicity and supplying carbon substrates for growth and mcl-PHA accumulation [260].

A two-step fed-batch cultivation of *P. putida* was performed with glucose and octanoate as the main carbon source for cell growth and PHA accumulation, respectively. Under nitrogen-and oxygen-limiting conditions 18.6 g L^{-1} PHA were obtained with a yield of about 0.4 g PHA g^{-1} octanoate. By supplying octanoate in the first step, production of mcl-PHAs was significantly enhanced; it yielded 35.9 g PHA L^{-1} (65.5% of cell dry mass) after 39 h of the fed-batch operation. This indicated that octanoate addition during growth stimulated quite efficiently the biosynthesis of mcl-PHAs [261].

When cultivated in chemostat with octanoate as sole carbon source and nitrogen limitation [258], *P. oleovorans* produced a PHA with a 3HB/3HHx/3HO/3HD ratio of 0.1:1.7:20.7:1.0 which was relatively independent of the octanoate concentration of the feed. The maximum productivity in copolymer was approx. 0.14 g L^{-1} h^{-1} from 7 g octanoate L^{-1} (all used) at the dilution rate investigated (0.24 h^{-1}).

P. oleovorans was also used by Preusting et al. to produce a similar medium-side-chain copolymer from high-cell-density, 2-liquid phase fed-batch fermentations with n-octane as the sole carbon source [262]. After 48 h in a 2.6-L, custom-made fermentor, 12.1 g copolymer L^{-1} (33% of the CDM) were produced with a productivity of 0.25 g L^{-1} h^{-1}. The preceding growth phase was supported by dosed additions of ammonium and magnesium source. The authors noted that the optimization of the medium composition and strategy of nutrient dosing should lead to even higher productivities.

14.4. THE USE OF BURKOLDERIA CEPACIA

B. cepacia (the new designation for *P. cepacia*) was used by Ramsay's group to produce PHB from fructose in batch fermentation [250]. *B. cepacia* accumulated in about 80 h 2.6 g PHB L^{-1} (47% of the CDM) in nitrogen-limited cultures. However, since about 35 g fructose L^{-1} were consumed by the micro-organism to achieve this, the polymer-from-sugar yield was only slightly above 0.074 g g^{-1}. Although this is low, the authors argued that since *B. cepacia* is likely to be capable of utilizing a variety of industrial or food wastes for growth, profitable PHA-producing processes might be conceivable with this organism.

The poly(3-HB-*co*-3H4PE) produced by strains of *B. cepacia* was obtained in substantial quantities by fermentation [113]. In an 10-L reactor, the new soil isolates produced in 72 h 17 to 34 g CDM L^{-1} (depending on the strain) from gluconic acid. The cells contained up to 70% wt/wt of a 3-mol%-3H4PE copolymer. One strain was also used in a 250-L culture to produce a 3.5-mol%-3H4PE polyester, albeit at concentrations below 2 g L^{-1} after 46 h.

14.5. THE USE OF AZOTOBACTER VINELANDII UWD

Since its first production by the transformation of *A. vinelandii* UWD in 1989 by Page and Knosp [263], *A. vinelandii* UWD has grown from a microbiological curiosity to a serious contender for the most interesting potential organism for the production of PHAs from inexpensive carbon sources. Early on, Page's group detected a very low NADH oxidase activity in the new strain. They concluded that this defect could explain the bacterium's habit of accumulating PHB in high quantities (65 to 75% of the CDM) during exponential growth: the polymer-synthesizing process is one (but not the only) way NAD can be regenerated [263]. This was the beginning of the work that eventually inspired Genser et al. [157] to search for a similar cause for the constitutive PHA accumulation in *A. latus*. Page and Cornish [264, 195] also found that while nitrogen-fixation in strain UWD interfered with PHB synthesis from glucose during nitrogen-depleted accumulation conditions, the addition to the medium of fish peptone, a complex nitrogen source, restored even greater polymer production rates and yields.) Since *A. vinelandii* UWD can utilize unrefined, complex carbon sources, such as beet molasses, for PHA production [265], Page et al. have investigated the potential of the micro-organism for profitable biopolymer-producing processes.

After establishing that valerate was the best 3HV precursor (and that propionate was no precursor), Page and co-workers [266] used the salt in combination with beet molasses to produce PHBV from strain UWD in a 2.5-L fermentor. Thirty-eight to 40 h of fed-batch regime with various valerate-addition strategies yielded 18 to 22 g copolymer L^{-1} (59 to 71% of the CDM) containing 8.5 to 23 mol% 3HV. In the absence of valerate, the cells produced only PHB to 23 g L^{-1} (66% of the CDM). The maximum PHA production rate was 1.1 g L^{-1} h^{-1}. Small amounts of P(3HB-*co*-45-mol% 3HV) were produced when valerate was the only carbon source. 4-Pentenoate was also used as 3HV precursor in fermentor. The authors claim that the utilization of beet molasses as carbon source should reduce by more than half the cost of PHAs formed from glucose.

A simpler polymer-extraction process using the highly fragile pleomorphic cells of *A. vinelandii* UWD formed after supplementing the nutrient broth with fish peptone [195] can be envisaged to also contribute to lower production costs.

14.6. THE USE OF OTHER MICROORGANISMS

The halobacterium *Haloferax mediterranei* is an organism of biotechnological potential due to its capacity as polymer producer [267]. This microorganism accumulates PHB as intracellular granules in very large amounts. The culture conditions for PHB production have been studied and optimized in batch as well as in continuous culture. Phosphate limitation is an essential condition for PHB accumulation in important quantities. Glucose and starch were the carbon sources giving the highest productivities. Under favorable circumstances in batch culture a concentration of 6.5 g/L PHB was reached, being 67% of the total biomass dry weight, and with a yield coefficient of 0.33 gg^{-1} for the carbon source. There is evidence indicating that the PHB produced by *H. mediterranei* is a copolymer with monomers of more than four carbon atoms, which is favorable for the manipulation of the physical properties of the polymer. *H. mediterranei* accumulates another polymer extracellularly. It is a sulfated heteropolysaccharide that has interesting rheological properties that are very resistant to environmental stress including, as may be expected, salinity. High productivities of this polysaccharide require sugars as carbon source, a condition that also favors PHB accumulation.

A mixed culture of polyphosphate accumulating bacteria was investigated with respect to PHA production by Lemos et al. [268]. Attention was devoted to understand how different carbon substrates and their concentrations can influence the production of PHA by polyphosphate-accumulating bacteria. Acetate, propionate, and butyrate were tested independently. The composition of the polymers formed was found to vary with the substrate used. Acetate leads to the production of a copolymer of hydroxybutyrate (HB) and hydroxyvalerate (HV) with the HB units being dominant. With propionate, HV units are mainly produced and only a small amount of HB is synthesized. When butyrate is used, the amount of polymer formed is much lower with the HB units being produced to a higher extent. The yield of polymer produced per carbon consumed ($Y_{P/S}$) was found to diminish from acetate (0.97) to propionate (0.61) to butyrate (0.21). Using a mixture of acetate, propionate, and butyrate and increasing the carbon source concentration, although maintaining the relative concentration of each substrate, propionate is primarily consumed and consequently, PHA synthesized was enriched in HV units. All the polymers synthesized were found to be quite homogeneous and their average molecular weight is of the same order of magnitude as the ones commercially available.

14.7. RECOMBINANT STRAINS FOR PHA PRODUCTION

As early as 1988 the poly-3-hydroxybutyrate (PHB) biosynthetic pathway from *Alcaligenes eutrophus* H16 has been cloned and expressed in *Escherichia coli* [269]. Initially, an *A. eutrophus* H16 genomic library was constructed by using cosmid pVK102, and cosmid clones that encoded the PHB biosynthetic pathway were sought by assaying for the first enzyme of the pathway, ß-ketothiolase. Six enzyme-positive clones were identified. Three of these clones manifested acetoacetyl coenzyme A reductase activity, the second enzyme of the biosynthetic pathway, and accumulated PHB. PHB was produced in the cosmid clones at approximately 50% of the level found in *A. eutrophus*. One cosmid clone was subjected to subcloning experiments, and the PHB biosynthetic pathway was isolated on a 5.2-kilobase KpnI-EcoRI fragment. This fragment, when cloned into small multicopy vectors, can direct the synthesis of PHB in *E. coli* to levels approaching 80% of the bacterial cell dry weight.

Similar experiments have been performed and published by Schubert et al.[270]. Eight mutants of *A. eutrophus* defective in the intracellular accumulation of poly-b-hydroxybutyric acid (PHB) were isolated after transposon Tn5 mutagenesis with the suicide vector pSUP5011. EcoRI fragments which harbor Tn5-mob were isolated from pHC79 cosmid gene banks. One of them, PPT1, was used as a probe to detect the intact 12.5-kilobase-pair EcoRI fragment PP1 in a lL47 gene bank of *A. eutrophus* genomic DNA. In six of these mutants (PSI, API, GPI, GPIV, GPV, and GPVI), the insertion of Tn5-mob was physically mapped within a region of approx. 1.2 kilobase pairs in PP1; in mutant API, cointegration of vector DNA has occurred. In two other mutants (GPII and GPIII), most probably only the insertion element had inserted into PP1. All PHB-negative mutants were completely impaired in the formation of active PHB synthase, which was measured by a radiometric assay. In addition, activities of ß-ketothiolase and of NADPH-dependent acetoacetyl CoA (acetoacetyl-CoA) reductase were diminished, whereas the activity of NADH-dependent acetoacetyl-CoA reductase was unaffected. In all PHB-negative mutants, the ability to accumulate PHB was restored upon complementation in trans with PP1. The PHB-synthetic pathway of *A. eutrophus* was heterologously expressed in *E. coli*. Recombinant strains of *E. coli* JM83 and K-12, which harbor pUC9-1::PP1, pSUP202::PP1, or pVK101::PP1, accumulated PHB up to 30% of the cellular dry wt. Crude extracts of these cells had significant activities of the enzymes PHB synthase, ß-ketothiolase, and NADPH-dependent acetoacetyl-CoA reductase. Therefore, PP1 most probably encode all three genes of the PHB-synthetic pathway in *A. eutrophus*. In addition to PHB-negative mutants, mutants were isolated which accumulate PHB at a much lower rate than the wild type does. These PHB-leaky mutants exhibited activities of all three PHB-synthetic enzymes; Tn5-mob had not inserted into PP1, and the phenotype of the wild type could not be restored with fragment PP1. The rationale for this mutant type remains unknown.

Alcaligenes eutrophus transformants AER3, AER4 and AER5 harboring cloned phbCAB, phbAB and phbC genes (from *A. eutrophus* encoding acetyl-CoA-acetyltransferase (EC-2.3.1.16), aceto-acetyl-CoA-reductase and poly-hydroxybutyrate-synthase) introduced via shuttle vector plasmid pKT230) were cultured under various different culture conditions to elucidate the optimal culture conditions for accumulation of PHB. The transformants showed increased total cell growth and PHB accumulation due to the recombinant enzymes. In batch culture, the transformant synthesized PHB more effectively at the high C/N molar ratio and low C compared to the parent strain. Fed-batch culture was more effective for maximizing PHB biosynthesis compared to the batch culture mode. The plasmid stability was maintained at about 85% after 36 hr and elongated morphological changes of transformant at the early growth stage were noticed. The gene amplification through the transformation of cloned PHB biosynthesis genes in *A. eutrophus* appears to be an excellent method for strain improvement to achieve an effective accumulation of PHB [271].

The increase of gene dosage of the poly-3-hydroxybutyrate biosynthesis operon in *Ralstonia eutropha* to test whether PHB synthesis rates may be increased by recombinant methods was studied by Jackson and Srienc [272]. The native *R. eutropha* phbCAB operon was inserted into the broad-host-range vector pKT230. This PHB operon-containing plasmid, and a control plasmid containing the identical broad-host-range replicon but not the PHB genes, was transferred to *R. eutropha* H16. Analysis of whole-cell lysates indicated that the strain harboring the operon-containing plasmid possessed ß-ketothiolase and acetoacetyl-CoA reductase specific activities that were 6.0 and 6.2 times elevated, respectively, as compared to the control strain with a single operon. After growth on fructose, PHB synthesis rates were sharply dependent on the type of carbon source offered during the PHB accumulation phase under nitrogen limitation. In the case of the strain harboring the control plasmid, and in comparison to fructose as carbon source, PHB accumulation was 2.15, 2.83, and 2.60 times faster when resuspended in nitrogen-free medium with lactate, acetate, or 3-hydroxybutyrate, respectively. The strain harboring the PHB operon-containing plasmid synthesized PHB at a lower specific rate in each case. During exponential growth on fructose, the strain harboring the control plasmid was again more efficient at forming PHB. These results suggest that increasing the intracellular concentration of PHB precursors may be a superior alternative to raising the levels of PHB enzymes for enhancing PHB productivity in *R. eutropha*.

Recombinant PHA producers seem to have several advantages as PHA producers compared with wild-type PHA-producing bacteria. However, the PHA productivity (amount of PHA produced per unit volume per unit time) obtained with these recombinant *E. coli* strains has been lower than that obtained with the wild-type bacterium *Alcaligenes latus*. To endow the potentially superior PHA biosynthetic machinery to *E. coli*, the PHA biosynthesis genes from *A. latus* have been cloned [273]. The three PHA biosynthesis genes formed an operon with the order PHA synthase, ß-ketothiolase, and acetoacetyl-CoA genes and were constitutively expressed from the natural promoter in *E. coli*. Recombinant *E. coli* strains harboring the *A. latus* PHA biosynthesis genes accumulated poly-3-hydroxybutyrate (PHB), a model PHA product, more efficiently than those harboring the *R. eutropha* genes. With a pH-stat fed-batch culture of recombinant *E. coli* harboring a stable plasmid containing the *A. latus* PHA biosynthesis genes, final cell and PHB concentrations of 194.1 and 141.6 gL^{-1}, respectively, were obtained, resulting in a high productivity of 4.63 g of PHB/liter/h. This improvement should allow recombinant *E. coli* to be used for the production of PHB with a high level of economic competitiveness.

Two types of polyhydroxyalkanoate (PHA) biosynthesis gene loci (phb and pha) of *Pseudomonas sp.* strain 61-3, which produces a blend of poly-3-hydroxybutyrate [PHB] homopolymer and a random copolymer poly(3-hydroxybutyrate-co-3-hydroxyalkanoate) [P(3HB-co-3HA] consisting of 3HA units of 4 to 12 carbon atoms, were cloned and analyzed at the molecular level [274]. In the phb locus, three open reading frames encoding polyhydroxybutyrate (PHB) synthase (PhbCPs), ß-ketothiolase (PhbAPs), and NADPH-dependent acetoacetyl coenzyme A reductase (PhbBPs) were found. The genetic organization showed a putative promoter region, followed by phbBPs-phbAPs-phbCPs. Upstream from phbBPs was found the phbRPs gene, which exhibits significant similarity to members of the AraC/XylS family of transcriptional activators. The phbRPs gene was found to be transcribed in the opposite direction from the three structural genes. Cloning of phbRPs in a relatively high-copy vector in *Pseudomonas sp*. strain 61-3 elevated the levels of ß-galactosidase activity from a transcriptional phb promoter-lacZ fusion and also enhanced the 3HB fraction in the polyesters synthesized by this strain, suggesting that PhbRPs is a positive regulatory protein controlling the transcription of phbBACPs in this bacterium. In the pha locus, two genes encoding PHA synthases (PhaC1Ps and PhaC2Ps) were flanked by a PHA depolymerase gene (phaZPs), and two adjacent open reading frames (ORF1 and phaDPs), and the gene order was ORF1, phaC1Ps, phaZPs, phaC2Ps, and phaDPs. Heterologous expression of the cloned fragments in PHA-negative mutants of *Pseudomonas putida* and *Ralstonia eutropha* revealed that PHB synthase and two PHA synthases of *Pseudomonas* sp. strain 61-3 were specific for short chain length and both short and medium chain length 3HA units, respectively.

In order to scale up medium-chain-length polyhydroxyalkanoate (mcl-PHA) production in recombinant microorganisms, Prieto et al. [275] generated and investigated different recombinant bacteria containing a stable regulated expression system for phaC1, which encodes one of the mcl-PHA polymerases of *Pseudomonas oleovorans*. The mini-Tn5 system was used as a tool to construct *Escherichia coli* 193MC1 and *P. oleovorans* POMC1, which had stable antibiotic resistance and PHA production phenotypes when they were cultured in a bioreactor in the absence of antibiotic selection. The molecular weight and the polydispersity index of the polymer varied, depending on the inducer level. *E. coli* 193MC1 produced considerably shorter polyesters than *P. oleovorans* produced; the weight average molecular weight ranged from 67, 000 to 70, 000, and the polydispersity index was 2.7. Lower amounts of inducer added to the media shifted the molecular weight to a higher value and resulted in a broader molecular mass distribution. In addition, it was found that *E. coli* 193MC1 incorporated exclusively the R configuration of the 3-hydroxyoctanoate monomer into the polymer, which corroborated the enantioselectivity of the PhaC1 polymerase enzyme.

Interesting results where published by Dennis et al. [276] on the formation of poly(3-hydroxybutyrate-co-3-hydroxyhexanoate). The acetoacetyl-CoA reductase and the polyhydroxy-alkanoate (PHA) synthase from *Ralstonia eutropha* were expressed in *Escherichia coli*, *Klebsiella aerogenes*, and PHA-negative mutants of *R. eutropha* and *Pseudomonas putida*. While expression in *E. coli* strains resulted in the accumulation of poly-3-hydroxybutyrate [PHB], strains of *R. eutropha*, *P. putida* and *K. aerogenes* accumulated poly(3-hydroxybutyrate-co-3-hydroxyhexanoate) [poly(3HB-co-3HHx)] when even chain fatty acids were provided as carbon source, and poly(3-hydroxybutyrate-co-3-hydroxyvalerate) [poly(3HB-co-3HV)] when odd chain fatty acids were provided as carbon source. This suggests that fatty acid degradation can be directly accessed employing only the acetoacetyl-CoA reductase and the PHA synthase. This is also the first proof that the PHA synthase from *R. eutropha* can incorporate 3-hydroxyhexanoate (3HHx) into PHA and has, therefore, a broader substrate specificity than previously described.

Recombinant strains of *Ralstonia eutropha* PHB-4, which harbored *Aeromonas caviae* polyhydroxyalkanoates (PHA) biosynthesis genes under the control of a promotor for *R. eutropha* phb operon, were examined for PHA production from various alkanoic acids [277]. The recombinants produced poly(3-hydroxybutyrate-co-3-hydroxyhexanoate) [P(3HB-co-3HHx)] from hexanoate and octanoate, and poly(3-hydroxybutyrate-co-3-hydroxyvalerate-co-3-hydroxypentanoate) [P(3HB-co-3HV-co-3HHp)] from pentanoate and nonanoate. One of the recombinant strains, *R. eutropha* PHB-4/pJRDBB39d3 harboring ORF1 and PHA synthase gene of *A. caviae* (phaCAc) accumulated copolyesters with much more 3HHx or 3HHp fraction than the other recombinant strains. To investigate the relationship between PHA synthase activity and in vivo PHA biosynthesis in *R. eutropha*, the PHB-4 strains harboring pJRDBB39d13 or pJRDEE32d13 were used, in which the heterologous expression of phaCAc was controlled by promoters for *R. eutropha* phb operon and *A. caviae* pha operon, respectively. The PHA contents and PHA accumulation rates were similar between the two recombinant strains in spite of the quite different levels of PHA synthase activity, indicating that the polymerization step is not the rate-determining one in PHA biosynthesis by *R. eutropha*. The molecular weights of poly-3-hydroxybutyrate produced by the recombinant strains were also independent of the levels of PHA synthase activity. It was suggested that a chain-transfer agent is generated in *R. eutropha* cells to regulate the chain length of polymers.

In order to produce poly(3-hydroxybutyrate-co-3-hydroxyhexanoate) and poly(3-hydroxyvalerate-co-3-hydroxyheptanoate), the PHA synthase gene (phaCNc) from *Nocardia corallina* was identified in a lambda library on a 6-kb BamHI fragment. A 2.8-kb XhoII subfragment was found to contain the intact PHA synthase. This 2.8-kb fragment was subjected to DNA sequencing and was found to contain the coding region for the PHA synthase and a small downstream open reading frame of unknown function. On the basis of DNA sequence, phaCNc is closest in homology to the PHA synthases (phaCPaI and phaCPaII) of *Pseudomonas aeruginosa* (approximately 41% identity and 55% similarity). The 2.8-kb XhoII fragment containing phaCNc was subcloned into broad host range mobilizable plasmids and transferred into *Escherichia coli*, *Klebsiella aerogenes* (both containing a plasmid bearing phaA and phaB from *Ralstonia eutropha*), and PHA-negative strains of *R. eutropha* and *Pseudomonas putida*. The recombinant strains were grown on various carbon sources and the resulting polymers were analyzed. In these strains, the PHA synthase from *N. corallina* was able to mediate the production of poly(3-hydroxybutyrate-co-3-hydroxyhexanoate) containing high levels of 3-hydroxyhexanoate when grown on hexanoate and larger even-chain fatty acids and poly(3-hydroxyvalerate-co-3-hydroxyheptanoate) containing high levels of 3-hydroxyheptanoate when grown on heptanoate or larger odd-chain fatty acids [278].

Another approach towards production of mcl PHAs from recombinant *E. coli* was shown by Doi's group [279]. The *Escherichia coli* 3-ketoacyl-ACP reductase gene (fabGEc) was cloned using a PCR technique to investigate the metabolic link between fatty acid metabolism and polyhydroxyalkanoate (PHA) production. Three plasmids respectively harboring fabGEc and the poly-3-hydroxyalkanoate synthesis genes phaCAc and phaC1Ps from *Aeromonas caviae* and *Pseudomonas sp.* 61-3 respectively were constructed and introduced into *E. coli* HB101 strain. On a two-stage cultivation using dodecanoate as the sole carbon source, recombinant *E. coli* HB101 strains harboring fabGEc and phaC genes accumulated PHA copolymers (about 8 wt% of dry cell weight) consisting of several (R)-3-hydroxyalkanoate units of C4, C6, C8, and C10. It was suggested that overexpression of the fabGEc gene leads to the supply of (R)-3-hydroxyacyl-CoA for PHA synthesis via fatty acid degradation.

14.8. IN VITRO PRODUCTION OF PHAS

Beside the studies of PHA production in fermentation processes applying living microorganisms also in vitro systems may be used in future. A combined chemical and enzymatical procedure has been developed to synthesize macroscopic poly[(R)-3-hydroxybutyrate] (PHB) granules in vitro. The granules form in a matter of minutes when purified polyhydroxyalkanoate (PHA) synthase from *Alcaligenes eutrophus* is exposed to synthetically prepared (R)-3-hydroxybutyryl CoA, thereby establishing the minimal requirements for PHB granule formation. The artificial granules are spherical with diameters of up to 3 mm and significantly larger than their native counterparts (0.5 mm). The isolated PHB was characterized by ^{1}H and ^{13}C NMR, gel-permeation chromatography, and chemical analysis. The in vitro polymerization system yields PHB with a molecular mass > 10.10^{6} Da, exceeding by an order of magnitude the mass of PHBs typically extracted from microorganisms. It was demonstrated that the molecular mass of the polymer can be controlled by the initial PHA synthase concentration. Preliminary kinetic analysis of de novo granule formation confirms earlier findings of a lag time for the enzyme but suggests the involvement of an additional granule assembly step. Minimal requirements for substrate recognition were investigated. Since substrate analogs lacking the adenosine 3', 5'-bisphosphate moiety of (R)-3-hydroxybutyryl CoA were not accepted by the PHA synthase, the authors provide evidence that this structural element of the substrate is essential for catalysis [280].

Additional work in this field was performed by Lenz et al. [281], showing the effectivity of glycerol on stabilizing the polymerase after purification and on eliminating the lag phase in in vitro polymerization reactions of 3-hydroxybutyryl CoA (HBCoA), and 3-hydroxyvaleryl CoA (HVCoA). K_M values were determined for the activity of the polymerase with both HBCoA and HVCoA, and the rates of propagation for both monomers were estimated. With a racemic mixture of HBCoA, the enzyme polymerized only the [R] monomer.

14.9. PRODUCTION OF PHAS WITH TRANSGENIC PLANTS

The production of polyhydroxyalkanoates from genetically modified crop plants represents a drastic change in methodology. With this strategy, the steps necessary to procure the substrates used in a fermentative process are no longer required, as naturally occurring carbon dioxide and sunlight serve as carbon and energy sources, respectively. While this field of research is still in its infancy, progress since the initial trials has shown the concept to be promising. The first investigations reported on the use of the plant *Arabidopsis thaliana* harboring the PHA genes of *R. eutropha*.

Poirier et al. [282] reported the successful expression of the *R. eutropha* genes encoding acetoacetyl-CoA reductase and PHA synthase in the cytoplasm of *A. thaliana*. The 3-ketothiolase gene is endogenous in plant cytoplasm. These experiments resulted in PHB synthesis in the cytoplasm, nucleus and vacuoles of all plant tissue, but in low amounts and at the cost of stunted growth and poor seed production. This was attributed to the diversion toward polymer accumulation of acetyl-CoA normally channeled into essential metabolic pathways.

The second phase of research [283] has focused on the targeting of the PHA pathway to a specific subcellular compartment, the plastid, where biosynthesis of triglycerides from acetyl-CoA normally occurs. All three genes needed be cloned in this case, and this led to the accumulation of high levels of PHB with few deleterious effects on the growth or fertility of the hosts. The homopolymer was stored within plastids to up to 14% of the dry mass of the plants (a 100-fold increase from expression in the cytoplasm) in the form of granules of size and appearance similar to those of bacterial PHA inclusions.

The genes encoding acetoacetyl-CoA reductase and PHA synthase from *R. eutropha* were expressed in cotton (*Gossypium barbadense* L. cv Sea Island) fibers. Transgenic plants containing both enzymes produced PHA in the fibers, since β-ketothiolase activity is present in cotton fibers [284]. The presence of PHB granules in transgenic fibers resulted in measurable changes of thermal properties, the fibers exhibited better insulating characteristics. The rate of heat uptake and cooling was slower in transgenic fibers, resulting in higher heat capacity [285].

Attempts to demonstrate the feasibility of profitable production on an agricultural scale are the next steps [282]. Poirier's group has proposed a number of oilseed crops that could be targeted for seed-specific PHA production, like rapeseed (closely related to *A. thaliana*), sunflower and soybean. Some of these are already under investigation by major companies. Depending on whether accumulation levels can be further increased; PHAs stored in plants have any deleterious effects on crop value in other respects; synthesis of PHAs other than PHB can be induced; and extraction of the biopolyesters is feasible at reasonable costs, the cost of PHAs produced in plants might be lowered enough to make them competitive with conventional plastics. But the tendency of arable land to become one of the most precious commodities on Earth [286] will present a formidable obstacle to applications in this field.

15. Extraction and purification of PHAs

While other physical methods have been described [63], PHAs are usually extracted from the producing cells with solvents or mixtures thereof. Mild polar compounds like acetone and alcohols [64] weaken or break down non-polymer cell material (NPCM), leaving PHB granules intact, although some longer-side-chain PHAs are soluble in acetone [191]. NPCMs mostly consist of nucleic acids, lipids and phospholipids, peptidoglycan and proteinaceous materials. In contrast, chloroform [186] and other chlorinated hydrocarbons [66] dissolve all PHAs. Methods employing both types of solvents (i.e., lipid extraction with PHA non-solvent followed by polymer dissolution) are therefore usually applied. The dissolved polymer is then separated from the solvent, usually by evaporation or precipitation with acetone or an alcohol, such as methanol or ethanol. Drying the cells prior to the extraction steps [65] can facilitate the subsequent polymer recovery, as can changing the pH [67] or temperature [62, 67] of the polymer-solvent mixture.

Differential digestion of NPCM can also be achieved with alkaline solutions of sodium hypochlorite, a method developed by Williamson and Wilkinson [26]. The undigested polymer granules can then be separated from the aqueous phase by centrifugation. While this method was at first reported to cause severe damage to the granules, mostly by an important loss of molecular mass of the polymer [35, 123], Ramsay's group showed that optimization of the separation conditions (pH, temperature, duration and biomass-to-aqueous phase ratio) could reduce degradation [192], and that treatment of the cells with a surfactant prior to washing with hypochlorite led to further improvements in the degree of purity and molecular mass of the final product [193].

The use of dispersions of sodium hypochlorite and chloroform for simultaneous differential digestion of NCPM and migration of the released PHB, which is hydrophobic, into the organic phase. [194]. The authors used the method on recovery of PHB from both *R. eutropha* and a PHB-producing recombinant strain of *E. coli* [127].

Flocculation of the cells renders subsequent separation of the PHB solution from the cell debris more facile. Preferably lipids are extracted from the flocculated cells before contact with the PHB extraction solvent. [287]

Page and Cornish [195] obtained a high-molecular-mass PHB by treating post-fermentation *A. vinelandii* cells with 1 N aqueous NH_3 in a process substantially simpler than enzymatic recovery. The addition of fish peptone to the culture to enhance polymer production rendered the cells osmotically sensitive and fragile, thus susceptible to disruption by NH_3.

Non-solvent processes have been developed in answer to the high cost of large-scale solvent extraction. Holmes and Lim [61] described the enzymatic process used at Zeneca for the recovery of PHB and PHBV. First, a high-temperature (100 to 150 °C) treatment of the cells provokes cell lysis and denaturation of nucleic acids, which could otherwise interfere with the subsequent steps. Non-PHA biomass is then solubilized with proteolytic enzymes (pepsin, trypsin, papain, others, and mixtures thereof) and anionic surfactants. Concentration of PHA by centrifugation is finally followed by bleaching with H_2O_2.

Another method was found by Findley and White [288]. In this study the determination of PHA after extraction with a one-phase chloroform:methanol solvent is shown to be as effective as the boiling chloroform method which is quantitative for PHA added to environmental samples. The one-phase chloroform:methanol extraction also quantitatively recovers the ester-linked fatty acids of the phospholipids (PLFA). The lipid extract is then partitioned on a disposable silicic acid column with quantitative elution of neutral lipids with chloroform, glycolipids and PHB with acetone, and phospholipids with methanol for analysis of each component. This extraction and simple column fractionation method for determination of PHA and PLFA simplifies previous methods for the assessment of the ratio of rates of formation of PHA and PLFA after a brief exposure to [14C] acetate which has been shown to be a sensitive measure of the nutritional status of bacteria in the environment.

Poly-3-hydroxybutyrate (PHB) or poly-3-hydroxyalkanoate (PHA) co-polymers are extracted from *Alcaligenes eutrophus* ATCC 17699 (KCTC 1006) using alkali and protease, in a method that involves: (1) washing strain KCTC 1006 in water at 40-120 deg or acetone; (2) grinding the dried biomass; (3) treating the biomass powder with protease in 0.2-2.0 M sodium hydroxide or potassium hydroxide solution at 20-60 °C; and (4) washing with a non-polar solvent and drying. This method prevents workers from exposing themselves to harmful solvents [289].

Poly-3-hydroxyoctanoate is extracted from the predried biomass with a non-chlorinated organic solvent such as acetone or THF. The biomass may be dried by suspension in acetone or isopropanol and agitation until a uniform mixture is obtained, separation of the cells, and air-drying. The extract is evaporated to obtain the polymer as a film [290].

PHB can also be separated from the biomass by heating to above 100 °C under pressure, releasing the pressure, and separating PHB granules from the cell debris [291] or by drying a finely divided stream or spray of an aqueous suspension of the cells with a gas heated to above 100 °C. and then extracting the PHB, preferably after a lipid extraction step with a solvent such as a partially halogenated hydrocarbon such as 1,2-dichloroethane or chloroform [292]. Brake used heating under pressure in the presence of a C1-C6 alcohol, and optionally also water [293].

Lafferty and Heinzle [294] described an extraction method using ethylene or propylene carbonate as extracting solvent for PHB from *Alcaligenes eutrophus H 16*. PHB is soluble in concentrations up to nearby 200 g/L at a temperature of 120 °C in these solvents, whilst at 100 °C solubility is as low as 2 g/L only. Biomass was contacted in a plug flow extractor at 120 °C, residual biomass was removed by a centrifugation step at the same temperature, and PHB was precipitated then by lowering the temperature to 95 °C and removed fy pressure filtration. The extraction solvent was recirculated back to the extractor, remaining solvent was removed from the product by a simple hot water extraction.

Another quite similar method for PHA extraction from biomass was developed by Kurdikar et al. [295] by dissolving the PHA in a non-halogenated solvent which comprises a PHA-poor solvent that dissolves less than 1 % of the PHA at temperatures less than the solvent boiling point, or a mixture of a PHA-poor solvent and a PHA-good solvent. Following extraction of PHA under pressure at a temperature above about 80 °C, typically above the boiling point of the PHA-poor solvent, PHA polymer is precipitated by cooling the PHA-enriched solvent mixture. Suitable PHA-poor solvents can include linear and branched R1-OH alcohols and R2-COOR3 esters where R1 = C1-C4, R2 = H, C1, C2, or C3, and R3 = C1-C5.

PHB recovery from fed-batch cultured *Alcaligenes latus*, ATCC 29713, was examined by Tarner et al. using combinations of chemical and mechanical treatments to disrupt the cells. Chemical pretreatments used sodium chloride and sodium hydroxide. For salt pretreatment the cells were exposed to NaCl (8 kg m^{-3}) and heat (60°C, 1 h), cooled to 4°C, and mechanically disrupted. For alkaline treatments, the cells were exposed to sodium hydroxide (0.025-0.8 kg NaOH per kg biomass) and mechanically disrupted at ambient temperature. A combined treatment with sodium chloride (8 kg m^{-3}), heat (60 °C, 1 h), and alkaline pH shock (pH 11.5, 1 min) was also tested. Mechanical disruption employed a continuous flow bead mill (2,800 rpm agitation speed, 90 ml min^{-1} slurry flow rate, 512 m mean bead diameter, bead loadings of 80% or 85% of chamber volume). Disruption was quantified by protein release. Over most of the disruption period, the release of PHB was approximately proportional to protein release. Regardless of the pretreatment or bead load, the disruption obeyed first order kinetics; hence, the rate of protein release was directly proportional to the amount of unreleased protein.Relative to untreated biomass, pretreatment always produced earlier protein release during milling. Pretreatment with a minimum of 0.12 kg NaOH per kg biomass was necessary to enable complete disruption within three passes (85% bead load). Untreated biomass required more than twice as many passes. Irrespective of the chemical pretreatment, the bead loading strongly influenced the disruption rate which was higher at the higher loading. Alkaline hydrolysis associated PHB loss was observed, but it could be limited to insignificant levels by immediate neutralization of disrupted homogenates [296].

Hypochlorite digestion of bacterial biomass from intracellular poly-3-hydroxybutyrate has not been used on a large scale since it has been reported to severely degrade PHB. In their study Berger et al., to minimize degradation, the initial *Alcaligenes eutrophus* biomass concentration, digestion time, and pH of NaOCl solvent were optimized to minimize degradation of poly-3-hydroxybutyric acid. Consequently, a PHA of 95% purity with a Mw of 600,000 and polydispersity index (PI) of 4.5 was recovered from biomass initially containing a polymer with Mw of $1.2*10^6$ and a PI of 3 [192].

Hahn et al. studied the recovery of PHB from *Alcaligenes eutrophus* and a recombinant *Escherichia coli* strain harboring the *A. eutrophus* poly(3-hydroxyalkanoic acid) biosynthesis genes. The amount of PHB degraded to a lower-molecular-weight compound in *A. eutrophus* during the recovery process was significant when sodium hypochlorite was used, but the amount degraded in the recombinant *E. coli* strain was negligible. However, there was no difference between the two microorganisms in the patterns of molecular weight change when PHB was recovered by using dispersions of a sodium hypochlorite solution and chloroform. To understand these findings, they examined purified PHB and lyophilized cells containing PHB by using a differential scanning calorimeter, a thermogravimetric analyzer, and nuclear magnetic resonance. The results of their analysis of lyophilized whole cells containing PHB with the differential scanning calorimeter suggested that the PHB granules in the recombinant *E. coli* strain were crystalline, while most of the PHB in *A. eutrophus* was in a mobile amorphous state. The stability of the native PHB in the recombinant *E. coli* strain during sodium hypochlorite treatment seemed to be due to its crystalline morphology. In addition, as determined by the thermogravimetric analyzer study, lyophilized cell powder of the recombinant *E. coli* strain containing PHB exhibited greater thermal stability than purified PHB obtained by chloroform extraction. The PHB preparations extracted from the two microorganisms had identical polymer properties [297].

Ling et al. [298] developed a new method of PHB extraction from recombinant *E. coli*, using homogenization and centrifugation coupled with sodium hypochlorite treatment. The size of PHB granules and cell debris in homogenates was characterized as a function of the noumber of homogenization passes. Simulation was used to develop the PHB and cell debris fractionation system, enabling numerical examination of the effects of repeated homogenization and centrifuge-feedrate variation. The simulation provided a good prediction of experimental performance. Sodium hypochlorite treatment was necessary to optimize PHB fractionation. A PHB recovery of 80% at a purity of 96.5% was obtained with the final optimized process. Protein and DNA contained in the resultant product were negligible. The developed process holds promise for significantly reducing the recovery cost associated with PHB manufacture.

16. Conclusions

Polyhydroxyalkanoates are thought to have an interesting future as sustainable polyesters. For medical applications their price is acceptable, but for broader use for example as packaging materials their high price is still the main hinderness for the application of such materials.

If a fermentation process is used for PHA synthesis this problem can partially be overcome by using cheap surplus and waste materials as renewable carbon sources (e.g. molasses, whey, cellulose hydrolysate) or other cheap carbon sources from fossil resources like methanol derived from natural gas, because roughly 50% of the total production costs derive from the carbon source costs. Unfortunately many of the well known production strains can not be used for PHA production from such substrates, because these microbial strains show either low yields or low production rates, when they grow on these substrates, or they simply cannot utilize these carbon sources at all. These drawbacks can be overcome either by isolating new microbial strains or by applying genetically modified strains for the production process.

Beside the main carbon source the cosubstrates for the production of copolyesters are another important cost factor for PHA production by fermentation. In some cases the yields for the comonomer in the copolyesters are rather low, because parts of the cosubstrates are completely metabolized for energy production instead of being inserted into the copolyester. Optimization of the fermentation conditions (e.g. PHA production at low DOC) can solve this problem.

Beside carbon source and cosubstrate for copolyester synthesis, bacterial growth rate and PHA production rate can be influenced by the nitrogen source needed for bacterial synthesis of biomass. In many cases a mixture of inorganic niterogen sources (e.g. ammonium sulfate) and protein hydrolizates can enhance as well the bacterial growth rate as the PHA production rate. The ideal composition of nutritional growth medium and PHA production medium have to be optimized for every interesting production strain.

After clarifying the physiological background for each combination of strains and nutrients, determination of growth and PHA production kinetics is of high importance for the fermentation part of the process development. Due to higher productivity a continuous fermentation process is of big advantage.Depending on the question if the PHA is synthesized associated to the bacterial growth or not, either a continuous stirred tank reactor (CSTR) or a continuous plug flow tubular reactor (PFTR) system or a cascade of several CSTRs will be the ideal fermentation system. This part of the process development has to be performed in strong cooperation with polymer scientists, because polymer quality can be strongly dependent from the type of fermentation process chosen. Having the feedback from polyester characterization, polyester quality can be controlled and influenced during microbial synthesis. Beside the upstream and the fermentation part of a PHA production process, also polymer recovery from the microbial biomass is of high interest and can strongly influence the polymer production costs.

In total it can be said, that the whole process for PHA production has to be optimized, and the interdisciplinary cooperation of microbial physiologists, biochemical engineers, polymer scientists, and chemical engineers is needed to solve all the problems in order to produce high quality polyesters within a price frame acceptable in comparison to oil-born plastics.

In the future PHA production by genetically modified plants can be a cheap way for their production, especially if not only PHAs but also other chemicals (e.g. starch) are produced by the same plant, and when it is used as a whole after being harvested. For poly-(R)-3-hdroxybutyrate production such a process should be reality within the coming decade, even though the polyester concentrations in plants are still too low to develop an economically feasible extraction process. For specialty copolyesters having a distinct composition such a plant production process seems to be rather doubtful. A way out of this problem might be the development of acceptable blends from lower quantities of high quality PHAs produced via fermentation with very cheap PHAs produced in plants.

References

1 Fuller, R. C., and R. W. Lenz. 1990. Natural History **5**:82-84.
2 Chowdhury, A. A. 1963. Archiv. Mikrobiol. **47**:167-200.
3 Lemoigne, M. 1927. Ann. Inst. Pasteur **41**:148-165.
4 Macrae, R. M., and J. F. Wilkinson. 1958. J. Gen. Microbiol. **19**:210-222.
5 Macrae, R. M., and J. F. Wilkinson. 1958. Proc. R. Phys. Soc. Edin. **27**:73-78.
6 Delafield, F. P., M. Doudoroff, N. J. Palleroni, C. J. Lusty, and R. Contopoulos. 1965. J. Bacteriol. **90**:1455-1466.
7 Doudoroff, M., and R. Y. Stanier. 1959. Nature **183**:1440-1442.
8 Stockdale, H., D. W. Ribbons, and E. A. Dawes. 1968. J. Bacteriol. **95**:1798-1803.
9 Schlegel, H. G., G. Gottschalk, and R. von Bartha. 1961. Nature **191**:463-465.
10 Schlegel, H. G. 1962. Arch. Mikrobiol. **42**:110-116.
11 Jensen, T. E., and L. M. Sicko. 1971. J. Bacteriol. **106**: 683 - 686.
12 Dawes, E. A., and P. J. Senior. 1973. Adv. Microbiol. Physiol. **10**:135-266.
13 Lundgren, D. G., R. Alper, C. Schnaitman, and R. H. Marchessault. 1965. J. Bacteriol. **89**:245-251.
14 Alper, R., and D. G. Lundgren. 1963. Biopolymers **1**:545-556.
15 Schlegel, H. G., and G. Gottschalk. 1962. Angew. Chem. **74**:342-347.
16 Cornibert, J., and R. H. Marchessault. 1972. J. Mol. Biol. **71**:735-756.
17 Wilkinson, J. F., and A. L. S. Munro. 1967. In Powell, E. O., C. G. T. Evans, R. E. Strange, and D. W. Tempest (eds.), Microbial physiology and continuous culture, H.M.S.O. London, pp.173-185.
18 Boatman, E. S. 1964. J. Cell Biol. **20**:297-311.
19 Ellar, D., D. G. Lundgren, K. Okamura, and R. H. Marchessault. 1968. J. Mol. Biol. **35**:489-502.
20 Griebel, R., Z. Smith, and J. M. Merrick. 1968. Biochem. **7**:3676-3681.

21 Lundgren, D. G., R. M. Pfister, and J. M. Merrick. 1964. J. Gen. Microbiol. **34** :441-446.
22 Merrick, J. M., and M. Doudoroff. 1961. Nature **189**:890-892.
23 Merrick, J. M., D. G. Lundgren, and R. M. Pfister. 1965. J. Bacteriol. **89**:234-239.
24 Law, J. H., and R. A. Slepecky. 1961. J. Bacteriol. **82**:33-36.
25 Norris, K. P., and J. E. S. Greenstreet. 1958. J. Gen. Microbiol. **19**:566-580.
26 Williamson, D. H., and J. F. Wilkinson. 1958. J. Gen. Microbiol. **19**:198-209.
27 Emeruwa, A. C., and R. Z. Hawirko. 1973. J. Bacteriol. **116**:989-993.
28 Kominek, L. A., and H. O. Halvorson. 1965. J. Bacteriol. **90**:1251-1259.
29 Oeding, V., and H. G. Schlegel. 1973. Biochem. J. **134**:239-248.
30 Ritchie, G. A. F., and E. A. Dawes. 1969. Biochem. J. **112**:803-805.
31 Senior, P. J., G. A. Beech, G. A. F. Ritchie, and E. A. Dawes. 1972. Biochem. J. **128**:1193-1201.
32 Senior, P. J., and E. A. Dawes. 1971. Biochem. J. **125**:55-66.
33 Slepecky, R. A., and J. H. Law. 1961. J. Bacteriol. **82**:37-42.
34 Schuster, E., and H. G. Schlegel. 1967. Arch. Mikrobiol. **58**:380-409.
35 Senior, P. J., and E. A. Dawes. 1973. Biochem. J. **134**:225-238.
36 Griebel, R. J., and J. M. Merrick. 1971. J. Bacteriol. **108**: 782-789
37 Ritchie, G. A. F., P. J. Senior, and E. A. Dawes. 1971. Biochem. J. **121**:309-316.
38 Gavard, R., A. Dahinger, B. Hauttecoeur, and C. Reynaud. 1966. C. R. Acad. Sci. Paris **263**:1273-1275.
39 Lusty, C. J., and M. Doudoroff. 1966. Biochem. **56**:960-96.
40 Merrick, J. M., and M. Doudoroff. 1964. J. Bacteriol. **88**:60-71.
41 Merrick, J. M., and C. I. Yu. 1966. Biochem. **5**:3563-3568.
42 Sierra, G., and N. E. Gibbons. 1962. Can. J. Microbiol. **8**:249-253.
43 Baptist, J. N. 1962. U.S. patent No. 3 036 959.
44 Baptist, J. N. 1962. U.S. patent No. 3 044 942.
45 Baptist, J. N. 1965. U.S. patent No. 3 182 036.
46 Baptist, J. N. 1963. U.S. patent No. 3 107 172.
47 W. R. Grace & Co. 1963. British patent specification No. 1 034 123.
48 Anonymous. 1995. Organization of Petroleum Exporting Countries. Britannica Online, version 1.2. Britannica Advanced Publishing, Inc. (World Wide Web information service.)
49 Anonymous. 1997. OPEC Annual Report 1996. OPECNA Information Department, Vienna, Austria.
50 Senior, P. G. 1984. In Dean, A. C. R., D. C. Ellwood, and C. G. T. Evans (eds.), Continuous culture, vol. 8, Ellis Horwood, Chichester, England, pp. 266-271.
51 King, P. 1982. J. Chem. Technol. Biotechnol. **32**:2-8.
52 Howells, E. R. 1982. Chem. Ind. **8**:508-511.
53 Fukada, E., and Ando, Y. 1988. Biorheology **25(1-2)**:297-302.
54 Wallen, L. L., and E. N. Davis. 1972. Environ. Sci. Technol. **6**:161-164.
55 Wallen, L. L., and W. K. Rohwedder. 1974. Environ. Sci. Technol. **8**:576-579.
56 De Smet, M. J., G. Eggink, B. Witholt, J. Kingma, and H. Wynberg. 1983. J. Bacteriol. **154**:870-878.
57 Findlay, R. H., and D. C. White. 1983. Appl. Environ. Microbiol. **45**:71-78.

58 Odham, G., A. Tunlid, G. Westerdahl, and P. Mrdén. 1986. Appl. Environ. Microbiol. **52**:905-910.
59 Hughes, L., and K. R. Richardson. 1981. European patent No. 46 344.
60 Powell, K. A., B. A. Collinson, and K. R. Richardson. 1980. European patent No. 15 669.
61 Richardson, K. R. 1984. European patent No. 114 086.
62 Barham, P. J., and A. Selwood. 1982. European patent No. 58 480.
63 Holmes, P. A., and E. Jones. 1981. European patent No. 46 335.
64 Holmes, P. A., and G. B. Lim. 1984. European patent No. 145 233.
65 Holmes, P. A., L. F. Wright, B. Alderson, and P. J. Senior. 1980. European patent No. 15 123.
66 Stageman, J. F. 1984. European patent No. 124 309.
67 Walker, J., J. R. Whitton, and B. Alderson. 1981. European patent No. 46 017.
68 Holmes, P. A., A. B. Newton, and F. M. Willmouth. 1981. European patent No. 52 460.
69 Holmes, P. A., L. F. Wright, and S. H. Collins. 1982. European patent No. 69 497.
70 Luzier, W. D. 1992. Proc. Natl. Acad. Sci. USA **89**:839-842.
71 Naylor, L., Zeneca Bio Products. 1995. Personal communication, March 16.
72 Lenz, R. W. 1995. National Technical Information Service (NTIS) report # PB95-199071 (JTEC monograph), U.S. Dept. of Commerce.
73 Lafferty, R. M., and G. Braunegg. 1988. Canadian patent No. 1 236 415.
74 Lafferty, R. M., and G. Braunegg. 1983. German patent No. 379 613.
75 Hrabak, O. 1992. FEMS Microbiol. Rev. **103**:251-256.
76 Poirier, Y., C. Nawrath, and C. Somerville. 1995. Bio/technol. **13**:142-150.
77 Naylor, L., Zeneca Bio Products. 1995. Personal communication, May 9.
78 Anderson, A. J., and E. A. Dawes. 1990. Microbiol. Rev. **54**:450-472.
79 Braunegg, G., and G. Lefebvre. 1993. Kem. Ind. **42**:313-322.
80 Lee, S.Y. 1996. Biotechnol. Bioeng. **49**:1-14.
81 Doi, Y. 1990. Microbial polyesters, VCH Publishers Inc., New York.
82 Steinbüchel, A., and H. E. Valentin. 1995. FEMS Microbiol. Lett. **128**:219-228.
83 Reusch, R. N. 1992. FEMS Microbiol. Rev. **103**:119-130.
84 Seebach, D., A. Brunner, H. M. Bürger, J. Schneider, and R. N. Reusch. 1994. Eur. J. Biochem. **224**:317-328.
85 Doi, Y., M. Kunioka, Y. Nakamura, and K. Soga. 1988. Macromolecules **21**:2722-2727.
86 Saito, Y., S. Nakamura, M. Hiramitsu, and Y. Doi. 1996. Polym. Int. **39**:169-174.
87 Doi, Y., A. Tamaki, M. Kunioka, and K. Soga. 1987. Makromol. Chem., Rapid Commun. **8**:631-635.
88 Kunioka, M., Y. Nakamura, and Y. Doi. 1988. Polym. Commun. **29**:174-176.
89 Palleroni, N. J., and A. V. Palleroni. 1978. Int. J. Syst. Bacteriol. **28**:416-424.
90 Chen, G.-Q., K. H. König, and R. M. Lafferty. 1991. Antonie van Leeuwenhoek **60**:61-66.
91 Ramsay, B. A., K. Lomaliza, C. Chavarie, B. Dubé, P. Bataille, and J.A. Ramsay. 1990. Appl. Environm. Microbiol. **56**: 2093-2098
92 Nakamura, S., Y. Doi, and M. Scandola. 1992. Macromolecules **25**:4237-4241.

93 Nakamura, K., Goto, Y., Yoshie, N., Inoue, Y., and R. Chujo. 1992. Int. J. Biol. Macromol. **14(2)**:117-118.

94 Steinbüchel, A., H. E. Valentin, and A. Schönebaum. 1994. J. Environ. Polym. Degrad. **2**:67-74.

95 Valentin, H. E., A. Schönebaum, and A. Steinbüchel. 1992. Appl. Microbiol. Biotechnol. **36**:507-514.

96 Nakamura, S., M. Kunioka, and Y. Doi. 1991. Macromol. Rep. **A 28**:15-24.

97 Hiramitsu, M., and Y. Doi. 1993. Polymer **34**:4782-4786.

98 Hiramitsu, M., N. Koyama, and Y. Doi. 1993. Biotechnol. Lett. **15**:461-464.

99 Huisman, G. W., O. de Leeuw, G. Eggink, and B. Witholt. 1989. Appl. Environ. Microbiol. **55**:1949-1954.

100 Huijberts, G. N. M., G. Eggink, P. de Waard, G. W. Huisman, and B. Witholt. 1992. Appl. Environ. Microbiol. **58**:536-544.

101 Lee, E. Y., D. Jendrossek, A. Schirmer, C. Y. Choi, and A. Steinbüchel. 1995. Appl. Microbiol. Biotechnol. **42**:901-909.

102 Schirmer, A., D. Jendrossek, and H. G. Schlegel. 1993. Appl. Environ. Microbiol. **59**:1220-1227.

103 Abe, H., Doi, Y., Fukushima, T., and H. Eya. 1994. Int. J. Biol. Macromol.16(3):115-119.

104 de Koning, G. J. M., H. M. M. van Bilsen, P. J. Lemstra, W. Hasenberg, B. Witholt, H. Preusting, J. G. van der Galiën, A. Schirmer, and D. Jendrossek. 1993. Polymer **35**:2090-2097.

105 Kim, Y. B., R. W. Lenz, and R. C. Fuller. 1991. Macromolecules **24**:5256-5260.

106 Lenz, R. W., Y. B. Kim, and R. C. Fuller. 1992. FEMS Microbiol.Rev. **103**:207-8.

107 Abe, C., Y. Taima, Y. Nakamura, and Y. Doi. 1990. Polym. Commun. **31**: 404-406.

108 Doi, Y., and C. Abe. 1990. Macromolecules **23**:3705-3707.

109 Kim, O., R. A. Gross., W. J. Hammar, and R. A. Newmark. 1996. Macromolecules **29**:4572-4581.

110 Kim, Y. B., R. W. Lenz, and R. C. Fuller. 1992. Macromolecules **25**:1852-1857.

111 Bear, M. M., M. A. Leboucherdurand, V. Langlois, R. W. Lenz, S. Goodwin, and P. Guerin. 1997. React. Funct. Polymers **34**:65-77.

112 Ulmer, H. W., R. A. Gross, M. Posada, P. Weisbach, R. C. Fuller, and R. W. Lenz. 1994. Macromolecules **27**:1675-1679.

113 Rodrigues, M. F. A., L. F. da Silva, J. G. C. Gomez, H. E. Valentin, and A. Steinbüchel. 1995. Appl. Microbiol. Biotechnol. **43**:880-886.

114 Fuchtenbusch, B., D. Fabritius, and A. Steinbüchel. 1996. FEMS Microbiol. Lett. **138**:153-160.

115 Mitomo, H., Takahashi, T., Ito, H., and T. Saito. 1999. Int. J. Biol. Macromol. **24(4)**:311-318.

116 Steinbüchel, A., E.-M. Debsi, R. H. Marchessault, and A. Timm. 1993. Appl. Microbiol. Biotechnol. **39**:443-449.

117 Kang, C. K., H. S. Lee, and J. H. Kim. 1993. Biotechnol. Lett. **15**:1017-1020.

118 Caballero, K.P., Karel, S.F., and R.A. Register. 1995. Int. J. Biol. Macromol. **17(2)**:86-92.

119 Takeda, M., Matsuoka, H., Ban, H., Ohashi, Y., Hikuma M., and J.. Koizumi. 1995. Appl. Microbiol. Biotechnol. **44**: 37-42.
120 Ballard, D. G. H., P. A. Holmes, and P. J. Senior. 1987. In Fontanille M., and A. Guyot (eds.), Recent advances in mechanistic and synthetic aspects of polymerization, vol. 215, Reidel (Kluywer) Publishing Co., Lancaster, U.K., pp. 293-314.
121 McCool, G. J., T. Fernandez, N. Li, and M. C. Cannon. 1996. FEMS Microbiol.Lett. **138**:41-48.
122 Dunlop, W. F., and A. W. Robards. 1973. J. Bacteriol. **114**:1271-1280.
123 Nuti, M. P., M. de Bertoldi, and A. A. Lepidi. 1972. Can. J. Microbiol. **18**:1257-1261.
124 Wang, W. S., and D. G. Lundgren. 1969. J. Bacteriol. **97**:947-950.
125 Barnard, G. N., and J. K. M. Sanders. 1989. J. Biol. Chem. **264**:3286-3291.
126 Barnard, G. N., and J. K. M. Sanders. 1988. FEBS Lett. **231**:16-18.
127 Hahn, S. K., Y. K. Chang, and S. Y. Lee. 1995. Appl. Environ. Microbiol. **61**:34-39.
128 Mas, J., C. Pedrós-Alió, and R. Guerrero. 1985. J. Bacteriol. **164**: 749-756
129 De Koning, G. J. M., and P. J. Lemstra. 1992. Polym. Commun. **33**:3292-3294.
130 Lauzier, C. A., and R. H. Marchessault. 1994. Symposium on Physiology, Kinetics, Production and Use of Biopolymers, Seggau, Austria. Proceedings, 59-69.
131 Haywood, G. W., A. J. Anderson, and E. A. Dawes. 1989. FEMS Microbiol. Lett. **57**:1-6.
132 Stuart, E. S., L. J. R. Foster, R. W. Lenz, and R. C. Fuller. 1996. Int. J. Biol. Macromol. **19**:171-176.
133 Pranamuda, H., Y. Tokiwa, and H. Tanaka. 1995. Appl. Environ. Microbiol. **61**:1828-1832.
134 Foster, L. J. R., E. S. Stuart, A. Tehrani, R. W. Lenz, and R. C. Fuller. 1996. Int. J. Biol. Macromol. **19**:177-183.
135 Byrom, D. 1987. Tibtech **5**:246-250.
136 Repaske, R., and C. Repaske. 1976. Appl. Environ. Microbiol. **32**:585-591.
137 Henderson, R. A., and C. W. Jones. 1997. Arch. Microbiol. **168**:486-492.
138 Shi, H. D., M. Shiraishi, and K. Shimizu. 1997. J. Ferm. Bioeng. **84**:579-587.
139 Lee, I. Y., M. K. Kim, Y. H. Park, and S.Y.Lee. 1996. Biotechnol. Bioeng. **52**:707-712.
140 Doi, Y., A. Tamaki, M. Kunioka, and K. Soga. 1988. Appl. Microbiol. Biotechnol. **28**:330-334.
141 Doi, Y., M. Kunioka, Y. Nakamura, and K. Soga. 1987. Macromolecules **20**:2988-2991.
142 Doi, Y., A. Tamaki, M. Kunioka, and K. Soga. 1987. J. Chem. Soc., Chem. Commun. **1987**:1635-1636.
143 Kunioka, M., Y. Kawagushi, and Y. Doi. 1989. Appl. Microbiol. Biotechnol. **30**:569-573.
144 Valentin H. E., G. Zwingmann, A. Schönebaum, and A. Steinbüchel. 1995. Eur. J. Biochem. **227**:43-60.
145 Akiyama, M., and Y. Doi. 1993. Biotechnol. Lett. **15**:163-168.

146 Valentin, H. E., and D. Dennis. 1996. Appl. Environm. Microbiol. **62**:372-379.
147 Haywood, G. W., A. J. Anderson, L. Chu, and E. A. Dawes. 1988. FEMS Microbiol. Lett. **52**:91-96.
148 Doi, Y., Y. Kawaguchi, N. Koyama, S. Nakamura, M. Hiramitsu, Y. Yoshida, and H. Kimura. 1992. FEMS Microbiol. Rev. **103**:103-108.
149 Sim, S. J., K. D. Snell, S. A. Hogan, J. Stubbe, C. K. Rha, and A. J. Sinskey. 1997. Nature Biotechnol. **15**:63-67.
150 Wodzinska, J., K. D. Snell, A. Rhomberg, A. J. Sinskey, K. Biemann, and J. Stubbe. 1996. J. American Chem. Soc. **118**:6319-6320.
151 Steinbüchel, A., E. Hustede, M. Liebergesell, U. Pieper, A. Timm, and H. Valentin. 1992. FEMS Microbiol. Rev. **103**:217-230.
152 Peoples, O. P., and A. J. Sinskey. 1989. J. Biol. Chem. **264**:15293-15297.
153 Peoples, O. P., and A. J. Sinskey. 1989. J. Biol. Chem. **264**:15298-15303.
154 Schubert, P., A. Steinbüchel, and H. G. Schlegel. 1988. J. Bacteriol. **170**:5837-5847.
155 Slater, S. C., W. H. Voige, and D. E. Dennis. 1988. J. Bacteriol. **170**:4431-4436.
156 Schubert, P., N. Krüger, and A. Steinbüchel. 1991. J. Bacteriol. **173**:168-175.
157 Genser, K. F., G. Renner, and H. Schwab. 1998. J. Biotechnol. **64**:125-135.
158 Doi, Y., A. Segawa, Y. Kawaguchi, and M. Kunioka. 1990. FEMS Microbiol. Lett. **67**:165-170.
159 McInerney, M. J., D. A. Amos, K. S. Kealy, and J. A. Palmer. 1992. FEMS Microbiol. Rev. **103**:195-206.
160 Urmeneta, J., J. Mas-Castellà, and R. Guerrero. 1995. Appl. Environ. Microbiol. **61**:2046-2048.
161 Jackson, F.A., and E. A. Dawes. 1976. J. Gen. Microbiol. **97**:303-312.
162 Ward, A. C., B. I. Rowley, and E. A. Dawes. 1977. J. Gen. Microbiol. **102**:61-68.
163 Morinaga, Y., S. Yamanaka, A. Ishizaki, and Y. Hirose. 1978. Agric. Biol. Chem. **42**:439-444.
164 Sonnleitner, B., E. Heinzle, G. Braunegg, and R. M. Lafferty. 1979. Eur. J. Appl. Microbiol. Biotechnol. **7**:1-10.
165 Steinbüchel, A., and H. G. Schlegel. 1989. Appl. Microbiol. Biotechnol. **31**:168-175.
166 Lefebvre, G., M. Rocher, and G. Braunegg. 1997. Appl. Environm. Microbiol. **63**:827-833.
167 Braunegg, G., and B. Bogensberger. 1985. Acta Biotechnol. **4**:339-345.
168 Manchak, J., and W. J. Page. 1994. Microbiology **140**:953-963.
169 Briese, B. H., B. Schmidt, and D. Jendrossek. 1994. J. Environ. Polym. Degrad. **2**:75-87.
170 Saito, T., A. Iwata, and T. Watanabe. 1993. J. Environ. Polym. Degrad. **1**:99-105.
171 Saito, T., K. Suzuki, J. Yamamoto, T. Fukui, K. Miwa, K. Tomita, S. Nakanishi, S. Odani, J.-I. Suzuki, and K. Ishikawa. 1989. Appl. Environ. Microbiol. **171**:184-189.
172 Tanio, T., T. Fukui, Y. Shirakura, T. Saito, K. Tomita, T. Kaiho, and S. Masamune. 1982. Eur. J. Biochem. **124**:71-77.
173 Kita, K., K. Ishimaru, M. Teraoka, H. Yanase, and N. Kato. 1995. Appl. Environ. Microbiol. **61**:1727-1730.

174 Kita, K., S. Mashiba, M. Nagita, K. Ishimaru, K. Okamoto, H. Yanase, and N. Kato. 1997. Biochim. Biophys. Acta - Gene Structure and Expression. **1352**:113-122.
175 Jendrossek, D., I. Knoke, R. B. Habibian, A. Steinbüchel, and H. G. Schlegel. 1993. J. Environ. Polym. Degrad. **1**:53-63.
176 Jendrossek, D., A. Friese, A. Behrends, M. Andermann, H. D. Kratzin, T. Stanislawski, and H. G. Schlegel. 1995. J. Bacteriol. **177**:596-607.
177 Müller, B., and D. Jendrossek. 1993. Appl. Microbiol. Biotechnol. **38**:487-492.
178 Schirmer, A., and D. Jendrossek. 1994. J. Bacteriol. **176**:7065-7073.
179 Ramsay, B. A., I. Saracovan, J. A. Ramsay, and R. H. Marchessault. 1994. J. Environ. Polym. Degrad. **2**:1-7.
180 Ostle, A., and J. G. Holt. 1982. Appl. Environ. Microbiol. **44**:238-241.
181 Braunegg, G., B. Sonnleitner, and R. M. Lafferty. 1978. Eur. J. Appl. Microbiol. Biotechnol. **6**:29-37.
182 Riis, V., and W. Mai. 1988. J. Chromatogr. **445**:285-289.
183 Morikawa, H., and R. H. Marchessault. 1981. Can. J. Chem. **59**:2306-2313.
184 Doi, Y., M. Kunioka, Y. Nakamura, and K. Soga. 1986. Macromolecules **19**:1274-1276.
185 Doi, Y., M. Kunioka, Y. Nakamura, and K. Soga. 1986. Macromolecules **19**:2860-2864.
186 Jüttner, R.-R., R. M. Lafferty, and H.-J. Knackmuss. 1975. Eur. J. Appl. Microbiol. **1**:233-237.
187 Braunegg, G., G. Lefebvre, G. Renner, A. Zeiser, G. Haage, and K. Loidl-Lanthaler. 1995. Can J. Microbiol **41**:239-248.
188 Degelau, A., T. Scheper, J. E. Bailey, and C. Guske. 1995. Appl. Microbiol. Biotechnol. **42**:653-657.
189 Jacob, G.S., Garbow, J.R., and J. Schaefer. 1986. J. Biol. Chem. **261(36)**:16785-16787.
190 Cook, P. D., Gao, Y., Smith, R., Toube, T. P., and J.H.P. Utley. 1995J. Mater. Chem. **5(3)**: 413-415.
191 Brandl, H., R. A. Gross, R. W. Lenz, and R. C. Fuller. 1988. Appl. Environ. Microbiol. **54**:1977-1982.
192 Berger, E., B. A. Ramsay, J. A. Ramsay, C. Chavarie, and G. Braunegg 1989. Biotechnol. Tech. **3:**227-232.
193 Ramsay, J. A., E. Berger, B. A. Ramsay, and C. Chavarie. 1990. Biotechnol. Tech. **4**:221-226.
194 Hahn, S. K., Y. K. Chang, B. S. Kim, and H. N. Chang. 1994. Biotechnol. Bioeng. **44**:256-261.
195 Page, W. J., and A. Cornish. 1993.. Appl. Environ. Microbiol. **59**:4236-4244.
196 Marchessault, R. H., T. L. Bluhm, Y. Deslandes, G. K. Hamer, W. J. Orts, P. R. Sundarajan, M. G. Taylor, S. Bloembergen, and D. A. Holden. 1988. Makromol. Chem., Macromol. Symp. **19**:235-254.
197 Holmes, P. A. 1985. Phys. Technol. **16**:32-36.
198 Bluhm, T. L., G. K. Hamer, R. H. Marchessault, C. A. Fyfe, and R. P. Veregin. 1986. Macromolecules **19**:2871-2876.

199 Song, J.J., Yoon, S.C., Yu, S.M., and R.W. Lenz. 1998. Int. J. Biol. Macromol. **23(3)**: 165-173.

200 Scandola, M., Ceccorulli, G., and Doi Y. 1990. Int.J. Biol. Macromol. **12**: 112-117.

201 Terada, M., and R.H. Marchessault. 1999. Int. J. Biol. Macromol. 25(1-3): 207-215.

202 Chen, G.-Q., and W. J. Page. 1994. Biotechnol. Lett. **16**:155-160.

203 Anderson, A. J., D. R. Williams, B. Taidi, E. A. Dawes, and D. F. Ewing. 1992. FEMS Microbiol. Rev. **103**:93-102.

204 Taidi, B., A. J. Anderson, E. A. Dawes, and D. Byrom. 1994. Appl. Microbiol. Biotechnol. **40**:786-790.

205 Koyama, N., and Y. Doi. 1995. Biotechnol. Lett. **17**:281-284.

206 Asenjo, J. A., A. S. Schmidt, P. R. Andersen, and B. A. Andrews. 1995. Biotechnol. Bioeng. **46**:497-502.

207 Kawaguchi, Y., and Y. Doi. 1992. Macromolecules **25**:2324-2329.

208 Suzuki, T., H. Deguchi, T. Yamane, S. Shimizu, and K. Gekko. 1988. Appl. Microbiol. Biotechnol. **27**:487-491.

209 Kang, C.-K., S. Kusaka, and Y. Doi. 1995. Biotechnol. Lett. **17**:583-588.

210 Doi, Y., Y. Kanesawa, M. Kunioka, and T. Saito. 1990. Macromolecules **23**:26-31.

211 Nishida, H., and Y. Tokiwa. 1993. J. Environ. Polym. Degrad. **1**:65-80.

212 Day, M., K. Shaw, and D. Cooney. 1994. J. Environ. Polym. Degrad. **2**:121-128.

213 Mergaert, J., C. Anderson, A. Wouters, and J. Swings. 1994. J. Environ. Polym. Degrad. **2**:177-183.

214 Gilmore, D. F., S. Antoun, R. W. Lenz, and R. C. Fuller. 1993. J. Environ. Polym. Degrad. **1**:269-274.

215 Briese, B. H., D. Jendrossek, and H. G. Schlegel. 1994. FEMS Microbiol. Lett. **117**:107-112.

216 Mayer, J. M., D. L. Kaplan, R. E. Stote, K. L. Dixon, A. E. Shupe, A. L. Allen, and J. E. McCassie. 1996. ACS Symposium Series **627**:159-170.

217 Ramsay, B.A., Langlade, V., Carreau, P.J., and J.A. Ramsay. 1993. Appl. Environ. Microbiol. **59(4)**:1242-1246.

218 Imam, S.H., Gordon, S.H., Shogren, R.L., Tosteson. T.R., Govind, N.S., and R.V. Greene. 1999. Appl. Environ. Microbiol. **65(2)**: 431-437.

219 Koller, I., and Owen, A. J. 1996. Polym. Int. **39**(3), 175-81.

220 Cyras, V., Fernandez, P., Galego; N., and A. Vazquez. 1999. Polym. Int. **48(8)**: 705-712.

221 Dubini Paglia, E., Beltrame, P. L., Canette, M., Seves, A., Marcandalli, B., and E. Martuscelli. 1993. Polymer **34(5)**: 996-1001.

222 Sadocco, P., Bulli, C., Elegir, G., Seves, A., and E. Martuscelli. 1993. Makromol. Chem.**194(10)**: 2675-2686.

223 Gatenholm, P.; and A. Mathiasson. 1994. J. Appl. Polym. Sci., **51(7)**:1231-1237.

224 Brandl, H., Aeberli, B., Bachofen, R., Schwegler, I., Muller, H.M., Burger M.H., Hoffmann, T., Lengweiler, U.D., and D. Seebach. 1995. Can.J. Microbiol. **41** Suppl 1:180-186.

225 Yoshie, N., Azuma, Y., Sakurai, M., and Y. Inoue. 1995. J. Appl. Polym. Sci., **56(1)**: 17-24.

226 Al-Salah, Hasan A. 1998. Polymer Bulletin **41**: 593-600.
227 Byrom, D. 1992. FEMS Microbiol. Rev. **103**:247-250.
228 Lee, S. Y., and H. N. Chang. 1995. In Fiechter, A. (ed.), Advances in biochemical engineering/biotechnology, vol. 52, Springer-Verlag, Berlin, pp. 27-58.
229 Beaulieu, M., Y. Beaulieu, J. Mélinard, S. Pandian, and J. Goulet. 1995. Appl. Environ. Microbiol. **61**:165-169.
230 Seki, H., M. Karita, and A. Suzuki. 1993. J. Chem. Eng. Jap. **27**:423-425.
231 Lee, S. Y. 1996. Trends Biotechnol. **14**:431-438.
232 Kim, B. S., S. C. Lee, S. Y. Lee, H. N. Chang, Y. K. Chang, and S. I. Woo. 1994. Enzyme Microb. Technol. **16**:556-561.
233 Suzuki, T., T. Yamane, and S. Shimizu. 1986. Appl. Microbiol. Biotechnol. **23**:322-329.
234 Koyama, N., and Y. Doi. 1993. J. Environ. Polym. Degrad. **3**:235-240.
235 Aragao, G. M. F., N. D. Lindley, J. L. Uribelarrea, and A. Pareilleux. 1996. Biotechnol. Lett. **18**:937-942.
236 Bitar, A., and S. Underhill. 1990. Biotechnol. Lett. **12**:563-568.
237 Lee, I. Y., M. K. Kim, G. J. Kim, H. N. Chang, and Y. H. Park. 1995. Biotechnol. Lett. **17**:571-574.
238 Fukui, T., and Y. Doi. 1998. Appl. Microbiol. Biotechnol. **49(3)**: 333-336.
239 Cho, G., Yoon, J., Oh, J. T., and W. Kim. 1997. Hwahak Konghak **35(3)**: 412-418.
240 Tanaka, K., K. Katamune, and A. Ishizaki. 1993. Biotechnol. Lett. **15**:1217-1222.
241 Rusendi, D., and J. D. Sheppard. 1995. Bioresource Technol. **54**:191-196.
242 Ishizaki, A., and K. Tanaka. 1991. J. Ferment. Bioeng. **71**:254-257.
243 Tanaka, K., A. Ishizaki, T. Kanamaru, and T. Kawano. 1994. Biotechnol. Bioeng. **45**:268-275.
244 Sheppard, J. D., P. Marchessault, T. Whalen, and S. F. Barrington. 1994. J. Chem. Tech. Biotechnol. **59**:83-89.
245 Katoh, T., Yuguchi, D., Yoshii, H., Shi, H., and K. Shimizu. 1999. J. Biotechnol. **67(2-3)**:113-134.
246 Kim, B. S., S. Y. Lee, and H. N. Chang. 1992. Biotechnol. Lett. **14**:811-816.
247 Wang, F. L., and S. Y. Lee. 1997. Appl. Environm. Microbiol. **63**:4765-4769.
248 Yim, K. S., S. Y. Lee, and H. N. Chang. 1996. Biotechnol. Bioeng. **49**:495-503.
249 Taga, N., Tanaka, K., and A. Ishizaki. 1999. Biotechnol. Bioengng. **62**: 546-553.
250 Ramsay, B. A., J. A. Ramsay, and D. G. Cooper. 1989. Appl. Microbiol. Microbiol. **55**:584-589.
251 Alderete, J. A., D. W. Karl, and C.-H. Park. 1993. Biotechnol. Prog. **9**:520-525.
252 Park, C.-H., and V. K. Damodaran. 1994. Biotechnol. Bioeng. **44**:1306-1314.
253 Ueda, S., S. Matsumoto, A. Takagi, and T. Yamane. 1992. Appl. Environ. Microbiol. **58**:3574-3579.
254 Yamane, T., X. F. Chen, and S. Ueda. 1996. FEMS Microbiol. Lett. **135**:207-211.
255 Bourque, D., B. Ouellette, G. André, and D. Groleau. 1992. Appl. Microbiol. Biotechnol. **37**:7-12.
256 Daniel, M., J. H. Choi, J. H. Kim, and J. M. Lebeault. 1992. Appl. Microbiol. Biotechnol. **37**:702-706.
257 Bertrand, J.-L., B. A. Ramsay, J. A. Ramsay, and C. Chavarie. 1990. Appl. Environ. Microbiol. **56**:3133-3138.

258 Ramsay, B. A., I. Saracovan, J. A. Ramsay, and R. H. Marchessault. 1991. Appl. Environ. Microbiol. **57**:625-629.
259 Huijberts, G. N. M., and G. Eggink. 1996. Appl. Microbiol. Biotechnol. **46**:233-239.
260 Kellerhals, M.B., Kessler, B., and B Witholt. 1999. Biotechnol. Bioengng. **65**: 306-315.
261 Kim, G. J., I. Y. Lee, S. C. Yoon, Y. C. Shin, and Y. H. Park. 1997. Enz. Microbial Technol. **20**:500-505.
262 Preusting, H., R. van Houten, A. Hoefs, E. K. van Langenberghe, O. Favre-Bulle, and B. Witholt. 1992. Biotechnol. Bioeng. **41**:550-556.
263 Page, W. J., and O. Knosp. 1989. Appl. Environ. Microbiol. **55**:1334-1339.
264 Page, W. J. 1992. Appl. Microbiol. Biotechnol. **38**:117-121.
265 Page, W. J. 1989. Appl. Microbiol. Biotechnol. **31**:329-333.
266 Page, W. J., J. Manchak, and B. Rudy. 1992. Appl. Environ. Microbiol. **58**:2866-2873
267 Rodriguez-Valera, F., Garcia Lillo, J. A., Anton, J., and I. Meseguer. 1991. NATO ASI Ser., Ser. A, **201**(Gen. Appl. Aspects Halophilic Microorg.): 373-380.
268 Lemos, P. C., Viana, C., Salgueiro, E. N., Ramos, A. M., Crespo, J.P.S.G., and M.A.M. Reis. 1998. Enzyme Microb. Technol. **22(8)**: 662-671.
269 Slater, S.C., Voige, W.H., and D.E. Dennis. 1988. J Bacteriol.**170(10)**: 4431-4436.
270 Schubert, P., Steinbüchel, A., and H.G. Schlegel. 1988. J. Bacteriol. **170(12)**: 5837-5847.
271 Park, J. S., Huh, T. L., and Y.H. Lee. 1997. Enzyme Microb.Technol. **21**: 85-90.
272 Jackson, J. K., and F. Srienc. 1999. J. Biotechnol. **68**: 49-60.
273 Choi, J.I., Lee, S.Y., and K. Han. 1998. Appl. Environ. Microbiol. **64(12)**:4897-4903.
274 Matsusaki, H., Manji, S., Taguchi, K., Kato, M., Fukui, T., and Y. Doi. 1998. J Bacteriol. **180(24)**:6459-6467.
275 Prieto, M.A., Kellerhals, M.B., Bozzato, G.B., Radnovic, D, Witholt, B., and B. Kessler. 1999. Appl. Environ. Microbiol. **65(8)**:3265-3271.
276 Dennis, D., McCoy, M., Stangl, A., Valentin, H. E., and Z. Wu. 1998. J. Biotechnol. **64(2, 3)**: 177-186.
277 Kichise, T., Fukui, T., Yoshida, Y., and Doi, Y. 1999. Int. J. Biol. Macromol. **25(1-3)**: 69-77.
278 Hall, B., Baldwin, J., Rhie, H.G., and D. Dennis. 1998. Can. J. Microbiol. **44(7)**: 687-691.
279 Taguchi, K., Aoyagi, Y., Matsusaki, H., Fukui, T., and Y. Doi. 1999. FEMS Microbiol. Lett.**176**: 183-190.
280 Gerngross, T. U., and D.P. Martin. 1995. Proc. Natl. Acad. Sci. U. S. A., 92(14): 6279-6283.
281 Lenz, R.W., Farcet, C., Dijkstra, P.J., Goodwin, S., and S. Zhang. 1999. Int. J. Biol. Macromol. **25**: 55-60.
282 Poirier, Y., D. Dennis, K. Klomparens, C. Nawrath, and C. Somerville. 1992. FEMS Microbiol. Rev. **103**: 237-246.
283 Nawrath, C., Y. Poirier, and C. Somerville. 1994. Proc. Natl. Acad. Sci. USA **91**:12760-12764.

284 Rinehart, J. A., M. W. Petersen, and M. E. John. 1996. Plant Physiol. **112**:1331-1341.
285 John, M. E., and G. Keller. 1996. Proc. Natl. Acad. Sci. USA **93**:12768-12773.
286 Skow, J. 1995. Time International **146**:64-66.
287 Walker, J., and J.R. Whitton. 1982. United States Patent US4358583, 09-11-1982.
288 Findley, R.H.; White, D.C. 1989. J. Microbiol. Methods **6**:113-120
289 Im, K. B., and K.J. Kim. Korean Patent KR 9502866, 27-03-1995.
290 Ohleyer, Eric, 1993, PCT Int. Appl. WO 9311656 Al
291 Holmes P.A. and Jones E., 1982. EP0046335 1982-02-24
292 Senior P. J. et al., 1982, Extraction process, US4324907 1982-04-13
293 Brake L.D., 1993, Recovery of polyhydroxy acids, US5264614 1993-11-23
294 Lafferty, R.M., and Heinzle, E., 1977. Chem. Rundschau **30**:14-16
295 Kurdikar, D.L., Strauser, F.E., Solodar, A.J., Paster, M.D. 1988. US patent WO9846783, 1998-10-22
296 Tamer I.M., Moo-Young, M., and Chisty,Y., 1998. Bioprocess Engineering Abstract **19** (6), 459-468
297 Hahn S.K., Chang, Y.K., and Lee, S.Y., 1995. Appl. Environ. Microbiol. **61**(1):34-39
298 Ling, Y., Wong, H. H., Thomas, C. J., Williams, D. R. G., Middelberg, A. P. J., 1997. Bioseparation **7**(1), 9-15

9

POLYHYDROXYALKANOATES: PROPERTIES AND MODIFICATION FOR HIGH VOLUME APPLICATIONS

IVAN CHODAK
Polymer Institute of the Slovak Academy of Sciences,
842 36 Bratislava

1 Introduction

Polyhydroxy alkanoates (PHAs) comprise a group of biodegradable polyesters produced by bacteria. Their primary function is energy storage and they are used as an energy reserve for bacteria, similar to the role of polysaccharides or polyphosphates in living cells. PHAs can be produced by relative simple and efficient procedure based completely on biotechnology utilizing fully renewable resources (see Chapter 8). Variations in this procedure, mainly consisting in changes in the composition of the food supplied to the bacteria, lead to a production of modified PHAs as homo or copolymers, containing different functional groups. The original or modified PHAs seem to be ideal for applications in various short-term plastic products, especially packaging. In spite of their attractive potential, the applications of PHAs are at present marginal, because they possess several serious drawbacks that prevent high volume production and application.

Several excellent reviews have appeared, dealing with various aspects of the preparation, production, and properties of polyhydroxyalkanoates [1-5]. In the present review, the potential of PHAs will be discussed for high volume application, packaging being the main target. The drawbacks will be shown and possible ways of remediation will be outlined. A major part of the review is devoted to poly(3-hydroxybutyrate), which is the simplest species in PHA homologous series.

2 Poly(3-hydroxybutyrate) and related PHAs

Poly(3-hydroxybutyrate) (PHB) can be characterized as rather controversial polymer. PHB is a completely biodegradable, highly hydrophobic thermoplastics, containing almost 80 % crystallinity, with high melting temperature, resistance to organic solvents

G. Scott (ed.), Degradable Polymers, 2nd Edition, 295-319.

and possessing excellent mechanical strength and modulus resembling that of polypropylene [3]. The basic physical properties compared to those of polypropylene are summarized in the Table 1. In spite of excellent properties, especially strength parameters, an extensive application of this material in high volume range is hindered by several serious drawbacks so that at present PHB is used only exceptionally and in small quantities for special purposes. Pronounced brittleness, very low deformability, high susceptibility to a rapid thermal degradation, difficult processing by conventional thermoplastic technologies (mainly due to fast thermal degradation) and rather high price compared to other high volume plastics may be named as the major factors hindering a wider application of PHB. Additional problems related to processing are connected mainly with low shear strength of the melt, which needs to be addressed if considering certain applications.

2.1 MECHANICAL PROPERTIES

When considering high volume application opportunities, packaging is the obvious primary target, especially due to the advantage of biodegradation as an alternative plastic waste management procedure. From this point of view, the comparison with the polyolefins given in Table 1, is the most appropriate.

Table 1 : A comparison of physical properties of PHB, copolymers of PHB with higher PHAs, polypropylene (PP), and low-density polyethylene (LDPE). Most data taken from [3].

	PHB	20V[1)]	6HA[2)]	PP	LDPE
melting temperature, °C	175	145	133	176	110
glass transition temp., °C	4	-1	-8	-10	-30
crystallinity, %	60	ng	ng	50	50
density, g / cm^3	1,25	ng	ng	0,91	0,92
E modulus, MPa	3,5	0.8	0.2	1,5	0,2
tensile strength, MPa	40	20	17	38	10
elongation at break, % 5	50	680	400	600	

[1)] poly(3-hydroxybutyrate-co-20 mol % hydroxyvalerate)
[2)] poly(3-hydroxybutyrate-co-6 mol % HAs, HAs (hydroxyalkanoates) = 3 % 3-hydroxydecano-ate, 3 % 3-hydroxydodecanoate, < 1 % 3-hydroxyoctanoate, < 1 % 5-hydroxy-dodecanoate, ng = negligible

It is seen that properties of PHB are rather close to those of polypropylene, outperforming those of polyethylene in most parameters. The factor of primary importance seems to be a low deformation at break, related to low film toughness and unacceptable rigidity and brittleness. The reason for the brittleness consists mainly in a presence of large crystals in the form of spherulites. On the other hand, high strength and E modulus represent a suitable starting point for modification since there is a large margin in strength parameters to increase deformability and toughness. A more fundamental study was directed to explaining mechanical behavior of PHB using linear elastic fracture mechanics techniques [6].

Copolymers exhibit properties much closer to those of LDPE; however, their availability and price represent a hindrance for these materials to be considered as a serious competitor for commodity polyolefins.

2.2 THERMAL DEGRADATION

A very low resistance to thermal degradation seems to be a serious problem. Since the melting temperature of PHB is around 180 °C, the processing temperature should be at least 190 °C. At this temperature thermal degradation proceeds rapidly so that the acceptable residence time in the processing equipment is only few minutes. The extent of the degradation is illustrated by the changes in molecular weight after annealing at various temperatures as shown in Table 2 [7].

Table 2 GPC molecular weights (M_p = MW in the peak) after thermal treatment (10, 80 minutes) of PHB and PHB with 20 wt.% of plasticizers; glycerol triacetate (TAC) and glycerol. Dependence on annealing temperature [7].

T, °C	time, min	M_p		
		no plasticizer	TAC	Glycerol
no treatment*		270 000	240 000	170 000
180	10	97 000	-	-
200	10	29 000	30 000	9 000
220	10	4 500	4 500	**
190	80	6 000	6 000	**

* Compression moulding at 170 °C for 1 minute
** M_p is below the threshold sensitivity of the instrument, less than 3000.

The degradation of aromatic polyesters has been studied for many years and the principal problem was identified as hydrolysis under the high temperature conditions. However, free radical autoxidation process also occurs when aliphatic groups are present [1]. It can be expected that similar mechanisms will be active also in the case of bacterial polyesters.

The mechanism of PHB degradation consists in a β - elimination reaction during a random chain scission with the formation of unsaturated end products of polymer fragments [8] as seen in Scheme 1 with a formation of crotonic acid [8]. However, the changes in molecular weight indicate that more complex side reactions occur. GPC measurements of PHB heated at various temperatures as a function of time revealed that molecular weight after an initial decrease increased slightly before continuing to decline [9]. The effect was attributed to polycondensation of the initial hydroxyl and carboxyl groups formed by the elimination process. Under severe degradation conditions (e.g. up to 300 °C under vacuum) leads to a formation of crotonic and isocrotonic acids and the dimer, trimer and tetramer of PHB [10].

Thermogravimetric analysis also reveals quite the low thermal stability of PHAs, compared to most of synthetic polymers but also regarding other biodegradable polymers of polyester or similar structures. Major degradation temperature range with steep weight loss started at 263 °C for PHB, which is much lower temperature compared to other biodegradable polymers with potential use for packaging (e.g. 355, 317 and 440 for polycaprolactone, poly lactic acid and modified polyethylene, respectively) [11].

The presence of a second monomer in PHAs copolymers seems to have a certain stabilizing effect in thermal degradation, as revealed by thermogravimetry for PHB, and its copolymers with hydroxyvalerate and hydroxyhexanoate [12]. Moreover, in the case of

Scheme 1 : Mechanism of the β-elimination reaction

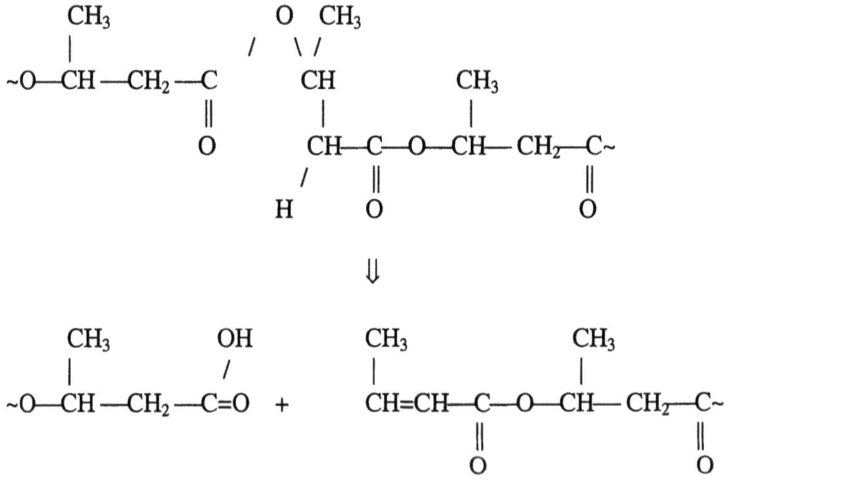

```
    CH3                      O  CH3
    |                    /    \ /
~O—CH—CH2—C               CH           CH3
           ||              |            |
           O              CH—C—O—CH—CH2—C~
                          /   ||         ||
                         H    O          O

                              ⇓

    CH3         OH            CH3           CH3
    |           /             |             |
~O—CH—CH2—C=O   +            CH=CH—C—O—CH—CH2—C~
                                  ||         ||
                                  O          O
```

PHBV copolymer, two peaks were observed in thermograms, indicating that two different monomers are evaporating at different temperatures, unlike for PHB where a single peak was found [13].

The presence of various additives results in a change of thermal degradation kinetics. Many species act as prodegradants. Aluminum compounds and fumed silica have been reported to have slight stabilizing effect followed by a prodegradant activity [1]. Other inorganic compounds have been shown to be prodegradants as revealed by dynamic TG experiments [14] and the same effect was observed for impurities present in technical PHB when compared to carefully purified PHB [15]. As expected, degradation has a negative influence on the mechanical properties of PHB [16]. From this point of view it is important to estimate the effect of plasticizers on thermal degradation. As seen in Table 2, glycerol triacetate has hardly any negative influence on thermal stability of PHB, while glycerol has a detrimental effect on molecular weight during processing [7]. The effect is attributed to a trans-esterification where hydroxyl groups of the glycerol play a crucial role.

Trans-esterification must also be considered when blends of PHB with other polymers are investigated, especially if they contain hydroxyl, carboxyl or other reactive moieties. A significant decrease in activation energy of thermal decomposition of PHB was observed for PMMA as a second component in the blend, while marginal changes were registered if PHB was mixed with polypropylene, as revealed by thermogravimetry [17], although the changes were not attributed to any particular PHB / PMMA interaction.

It is important to note that although PHB is a highly hydrophobic material, it takes up minor quantities of water upon storage [18]. Although the amount of moisture is low (about 0.2 wt %), hydrolysis has also to be considered as a reason for degradation of PHB. Hydrolysis of PHAs proceeds in water at a very low rate at ambient temperatures. However, at higher temperature or in alkaline medium the process is much faster [19]. The topic of hydrolysis of aliphatic polyesters was reviewed recently by Tsuji [20].

As shown above, thermal degradation of PHB and other PHAs is fast. In spite of low thermostability, PHB can be processed by injection molding or extrusion if care is taken to keep the processing temperature as low as possible and to minimize the residence time [21,22].

2.3 PHYSICAL AGEING

The phenomenon of an extensive physical ageing is an interesting feature of PHB behavior. It manifests itself in a substantial increase in E-modulus and tensile strength during storage. This effect is not accomplished in a short time period but can be observed over several months as demonstrated in Fig. 1.

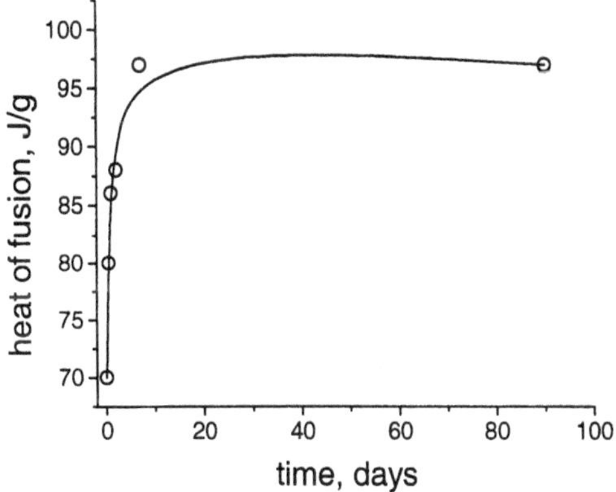

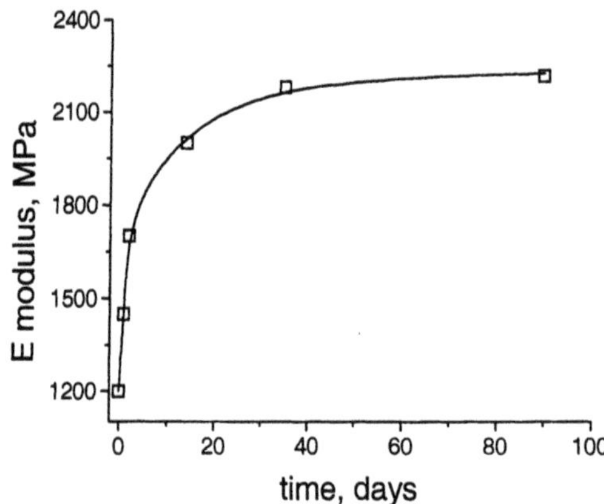

Fig. 1 : Changes in heat of fusion and E-modulus of PHB due to the time of storing at room temperature.

The effect is generally ascribed to recrystallization leading to an increase in crystalline portion and consequently to the increase in E modulus and tensile or flexural strength values. At the same time, a decrease in amorphous region results in a drop in ductility and enhanced brittleness of the material. However, as seen in Fig 1, the increase in crystallinity occurs much faster compared to the increase of strength, since the crystallinity, calculated from DSC heat of fusion , levels off after few days, when modulus has reached only about 70 % of its ultimate equilibrium value. This behavior, of course does not exclude the possibility of changing the crystalline structure while keeping the crystalline portion at the same level, but the extent of variation of strength indicates that some additional mechanism may be responsible for the phenomenon. A detailed study of the problem by de Koning et al [23] suggested a model explaining the process quantitatively. The model is based on a large specific crystalline–amorphous interface. Due to large interface area, subsequent crystal perfection during ageing results in a significant increase in a number of macromolecules in amorphous phase being under residual stress. Thus, relatively minor changes in the morphology of the crystalline part lead to much higher ageing effect compared to material with a lower crystalline-amorphous interface area, as is usual for the majority of synthetic polymers.

A detailed investigation of the ageing behavior of PHB containing 8 or 12 % of valerate as a comonomer has been described [24]. Both extruded and compression molded samples were investigated by DSC, DMTA, dielectric spectroscopy, and thermally stimulated discharge technique. Besides changes in the amorphous and crystalline phases, the authors considered the importance of the interphase region on the ageing process, consisting in relaxation above T_g due to morphological reorganization in the interphase [24]. This opinion is supported by fracture mechanics data indicating that the ageing process does not consist of a simple embrittlement but rather in a reduction of the energy dissipating properties of the material accompanied with the ability to survive high stress levels [6].

From this point of view, the changes in FTIR spectra of PHB during storage may be of interest. Two IR peaks have been found [25] to be sensitive to storage, namely 1685 cm^{-1}, and 3435 cm^{-1}, corresponding to carbonyl and hydroxyl stretch, respectively. Both peaks grow during storage with no leveling off after 14 days, indicating that some interactions between the functional groups of PHB may result from storage. It is not clear to what extent these interactions are related to the physical ageing. Nevertheless, it is worth considering the importance of these effects in addition to changes in crystalline structure, especially in the interphase region.

4 Procedures suggested for improving toughness

The extremely low deformability of PHB seems to be the most important parameter negatively influencing the ultimate properties of the polymer. Therefore research activities are aimed, not only at increasing thermal stability, but also at the improvement of the ductility and, consequently, toughness of PHB. For this purpose, many different approaches have been described and published that lead to a partial improvement of toughness and flexibility. The procedures include synthesis of new copolymer-based PHAs, physical modification either by additives or by changing the morphology and chemical modification. Of course, brittleness and thermal degradation are related parameters since a decrease in molecular weight due to degradation leads to an increase in brittleness. Thus, in some cases a modest improvement in ductility may be achieved by careful and fast processing at the lowest possible temperature.

4.1 POLYHYDROXYALKANOATE ADDITIVES

The only modification, that results in a really tough material in some cases, seems to be a bacterial preparation of copolymers of PHB with higher polyhydroxyalkanoates. The introduction of higher alkanoates into the polymer backbone results in the formation of defects in the crystalline structure leading to lower crystallinity and much higher mobility in the amorphous portion, as revealed by DMTA measurements. Polyhydroxybutyrate/ valerate is the most common species of this kind, possessing a good toughness and flexibility in films similar to that of polyethylene.

In order to improve properties of PHB, especially toughness, random copolymers have been prepared by replacing the methyl group on the PHB main chain by ethyl or longer substituents. This can be achieved by changing the substrate on which the bacteria grow (Chapter 8), the most straightforward way being a change in the ratio of glucose and propionic acid [26]. The mechanical properties are improved by such a procedure, compared with PHB homopolymer [27], especially with respect to higher deformability and, consequently toughness. Apparently, more bulky substituent on the chain results in more flexible chains and the formation of a higher proportion of mobile phase due to higher free volume, as indicated in Fig. 2.

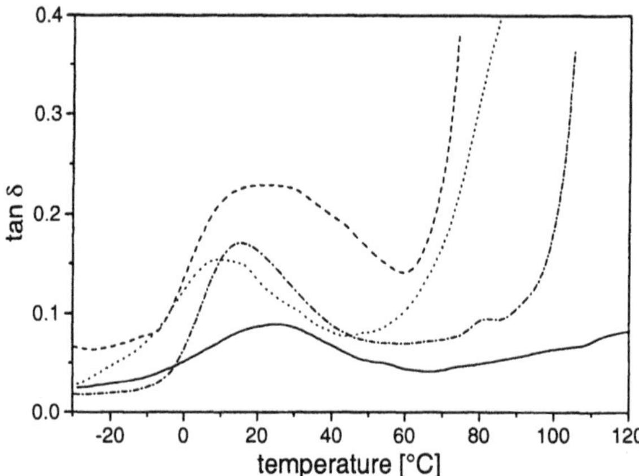

Fig. 2 : DMTA curves of poly}3-hydroxybutyrate-co-valerate) in dependence on valerate content. the area under the curve corresponds roughly to the mobile (amorphous) portion. Vaterate content 4.5 (solid line), 32 (dots), 39 (dash-dots), and 75 % (dash).

A comparison of the mechanical properties of two copolymers with PHB and common polyolefins are given in Table 1. In spite of interesting properties, the PHB/V copolymer seems to be too expensive for high – volume applications. Copolymers with more bulky substituents are even less economically acceptable, thus their application can be only considered for special purposes. Moreover, the presence of a comonomer affects the crystallization kinetics resulting in a longer processing cycles [28].

Bacterial synthesis using sodium octanoate as a sole carbon source was reported to result in a formation of copolymer with majority of polyhydroxyoctanoate segments [29]. The mechanical properties of this thermoplastic material are within the range typical for thermoplastic elastomers, with ultimate elongation 380 %, tensile strength and modulus 9 and 8 MPa, respectively, however, the tensile set was rather high, reaching 35 % at 100 % elongation.

4.2. PHYSICAL MODIFICATION

A direct and simple approach to obtain a more ductile material consists in the addition of plasticizers. A decrease in processing temperature and a more convenient processing regime can be achieved in this way, resulting from a decrease in melting temperature [1]. Some decrease in strength and modulus must be accepted in this case, depending on the amount of plasticizer added. However, no substantial improvement in toughness has been reported by using low molar mass plasticizers. Two to three-fold increase in deformability can be achieved by the addition of 10 to 20 wt % of glycerol triacetate, which seems to be rather good plasticizer for PHB, considering its biodegradability and miscibility with PHB (Fig. 3, [30]). Further increase in the plasticizer content results not only in a decrease of strength and modulus but at the same time to a drop in deformation at break. Apparently, the cohesion of the material is not high enough after absorbing such a large content of low molar mass additive. Better properties can be obtained by using oligomeric or polymeric additives. Blending with other polymers may result in good mechanical properties but the number of polymers is limited to biodegradable species if the biodegradable nature of the resulting material is to be maintained.

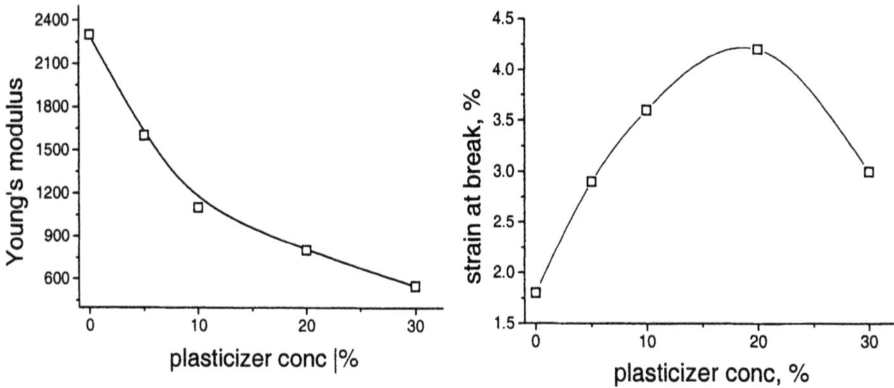

Fig. 3 :Young's modulus and deformation at break of PHB in dependence on plasticizer content (triacetine)

An application of epoxidized soybean oil has been reported recently for plasticizing blends of PHBV and cellulose acetate butyrate. The material is claimed to be suitable for such applications as packaging [31].

A hot rolling treatment has been shown to substantially improve the ductility of PHB. It is proposed that rolling results in the healing of cracks present inside the spherulites leading to more ductile films [32]. Luepke et al, using solid–state processing of PHB powders, applied a more sophisticated approach. To prevent thermal degradation, extrusion proceeded far below the melting temperature of PHB. Thus, materials with improved thermal properties have been obtained. In particular, the ductility of the PHB increased substantially compared to traditionally melt–processed PHB. The improved mechanical properties were attributed to structural differences at the molecular and supramolecular level [33]. The changes are apparently similar to those induced by hot–rolling; just the extent should be much higher due to higher stress. Avoiding thermal treatment contributes to the improved properties significantly due to negligible thermal degradation during processing, although some degree of mechano-destruction might be expected in this case.

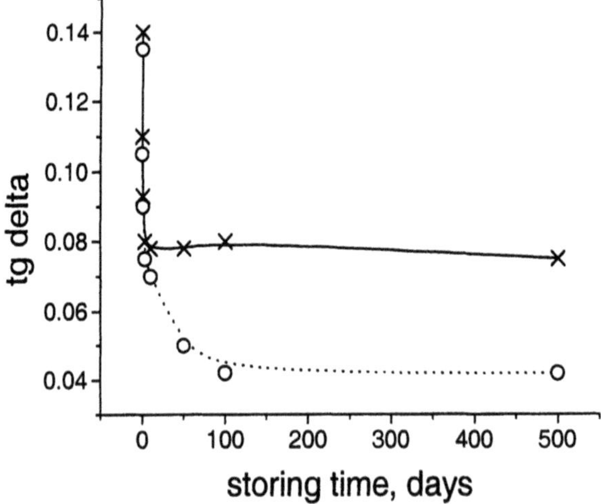

Fig. 4. : A change in height of DMTA peak with time of storing for as – moulded PHB (circles) and PHB annealed at 110 °C for 16 h [data according to [23])

An interesting and rather simple option for toughness improvement consists in annealing the material at 120 °C or above [34]. The elongation at break has been found to increase up 30 % [23], although an increase up to 60 % was reported in individual cases [34]. A change in lamellar morphology during annealing was suggested as an explanation of the effect, resulting in a substantial increase of the mobile phase as indicated by the area under tg δ peak measured by DMTA (Table 3). Subsequent ageing leads to a decrease in the area under DMTA curve with time; the rate for melt-crystallized being almost the same as for the annealed materials. However, the leveling off occurs at the height of the peak and is about a double for annealed PHB (Fig 4 [23]). The reason for the morphology changes is explained suggesting that the annealing could be considered as if crystallization proceeded from the melt at much lower rate. Fracture mechanics analysis indicates that annealing results in an improvement of both the critical stress intensity factor and strain energy release rate. These factors, unlike elongation at break, do not change significantly with reageing [35].

Table 3 : Change in the area under the DMTA peak due to annealing [23].

Annealing	area under tg δ peak a.u.
0 min, RT	0.8
60 min, 90 °C	1.4
60 min, 115 °C	1.8
10 min, 147 °C	2.0

It is seen that a significant improvement of mechanical properties can be achieved via physical modification. However, the degree of the changes is not sufficient for satisfactory long–term properties of PHB.

4.3. CHEMICAL MODIFICATION

Trans-esterification seems to be quite frequently used to modify PHB. In many cases transesterification is undesirable process, as discussed above concerning the prodegradation effect of glycerol on PHB thermal degradation. In other cases, however, transesterification may contribute to an improvement of properties due to influencing the crystallinity or morphology and introducing flexible segments. Various materials and procedures have been described. For example a poly(hydroxy ether) prepared from Bisphenol A was reported to react with PHB by trans-esterification during annealing at 180 °C [36]. The number of free phenoxy groups, determined by FTIR, was found to pass through a maximum with the increase of the annealing time, becoming negligible once a single phase was formed.

Other chemical modifications have been reported in the literature. Among these, graft polymerization of various monomers onto the PHA 'backbone' may be of some interest. Grafting of acrylic acid onto PHB and PHBV was initiated by gamma irradiation [37]. An increase in biodegradability was reported as a result of small amounts of grafts due to improved wetability. Grafting of styrene or methylmethacrylate initiated by gamma irradiation was also reported [38].

Crosslinking is considered as a good option for modification of properties of many polymers due to the formation of an additional number of tie molecules, the formation of an additional number of tie molecules, and changes in crystallinity and morphology which influence properties. Nevertheless, in the case of PHAs the attempts to crosslink the material were only partially successful. It was found to be necessary to add a crosslinking coagent (e.g. a polyfunctional monomer such as triallyl cyanurate) in order to obtain some crosslinked gel if the process is initiated by thermal decomposition of organic peroxides. Co - gamma irradiation was shown to cause degradation of PHB [39] and both thermal and tensile properties were significantly affected. Irradiation resulted in a decrease of the melting temperature and a dramatic decrease in tensile strength and strain at break, indicating pronounced brittleness of the degraded material. The observed effect was very extensive The molecular weight decreased from original 245000 down to 77000 and 37000 for radiation dose 100 and 250 kGy, respectively..

On the other hand, rather effective crosslinking can be achieved with PHA containing unsaturated double bonds in the side chain. The polymer was prepared by feeding the appropriate bacteria (Pseudomonas oleovorans) with soybean oily acids under formation of polyhydroxyoctanoate with about 10 % mol of unsaturated chains. This polymer was crosslinked, either by uv irradiation or by thermal decomposition of benzoyl peroxide, leading to a formation of an insoluble crosslinked gel content between 81 and 93 wt %, as shown in Table 4. The glass transition temperature (–60 °C) was affected significantly compared to the Tg for unmodified PHA with unsaturated chains [40]. Nevertheless, it is of interest that highest degree of crosslinking is achieved without the use of a crosslinking initiator, although the time of reaction was much longer than for the other samples. This indicates that in this case, there is some kind of ambient atmosphere- initiated reaction, possibly oxidative, similar to the slow hardening of unsaturated oils or lacquers.

Table 4 : Crosslinked insoluble gel content and glass transition temperature (T_g)of PHA containing double bonds. Crosslinked by UV irradiation or thermally in the presence of benzophenone (BPh) or benzoyl peroxide (BPx) and ethylene glycol dimethacrylate (EGDM).

Initiation	BPh wt%	BPx wt%	EGDM wt%	time days	gel %	Tg °C
UV	-	-	-	2,4	93	-45
UV	0,4	-	-	1,5	81	-33
UV	-	-	1,0	1,3	84	-42
UV	-	0,3	1,0	1,0	89	-46
thermal	-	3,3	-	0,1	87	-35
none	-	-	-	120	85	-40

It is not clear what is the effect of thermal decomposition of benzoyl peroxide since the decomposition half-life for this initiator in solution is around 10 hours and it must be at least as long or even longer in solid polymer. Thus, although the demonstration of efficient crosslinking of PHAs containing unsaturation is very interesting, several questions arise for the special case described [40].

Some ideas for more sophisticated chemical modifications may be derived from other procedure described in the literature. For example, PHB can be modified by alcoholysis to prepolymer and then selectively end–capped by maleation [41] . The molecular weight of the macromer can be easily controlled by the alcoholysis time and the molecular weight distribution of the final product with double bonds in the molecule is narrow. The macromer can be used for synthesis of new macromolecules with PHB segments.

Biodegradable PHB with modified properties can be prepared synthetically, by ring-opening polymerization of butyrolactone. Under certain conditions a telechelic polymer may be prepared with oxytetramethylene groups in the main-chain. This polymer reacts with L-lactide in the presence of a catalyst with the formation of an A-B-A triblock copolymer, resulting in the formation of a biodegradable thermoplastic elastomer [42].

4.4. BLENDS WITH OTHER POLYMERS

Blending with other polymers is apparently a feasible way to affect the properties of PHAs and, in some cases, to decrease the price of the material. Both miscible and immiscible pairs may be considered although, in the latter case, an efficient compatibilizer has to be found if the two polymers forming the two-phase blend are incompatible. Biodegradability of the second polymeric component is an obvious requirement if the blend is to be claimed to be biodegradable. However, PHB blends with non–biodegradable polymers may also be of important.

Blends of PHAs with polyvinyl alcohol (PVOH) have been attracting the attention of scientists. The two polymers are claimed to be compatible, even miscible. However, the scatter in the data presented in scientific literature may be caused by the fact that polyvinyl alcohol is in fact a copolymer of vinyl alcohol and vinyl acetate prepared from polyvinyl acetate by hydrolysis. The hydrolysis is rarely complete, so that up to 30 % of vinyl acetate may be present in "polyvinyl alcohol" of different commercial grades. The properties, especially the solubility in water depends very much on the residual acetate content and will definitely influence many properties, including miscibility/compatibility with PHAs.

Yoshie et al [43] found that PVOH/PHB blends are compatible only when the blend contains a large content of PVOH. Compatible blends exhibit low crystallinity for both PHB and PVOH. The NMR spectra indicate that compatibility occurs due to hydrogen–bonding interactions in the amorphous phase and is affected also by tacticity of PVOH. The compatibility concentration range is wider for syndiotactic–rich PVOH/PHB compared to atactic PVOH/PHB blends [43]. The dependence of tensile strength on the PHB/PVOH ratio shows a very pronounced minimum at a ratio about 1:1, as seen in Fig. 5 [30]. In this case an extensive prodegradant effect was proposed to contribute significantly to poor properties due to a presence of glycerol used as a plasticizer for PVOH. In this connection, it may be of interest that the miscibility of PHB with ethylene- vinyl actetate depends on the EVA composition. Copolymer containing 70 % of acetate was found to be immiscible with PHB while for the PHB blend with 85 % of acetate the Flory-Huggins parameter was determined to be negative, indicating thermodynamic miscibility [44].

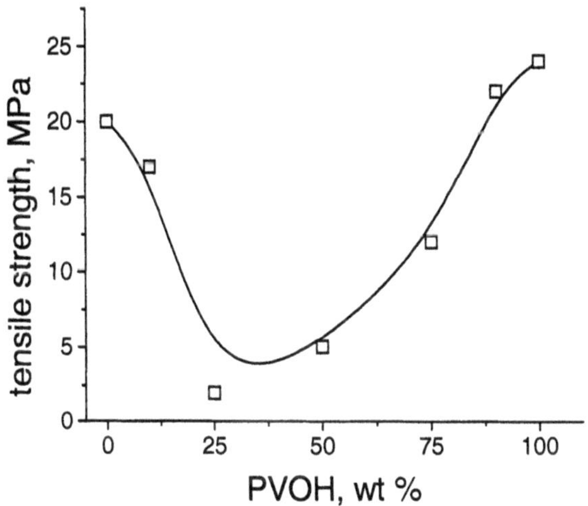

Fig. 5 : Tensile strength of the blend PHB / PVOH in dependence on the blend composition

Polycaprolactone (PCL) is another candidate for blending with PHB, since it is biodegradable and a rather ductile material. Blends of PHB and PCL were found to be immiscible, however, their toughness has been reported to be higher compared to PHB [45] and mechanical properties are claimed to be excellent [46]. PHB/PCL (60:40) blends exhibit mechanical properties superior to PHBV copolymer or low-density polyethylene. The oxygen permeability of PCL decreased due to addition of PHB at 25 % compared to the value for PCL and an improvement in heat stability and impact strength was recorded [47]. The miscibility was improved if a copolymer of caprolactone and ethylene glycol (PECL) was used instead of PCL. The morphology of PHB/PECL also changed compared with PHB/PCL blends. An attempt was made to improve PHB/PCL compatibility by crosslinking, initiated by organic peroxide [48, 49]. As seen in Fig. 6, the deformability of the blends is rather high if the PCL content is at least 30 %, while decrease in strength parameters was observed with decreasing PHB in the blend [49]. Two-phase morphology was also found for blends of polycaprolactone with PHBV (7% valerate content) [50].

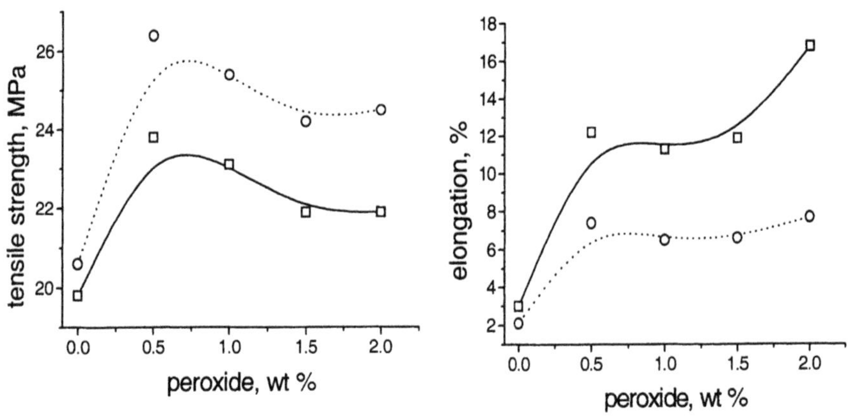

Fig. 6 : The effect of crosslinking (degree given by content of decomposed peroxide) on mechanical properties of PHB / PCL blends. The effect of ageing is shown by testing after 2 (full line) or 14 days (dashed line) after compresson moulding.

Blends of PHB and polyethylene glycol have been reported to be miscible as indicated by DSC data. A decrease in PHB melting temperature was reported as a result of increasing content of PEG in the blend. This improves the processing of PHB and melt flow index data support this conclusion [51]. Semi-interpenetrating PHB/EG networks have been prepared by crosslinking PEG with acryloyl chloride and the resulting hydrogel had improved mechanical properties [52]. Miscibility with PHB was also found for poly-(epichlorohydrin) for the full concentration range in the melt. The blends show a single glass transition temperature of the value corresponding to the addition law [53]. A presence of specific molecular interactions was suggested as an explanation for observed miscibility of PHB with poly(vinylidene chloride-co-acrylonitrile) [54]. Blends of PHB with cellulose acetate butyrate have been reported to be miscible in the melt. Flory- Huggins parameter for the blend is composition – dependent and always negative [55].

Polylactic acid (PLA) forms immiscible blends with PHB, obviously due to a lack of specific interactions between the two polymers. However, a copolymer of PLA with polyethylene glycol is miscible with PHB in the melt and amorphous states, similar to PEG [56].

Blends of PHB with poly(ethylene oxide) are claimed to be miscible [57], or partially immiscible in the melt depending on the blend composition [58]. On the other hand, PHB was reported to be immiscible with poly(methylene oxide), with two distinct spherulitic phases in the solid state [59]. Two-phase morphology is present in PHBV (20 % HV) in poly-L-lactide as indicated by two T_g values [60]. Stiffer but also more brittle material was obtained by blending, compared to PHBV, as seen in Table 5. Blends of PHB with poly(amino acids) were also investigated. Immiscible but compatible blends were reported for PHB with poly(γ-benzyl-L-glutamate) [61], poly(butylene succinate-co-butylene adipate) and poly(butylene succinate-co-ε-caprolactone) [62].

Two phase morphology and positive Flory Huggins parameters χ_{12} were observed for PHB/poly (vinyl butyral) (PVB), a random copolymer of vinyl butyral and vinyl alcohol of different compositions). It was found that a degree of immiscibility (as characterized by χ_{12} values) depends on PVB composition with partial miscibility and co-continuous morphology observed for a 50/50 blend and 25 – 36 vinyl alcohol content in PVB. The results are interpreted in terms of repulsive effect of hydrophobic and hydrophilic component in PVB [63]

Table 5 : Mechanical properties (Young's modulus E, tensile strength σ, elongation at break ε) of solvent cast PHBV/poly-L-lactide (PLLA) blends (average ± standard deviation) [60]

PLLA %	E MPa	σ MPa	ε %
100	2415 ± 140	71 ± 3	5.6 ± 1.0
80	2083 ± 45	54 ± 3	6.2 ± 0.5
60	1552 ± 56	39 ± 2	6.7 ± 0.7
40	1258 ± 17	29 ± 2	4.1 ± 0.4
20	1076 ± 17	24 ± 1	6.9 ± 0.9
0	882 ± 62	25 ± 3	13.8 ± 1.2

Obviously, blends of two different PHAs may be of interest from both a scientific and an applicational point of view. The miscibility of PHB with PHBV depends on the content of valerate in the copolymer. For low valerate compositions the blends are miscible and even co-crystallization was observed [64] for blends with a valerate content in the copolymer of 9 % [65]. Miscibility of blends based on copolymers of poly(3-hydroxybutyrate-co-3-hydroxypropionate) with compositions varying in a wide range was investigated in [66]. The blends were found to be miscible in amorphous phase if the difference in HB content between two co-polyesters was less than 30 – 40 mol %.

From a scientific point of view, materials without a hydroxybutyrate component are of interest. The blends of poly(3-hydroxyvalerate) with poly(p-vinyl phenol) (PVPh) were investigated. The blends were found to be miscible over the whole concentration range as indicated by a single T_g. The Flory – Huggins parameter calculated from the melting point depression data was found to be –1.2, which is a value well in negative range demonstrating good thermodynamic miscibility. The good miscibility observed may be attributed to hydrogen bonding interactions between carbonyl groups of PHV and hydroxyl groups of PVPh. The addition of 50 % of the PVPh leads to a complete suppression of cold crystallization of the PHV [67].

Among blends of PHAs with synthetic non-biodegradable polymers, the system PHBV/PVC are of interest. Two T_gs were observed for a blend of PVC and PHBV with 8 % of valerate, while a thermodynamically miscible blend with a negative Flory Huggins parameter was formed if the valerate content in PHBV increased to 18 % [68]. Interesting behavior was reported for blends of chemosynthetic atactic poly(R,S-3-hydroxybutyrate) with polymethyl methacrylate (PMMA). These blends are partially miscible at PMMA content ranging from 10 to 50 %. Enzymatic degradation of PHB was enhanced in the presence of the amorphous non-biodegradable polymer [69].

Data on the phase behavior of blends of PHAs with other polymers can be interesting for the design of industrially important materials. However, most of the papers are based on the investigation of thermal properties such as crystallization and melting behaviors, and structural data. Little or no information is given on mechanical and rheological properties of the blends, possibly indicating that the authors are not confident about the technological applicability of the materials. The data on miscibility of PHAs with various polymers are summarized in Table 6.

Table 6 : Miscibility of PHAs with various polymers

PHA	Second polymer	Miscibility		ref
PHB	PVOH	yes ?	blend composition dependent	[44]
PHB	EVA	yes	EVA composition dependent	[44]
PHB	PCL	no		[45]
PHB7V	PCL	no		[45]
PHB	PECL	yes		[48]
PHB	PEG	yes		[51]
PHB	PECH	yes		[53]
PHB	PVdC-AN	yes		[54]
PHB	CAB*	yes ?		[55]
PHB	PLA	no		[56]
PHB	PLAcoPEG	yes		[56]
PHB	PEO	yes	blend composition dependent	[58]
PHB	PMO	no		[59]
PHB20V	PLLA	no		[60]
PHB	PVB	no	partially, PVB composition dependent	[63]
PHB	PHBV	?	dependent on valerate content	[64]
PHBP	PHBP	?	copolymer composition dependent	[66]
PHV	PVPh	yes		[67]
PHBV	PVC	yes, if V > 18 %		[68]
PHBsyn	PMMA	yes	blend composition dependent	[46]

* cellulose acetate butyrate

5. Composites – incorporation of fillers

The main reason for mixing PHB with fillers is the need to improve certain properties but also to decrease the price of the material. For the latter reason, inexpensive fillers are looked for, although usually biodegradability of the filler is required.

Starch seems to be an obvious option for blending with PHAs. The advantages sought are better mechanical properties and a decrease in a price. Biodegradation is maintained; the presence of starch even accelerates the degradation of PHAs in composting conditions [70]. Destructured starch is claimed to be superior to native starch when comparing reinforcing ability, presumably due to the surface area, shape irregularity and smaller particles of the former, all of which contribute to higher surface area [71].

Nucleation of PHBV on wood fibers has been reported as revealed by hot–stage microscopy, modulated DSC and other techniques [72]. Crystallization kinetics analysis has led to somewhat controversial conclusions regarding the rate and mechanism of crystal growth based on the Avrami exponent n [72]. An increase in Young's modulus and tensile strength of PHB was found for cast material due to blending with cellulose propionate. On the other hand, elongation at break and toughness increased for melt–quenched blends. The difference was attributed to depression of crystallinity of PHB in the presence of cellulose derivative during cooling from the melt [73].

Similarly, the addition of straw fibers leads to an increase in the crystallization rate of PHBV, without affecting the crystallinity content. The composite was found to be stiffer compared to the matrix biopolyester. The presence of the filler does not influence the biodegradability rate but in composting tests the rate of biodegradation is reduced for composites containing more than 10 % of straw [74].

6. Current state-of-the-art regarding commercial production

Current production of PHB is marginal. After initial pilot production starting in ICI and later transfer to Zeneca and Monsanto, the PHBV copolymer is available from this source in small quantities under trademark BIOPOL supplied by the Biopol Business Unit of Monsanto [39]. Production capacity is about 500 t/year [75]. There are a few companies producing the material commercially in small quantities with an estimated production in the range of a few tons/year. Biomer (Germany) is apparently the oldest active small player on the market. They offer compositions based on PHB of high purity, however, the products have no FDA approval and are not used for food or medical applications. The quoted price range is high, being around 17 – 20 EURO.

Brazilian Copersucar used to supply samples of PHB in quantities of few hundreds of kilograms produced in a pilot plant connected with a sugar production. Although the purity of the product was not the best; the material was suitable for many purposes. After some interruption of production in 2000, they stopped supplying free samples and they seem to be more or less sold out, supplying only to various small customers, apparently to research laboratories and some companies using the material for special purposes [55]. The purity of the product and the quality standard over a long term time scale should be a matter of concern for potential customer. Metabolics (Cambridge, Mass) and Novartem (Canada) are reported to be producing PHB and are considering increasing production. Recently Huayi Biotech in China also started production. Production of PHB by Chemie Linz in Austria was reported in 1992 [76].

The main limiting factor in wide spread application seems to be the price of PHB and especially of PHBV. Since the price is relatively high, the demand is low; therefore the production is low keeping the price high. It seems that investors are rather wary about starting massive production, which could result in a lower price. The targeted price for high volume applications seems to be around 2 $ per kg, which is considered to be competitive with commodity plastics in the packaging industry. Although the quoted price is about double that of the polyolefins, other advantages of PHAs are believed to compensate for the increase. Obviously, the investors consider the target 2 $ per kilogram to be unrealistic in the near future, in view of the cost of raw materials (i.e. substrates) and the efficiency of the biotechnological process. A comparison of price of several biodegradable polymers and synthetic commodity plastics is shown in Table 7 [77].

Table 7 : Comparison of some properties and prices of biodegradable polymers and synthetic plastics [77].

	cost ($/kg)	cost ($/kg) finished film	moisture barrier	oxygen barrier	mechanical properties
PHBV	6.00-12.00	not available	good	good	moderate
polylactic acid	2.00-10.00	not available	moderate	poor	good
cellulose acetate	3.30-4.50	8.00	moderate	poor	moderate
LDPE	1.00	2.00	good	good	moderate
polystyrene	1.10	4.00	good	good	moderate
PET	1.50	6.00	good	good	good

Three main components make up the cost of production of PHAs. Of these, the fermentation process seems to be optimized enough to be competitive, so that the main problems in economics is seen in the costs of the substrate and of the separation process [27]. Alternative lower cost substrates include methanol, molasses, and hemicellulose hydrolysate [2]. For instance, replacing glucose by lactose could reduce the substrate cost by more than 80 %, while the efficiency of the process remains similar to that with glucose at about 0.33 kg PHB per kg of the substrate [2,27]. Scientists and engineers are thus challenged to search for approaches to increase the efficiency and consequently the productivity of the process. Mixing substrates seems to be a feasible possibility. Using a mixture of acetate and glucose or adding a formate to methanol can be cited as successful examples of this approach. Continuous synthesis of PHB in one step process is another possibility, aimed at increasing productivity. The topic was reviewed recently by Ackermann and Babel [78]. See also Chapter 8.

The possibility of producing PHAs by genetically engineered microorganisms or even transgenic plants is being investigated and some optimistic opinions have been published regarding the possibilities of large-scale production in the future [3]. It should be mentioned that PHB and other PHAs could also be produced also synthetically via ring opening polymerization of respective lactones. This way however is not commercially viable on industrial scale at present due to the cost of producing the lactone monomers and polymerization catalyst [2].

7 Applications

PHB has been suggested as a biodegradable substitute for polyolefin containers, plastic films and bags. Several examples of commercial applications of PHB or PHBV are being evaluated, such as injection blow molded bottles for packaging of a biodegradable hair shampoo, produced by Wella AG (Darmstadt, Germany) [79], or motor oil containers and disposable razor handles [80]. Excellent gas barrier properties may be considered for application in foodstuffs packaging [81]. Another possibility is for coating paper and films [79]. Rather detailed study of the transport properties of various liquids, vapors and gases through PHB was carried out by Miguel et al [82]. Unlike paper with polyethylene coating, PHB or PHBV coated paper has been shown to be completely biodegradable and also easier to recycle than conventionally coated paper [83]. A direct electrostatic coating technique is possible to use for depositing PHB on a low dielectric substrate such as paper [84].

Quite a number of applications are suggested, tested or used in medicine. PHBV is non-toxic and compatible with living cells, producing an extremely mild foreign body response and the biodegradation rate in vivo is low. The sole degradation product is R-β-hydroxy-butyric acid, which is a common mammalian metabolite [2,85]. Applications such as controlled drug release microcapsules [86], surgical sutures, surgical swabs, wound dressings, lubricating powders for surgeon's gloves [87] and even blood–compatible membranes [88] can be quoted as typical applications for consideration in hospitals. In this connection, water–transport behavior and sorption was also examined for PHB [89] and a texture of PHB membranes was investigated by wide angle X-ray scattering (WAXS) regarding the orientation of crystallites and the effect on transport properties [90]. The advantage of using PHB or PHBV is that, unlike cotton, small pieces of the material from swab or dressing can be left in the wound without danger of inflammation.

A detailed discussion on possible applications of PHB has been published [2], especially in medicine but considering also its optical activity and piezoelectric properties (by one order of magnitude lower than polyvinylidene fluoride but without interference from pyroelectricity due to temperature changes). Applications in various areas have been discussed not only for PHB but also for other important biodegradable plastics [77].

Many fibre applications may be envisaged for PHAs. However, it is not a simple procedure to draw PHB to high draw ratios. A rather sophisticated procedure for the preparation of PHB fibers by spinning and drawing has recently been described in several papers. The prerequisite seems to be the synthesis of PHB with ultra-high molecular weight. This should be well above 500 000 [91] and preferably in the range of 1 to 10 million [92], although some authors do not refer to the molecular weight of the PHB used at all [93]. Solvent cast films were reported to be easily stretched at 160 °C to a draw ratio 400–650 %. Stretched films were highly oriented as indicated by X-ray diffraction and their mechanical properties were improved compared to isotropic PHB. Annealing results in further improvement of properties, as seen in Table 8 [92]. The comparison of stretched and undrawn material indicates a somewhat different behavior compared to conventional thermoplastics, e.g. polypropylene. Drawing in the latter case results in a substantial increase in modulus and a dramatic decrease in elongation at break, while for PHB quite opposite tendency was observed (slight decrease in modulus and an increase in deformability). Obviously, the changes in morphology leading to an increase in deformability result in a lower stiffness of the material, in spite of the orientation and crystallinity increase. Orientation leads to an increase in tensile strength for both materials.

Other papers describe melt spinning and subsequent cold or hot drawing [93]. The essential factor seems to be immediate post-extrusion stretching of the material. This process, which suppresses the formation of large crystals during cooling, achieves drawability via necking even at room temperature [94]. Impressive values of tensile strength and modulus were reported, namely 190 MPa and 5.6 GPa, respectively. Using a similar, slightly modified procedure, a drawing ratio 6 and tensile strength 250 MPa resulted [93]. However, modulus was found to be only 2.5 GPa, while elongation at break of 40% was measured. Annealing was carried out for 2.5 minutes in heated air. The application of various temperatures (75 to 150 °C) and constant tensile stresses during annealing (0, 50 or 100 MPa) had only marginal effect on the mechanical properties [93] in contrast to the data of other authors.

Table 8 : The effect of stretching on the mechanical properties of PHB (tensile strength σ, elongation at break ε, and Young's modulus E)

	σ MPa	ε %	E MPa
PHB	40	2	2.0
PHB UHMW stretched	62	58	1.1
PHB UHMW stretched and annealed	77	67	1,8

In a recent paper, Gordeyev et al. addressed the problem of thermal stability by introducing a gel-spun technique [95]. The gel was prepared by dissolving PHB in an appropriate solvent (1,2-dichloroethane was found to be the best) followed by evaporating part of the solvent to form a solid gel with polymer content around 30 wt %. The gel was extruded, pre-stretched to a draw ratio (DR) about 2 and hot drawn to DR around 10. The last step consisted in stretching at room temperature to 180 % and annealing in a fixed position at 150 °C for 1 hour. The ultimate properties of the fibres were impressive, as seen in Table 9 [95]. An important feature of the fibres was the negligible effect of physical ageing. Both tensile strength and Young's modulus values have been maintained within the experimental error during 6 months period with no tendency to change, if the material was annealed after stretching. However, the hot drawn fibres without subsequent annealing showed an increase in modulus and, rather surprisingly, a decrease in tensile strength [95]. Considering the gel–spinning technique, the formation of thermo-reversible gels may be of interest. This process was described recently for PHB solutions in dimethyl formamide [96] or N-methyl-2-pyrrolidone [97] in which the polymer is soluble above 100 °C. A self–supporting gel is formed by cooling from solutions containing 0.1 – 20 % of PHB

The biodegradation of fibres was investigated and compared with isotropic PHB. An interesting transformation of the crystalline phase was revealed by WAXS due to biodegradation. In general, a decrease in enzymatic degradation rate was reported [98]. However, this conclusion was based only on changes of mechanical properties and no weight changes or CO_2 emissions were measured. PHBV was blended also with a polyester–polyurethane matrix at 5–15 % and the possibility of a formation of fibres from the blend was demonstrated [99].

Obviously, the option of producing biodegradable fibres with excellent mechanical properties will broaden the possible applications for PHAs, although the possibility of a decrease in biodegradability arises as a result of substantial changes in morphology.

Table 9 : Tensile properties (tensile strength σ, elongation at break ε, and Young's modulus E) of gel-spun fibres measured at room temperature

Sample	DR	σ, MPa	ε, %	E, MPa static	E, MPa dynamic
as spun	2	103	250	2.0	4.6
hot drawn	10	332	104	3.8	5.8
annealed	10	360	37	5.6	7.5

8 Conclusions

Polyhydroxyalkanoates and polyhydroxybutyrate in particular may be considered as rather controversial polymers. PHB is sometimes referred to as a material that appeared too early, before its time has came. To summarize, PHAs are prepared from renewable resources and are fully biodegradable. They outperform most of the other biodegradable polymers and many synthetic plastics in properties, notably mechanical strength and modulus, resistance to water and moisture due to high hydrophobicity, high crystallinity and several other physical properties such as barrier behavior or piezoelectricity.

On the other hand, severe drawbacks hinder high volume applications in many prospective fields. These include low thermal stability, difficult processing, unacceptable brittleness, and extensive physical ageing, together with high price. Some scientists believe that PHAs and especially PHB are highly promising materials and that their broad application is inevitable in the near future. In the academic community, this opinion is illustrated by the increasing number of scientific papers and patents dealing with various topics related to PHAs. Others, especially investors, are more skeptical about the future of PHAs and they consider other plastics as powerful competitors for PHAs. Among these, especially polylactic acid and polycaprolactone should be mentioned. The production of the first high volume bio-based synthetic polymer, polylactic acid, was initiated by Cargill Dow and a number of new applications and broadening of the market have been reported for polycaprolactone. Biodegradable/renewable materials based on starch, especially in its thermoplastic modification, are strong competitors too, especially regarding price, although a low resistance to moisture disqualify these polymers from may applications.

Although biodegradable plastics are apparently new materials in high volume market, conventional synthetic polymers, modified to biodegradable version, should be considered as well. There are now a number of commercial products based on polyolefins used in packaging and in commercial agriculture that are claimed to be biodegradable [100]. If the claims made for these materials are justified, then they must be considered as major competitors to the natural or synthetic bio-based products. Since the properties of the degradable polyolefins are already well understood and they are based on relatively cheap commodity plastics, they will be very difficult to displace from packaging and agricultural applications.

Thus, the future of PHAs as high volume plastics depends on the ability to improve the properties and at the same time to decrease the price. It seems that the final goal can be only reached by complementary approaches. The modification of properties should be based on a combination of physical and chemical procedures aimed at a modification of crystalline structure and crystallization kinetics without substantial decrease in the crystalline portion. Another prospective approach may consist in a substantial reduction of the crystalline domain, possibly by an appropriate chemical procedure and subsequent reinforcement by introduction of fillers. The latter idea might also result in a decrease in price of the material, especially if a one – step chemical modification would be accomplished by reactive processing using inexpensive chemicals and fillers. The most feasible process could involve property modification in the first stage, i.e. during the bacterial synthesis of the polymer. At the same time the optimization of the production process could contribute substantially to the economics, mainly via the development of technology utilizing of cheap but efficient carbon source as a feed for bacteria. These or similar sophisticated products may possibly result in a biodegradable PHA - based materials which could take over a part of the high volume plastics market in the near future in competition with synthetic petrochemical plastics.

References

1. Bilingham N.C., Henman T.J., Holmes P.A. (1987) *Development in Polymer Degradation 7*, chapter 3, Ed. N. Grassie, Elsevier Sci. Publishers
2. Hocking P.J., Marchessault R.H., Biopolyesters, in : *Chemistry and Technology of Biodegradable Polymers*, Blackie Academic & Professional, Ed. G.J.L. Griffin, London, Glasgow, NY, Tokio, Melbourne, Madras 1994
3. Sudesh K., Abe H., Doi Y. (2000) *Progr. Polymer Sci.* **25**, 1503 – 1555
4. Braunegg G., Lefebvre G., Genser K.F. (1998) J. Biotechnology **65**, 127 – 161
5. Hammond T., Liggat J.J. (1995) *Degradable Polymers: Principles and Applications*, Eds. G.Scott and D.Gilead, Chapman & Hall (Kluwa), Chapter 5
6. Hobbs J.K (1998) *J. Mater. Sci.* **33**, 2509 – 2514

7. Janigová I., Lacík I., Chodák I. *Polymer Degrad. Stability* **77**, 35 – 41 (2002)
8. Taylor R. In : *Chemistry of Functional Groups,* Chap. 15, Ed. S. Patai, Appl. Sci. Publishers, London 1979.
9. Grassie N., Murray E.J., Holmes PA..(1984) *Polymer. Degrad. Stab*ility. **6,** 95.
10. Grassie N., Murray E.J., Holmes P.A. (1984) *Polymer Degrad. Stability* **6**, 47
11. Day M., Cooney J.D., Shaw K., Watts J. (1998) *J. Thermal Analysis* **52**, 261 – 274
12. He J.-D., Cheung M.K., Yu P.H., Chen G.-Q. (2001) *J. Appl. Polymer Sci.* **82**, 90 – 98
13. Li S.-D., Yu P.H., Cheung M.K. (2001) *J.Appl. Polymer Sci.* **80**, 2237 – 2244
14. Csomorova K, Rychly J, Bakos D, Janigova I. (1994) Polym. Degrad. Stability. **43** 441 - 446
15. Kopinke F.D, Remmler M, Mackenzie K. (1996) *Polymer. Degrad. Stability* **52** 25 - 38
16. Hoffmann A, Kreuzberger S, Hinrichsen G. (1994) *Polymer Bull.*, **33,** 355 – 362.
17. Rychly J., Csomorova K., Janigova I., Broska R., Bakos D., (1995) *Iranian J. Polymer Sci. Technol.* **4**, 274 – 282
18. Weber E.J., Heusinger H. (1965) *Radiochimica Acta* **4**, 92
19. Albertsson A.-C., Karlsson S. (1995) *Acta Polymerica* **46**, 114 – 123
20. Tsuji H. (2000) *Recent Res. Devel. Polymer Sci.* **4**, 13 – 37
21. Tormala P., Rokkanen P., Vainiopaa S., Laiho J., Hepponen V.-P., Pohjonen T., (1990) *US Pat. Applic.* 4,968,317
22. Wang Y.D., Yamamoto T., Cakmak M. (1996) *J. Appl. Polymer Sci.* **61**, 1957
23. de Koning G.,J.,M., Scheeren A.H.C., Lemstra P.J., Peeters M., Reynaers H. (1994) *Polymer* **35**, 4598 – 4605
24. Chambers R., Daly J.H., Hayward D., Liggat J.J. (2001) *J. Mater. Sci.* **36**, 3785 – 3792
25. Karpátyová A., Chodák I., unpublished results
26. Doi Y., Kunioka M., Nakamura Y., Soga K (1987) *Macromolecules* **20**, 2988 – 2991
27. Byrom D. (1987) *Trends Biotechnol.* **5**, 246-250,
28. Organ S.J., Barham P.J. (1991) *J. Mater. Sci.* **26**, 1368,
29. Gagnon K.D., Lenz R.W., Farris R.J. (1992) *Rubber Chem. Technol.* **65**, 761 – 777
30. Chodak I., Nogellova Z., Alexy P., *Proc. Int. Workshop Environmentally Degradable Polymers,* Doha, March 1999, 66 – 72.
31. Yasin M., Amass A.J., Tighe B.J. (1995) *Adv. Mater. - 95*, Proc. Int. Symp. 4th Meeting Date 1995, 331 - 336.
32. Barham P.I., Keller A, *J.* (1986) *Polymer Sci., Polym. Phys. Ed.* **24**, 69
33. Luepke T., Radusch H.J., Metzner K. (1998) *Macromol. Symp.* **127**, 227 – 240
34. de Koning G.J.M., Lemstra P.J., (1993) *Polymer* **34**, 4089
35. Hobbs J.K., Barham P.J. (1998) *J. Mater. Sci.* **33**, 2515 – 2518
36. Yuan Y., Ruckenstein E., (1998) *Polymer* **39**, 1893 – 1897
37. Mitomo H., Sasaoka T., Yoshii F., Makuuchi K., Saito T. (1996) *Sen'i Gakkaishi* **52**, 623 – 626 (English)
38. Bahari M., Mitomo H., Enjoji T., Hasegawa S., Yoshii F., Makuuchi K. (1997) *Angew. Makromol. Chem.* **250**, 31 – 44
39. Luo S., Netravali A.N. (1999) *J. Appl. Polymer Sci.* **73**, 1059 - 1067

40. Hazer B., Demirel S.I., Borcakli M., Eroglu M.S., Cakmak M., Erman B. (2001) *Polymer Bull.* **46**, 389 – 394
41. Deng X.M., Hao J.Y. (2001) *Eur. Polymer J.* **37**, 211 – 214
42. Hiki S., Miyamoto M., Kimura Y. (2000) *Polymer* **41**, 7369 – 7379
43. Yoshie N., Azuma Y., Sakurai M., Inoue Y (1995) *J. Appl. Polymer Sci.* **56**, 17 – 24
44. Yoon J.-S., Oh S.-H., Kim M.-N. (1998) *Polymer* **39**, 2479 – 2487
45. Zhang L. Deng X. (1994) *Gaofenzi Cailiao Kexue Yu Gongcheng* **10**, 64 – 68 (Chinese)
46. *Eur. Patent Appl.* 93302847.4 (13.04.93, Mitsubishi Gas Chemical Co)
47. Urakami T., Imagawa S., Harada M., Iwamoto A., Tokiwa Y. (2000) *Kobunshi Ronbunshu* **57**, 263 – 270
48. Immirzi B., Malinconico M, Orsello G., Portofino S., Volpe M.G. (1999) *J. Mater. Sci.* **34**, 1625 – 1639
49. Chodak I., (2001) *Proc. Int. Workshop Environmentally Degradable Polymers*, Lodz – Pabianice, June 2001
50. Chun Y.S., Kim W.N. (2000) *Polymer* **41**, 2305 – 2308
51. Zhang Q., Zhang Y., Wang F., Liu L., Wang C. (1998) *J. Mater. Sci. Technol.* **14**, 95 – 96
52 Hao J., Deng X., (2001) *Polymer* **42**, 4091 – 4097
53. Finelli L., Sarti B., Scandola M. (1997) *J. Macromol. Sci., Pure Appl. Chem.* **A34**, 13 – 33
54. Lee J.-C., Nakajima K., Ikehara T., Nishi T. (1997) *J. Polymer Sci., Part B: Polym. Phys* **35**, 2645 – 2652
55. El-Shafee E., Saad G.R., Fahmy S.M. (2001) *Eur. Polymer J.*, **37**, 2091 – 2104
56. Deng X.M., Zhang L.L., Xiong C.D. (1993) *Chinese Chem. Letters* **4**, 265 – 268 (in English)
57. H.J., Park S.H., Yoon J.S., Lee H.-S., Choi S.J. (1995) *Polymer Engn. Sci.* **35**, 1636 –1642
58. He Y., Asakawa N., Inoue Y. (2000) *Polymer Int.* **49**, 609 – 617
59. Avella M., Martuscelli E., Orsello G., Raimo M. (1997) Pascucci B., *Polymer* **38**, 6135 – 6143
60. Iannace S., Ambrosio L., Huang S.J., Nicolais L. (1994) *J. Appl. Polymer Sci.* **54**, 1525 – 1536
61. Deng X., Hao J., Yuan M., Xiong C., Zhao S., (2001) *Polymer Int.* **50**, 37 – 44
62. He Y., Asakawa N., Masuda T., Cao A., Yoshie N., Inoue Y. (2000) *Eur. Polymer J.* **36**, 2221 – 2229
63. Chen W., David D.J., MacKnight W.J., Karasz F.E. (2001) *Polymer* **42**, 8407 – 8414
64. Yoshie N., Fujiwara M., Ohmori M., Inoue Y. (2001) *Polymer* **42**, 8557 – 8561
65. Saito M., Inoue Y., Yoshie N. (2001) *Polymer* **42**, 5573 – 5580
66. Na Y.-H., Arai Y., Asakawa N., Yoshie N., Inoue Y. (2001) *Macromolecules* **34**, 4834 – 4841
67. Zhang L., Goh S.H., Lee S.Y. (1999) *J. Appl. Polymer Sci.* **74**, 383 – 388
68. Choe S., Cha Y.-J., Lee H.-S., Yoon J.S., Choi H.J. (1995) *Polymer* **26**, 4977 – 4982
69. He Y., Shuai X., Cao A., Kasuya K., Doi Y., Inoue Y. (2001) *Polymer Degrad. Stability* **73**, 193 - 199

70. Imam S.H., Chen L., Gordon S.H., Shogren R.L., Weisleder D., Greene R.V., (1998) *J. Environ. Polymer Degrad.* **6**, 91 – 98
71. Koller I., Owen A.J. (1996) *Polymer Int.* **39**, 175 - 181
72. Reinsch V.E., Kelley S.S. (1997) *J. Appl. Polymer Sci.* **64**, 1785 – 1796
73. Maekawa M., Terahata Y., Manley R.S.J. (1997) *Nippon Kasei Gakkaishi* **48**, 981 – 987
74. Avella M., La Rota G., Martuscelli E., Raimo M., Sadocco P., Elegir G., Riva R., (2000) *J. Mater. Sci.* **35**, 829 – 836
75. Bastiolli C. (1998) *Macromol. Symp.* **135**, 193 – 204
76. Hrabak O., (1992) *FEMS Microbial Rev.*, 251 – 256
77. Ikada Y., Tsuji H. (2000)*Macromol Rapid. Commun.* **21**, 117 – 132
78. Ackermann J.-U., Babel W. (1998) *Polymer Degrad. Stability* **59**, 183 – 186
79. Ramsay J.A., Ramsay B.A. (1990) *Applied Phycology Forum* **7**, 1 – 5
80. McCarthy-Bates L. *Plastic World March*, 22 – 27 (1993)
81. Basta N., (1984) *High Technology,* February 1984,. 67 – 71
82. Miguel O., Fernandez-Berridi M.J., Iruin J.J (1997) *J. Appl. Polymer Sci.* **64**, 1849 – 1859
83. Marchessault R.H., Lepoutre P., Wrist P. (1991) *PCT International Application* WO 31 13 207
84. Marchessault R.H., Rioux P., Saracovan I. (1993) *Nordic Pulp and Paper Research J.* 211 – 216
85. ICI, Biopol, the unique biodegradable thermoplastic from ICI, *promotional material*
86. Trau M., Truss R.W (1988) *European Patent Application* EP 293 172
87. Holmes P.A. (1985) *British UK Patent Application* GB 2 160 208
88. Kamata T., Numazawa R., Kamo J. (1983) *Jpn. Kokai Tokyo Koho Jp,* 137 402
89. Iordanskii A.L., Kamaev P.P., Hanggi U.J (2000) *J. Appl. Polymer Sci.* **76**, 475 – 480
90. Krivandin A.V., Shatalova O.V., Iordanskii A.L (1997) *Polymer Sci. Ser. B*, **37**, 102 – 105
91. Doi Y., Iwata T., Kusaka S., (1998) *Eur. Patent Appl.* EP 849311 A2, 24
92. Kusaka S., Iwata T., Doi Y., (1998) *J. Macromol. Sci., Pure Appl. Chem.* **A35**, 319 – 335
93. Yamane H., Terao K., Hiki S., Kimura Y. (2001) *Polymer* **42**, 3241 – 3248
94. Gordeyev S.A., Nekrasov Y.P. (1999) *J. Mater. Sci. Lett.* 18, 1691 - 1692
95. Gordeyev S.A., Nekrasov Yu.P., Shilton S.J. (2001) *J. Appl. Polymer Sci.* **81**, 2260 - 2264
96. Cesaro A., Fabri D., Sussich F., Paradossi G. (1999) *Macromol. Symp.* **138**, 165 – 174
97. Fabri D., Guan J., Cesaro A. (1998) *Thermochimica Acta* **321**, 3 – 16
98. Yamane H., Terao K., Hiki S., Kawahara Y., Kimura Y., Saito T. (2001) *Polymer* **42**, 7873 – 7878
99. Kostic M., Skundric P., Divjakovic V., Zlatanic A., Medovic A. (1998) *Hem. vlakna* 38, 12 – 15 (Serbian)
100. Scott G. (1999) *Polymers and the Environment,* Royal Society of Chemistry, Chapter 5

10

BIODEGRADABLE POLYMERS IN MEDICINE

E. PİŞKİN
Chemical Engineering Department and Bioengineering Division,
Hacettepe University,
Beytepe, 06532 Ankara,
Turkey

1. Biomaterials: definition and major classes [1-23]

Biomaterials are substances that are used in prostheses or in medical devices designed for treatment, augmentation, or replacement any tissue, organ or function of the body. Both natural and synthetic materials are used as biomaterials.

Natural tissues, as auto-, allo- and xenografts (e.g., porcine skin and heart valves, bovine arteries) and their modified forms (e.g., catgut sutures) have been used as biomaterials in medicine for soft and hard tissue repair and replacement for centuries. They may have good biophysical properties, and may be used in blood-contacting systems without anticoagulants. However, they may exhibit some important problems including: risk of disease transmission; limited amount of tissue that may be used as autografts, donor site morbidity; unpredictable resorption characteristics; bio-mechanical degradation, and mineralization; possible immunogenic responses and ethical limitations with allo- and xenographs.

During the last 30 years, advances in material science have led to the development of synthetic materials that have unique properties for medical applications. Metals, ceramics, polymers, composites are the main classes of synthetic biomaterials. Metals and their alloys have been used in various forms as implants and for hard tissue repair (e.g., dental implants, joint replacement, fracture plates, screws, pins). They are mechanically strong, tough and ductile. They can be readily fabricated and sterilised. However, they may corrode in the biological media, their densities are high and their mechanical properties mismatch with bone, which may result undesirable destruction of the surrounding hard tissues.

Ceramic biomaterials have been given attention as implant materials due to some highly desirable properties for specific applications such as dental implants, hip sockets, joint implants, and heart valves. Their main advantage is that they have similar physical

G. Scott (ed.), Degradable Polymers, 2nd Edition, 321-377.

properties to bone. They can readily be sterilised, and can be inert (e.g., carbon or carbon-coated materials). Ceramics are generally hard and strong. Their compressive strengths are high, however, they are brittle, and they exhibit low impact strengths, therefore, they are very sensitive to notches or micro cracks that they can easily fracture. It is generally difficult to manufacture ceramic materials. They may also corrode in the body, which may be desirable in some cases.

Polymers, both synthetic and natural, are the most diverse class of biomaterials. Polymeric biomaterials are widely used in both medical and pharmaceutical applications, and contribute significantly to the quality and effectiveness of health care. They are available in a wide variety of compositions and properties. They can readily be processed to form complex shapes with any size according to their final application. In addition, their surface properties, which are important in biological applications, may be readily modified by physical, chemical, or biochemical means. Their main disadvantage is the extractables in their structures (remaining after synthesis or fabrication processes), which may leach out during the use, and may lead undesirable effects on the host.

Composite materials contain two or more distinct constituent materials or phases, on microscopic or macroscopic size scale. The use of composite materials is motivated by the fact that they can provide more desirable material properties than those of homogeneous materials. For example, mechanical properties of polymeric materials may be improved by reinforcement (e.g., silica particles in silicone implants) and the biocompatibilities of metallic implants may be improved by coating with carbon.

1.1 CRITERIA FOR SELECTION

All biomaterials must meet certain criteria and regulatory requirements. First of all biomaterials should be readily purified, fabricated and sterilised easily by conventional methods. Materials should be free from leachable impurities, such as initiators, stabilisers, emulsifiers, unreacted monomers or oligomers, and other additives (e.g., plasticisers, fragments of fillers, dyes) may leach out during application of polymeric biomaterials and may cause important side-effects.

Biomaterials should exhibit the biomechanical properties (in tension, compression and shear) necessary for the specific application. Therefore, they should have desirable physical structures (e.g., crystallinity, entanglement, equilibrium swelling). In addition, depending on the application, they should also have prerequisit permeability, elasticity, electrical properties.

Surface properties of biomaterials are among the most important issues to be considered. Hydrophobicity or hydrophilicity, wettability (wettable or non-wettable), surface charge (anionic or cationic), polarity (polar or apolar), heterogeneity in the distribution of reactive chemical groups (uniform or domain structure), surface energetics (high or low energy), type of sorbed water (oriented, structured or free), mobility of the surface molecules and smoothness (smooth, rough or porous) are considered important surface properties.

It is important that a biomaterial should maintain its properties, and function *in vivo* over desired time period (from hours to years). Material properties may change within the physiological environment during use. Movement, separation, adsorption/de-sorption, biodegradation, corrosion, etc. may lead to total failure of the biomaterial.

Finally, biomaterials should be biocompatible. In other terms, they and their degradation products (if they are biodegradable) should not induce undesirable host reactions (e.g., thrombosis, inflammatory reactions, tissue necrosis, toxicity, allergenic reactions, carcinogenesis). These are briefly introduced below.

1.2. BIOCOMPATIBILITY CONSIDERATIONS

Biomaterials in use come into contact either with the cardiovascular blood system (i.e., the intravascular system), or with the soft and hard tissues (i.e., the extravascular system), or with both. When a biomaterial is exposed to the living organism, there is a natural tendency to respond to this foreign object. The living organism is a highly complex system. Many interrelated local and systemic reactions, including various parameters may occur at the biomaterial-biological system interface in a complex and dynamic manner. Biocompatibility is a wide definition, which includes all the responses of the biological system to the biomaterial. Non-biocompatible materials may be rejected in time or may cause important problems, such as emboli, tissue necrosis or even tumor formation.

The blood is a highly complex fluid, which consists of water, ions, proteins and cells (i.e., erythrocytes, platelets and leukocytes). As expected, blood-material contact is also very complex, and there are many interrelated reactions and feedback networks including the cascade of coagulation and complementary system. The main undesirable event in blood-material interaction is thrombus formation, which may lead to emboli formation (which may even cause death) or failure of the biomaterial. When the blood first has contact with a biomaterial, small molecules (e.g., water and ions) reach the surface and may or may not be adsorbed. This is followed by plasma protein adsorption. Depending on the surface properties of the biomaterial, the adsorbed proteins may change their three-dimensional structures (i.e., orientations), may even be denatured, or may exchange places with other proteins competing for the same site(s) on the surface. This phenomenon is dynamic and is closely related to the biomaterial surface properties defined before. Furthermore, fluid dynamic factors, which depend on the design of the device (i.e., the final form of the biomaterial) and on the actual site of application, are involved in the process. The flow conditions (e.g., shear rates, turbulence and secondary flows), the residence time in the device and the size of the interface are important in this respect. The first protein layer adsorbed on the biomaterial surface determines the subsequent events of the coagulation cascade, and the complementary activation. Both mechanisms (i.e., coagulation and complement) are interrelated, and play a major role in activating the blood cells. Activation of plasma proteins/blood cells leads to the systemic inflammatory reactions and the generation of new molecules along with the release of substances from a multitude of cells. The plasma proteins, the blood cells deposited on the biomaterial surface or circulating in the blood stream and their released substances take part in the dynamic process of fibrinolysis and thrombus formation.

Materials implanted in tissues always generate a response. The major tissue response in the extravascular system is the inflammatory process, which starts as a local reaction to injury, insult or infection. Inflammation may be induced biologically, chemically or physically. Many proteins and cells are involved in this very complex process. The chemical characteristics of the biomaterial, or in the long-term, the released substances and/or the biodegradation products, may be responsible for foreign body reactions. Cell ingestion, fibrous encapsulation or fibrous ingrowth may occur depending on the geometry, configuration and size of the biomaterial. If the foreign body reaction in soft tissue to a biomaterial is a mild inflammatory response, healing occurs rapidly, and then the implant performs effectively. It may be difficult to understand the response of the biological process from the first contact of the biomaterial. In the long-term implantation, the formation of a fibrous sheath, a fibrous capsule, pseudo membranes/ligaments, and problem of interfacial instability, may occur. Severe inflammation or excessive fibrosis may even cause tissue necrosis, granulomas or tumorgenesis.

2. Biodegradable polymers [24-68]

Many types of surgically implantable devices and drug delivery systems that only function for a relatively short time *in vivo* can be made from polymers that are eliminated from the body by hydrolytic degradation and subsequent metabolism after serving their intended purpose. These types of polymers are commonly known as biodegradable polymers. Biomaterials made of biodegradable polymers are designed to degrade *in vivo* in a controlled manner over a pre-determined period. The application of degradable biomaterials for temporary artificial implants can have two major advantages. Firstly, degradable biomaterials do not have to remove after use by secondary surgery, because, as mentioned before, the degradation products formed can be excreted from the body via natural pathways. Secondly, the use of biodegradable materials may lead to a better recovery of the biological systems, because progressive loss of mechanical strength of the implant will lead to a continuous simulation of the healing tissues.

2.1 BIODEGRADATION

Degradation *in vivo* is a highly complex phenomenon. Both enzymatic and hydrolytic degradation may occur in the body. Mechanical stress in use can induce material fragmentation, therefore producing fragments or even particles with a larger surface/volume ratio, which contributes biodegradation in both cases. Oxidative biodegradation has also been evidenced for various polymers (ultrahigh molecular weight polyethylene in joint prostheses or in antioxidant-free polypropylene), in which superoxide anions, hydrogen peroxide and hypochlorous acid produced by polymorphonuclear leukocytes (neutrophils), macrophages and giant cells are involved.

Natural polymers, such as polyaminoacids, polysaccarides, polynucleotides, bacterial polyesters, etc. are generally degraded in the body by enzymatic degradation. Other chemical linkages in the polymer chains, such as amide, enamine, ester, urea and urethane are also susptible to biodegradation by several enzymes (proteases, amylases lysozyme, esterase, etc.). It is rather difficult to pre-evaluate enzymatic degradation rate and extent *in vivo* because of the differences in enzymatic concentration in different parts of the body. In this type of degradation, availability of the polymeric chains for the enzyme attacks is important. Enzymes large protein molecules and therefore cannot penetrate within the polymeric matrix, which leads only degradation from the surface (i.e., "surface erosion").

Flexible aliphatic polyester chains, that are flexible enough to fit into the active site of the enzyme, are readily degraded enzymatically, while more rigid aromatic polyesters are generally bioinert. In the crystalline regions, the polymer chains are densely packed, therefore they are not easily available for enzymatic attacks. However, smaller lateral crystallites size yields a higher crystallite edge surface, which leads to a higher rate of degradation. Note that enyzmatic degradation may also be as fast as hydrolysis or even faster. But is certainly more specific in some cases (i.e., collagenase for collagen degradation).

In hydrolytic degradation, there are a number of parameters that affect degradation. Water is the agent that causes hydrolysis. Therefore, water penetration rate is one of the important parameters in hydrolytic degradation. If water molecules penetrate easily within the polymeric matrix, as in the case of hydrophilic polymers, degradation occurs rapidly, and most probably, homogeneously within the bulk (i.e. so-called "bulk erosion"). More hydrophobic polymer matrices are degraded much more slowly and from the surface. The chemical structure of the matrix has a pronounced effect on degradation. Hydrolytically unstable bonds (e.g., anhydrides, ester and carbonate linkages) are degraded much faster. Polymers with lower molecular weights are usually degraded faster than those with higher molecular weights. The morphology of the polymer is important. Water penetration in crystalline regions is rather slow, therefore these regions are degraded much more slowly than the amorphous regions. Polymers that exhibit glass transition temperatures lower than body temperature (at 37°C) are rubbery under *in vivo* conditions, and are thus degraded faster, mainly due to the flexibility of the polymer chains. Other ingredients (e.g., monomers, drugs) within the polymeric structure may change the course of degradation. Geometrical factors (e.g., size, shape, and surface to volume ratio) influence the degradation. A high surface to volume ratio means high water penetration, which leads to high degradation rates. The degradation products (e.g., acidic compounds) may catalysis the degradation rate (so-called "autucatalytic degradation"). Environmental factors, e.g., the site of implantation or injection, and pH and ionic strength of the degradation site may have profound effect on both the rate and extent of hydrolytic degradation.

2.2 SYNTHETIC BIODEGRADABLE POLYMERS

2.2.1 Poly(α-hydroxy acids) [69-124]

During the last decade poly(α-hydroxy acids) (polyglycolide, polylactides and their related copolymers) have received special and much interest because of their potential and proven use in the medical and pharmaceutical field.

Alpha-hydroxy acids, being bifunctional molecules (one hydroxyl and one carboxylic acid group on each molecule), can be polymerized to form related polyesters. Such condensation polymerization occurs spontaneously in aqueous solution of the acids with concentration greater than 25%. The direct polycondensation reaction of α-hydroxy acids by heating with or without using a catalyst was first described by Filachione and Fisher in 1944 (Scheme1). Here, poly-condensation is affected by removal of water. An acid catalyst is beneficial to increase the reaction rate, but above 120°C the rate-limiting step is generally water removal. Even by exhaustive distillation using an effective agent one might expect to produce polymers with number average molecular weights not more than 10,000.

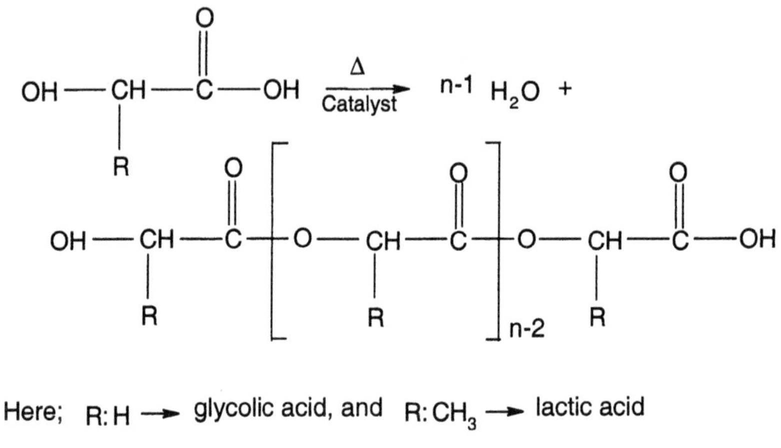

$$\mathrm{OH{-}CH(R){-}C({=}O){-}OH} \xrightarrow[\text{Catalyst}]{\Delta} \text{n-1}\ \mathrm{H_2O} +$$

$$\mathrm{OH{-}CH(R){-}C({=}O)}\left[\mathrm{{-}O{-}CH(R){-}C({=}O){-}}\right]_{n-2}\mathrm{O{-}CH(R){-}C({=}O){-}OH}$$

Here; R: H → glycolic acid, and R: CH_3 → lactic acid

Scheme 1. Polymerization of α-hydroxy acids.

In order to produce glycolic or lactic acid polymers with molecular weights usually higher than 40,000, the preferred route is based on the ring-opening polymerization of the respective cyclic dimers initiated by a catalyst. Several reaction mechanisms have been proposed for the ring-opening polymerization of glycolide or lactides. A broad spectrum of substances have been used as catalyst, including metallic, organometallic, inorganic and organic zinc and tin compounds, such as zinc, zinc chloride, zinc octoate, trialkyl aluminum, trialkyl tin, stannous chloride, stannous octoate, etc.

Glycolic acid, hydroxyacetic acid ($HO\text{-}CH_2\text{-}COOH$), is a naturally occurring substance found in sour milk, fruits (e.g., grapes), sugar beets and sugar cane, etc. Glycolic acid is isolated from natural materials or produced by a number of synthetic routes. Aqueous solutions of glycolic acid and higher molecular weight alkyl esters are used in medicine to trigger soft tissue regeneration, and also in personal care product formulations. For production of respective polyesters, glycolic acid is first converted to a cyclic dimer, "glycolide", and is then polymerized by a ring-opening mechanism as mentioned above. The preparation of glycolide is an old and much studied process. Here, the process has been conducted in two generally distinct batch steps involving first preparing an oligomer of the glycolic acid, then heating the oligomer under reduced pressure to generate the desired cyclic ester.

Poly(glycolic acid) (PGA), is the simplest linear aliphatic polyester. PGA was first synthesized in 1893, and was recognized to be a potential fibre-forming polymer, but with a high hydrolytic instability. PGA is a semi-crystalline polymer (45-55% crystallinity) with a melting point of 220-226°C and a glass transition temperature of 35-40°C. Due to its highly crystalline structure, PGA is soluble in only a few organic solvents (e.g. hexafluoroisopropanol and hexafluoroacetonesesquilhydrate). High melting temperature, solubility in only in few expensive solvents and sensitivity to humidity (especially at high temperatures) are considered to be important limitations in the processing of PGA.

Owing to their relatively hydrophilic nature, PGA materials are degraded *in vivo* rapidly, from hours to months depending mainly on the initial molecular weight, morphology (amorphous/crystalline phases) and surface-to-volume ratio of the sample (i.e. its shape). The PGA degradation mechanism is not yet fully understood but it is believed that degradation occurs mainly by way of hydrolysis, which starts in amorphous regions and then proceeds into the crystalline regions. Note that the hydro-lysis of ester bonds may be auto-catalyzed by the carboxylic acid end groups of the oligomers and monomer generated during degradation (see Chapter 5). Esterases have also been proposed to be most likely candidates for the enzymatic degradation of PGA materials. The degradation products are ultimately metabolized to carbon dioxide and water or are excreted via the kidneys.

PGA is mainly used as suture material in surgery. Sutures are threads, either as mono or multi-filament, and used to close wounds for a successful healing. They are made of both biodegradable and non-degradable polymers, either natural or synthetic origin. Biodegradable sutures undergo rather fast degradation in the living body (from 2-3 weeks to 3-4 months) to soluble products and disappear from the sutured site as the wound heals. These groups of sutures are usually preferred in suturing of the wounds those cannot be reached after healing. The first totally synthetic biodegradable surgical suture was produced from PGA in the 1960s by Davis and Geck, Inc. (Danbury, CT). PGA-based sutures gradually replaced the collagen based natural biodegradable sutures (catgut and its modified forms) due mainly to its better tissue compatibility, mechanical properties (e.g., high flexibility, high tensile strength and knot-tying and security, reliable strength retention) and predictable biodegradation. The most widely used absorbable sutures based on PGA and copolymers are Dexon® (Davis and Geck), a multi-filament polyglycolide; Vicryl® (Ethicon), a copolymer with composition of (poly(L-lactide) (PLLA)-*co*-PGA (8% PLLA and 92%PGA); Maxon®, PGA-*co*-TMC (65% glycolide and 35% trimethylencarbonate) (Davis and Greck); Polysorb®, PGA-*co*-PLLA (U.S. Surgical); Biosyn®, PDO-PGA-TMC (U.S. Surgical); PGA Suture®, PGA (Lukens); Suretak® (PGA-*co*-TMC, from Smith and Nephew).

Along with suture threads, PGA and its copolymers have been used as knitted or non-woven fibrous materials for curing burns, abrasion and skin damage; as temporary scaffolds for the regeneration of the arterial wall; as woven gauze, felt-like sponges and tampons for packing the surface of bleeding organs; as a guided-tissue-regeneration membrane peridontal surgery (e.g., Resolut® (PGA-*co*-PDLLA) from W.L.Gore, and Vicryl® (PGA-*co*-PLLA) mesh from Ethicon); and more recently as scaffolds for tissue engineered soft and hard tissue repair.

For many applications, when devices or prostheses have to bear loads, for example in orthopedics or maxillofacial surgery, mechanical properties of polymers are not sufficient, and higher mechanical properties may be attained through specific treatments, such as sintering starting from fibres to prepare self-reinforced rods or by the so called orientrusion process or even by introducing inorganic fillers and fibres as reinforcing components. Self-reinforced PGA, PGA-TMC and PGA-PLA copolymers have been used for fracture fixation (SmartPins®, by Bionx Implants), interference screws (Endo-Fix Screw®, by Smith and Nephew), in meniscus repair (SD sorb®, from Surgical Dynamics), and Cranionmaxillofacial fixation (LactoSorb® screws and plates, from Biomet).

In order to control biodegradation of homopolymers, glycolide and lactides are copolymerised. Owing to difference in reactivities of these two dimers, copolymers having broad composition ranges can be produced. Therefore they degrade *in vivo* by hydrolysis in different periods of time, from weeks to months. PLA/PGA copolymers have also been studied in many drug delivery systems for controllable drug release and biodegradation profiles.

Lactic acid is also a natural product. As its name implies, it is in milk that, in 1780, Carl Wilhem Scheele found an acid, which he separated by crystallizing a calcium salt. Scheele had discovered the "milk acid" but thought it to be a milk component and not a fermentation product of rancid milk. Lactic acid has thus been used for centuries as a natural preservative in many food products. Lactic acid is the smallest chemical molecule with an asymmetric carbon and therefore exists with two optical-isomers: the L (+) and the D (-). The L form is the natural one, and is naturally present in animal and human tissues as well as in numerous food products (meat, milk products, pickles, beer, etc.). Today, lactic acid, its salts and esters are extensively used in food, cosmetic and pharmaceutical industries. These industries show preference for the L-lactic acid and its related compounds because the D form cannot be metabolized by the human body.

Lactic acid can be produced by chemical processes, or by fermentation. There are several possible routes for the production by chemical processes. Today more than half of the total consumption is produced by fermentation in which several carbohydrate sources such as whey, barley, sugarcane, soybean, milk, corn, sulfite waste liquor and potatoes can be used as substrate. Careful selection of the fermentation bacteria (*Lactobacillus, Leuconostos, Lactococcus, Pediococcus, Carno-bacterium Listeria, Staphylococcus and Bacillus*) allows producing the desired isomers, the L, the D or both. Industrially used specie is *Lactobacillus Delbrueckii* with glucose or sucrose as substrate.

The bimolecular cyclic ester of lactic acid is also called "lactide". According to the configurations of the chiral carbons, lactide can be classified as L-, D-, and meso-lactide as shown in Fig.1. The racemic D,L-lactide consists of a mixture of L-, and D- forms.

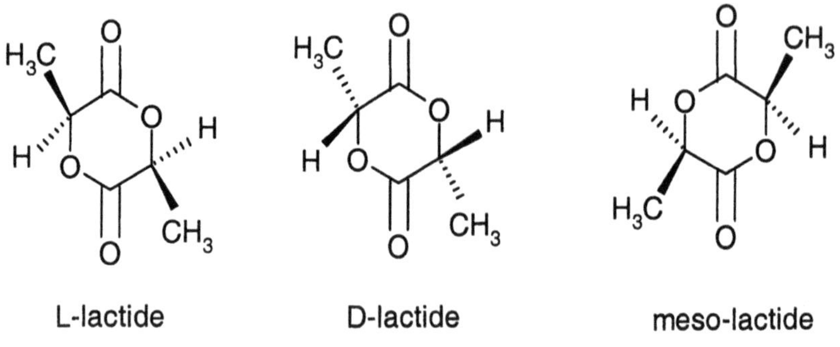

L-lactide D-lactide meso-lactide

Fig. 1. Stereoisomers of lactides.

Lactide enantiomers are synthesized by a two-step process similar to the production of the glycolide dimer. At the first step, lactic acid is converted to an intermediate, relatively low molecular weight, polylactic acid. Then, at the second cyclising step lactide, which is a cyclic ester of two molecules of lactic acid, is formed from this intermediate at a temperature of not lower than 180°C, in the presence of catalyst. The main lactide producers are Purac Biochem (The Netherlands), Boehringer (Germany), Cargill (USA) and DuPont (USA).

Similarly to glycolic acid, lactic acid can be polymerized directly into polylactic acid with low molecular weight by heating. Polylactides with high molecular weights therefore reliable mechanical properties can be synthesized from lactide dimers by ring-opening polymerisation. Poly(L-lactide) is a semi-crystalline polymer (up to 40% crystallinity) with a melting and glass transition temperatures of 175-184°C and 57-65°C, respectively, while poly(D,L-lactide) is fully amorphous and has a glass trans-ition temperature of 54-59°C. Poly(D,L-lactide) is soluble in most organic solvents such as tetrahydrofuran, acetone, chloroform, benzene, while poly(L-lactide) can be dissolved mainly in chloroform and methylene chloride.

Owing to the methyl groups in lactic acid, polylactides are more hydrophobic than PGA. Therefore they degrade much slower *in vivo* by hydrolysis, from weeks to years depending on the initial molecular weight, morphology, type of stereoisomers, and shape. Note that the carboxylic acid end groups formed during degradation may be auto-catalyzed the hydrolysis. It is generally accepted that enzymatic reactions are not responsible for the observed degradation of polylactides.

L-lactic acid, the final degradation product of poly(L-lactide), is a normal intermediate of carbohydrate metabolism in man. Therefore, it has attracted a great deal of attention as a biodegradable biomaterial. L-lactic acid is recycled by conversion to glycogen in the liver, while D-lactic acid is most probably excreted through the kidneys unchanged. In biomedical applications mostly poly(L-lactide) and poly(D,L-lactide) have been studied. With high mechanical strength and toughness, semi-crystalline poly(L-lactide) has been used in orthopaedic devices, such as fracture-fixation plates, clips, staples, pins, screws, meshes, and hard tissue substitutes. Examples are as follows: fracture fixation (SmartPins®, SmartScrew® and SmartTack® from Bion Implants); interference screws (Bioscrew®, from Linvatec, Sysorp®, from Sulzer Orthopedics); for meniscus repair (Meniscus Arrow®, from Bionx Implants, Clearfix®, from Innovasive Devizes, Meniscal Stringer®, from Linvatec) With amorphous structure poly(D,L-lactide) has been studied in drug delivery and targeting systems, as the basic carrier matrix in several forms, such as hollow fibers, films, microcapsules and microbeads.

On the other hand, a method to obtain biodegradable and less stiff polymers consists in the copolymerization of poly(ethyleneoxide) (PEO) and polylactic acid (PLA). These poly(ether-esters) are characterised by hard semi-crystalline PLLA domains and flexible, elastic and hydrophilic PEO regions. Degradation kinetics suggested a bio-medical application as replacement of soft tissue and drug delivery.

Poly(ester-urethane) and poly(L-lactide) mixtures were purposed as biodegradable vascular prostheses and biodegradable nerve guides. The two polymers are well known to degrade differently, and mechanical properties and degradation rate can be optimized as function of composition and geometric factors.

2.2.2. Poly(ε-caprolactone) [90,91,95,97,98,103,106-110,113,125-131]

Poly(ε-caprolactone) (PCL) is another biodegradable polyester that has been extensively investigated as a potential biomaterial. PCL was first synthesized by Carothers by ring-opening polymerization of ε-caprolactone. ε-caprolactone, a cyclic ester, which can be prepared by the Baeyer-Villiger reaction for the oxidation of cyclic ketones and lactones. ε-caprolactone can be polymerized by either a cationic or an anionic mechanism (Scheme 2). Cationic polymerization usually results in low molecular weight polymers because of proton transfer or to inter- or intramolecular chain transfer to polymer. The mechanism in which there is an alkyl-oxygen bond scission involves attack at hexocyclic oxygen atom. A wide variety of catalyst has been used to prepare homopolymers of ε-caprolactone. These include alkaline earth hex-ammoniates, alkyl and metal amides, alkoxides, alkaline metals and metal alkyls. Selectively functionalized polymers can be prepared with organic metallic salts such as stannous octanoate, dibutyltin dilaurate, zinc octanoate and similar compounds. Uranyl nitrate has been used as an initiator/sensitiser to photopolymerise ε-caprolactone. Poly(ε-caprolactone) is commercially available in a variety of molecular weights, that is usually controlled by use of dry monomer (water would act as initiatior) and addition of a specific amount of active-hydrogen initiator.

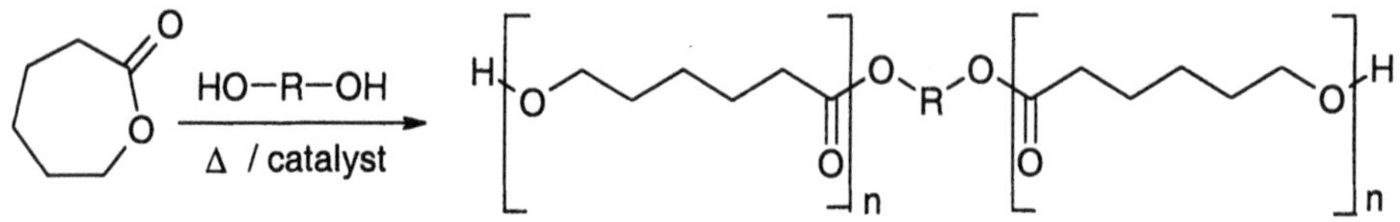

HO—R—OH

Δ / catalyst

Scheme 2. Polycaprolactone synthesis.

Although low molecular weight polycaprolactones range from liquids to hard waxes, the high molecular weight polymer is a strong, ductile polymer with excellent mechanical characteristics. It is a hydrophobic and semi-crystalline polymer with a melting point of 59-64°C and a glass-transition temperature of –60°C. PCL is in the rubbery state at room temperature and has a relatively low tensile strength (23 MPa) but very high ultimate elongation (>700%). PCL has the unique characteristic of being miscible with almost all other polymers (polyethylene, polypropylene, polystyrene, poly(methyl methacrylate), polycarbonates, polysulfone, poly(vinyl acetate), etc.). High molecular weight PCL is usually used as an additive to other polymers to obtain special effects, but it is used as the major ingredient in many formulations.

Polycaprolactone is degraded very slowly, much more slowly than poly(α-hydroxy acids), *in vivo* to yield ε-hydroxycaproic acid. Its *in vivo* degradation is initiated by non-enzymatic ester hydrolysis in the extracellular matrix. The final stage of degrad-ation, however, was found to involve phagocytosis of polymer fragments by macrophages and giant cells, and degradation within these cells by lysosome-derived enzymes. *In vitro* studies on polycaprolactone degradation have established its sensitivity to microbial enzymes and, as expected, increased degradability of amorph-ous regions relative to the crystalline phase.

The high permeability of PCL to various substances (such as contraceptive steroids) has made it an important candidate for the development of long-term implantable drug delivery systems. High loading and diffusion rates for hydrophobic drugs can be achieved in PCL matrices. Capronor® is a typical 1-year implantable contraceptive device prepared from PCL. Several groups have also investigated the use of lactide and ε-caprolactone copolymers with different chain structures (random, block, etc.) in controlled drug release systems. A series of tri-component copolymers by ring opening polymerization of glycolide, L-lactide and ε-caprolactone, using stannous octoate as a catalyst have been synthesized to modulate the degradation rate. Polyethylene glycol-coated biodegradable microspheres composed of polylactic acid/ poly(ε-caprolactone) blends were proposed for targeting antiproliferative agents for prolonged periods to treat retenosis. PCL microparticles for encapsulating of both lipophilic and hydrophilic drugs for oral delivery have been prepared by oil-in-water or water-in-oil-in-water solvent evaporation method. PCL microparticles with a mean size between 5 and 10 micron, obtained by a double emulsion-solvent evaporation technique, have been studied as potential oral vaccine delivery matrices.

Copolymerisation of polycaprolactone with especially poly(D,L-lactide) gives flexible matrices (more flexible than either polylactide or polycaprolactone) with lower crystallinity and, therefore, higher degradation rates are achieved. An 80/20 copolymer of D,L-lactide/caprolactone was used to produce flexible suturable films that were successfully tested *in vivo* for the prevention of post-operative pleural and pericardial adhesion. These films maintained integrity for about one month post-implantation and were estimated by *in vitro* tests to be absorbed in about five months. Histological examination of the copolymer films revealed no evidence of interference with the natural healing of adjacent tissues.

Caprolactone has also been copolymerized alternatively with glycolide and trimethylene carbonate to provide coating for bioabsorbable sutures. Sutures treated with these coatings exibited improved knot security and knot repositioning characteristics. A block copolymer of ε-caprolactone with glycolide is marketed by Ethicon (Monocryl®), which has reduced stiffness compared to pure PDA sutures.

Blends of poly(ε-caprolactone) and poly(D,L-lactic-co-glycolic acid), including also hydroxyapatite granules have been investigated as scaffolds (in the form of porous discs) for applications in bone tissue engineering. PDLLA-PCL copolymers and their composites with hydroxyapatite have been investigated as potential hard tissue filling material. Poly(L-lactide/ε-caprolactone) sponges containing chondrocytes have been evaluated as tissue-engineered cartilage.

2.2.3 Poly(α-amino acids) [132-150]

Proteins are among the essential molecules in all respect of the living organism. Hundreds of different proteins can be found in any single cell, and together they make up 50% or more of a cells's dry weight. They have several vital functions. For example, they catalyze biochemical reactions as enzymes). They transport several chemicals (in the blood stream and in and out from the cell wall). They have an important role in the vertebrate immune system as antibodies) and have regulatory functions as hormones).

Proteins are linear polypeptides, and composed of amino acids. There are twenty different amino acids, and each amino acid has the same basic formula as shown in Fig. 2. Amino acids contain at least one carboxyl (-COOH) group and one amino (-NH_2) group attached to the same carbon atom, called an alpha-carbon, and a variable R group. The side group (R group), which is different in different amino acids, can be a hydrogen atom, an unbranched or branched chain of atoms, aliphatic or aromatic

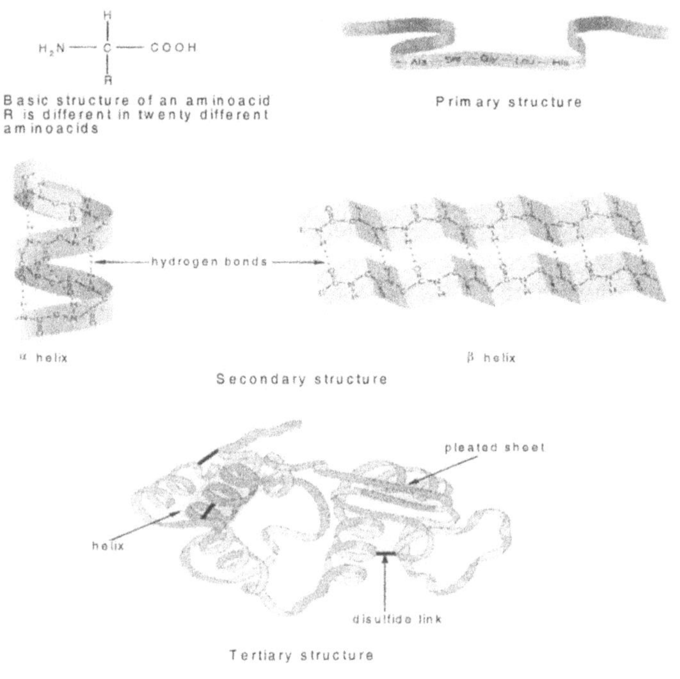

Fig. 2. Structure of amino acids and polypeptides

forms. The side groups can contain functional groups, such as the thiol group (-SH), the hydroxyl group (-OH), or additional carboxyl or amino groups, which are specially important in the formation of architectures and three-dimensional shapes of the proteins, which are directly related to their diverse functions. Amino acids exist in either of two stereoisomeric configurations designated D and L, which are mirror images of each other. The amino acids found in proteins are always L-isomer (except glycine, which has no stereoisomer). Carboxylic acid groups (attached to the alpha-carbon) of one amino acid react with the amino group (attached to the alpha-carbon) and form peptide bonds. Several amino acids with different sequences and numbers together form linear polypeptide chains (Fig.2). The sequence of amino acids on the chain is the primary structure of the protein. The order of the amino acids determines all of the higher order levels of structure of the protein. The polypeptide chain folds spontaneously to secondary structures (as helices or pleated sheets) and then to a final three-dimensional structure (the tertiary structure). Hydrogen bonds and other relatively week interactions between the side chain groups of the amino acids located on the backbone are responsible of these foldings. In addition, thiol (-SH) groups on two different amino acid subunits can form a covalent, disulfide link (-S-S-) by removal of the hydrogen atom, which further stabilizes the three-dimesional structure of the protein. Note that some proteins have also a quaternary structure, which consists of an aggregate of two or more individual polypeptide chains.

Since proteins are composed of amino acids, many researchers have tried to develop synthetic polypeptides, which can be manufactured by polymerization of the respective monomers (amino acids) or by fermentation. It should be noted that the functional groups on the side chain (e.g., ω-carboxylic groups) should be protected in different chemical forms (e.g., methyl, benzyl esters) during polymerization of amino acids. Several polypeptides have been synthesized to serve as models for structural, biological, and immunological studies. In addition, many different types of synthetic polypeptides have been investigated for use in biomedical applications.

It has been shown that poly(amino acids) are enzymatically degraded and the rate of *in vivo* degradation of these synthetic polypeptides can be controlled by varying the hydrophilicity of the side chain groups. The degradation was attributed to cleavage of the polypeptide chains by proteolytic enzymes, such as endopeptidase cathepsin B, released during acute and chronic stages of the inflammatory response. Because these polymers release naturally occuring amino acids as the primary products of polymer backbone cleavage, their degradation products may be expected to show a low level of systemic toxicity.

Synthetic polypeptides have several potential advantages in biomedical use. Many types of synthetic polypeptides have been prepared for biomedical applications, such as sutures, artifical skin substurates, and drug delivery systems. Side chains offer sites (e.g., ω-carboxylic acid group of glutamic acid or aspartic acid, amino group at the side chain of of lysine) for the attachment of drugs, crosslinking agents, or pendant groups that can be used to modify the physico-chemical properties of the polypeptide chain. In spite of their apparent potential as biomaterials, synthetic polypeptides have actually found few practical applications, mainly because of their low their insolubilities and difficulties in processing.

Among the potentials of biodegradable materials in medical applications, water-soluble polypeptides from amino acids have been attracted the most attention as carriers for drugs (for endocellular chemotherapy of tumors) and for genetic material (mostly plasmid DNA for gene theraphy). In the drug delivery approach, several drugs (e.g., methotrexate, 6-aminonicotinamid, netrexone, catecholamine, 14-bromodauno-rubicin, danuromycin, melphalan, mitomycin, etc.) have been covalently attached to the polypeptide chain (especially poly(L-glutamic and aspartic acids), poly(L-lysine), poly[(N^5-(2-hydroxyethyl)-L-glutamine], poly[(N^5-(3-hydroxypropyl)-L-glutamine or their copolymers with other poly(α-amino acids) or polyethylene glycol, etc.), through the ω-carboxylic acid or amino groups (directly or through a spacer arm). In gene theraphy, especially polylysine and its copolymers with polyethylene glycol, dextran, etc. have been used as carriers, in which the negatively charged DNA forms conjugates with the positively charged polylysine chains (in the form of nano particles, 40-100 nm in diameter). In both applications, the conjugates are injected into the blood stream, and it is expected that they will reach to the target cells and pass the cell wall and therefore directly exhibit its effect in the cytoplasma. The size, and surface charge of the conjugates are important to have an efficient transfection without caus-ing toxicity or other side effects. Several targeting agents (folic acid, mannose, anti-bodies against the membrane receptors of the target cells, etc.) have been also in-corporated in the carriers molecules in order to target the conjugates to the desired cell population.

Hydrophilic polypeptides have been processed into hydrogel type of membranes by using crosslinkers, which have potential as temporary artifical skin substitues in burn therapy. Synthetic polypeptide fibers (e.g., poly(N-hydroxyalkyl-L-glutamine-co-γ-methyl-L-glutamate) have been investigated for possible uses of biomedical applic-ations, such as a surgical suture.

Synthetic polypeptides from α-amino acids are well known as typical examples of synthetic polymers that can produce secondary structures such as α-helix and β-structure spontaneously and have been developed as model compounds of proteins. Immobilization of their functions expands their possible applications. It is very attractive to produce spherical particles from poly(α-amino acids), especially in the application field of packing materials for bioaffinity separation and carrier particles in drug delivery. The "suspension evaporation" method has been developed in order to obtain spherical particles directly from poly(amino acids). Macromolecular colloids of diblock poly(amino acids), poly(L-leucine-*block*-L-glutamate) have been prepared for insulin binding. Biodegradable copolymers containing alpha-hydroxy acid and alpha-amino acid units, have been produced for use as particulate carriers containing functionalizable amino subunits for coupling with targeting ligand. The detection of phosphopeptides by fluorescence polarization have been studied in the presence of cationic polyamino acids such as polyarginine and polylysine for possible application to kinase assays. Microparticles of novel branched copolymers of lactic acid and amino acids that composed of a poly(L-lactic acid-co-L-lysine) (PLAL) backbone, and poly(L-lysine), poly(D,L-alanine) or poly(L-aspartic acid) side chains extending from the lysine residues of PLAL have been proposed as carriers in controlled drug release. Crosslinked poly(amino acids) have been found as superabsorbent polymers. Use of polyaziridine and polyepoxide crosslinkers allows the production of superabsorbent polymers that is free of special handling steps required to process hydrogel materials using conventional preparation methods. Preferably, the crosslinked poly(amino acids) have a polyanionic backbone, such as poly(aspartic acid) or poly(glutamic acid), with the remainder of the polymer comprising crosslinking elements joined to backbone polymer via reaction with the side chain carboxyl groups.

In an attempt to circumvent the problems associated with conventional poly(amino acids), backbone-modified, the so-called "pseudo"-poly(amino acids) have been investigated. The pseudo-poly(amino acids), belonging to the polyamides group have been proposed to be utilized for medical use, ranging from biodegradable bone nails to implantable adjuvants. Recent studies indicate that the backbone modification of poly(α-amino acids) may be a generally applicable approach for improving the physico-mechanical properties of conventional poly(α-amino acids). For example tyrosine-derived polycarbonates are high-strenght materials that may be useful in the formulation of degradable orthopedic implants.

2.2.4 Polyanhydrides [151-159]

Polyanhydrides are among the most hydrolytically unstable polymers. These polymers were prepared first in 1909, subsequently investigated as potential textile fibers but found unstable melt-polycondensation. Here, starting from a dicarboxylic acid monomer, a prepolymer consisting of a mixed anhydride of the diacid with acetic acid is formed (Scheme 3). The polymer is obtained by subsequently heating the prepolymer under vacuum to eliminate the acetic anhydride by-product. Several aliphatic dicarboxylic acid compounds (sebacic acid (SA), fumaric acid (FA), etc.) and aromatic (isophthalic acid (Iph), p-carboxyphenoxy propane (CPP), p-carboxyphenoxy hexane (CPH), etc.) have been used to obtain polyanhydrates with different molecular weights (MW_n: 5000-20000) and polydispersity indices (P.I.=MW_w/MW_n: 3-10), therefore with different degradation rates (Fig.3). Since melt-polycondensation requires high temperature, a milder reaction is desirable for heat sensitive monomers. Several solution polymerization schemes have therefore been proposed. In a typical synthesis, polyanhydride is formed at room temperature by dehydrochlorination between a diacid chloride (e.g., sebacoil chloride) and a dicarboxylic acid (e.g., sebacic acid). Interfacial polycondensation has also been considered as an option for synthesis of polyanhydrides, however, the molecular weights obtained have been less than satisfying due to hydrolytic decomposition of the diacid chloride.

Poly(CPH) is amorphous, while the others exhibit quite high degree of crystallinity (e.g., up to 66% for poly(SA) and 62% for poly(CPP)). Note that the copolymers of SA and CPP have much lower crystallinities comparing to the homopolymers (e.g., 6% for poly(CPP-SA) 50:50). Little data exist on the mechanical properties of polyanhydrides. The majority of the polyanhydrides studied dissolve in common organic solvents such as dichloromethane and chloroform. For the highly aromatic structures, more polar solvents (e.g., dimethyl-formamide or m-cresol) are needed. Poly[1,2-bis(*p*-carboxyphenoxy)ethane anhydride] fibres obtained in a melt-spinning process showed a tensile strength of 392 MPa with an elongation of 17.2% and a Young Modulus of 4.95 GPa. In the case of CPP and SA copolymers, it has been shown that the tensile strength increases when either the CPP content in the copolymer or the molecular weight is increased.

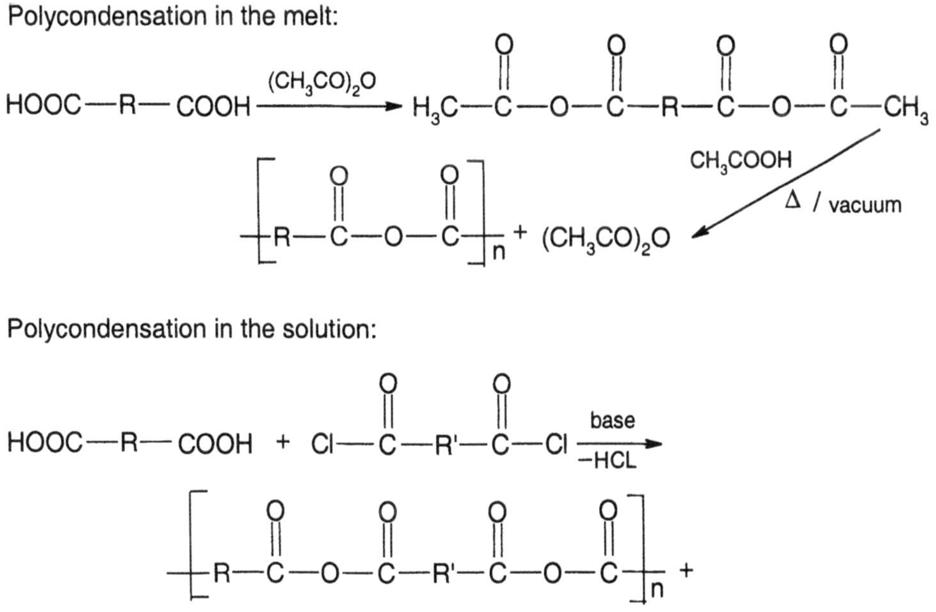

Polycondensation in the melt:

$$\mathrm{HOOC{-}R{-}COOH} \xrightarrow{(CH_3CO)_2O} \mathrm{H_3C{-}\overset{O}{\overset{\|}{C}}{-}O{-}\overset{O}{\overset{\|}{C}}{-}R{-}\overset{O}{\overset{\|}{C}}{-}O{-}\overset{O}{\overset{\|}{C}}{-}CH_3}$$

$$\xrightarrow[\Delta\ /\ \text{vacuum}]{CH_3COOH} \left[\mathrm{R{-}\overset{O}{\overset{\|}{C}}{-}O{-}\overset{O}{\overset{\|}{C}}}\right]_n + (CH_3CO)_2O$$

Polycondensation in the solution:

$$\mathrm{HOOC{-}R{-}COOH} + \mathrm{Cl{-}\overset{O}{\overset{\|}{C}}{-}R'{-}\overset{O}{\overset{\|}{C}}{-}Cl} \xrightarrow[-HCL]{\text{base}}$$

$$\left[\mathrm{R{-}\overset{O}{\overset{\|}{C}}{-}O{-}\overset{O}{\overset{\|}{C}}{-}R'{-}\overset{O}{\overset{\|}{C}}{-}O{-}\overset{O}{\overset{\|}{C}}}\right]_n +$$

Scheme 3. Synthesis of polyanhydrides by polycondensation in melt and in solution.

Polyanhydrides are among the most reactive and hydrolytically unstable polymers currently used as biomaterials. The hydrophobic character of anhydride polymers reflects a very low water sorption, whereas the rate of hydrolytic degradation results much faster. This gives a unique property to polyanhydrides that they exhibit surface erosion. The high chemical reactivity is both an advantage and a limitation of polyanhydrides. Aliphatic polyanhydrides are known to decompose on standing, while aromatic polyanhydrides are much more resistant to hydrolysis. In general the degradation rate decreases with the aromaticity and hydrophobicity of the backbone. Although hydrolysis of monomeric anhydrides is both acid and base-catalyzed, it is generally observed that the hydrolytic degradation rate of polyanhydrides increases with pH. In the solid state and at room temperature, polyanhydrides are unlikely to react with other compounds, for instance, with nucleophiles other than the hydroxide ion.

Because of their surface erosion characteristics, polyanhydrides have been used in controlled release of drugs with zero-order drug release kinetics. This surface mechanism erosion represent a potential advantage with respect to polyorthoesters that usually require the incorporation of additives to exhibit the same behaviour. Studies of polyanhydride copolymers demonstrated that degradation rates are strongly influenced by copolymer composition. Depending on the type and ratio of diacids, the degradation time of polyanhydrides can be properly ranged from few days (e.g., with the poly(SA) matrices) to few years (e.g., with the poly(CPP) matrices). The potential reactivity of the polymer matrix toward nucleophiles limits the types of drugs that can be successfully incorporated into a polyanhydride matrix by melt processing techniques.

Fig. 3. Examples of dicarboxylic acid compounds used for synthesis of polyanhydrides.

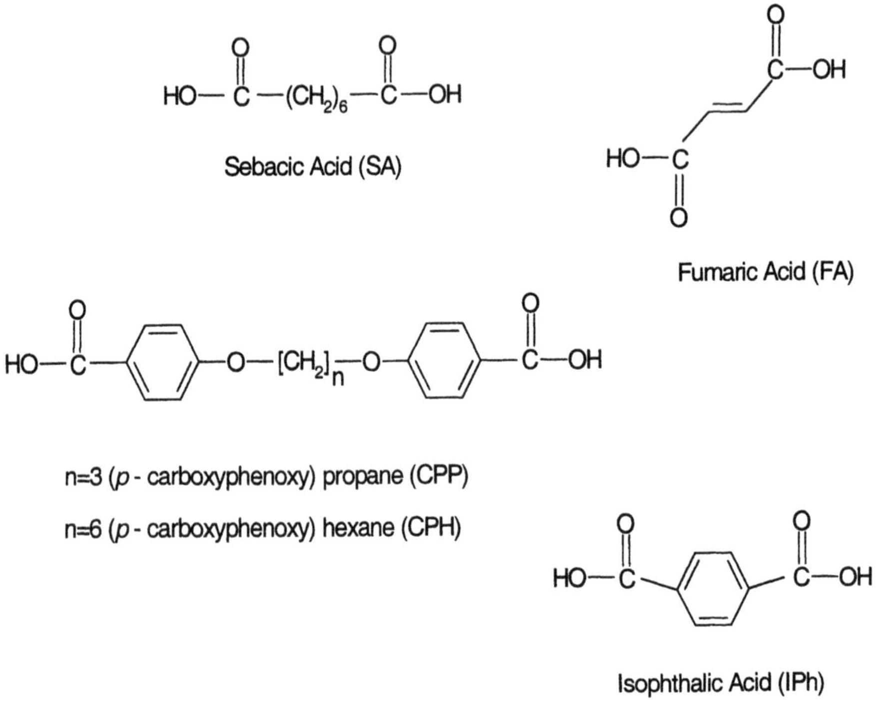

Sebacic Acid (SA)

Fumaric Acid (FA)

n=3 (*p* - carboxyphenoxy) propane (CPP)

n=6 (*p* - carboxyphenoxy) hexane (CPH)

Isophthalic Acid (IPh)

Applications of polyanhydrates in drug delivery includes implantable delivery systems for the chemotherapeutic agents (e.g., nitrosoureas, carmustine) for tumor theraphy, and for the release of other drugs and active agents (e.g., gentamicin, insulin, somatotropin, angiogenesis inhibitors, immunosuppressive agents) for treatment of other diseases

In addition, polyanhydrides have been considered as attractive materials for temporary biomaterials, such as sutures, and bioabsorbable prostheses. Degradation products of these polymers have anti-thrombotic/anti-inflammatory properties. Active polymers namely poly(anhydride esters) stimulate new tissue formation. The resulting combination features make polyanhydrides as ideal implantable materials.

2.2.5 *Polyorthoesters* [160-173]

Polyorthoesters are hydrophobic, but with hydrolytic linkages that are acid-sensitive. The first polyorthoesters have been reported in a series of patents by Choi and Heller, assigned to Alza Corporation. In the 1970s Alza Corporation firstly developed a polyorthoester matrix, designated with the trade name Chronomer® and later as Alzamer® (used for controlled release of contraceptives), which is produced by a transesterification reaction between a diol and diethoxytetrahydrofuran (Scheme 4A). The synthesis of such polymers, with minimal amount of crosslinkage, requires an orthoester starting material in which one alkoxy group has a greatly reduced reactivity, which can be achieved by using a cyclic structure.

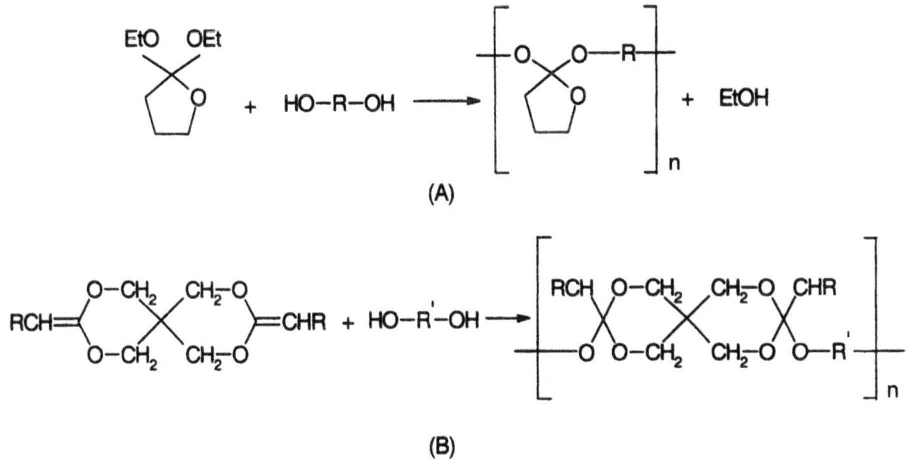

Scheme 4. Synthesis of polyorthoesters by:
(A) Transesterification; (B) addition of a polyol and diketene.

These polymers degrade by surface erosion, mainly due to the hydrophobicity of the matrix. At the first step of hydrolytic degradation, a diol and γ-butirrolactone are formed. Then, by hydrolysis, γ-butyrolactone is converted γ-hydroxybutyric acid, which accelerates polymer degradation (autocatalytic effect). The difference in the sensitivity of the hydrolysis of orthoester linkages in acid and in alkaline medium has been used an advantage in designing the orthoester-based delivery systems. This preferential sensitivity is used by incorporating acid anhydrides (e.g. phthalic anhydride) into the matrix to accelerate the rate of hydrolysis, whereas a base is used to stabilize the interior of the matrix.

A second family of polyorthoesters, derived from the addition of polyols to di-ketene acetals without forming condensation by-products, was developed by SRI International (Scheme 4B). Initial work was conducted with the monomer, diketene acetal (3,9-bis(ethylidene-2,4,8,10-tetraoxaspiro[5,5]undecane, DETOSU), derived from pentaerythritol and 1,6-hexane diol. This reaction is exothermic and proceeds to completion virtually instantaneously. Then, the diketene acetal is reacted with a polyol to synthesize the polyorthoester. Here, the hydrolytic products are the diol and pentaerythritol dipropionate; the further hydrolysis of this latter compound produces propionic acid, but no autocatalytic reaction is observed because this hydrolysis proceeds much more slowly than that of orthoester bonds.

Several diols, such as 1,6-hexanediol-trans-1,4-cyclohexanedimethanol, 1,6-cyclohexane-diol, ethylene glycol and bisphenol A, have been investigated to obtain polyorthoesters with different properties. The molecular weight of polyorthoesters is significantly dependent on the type of diol and catalyst used for synthesis. A linear, flexible diol like 1,6-hexanediol gives molecular weights greater than 200K, whereas bisphenol A in the presence of catalyst results in molecular weights around only 10,000. Mechanical properties of the linear polyorthoesters can be varied over a wide range by selecting various diol compositions. For instance, it was shown that the glass transition temperature of the polymer prepared from DETOSU can be varied from 25 to 110°C by simply changing the amount of 1,6-hexanediol in trans-1,4-cyclohexane dimethanol from 100 to 0%. There seems to be a linearly decreasing relationship between the T_g and percentage of 1,6-hexanediol.

For the synthesis of crosslinked polyorthoesters, a prepolymer is reacted with triols or a mixture of diols and triols. The prepolymer is a viscous liquid at room temperature. Thus, the compound of interest could be incorporated into the prepolymer (without using any solvent) along with the triol, and then the mixture can be crosslinked at temperatures as low as 40°C. This could be a good method for incorporating thermosensitive agents (e.g., drugs) within the matrix. However, one should be cautious when using compounds with hydroxyl functionality. This type of polyorthoesters has been derived from direct polymerization of a triol and an orthoester. Depending on the desired properties of the final polymeric matrix, flexible or rigid triols such as 1,2,6-hexanetriol or 1,1,4-cyclohexanetrimethanol can be used. In the hydrolysis of these types of crosslinked polyorthoesters, the first hydrolysis step occurs at the orthoester bond with the formation of hydroxyester, whose hydrolytic cleavage is much slower, and hence the overall reaction is not autocatalytic. The final degradation products are the triol and acetic or propionic acid.

The orthoester linkage is inherently unstable in the presence of water. However, because of their highly hydrophobic nature, polyorthoesters can be stored without careful exclusion of moisture. Even though the polymer is relatively stable in trace amounts of moisture, it is unstable to heat and undergoes degradation. The combination of moisture and heat can be fatal for the processing of polyorthoesters designed to degrade within days. Thus, if injection moulding is necessary to fabricate the device, then moisture must be rigorously excluded during fabrication. One should also consider the interaction between the incorporated anhydride as catalysts, the polymer and drug during the thermal processing. The preferred method for fabricating the devices would be under low thermal and shear stresses. This would include solution mixing or powder blending followed by compression moulding of the devices.

Polyorthoesters find their main application in drug delivery. Because, they degrade by surface erosion, which means that zero-order release kinetics can be achieved, and degradation rates can be controlled by incorporation of acidic or basic excipients. Polyorthoesters represents a unique polymer system that is a viscous, hydrophobic, paste-like material at room temperature even at fairly high molecular weights. The paste-like property allows incorporation of therapeutic agents under very mild conditions, at low temperature, and without using solvents. An additional objective is that the polymer erosion and the drug release should be concomitant so that no polymer residuals would be present in the tissue after all drug molecules have been released. This objective could be met only if the polymer was truly surface-eroding. For a surface-eroding polymer, the erosion process at the surface of the polymer should be much faster than in the interior of the device. To exhibit such a phenomenon, the polymer has to be extremely hydrophobic with very labile linkages. Hence, it was envisioned that polymeric devices with an orthoester linkage in the backbone, which is an acid-sensitive linkage, could provide a surface-croding polymer if the interior of the matrix is buffered with basic salts. Polyorthoesters have been used for the controlled delivery of several agents, such as contraceptive steroids (e.g., norethisterone), narcotic antagonists (e.g., naltrexone), polypeptides (e.g., insulin) and a significant number of publications describe the use of polyorthoesters for various drug delivery applications

2.2.6 *Polyphosphazanes [174-194]*

A relatively new class of biodegradable polymers is the family of polyphosphazenes. They have the general structure shown in Scheme 5A. They contain a long-chain backbone of altering phosphorous and nitrogen atoms, with two side groups attached to each phosphorous. Several methods have been proposed to synthesize a variety of polymers having the same backbone but different side groups, therefore exhibiting properties vary over a very broad range. A great deal of background information is now available on the relationship between side group structure and properties.

In polyphosphazene synthesis, molecular changes are achieved through substitution reactions carried out on a prepolymer, which is a highly reactive macromolecular intermediate, an inorganic macromolecule, i.e. "poly(dichlorophosphazene)". Its synthesis and typical reactions are shown in Scheme 5B. The starting material is the commercially available cyclic trimer, prepared from phosphorus pentachloride and ammonium chloride. A thermal ring-opening polymerization yields the reactive high polymeric intermediate, which reacts in solution with a wide variety of nucleophiles to give the organophosphazene species. These transformations generally yield hydrolytically stable polymers, but by substitution of specific side groups one may generate degradability. The reaction is carried out in general at room temperature in tetrahydrofuran or aromatic hydrocarbon solutions.

The weight-average molecular weights of these polymers are usually very high which correspond to 15,000 or more repeating units per chain.

Polyphosphazenes bearing fluoroalkoxy side groups are among the most hydrophobic synthetic polymers known, and are bioinert. These polymers are as hyrophobic as polytetrafluoroethylene but, unlike Teflon, are flexible or elastomeric, easy to prepare, and can be used as coatings for other materials. They are highly biocompatible and therefore have been proposed as good candidates for use in heart valves, heart pumps, blood vessel prostheses, or as coating materials for pacemakers or other implantable devices.

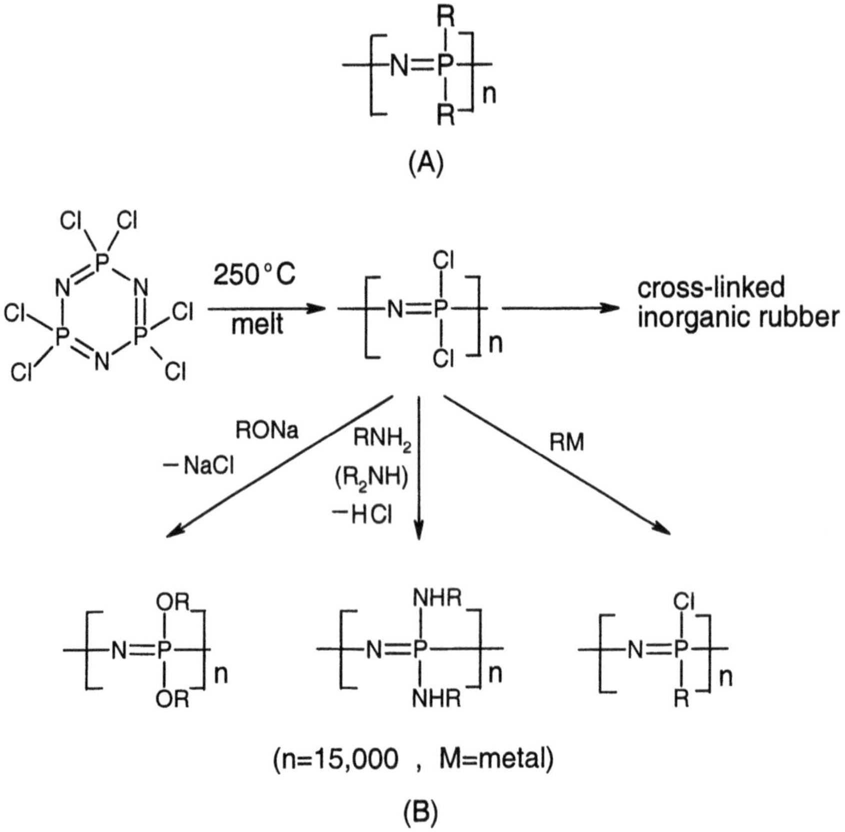

(A)

250 °C melt

cross-linked inorganic rubber

RONa −NaCl

RNH_2 (R_2NH) −HCl

RM

(n=15,000 , M=metal)

(B)

Scheme 5. Polyphosphazenes: (A) General structure; (B) synthesis and typical reactions.

In the degradation of polyphosphazenes, the inorganic backbone is hydrolyzed to phosphate and ammonia due to the presence of appropriate side groups. The phosphate can be metabolized and the ammonia excreted. Theoretically, if side groups attached to the polymer are released by the same process being excretable or metabolizable, then the polymer can be eroded under hydrolytic conditions without any toxic response.

Polyphosphazenes containing amino acid ester side groups were the first biodegradable polyphosphazenes synthesized. These polyphosphazene derivatives are prepared by reaction of poly(dichloro phosphazene) with ethyl esters of amino acids and dipeptides. They are degraded hydrolytically to ethanol, glycine, phosphate, and ammonia. The rate of degradation depends on the nature of the amino acids. Polyphosphazenes substituted with amino acid ester side groups are partially crystalline. The degree of crystallinity and the morphology of the crystalline phase depend on the storage conditions. It was reported that these polymers showed no evidence of irritation, cell toxicity, giant cell formation, or tissue inflammation.

Imidazolyl-substituted polyphosphazene is hydrolytically unstable and hydrolyzes in room moisture. The rate of hydrolysis can be slowed by the incorporation of hydrophobic side groups such as phenoxy or methylphenoxy. Imidazoyl groups linked to polyphosphazene chains are also hydrolyzed very easily, showing good biocompatibility.

Various polyphosphazene have been evaluated as drug delivery matrices. Amino acid ester-substituted polyphosphazene have been used for controlled release of the covalently bound anti-inflammatory agent naproxen. Three different poly(amino acid ester) phosphazenes, i.e., polydi(ethylglycinato) phosphazene, polydi(ethylalanato) phosphazene and polydi(benzylalanato) phosphazene have been examined in order to investigate their possible use as drug delivery vehicles. The *in vivo* and *in vitro* release of progesterone and bovine serum albumin from imidazole-4-methylphenoxy-substituted polyphosphazene matrices has been evaluated. Polyphosphazene microspheres, prepared with phenylalanine ethyl ester as a phosphorus substituent and loaded with succinylsulphathiazole or naproxen, have been studied. Polyphosphazene matrices with 50% ethyl glycinato and 50% p-methylphenoxy substitution have been investigated as vehicles for the controlled release of macromolecules. It was reported that these polymers can predictably release macromolecules, in which release can be modulated through changes in pH environment and drug loading. A biocompatible and biodegradable polyphosphazene bearing phenylalanine ethyl ester, imidazole and chlorine (10.7:1:2.5 molar ratio) as substituents of the phosphorus atoms of the polymer backbone was studied for the preparation of polymeric naproxen slow-release systems. Poly(bis(carboxylatophenoxy) phosphazene) forms hydrogels when treated in aqueous media with salts of di- (Ca^{2+}, Mg^{2+}, Zn^{2+}) or tri- (Al^{3+}) valent metal cations. These very soft, highly swellable hydrogels have been proposed to encapsulate drugs, enzymes, living cells, antigens, proteins, etc. Other phosphazene macromolecules functionalized with glycolic acid esters and lactic acid esters or new types of phosphazene-based blends have been also explored as bioerodible materials for drug delivery systems.

Membranes or microcapsules made from polyphosphazenes bearing amino acid side groups have been proposed for the treatment of periodontal diseases. Polyphosphazene membranes, prepared with alanine ethyl ester and imidazole in the molar ratio of 80:20 as phosphorus substitutents with degradation rates that corresponded to the healing of the bone defect have been prepared. These membranes have been found much more successful in promoting healing of rabbit tibia defects than Teflon membranes. It was proposed that antibacterial or anti-inflammatory drugs, useful in periodontal tissue regeneration, could be entrapped in the polyphosphazene membranes and released both *in vitro* and *in vivo* at a rate that ensured therapeutic concentrations in the surrounding tissue.

Polyphosphazenes have been investigated as potential candidates as for the production of biodegradable sutures. Nerve regeneration experiments have been performed using tubular nerve guides of polyolyphosphazenes, which have been prepared by deposition of the dissolved polymer on a glass capillary tube, followed by evaporation of the solvent (methylene dichloride). Successful nerve regeneration has been reported in animal studies. Fluorinated polyphosphazenes have been selected for evaluation as coronary stent coating. After applying the polymer film by dipcoating, the stents were implanted in porcine coronary arteries, in which promising results have been obtained.

2.2.7 *Polycyanoacrylates [195-211]*

Alkylcyanoacrylates are mainly used in the formulations of adhesives. The related products currently available contain methyl, ethyl, butyl, isobuthyl or octyl esters (Fig. 4). These monomers undergo very rapid anionic polymerization at room temperature, in which hydroxyl ions (in the presence of water) initiate the reactions. Polycyanoacrylates degrade very fast (few hours) in alkaline pH, while they are quite stable in acidic pH (usually around pH: 1-2). Lower homologs (e.g., methyl or ethyl) are degraded much faster than the higher homologs. It is generally accepted that due to very fast degradation rates, which means very fast accumulation of the degradation products, i.e., oligomers and formaldehyde at the surrounding tissue, those lower homologs may lead toxicity in medical uses as adhesives.

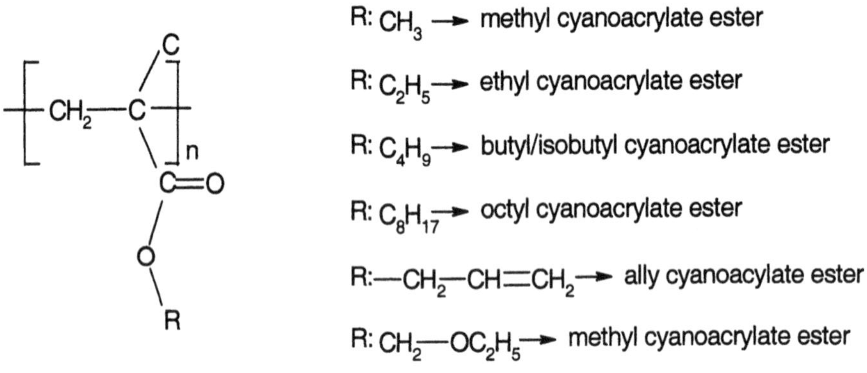

R: CH_3 → methyl cyanoacrylate ester

R: C_2H_5 → ethyl cyanoacrylate ester

R: C_4H_9 → butyl/isobutyl cyanoacrylate ester

R: C_8H_{17} → octyl cyanoacrylate ester

R: —CH_2—CH=CH_2 → ally cyanoacylate ester

R: CH_2—OC_2H_5 → methyl cyanoacrylate ester

Fig.4. Currently available alkylcyanoacrylates.

Cyanoacrylate base adhesives are solvent-free, one-part, and room-temperature-curable. They are available under several brand names, such as Super Glue®, Quick-Tite®, and Crazy Glue® brand names in a wide range of viscosities. When confined in a thin film between two surfaces or sprayed with a chemical activator, cyanoacrylates cure rapidly to form rigid thermoplastics with excellent adhesion to most substrates. Cyanoacrylates typically fixture within 1 minute and achieve full bond strength in 24 hours. Formulations based on methyl cyanoacrylates were the first commercially available cyanoacrylate products, and they lead to a rigid polymer matrix. Ethyl cyanoacrylate is the most commonly used monomer in cyanoacrylate adhesives, and offers superior performance on plastics and elastomeric substrates compared with methyl cyanoacrylates. Use of cyanoacrylate adhesives for high-speed medical device assembly has become a widely accepted pratice because of the adhesive's rapid curing time and high bond strength.

Due to the toxicities of the methyl and ethyl esters, such as inflammation or local body reaction, medical grade adhesives currently available contain either butyl, isobuthyl or octyl esters. As surgical adhesives, they provide an important option for wound closure. They are as effective as sutures in low tension-incisions and lacerations. A few minutes are enough to reach the maximum bonding strength. Faster repair time, better acceptance by patients, water resistance in coatings, no foreign-body reaction, better arrangement of collagen and tissue, antibacterial properties show some advantages over sutures. In addition, they are easier to apply, time saving, and more economical. A strong yet flexible coating is formed on the wound that protects the wound during the normal healing process. These adhesives have been applied successfully in cartilage and bone grafting, coating of corneal ulcer in ophthalmology, embolization of gastrointestinal varices and in neurolvascular surgery, topical skin adhesives and closure, facial plastic surgery, cerebrospinal fluid leak closure, bonding of dental materials such as crowns, caps and pins, etc.

Cyanoacrylates have also been evaluated as carriers in the forms of nano- or microparticles in intravascular drug delivery formulations, especially in cancer theraphy. Polyalkylcyanoacrylate nanocapsules have been successfully used for oral administration of insulin in diabetic rats. Polycyanoacrylate monolayers have been evaluated to produce coatings to immobilize various proteins, such as enzymes and antibodies, for the preparation of biologically active media.

2.3 NATURAL BIODEGRADABLE POLYMERS

2.3.1 Albumin [219-243]

Blood plasma is fluid that consists of 90% water and 10% dissolved fats, salts, sugars, and proteins called plasma proteins, which are divided into three types: albumins, globulins, and fibrinogen. In humans, albumin is the most abundant plasma protein, accounting for 55-60% of the total. The complete gene sequence of human albumin is known today. It consists of a single polypeptide chain of 585 amino acids with a molecular weight of 66.5 kDa (Fig. 5). The chain is characterized by having no carbohydrate moiety, a few tryptophan and methionine residues, but an abundance of charged residues, such as lysine, arginine, glutamic acid and aspartic acid. An albumin molecule has a series of alpha-helices held by disulphide bridges. The molecule is very flexible and changes shape readily with variations in environmental conditions and with binding of ligands. But these changes are reversible, and albumin molecules can regain shape easily, owing to the disulphide bridges, which provide strength, especially in physiological conditions. Denaturation occurs only with dramatic and non-physiological changes in temperature, pH and the ionic or chemical environment. The serum albumin concentration is a function of its rates of synthesis and degradation and its distribution between the intravascular and extravascular compartments. The total amount of body albumin is about 3.5-5.0 g/kg bodyweight. About 42% of the total is in the plasma compartment.

One of the main roles of albumin is the regulation of blood osmotic pressure. Albumin contributes up to 80% of the total osmotic pressure. Due to the large extravascular pool of albumin, its water-solubility and its negative charge, albumin also plays a significant role in the regulation of tissue fluid distribution.

Albumin incorporates many different substances. It has a strong negative charge, but there is little correlation between the charge of the compound and the degree of binding to albumin. The most strongly bound ones are medium-sized hydrophobic organic anions, including long-chain fatty acids, bilirubin and haematin. Less hydrophobic and smaller substances can be bound specifically but with lower affinity. This may be considered as an advantage, since there is easy off-loading at target sites. Monovalent cations do not bind, but divalent cations (e.g.,Ca^{2+}, Mg^{2+}, Zn^{2+}) do. There are a variety of binding sites on the albumin molecule for a variety of drugs. Drug binding strongly affects the delivery of bound drug to tissue sites and the metabolism and elimination of the drug. There are several factors that are important in drug-albumin interactions and may be responsible for the wide inter-individual variation seen. These include age; temperature, pH and ionic strength, which can affect the number of binding sites *in vitro*; and competition between drugs for binding sites. Displacement of drugs from their binding sites by other drugs or by endogenous substances occurs and may alter the distribution, pharmacological action, metabolism and excretion of the displaced drug.

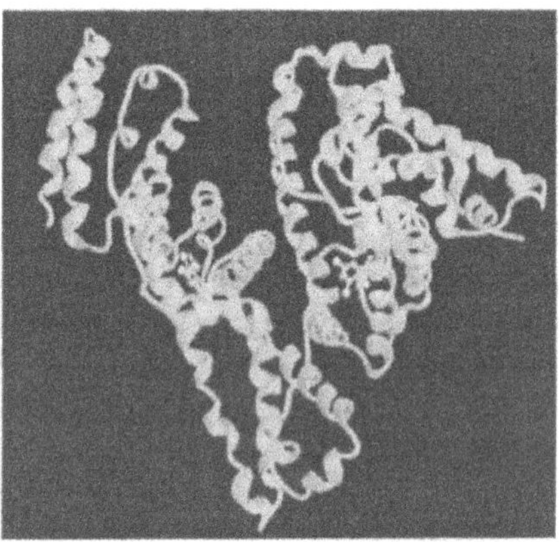

Fig, 5. Structure of albumin.

Apart from its vital roles in regulation of osmotic pressure and transporting drugs and endogenous compounds, albumin has some other functions. Albumin is involved in the inactivation of a small group of compounds. The presence of many charged residues on the albumin molecule and the relative abundance of albumin in plasma mean that it can act as an effective plasma buffer. Under physiological conditions, albumin may have significant antioxidant potential. It is involved in the scavenging of oxygen free radicals, which have been implicated in the pathogenesis of inflammatory diseases. It is possible that albumin has a role in limiting the leakage from capillary beds during stress-induced increases in capillary permeability. Albumin has effects on blood coagulation.

In biomedical applications, because of its very high blood-compatibility, albumin has mostly been evaluated as a potential carrier matrix in intravascularly injectable drug delivery systems. Albumin microspheres are usually prepared by emulsion stabilization (emulsion crosslinking). Crosslinking may be achieved thermally (at 100-180°C) or chemically (by using chemical crosslinkers, e.g., formaldehyde, glutaraldehyde) through disulphide bonds or formation of lysine-alanine crosslinks. Albumin particles in a wide range of size (1-1000 μm) have been produced by this method. Both hydrophobic and hydrophilic drugs, such as L-epinephrine, adrymycin, 5-fluorouracil, insulin, epirubicin, and prednisolon, have been incorporated into albumin particles. Low loadings, thermal decomposition of drugs during thermal crosslinking, strong interaction of drug molecules with albumin and crosslinking agent, and enyzmatic degradation of the matrix, which leads bulk erosion (means usually uncontrolable and unpredictable degradation) are noted as the main disadvantages of albumin carriers.

Radiolabelled albumin microspheres have also been used in dignostic nuclear medicine for imaging of organs and tissues. The use of immobilized human serum albumin as a stationary phase in affinity chromatography has been shown to be useful in resolving optical antipodes or to investigate interactions between drugs and protein. Albumin carrying sorbents have been investigated for removal of heavy metal ions. Blood-compatibilities of blood contacting biomaterials may be significantly improved by attaching albumin molecules on biomaterial surfaces. It has been shown that surfaces coated with albumin are suitable substrates for endothelial cell seeding.

2.3.2 Collagen and Gelatin [244-260]

Collagen is the most abundant naturally-occuring fibrous protein in all mammals, contributing about 30% of total body protein in vertebrate species. It is present throughout the body and provides strength and structural stability for skin, bones, teeth, tendons, ligaments and blood vessels.

The structural hierarchy of a collagen molecule is outline in Fig. 6. The collagen amino acid chains are unusually rich in glycine (about 33%, as often as every third amino acid in the chain), proline (20-30%) and hydroxy proline (20-30%). Another unusual amino acid, hydroxylysine is also found in similar quantities. In addition, small amount of carbohydrates, glucose and galactose (0.5-1.5%) are attached to these hydroxylysine residues.

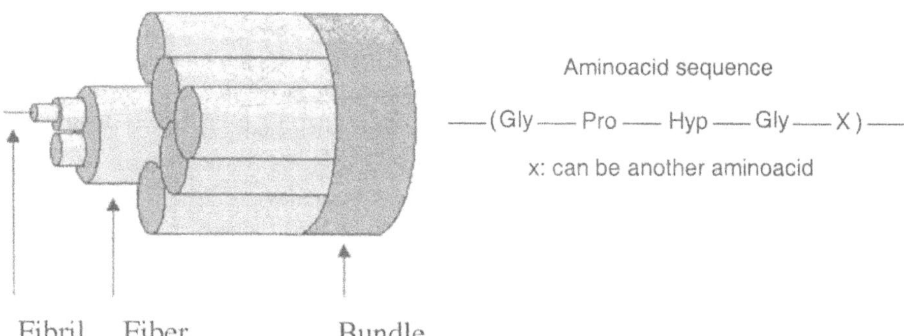

Fig. 6. Structure of collagen.

Three polypeptide chains (α-chains, each with an average molecular weight of 100,000) form a triple helix (lest-handed), the so-called "tropocollagen", a rigid rod-like molecule (length: 2800 Å and width: 15 Å). The -NH groups of glycine on one chain and the oxygen of the -C=O group of a residue on another chain forms hydrogen bonding and contributes this triplex structure. The free $-NH_2$ groups of the lysine residues become oxidatively deaminated to aldehyde groups (by lysine oxidase) in some chains, which may cause crosslinking by covalent bonding between different chains (by aldol condensation reactions), which further stabilise the triplex structure.

The tropocollagen molecules, as monomeric units, come together to assemble collagen filaments (0.01-0.02 μm), which form fibrils (0.1-0.2 μm). These fibrils are unit elements of the collagen fibres (1-2 μm), which usually hold together as fibre bundles (10-50 μm). This highly oriented and complex structure is unique and gives the extraordinary extremely high mechanical strength to the natural fibrous materials. Chemical crosslinking via aldehyde groups contributes also the stability of this structure. Formation of collagenous structure is probably influenced by other proteins like glycosaminoglycans, is different depending on the tissue to be supported. For instance, in tendon the fibres are closely packed parallel, while in skin, the fibres are randomly oriented and woven into a feltwork to provide some flexibility.

Different types of collageneous materials are known, from collagen Type I to X. Type I is the most abundant form (about 95% of all) and exists in bone, skin, tendon, cornea. In Type I collagen, two of three chains have identical composition, and the third is different. Type II (in hyaline cartilage, intervertebral discs, vitreous body of the eye), Type III (in skin, blood vessels, internal organs, etc.) Type IV (in the basement membranes around many types of cells) are compose of three identical polypeptide chains. Type V is present throughout the body in small amounts. It's tropocollagen chain structure is similar to Type I, but it is in non-fibrillar form.

Natural collageneous tissues can be directly used as biomaterials, after some purification steps to remove noncollageneous substances, but not loosing the original fibrous structure. For safety purposes, reconstituted collagen biomaterials are pre-pared, in which by enzymatic processes, the original structure is disintegrated down to tropocollagen units, and then after purification, these units are bring together by physical or chemical crosslinking ("tanning") to form the desired materials shape (fibers, films, non-woven matrices, sponges, etc.). Short wave length UV irradiation (254 nm) or dehydrothermal crosslinking (in vacuum for several days at temperature up to 100°C) are two physical crosslinking methods which have been used. Note that lower degree of crosslinking is achieved by these physical methods. Trivalent metal ions, such as aluminum or chromium are used in tanning of collagen fibres by forming ionic intra- and inter-molecular crosslinking. Chemical crosslinking through covalent bonds can be achieved by using a number of crosslinking agents (glutaraldehyde, formaldehyde, carbodiimide, etc.), which stabilise the structure, improve the fibre mechanical properties, suppresses the antigenicity of collagen, and more importantly allows better control the extend of biodegradation *in vivo*.

Collagen has been used in a variety of biomedical applications. Collagen fibres obtained from bovine intestine or the submucosal layer of sheep intestine, after purification but not loosing their original forms, has been used as sutures for closure of surgical or traumatical incisions for decades. Collagen materials in the form of films or nonwoven matrices (from natural materials or as crosslinked gels) have been marketed as temporary wound dressings to aid/confort wound healing. Injectable collagen solutions made from cows or pigs are used to treat wrinkles, scars and facial lines for cosmetic purposes, and also for occular surface lubrication. Sponge-like structures with primary cells have been utilized as scaffolds for tissue engineering, to restore damaged soft or hard tissues. Collageneous skin substitutes are produced by culturing keratinocytes on a matured dermal equivalent composed of fibroblasts included in a collagen gel and membranes. Tubular matrices from collagen and elastin have produced suitable small diameter vascular grafts from total or partially devitalised collagen:elastin matrix. As controlled drug delivery systems, injectable dispersions for local tumor treatment, sponges carrying antibiotics and minipellets loaded with protein drugs, a recombinant human bone morphogenetic protein soaked absorbable collagen sponge for bone regeneration. Other approved uses of collagen includes hemostatic agents, vascular sealents, tissue sealents, implant coatings (orthopedic and vascular), bone graft substitutes, corneal shields, dental implants, etc.

Collegen biomaterials are biocompatible and nontoxic. They are readly available. Their properties have been well-documented, and they can be processed into a variety of forms. However, collagen products suffer from poor reproducability, which highly depends on the source. They are biodegraded both hydrolytically and enzymatically (by collagenase, and other enzymes e.g., elastase, cathepsin G, etc.) and teherefore it is usually difficult to predict their *in vivo* degradation rates.

Breaking the secondary bonds between the polypeptide chains of fibrous collagen structures by heat or reagents such as urea gives an unoriented and water-soluble product, which is called "gelatin". Gelatin has extensively been employed for coating and also microencapsulation of various drugs, and for preparing biodegradable hydrogels (crosslinked films, sponges, etc.) for pharmaceutical and biomedical use.

2.3.4 Chitin and Chitosan [38,182,187,309-326]

Chitin is among the most abundant polysaccharides found in exoskeleton of crustacean like crubs and shrimps, in the crust of the insects like beetles, gold bugs, and grasshoppers, and in the cell wall of bacteria and fungi like moulds and mushrooms. It is found in association with proteins and minerals such as calcium carbonate. The different sources of chitin differ somewhat in their structure and percentage of chitin content.

It has a similar structure to cellulose; the difference is that the 2-hydroxy groups of the cellulose backbone has been replaced with acetamide groups, therefore the repeating unit in chitin is "N-acetylglucosamine" (Fig.7). Most commercial applications use the deacetylated derivative, the so-called "chitosan", which is formed of repeating units of D-glucosamine. The degree of deacetlylation is generally about 80% and changes depending on the chitin source and also processing conditions. Due to the primary amine groups, chitosan exhibits strong positive charge.

For chitin production, first the proteins are removed by treating with sodium hydroxide. Inorganic materials, such as calcium carbonate and calcium phosphate, are then extracted with hydrochloric acid. After rinsing, the chitin is dried as a flaked material. For chitosan manufacture, chitin flakes are first treated with very strong NaOH to hydrolyse the N-acetyl linkage, then rinsed, pH adjust, and dried. The chitosan flakes can be dissolved in acidic solutions (such as dilute acetic acid) for further processing to give the final form of the product.

Chitosan is linear polymer with reactive amino groups available for further derivitisation. It forms chelates with many transitional metal ions. It is a semi-crystalline polymer and the degree of crystallinity depends on the degree of deacetylation. Crystallinity is minimum at intermediate degree of deacetylation. Chitosan is soluble only in acidic solutions (pH<6.5), and insoluble in most organic solvents. Solutions are viscous and shear thinning. Viscous solutions can be extruded and gelled in aqueous solutions at high pH or in methanol (as non-solvent) to form fibres or films (membranes). Due to high positive charge density, chitosan forms also insoluble ionic complexes or complex coacervates with a variety of with negatively charged polyanions (alginic acid, polyacrylic acid, glycosaminoglycans, DNA, etc.).

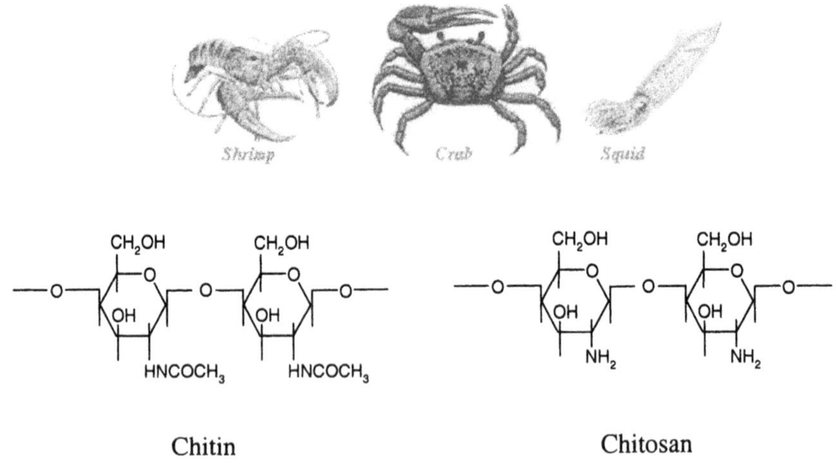

Fig. 7. Structure of chitin and chitosan.

Chitosan is biocompatible, biodegradable to normal body constituents, safe, non-toxic, hemostatic, fungistatic, spermicidal, antitumoral, anticholesteremic, immuno-adjuvant, central nervous system depressant, and good for cell attachment, and accelerates wound healing and the formation of osteoblasts reponsible for bone formation. With these highly favorable biological and chemical properties, chitin and especially chitosan have been employed as an important candidate/alternative material in many diverse biomedical and related applications in different forms such as "solution/gel" (e.g. bacteriostatic agent, fungistatic agent, cosmetics, flocculating agent, coating agent); "gel/paste" (drug delivery vehicle, spermicide, immobilization/ encapsulation agent, etc.); "beads" (immobilising enzymes/cells, protein separations, etc.); "films/membranes" (dialysis membrane, contact lens, wound-dressing, encapsulation of mammalian cells, etc.); "sponge" (mucosal hemostatic agent, wound dressings, etc.); "fibre" (suture, wound dressings, membrane drug delivery as hollow fiber, etc.); and "scaffolds" with controllable pore structures for tissue engineering.

In vivo, chitosan is degraded by enzymatic hydrolysis. The primary agent is lysozym, which appears to target acetylated residues. However, there is some evidence that some proteolytic enzymes show low levels of activity with chitosan. The degradation products are chitosan oligosaccharides of variable length. The degradation kinetics appears to be inversely related to the degree of crystallinity, which is controlled, mainly by the degree of deacetylation. Highly deacetylated forms (e.g. >85%) exhibit the lowest degradation rates and may last several months *in vivo,* whereas samples with lower levels of deacetylation degrade more rapidly. This issue has been addressed by derivatising the molecule with side chains of various types. Such treatments alter molecular chain packing and increase the amorphous fraction, thus allowing more rapid degradation. They also inherently affect both the mechanical and solubility properties.

2.3.5 Glycosaminoglycans [261-276]

"Glycosaminoglycans" (GAGs) are long unbranched polysaccharides containing a repeating disaccharide unit. The majority of GAGs in the body are linked to core proteins, forming "proteoglycans" (PGs), also called as "mucopolysaccharides". Both are the main components of extracellular matrix in connective tissue. They are gel-like substances in which the fibrous proteins (collagen, elastin, fibronectin and laminin) are embedded. These macromolecules are highly negatively charged with extended conformation leading high viscosity that helps hold the cells and tissues together and provides structural integrity. Note that the polysaccharide gel resists compressive forces on the tissue and the collagen fibres provide tensile strength. Along with the high viscosity of GAGs comes low compressibility, which makes these molecules ideal for a lubricating fluid in the joints. The aqueous phase of the polysaccharide gel also allows the rapid diffusion of nutrients, metabolites and hormones between the blood and the tissue cells.

The disaccharide units of GAGs contain either of two modified sugars; N-acetylgalactosamine or N-acetylglucosamine and a uronic acid (e.g., glucuronate or iduronate) (Fig. 8). In the PGs, the GAGs molecules extend perpendicularly from the core proteins in a brush-like structure. The linkage of GAGs to the protein core involves a specific trisaccharide linker, which is coupled to the protein core through an O-glycosidic bond to a S residue in the protein. Some forms of keratan sulfates are linked to the protein core through an N-asparaginyl bond. Many of these chains attach to a protein core that is so-called as "proteoglycan monomer" having an average molecular weight of one million. In articular cartilage, up to a hundred of these monomeric units link to a hyaluronic acid chain to form a PG aggregate with average molecular weights up to a hundred million.

The GAGs have numereous vital functions in the body, and the major ones are classified according to the saccharide-uronic acid subunits. "Hyaluronic acid" (HA) is a high molecular weight GAG about 2500 repeating disaccaride units, each containing one residue of D-glucuronic acid and N-acetyl-D-glucosamine linked by glycosidic bonds. HA's distribution in the body is as follows: 56% in the skin; 35% in the muscles/skeleton, 9% elsewhere. Note that HA is one of the major components in the synovial fluid of articular joints, and also exist in cartilage matrix as PG aggregates. HA is extracted usually from rooster combs (contains about 1% by weight HA) or fermented from bacteria. Fermentation is preferred due to not being associated with contaminants, and it can be produced on a large scale. Purified HA and its derivatives (especially esters) and also hydrogel complexes with polycations (particularly chitosan) have been used in diverse biomedical applications, including "viscous solutions" as lubricants for articulating surfaces to relieve pain associated with related diseases (e.g., osteoartritis), and for eye surgery and as skin-care products, "films/membranes" as wound dressing materials, and "swellable gels" as filling of the tissue spaces in directing and controlling tissue regeneration, as carrier of active agents (e.g., fibroblast growth factor) and as cartilage tissue engineering matrices.

"Chondroitin sulfate" (CS) is a very large molecule, often aggregates with HA. CS is the most abundant GAG in the body and predominant in cartilage, tendons and ligaments. CS has been demonstrated *in vitro* to inhibit several degradative enzymes that destroy cartilage and exhibit anti-inflammatory activity. CS is a sulphated GAG therefore can form hydrogel complexes with cationic polymers (e.g., chitosan).

"Dermatan suphate", a relatively small GAG, is widely distributed in the body (skin, blood vessels and heart valves). It is found in small amounts in cartilage and dense connective tissue. The high charge density and lack of crystallinity make the sulphated GAGs highly water-soluble. Due to their generally low molecular weight (5-40 kDa) they exhibit poor intrinsic gel formation properties. However, their complexes with polycations could be prepared in gel form and used as biomaterials similar to other GAGs.

Fig. 8. Structure of some selected GAGs.

"Heparin" is an intracellular component of mast cells and exists in the liver, lung and

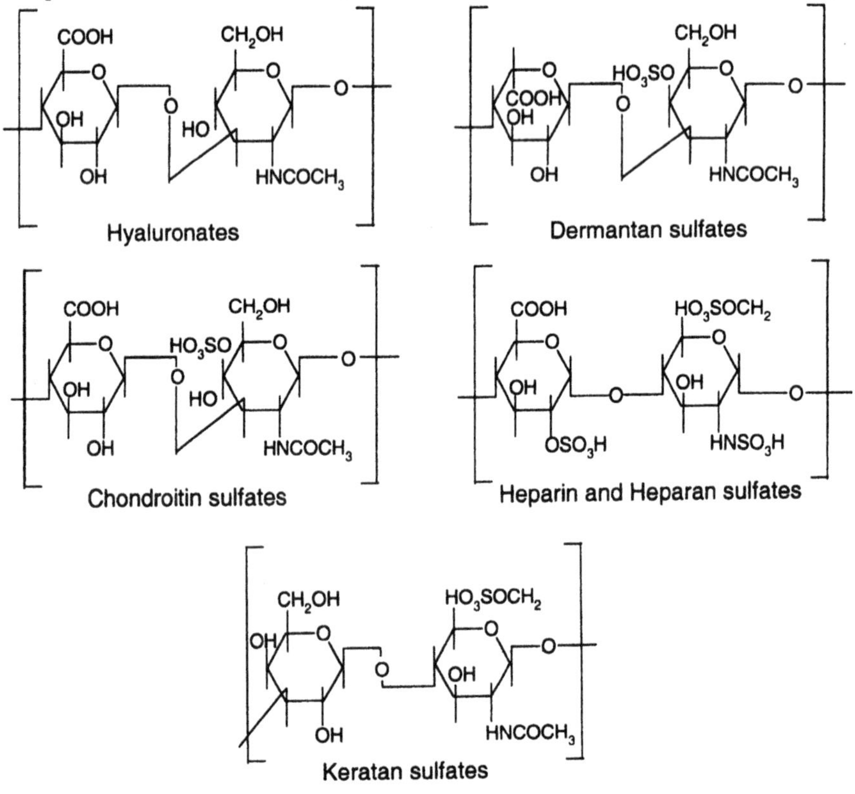

Hyaluronates

Dermantan sulfates

Chondroitin sulfates

Heparin and Heparan sulfates

Keratan sulfates

skin. Heparin is a highly sulphated polysaccharide attached to the core protein serglycin. It is stored in the granules of connective tissue type mast cells and released together with histamine, various mast cell proteases and cytokines upon mast cell activation. Heparan sulphate (HS), in contrast, has a more variable sulphation pattern and iduronic acid content than heparin. HS chains are attached to different core proteins (e.g., syndecans, glypicans, agrin and perlecan) and then secreted into the extracellular matrix or localized to the cell surfaces. Heparin is used clinically as an anti-coagulant and lipid-clearing agent. Numereous techniques have been developed to immobilise heparin onto blood-contacting devices to improve their use as anti-coagulants and blood compatibilisers. Experiments have shown that the use of immobilised heparin provides greater benefits than does administration of systemic heparin, for instance it helps reduce patient blood loss and transfusion requirements while guarding against thrombus formation in open-heart surgery.

"Keratan sulfate" is found along with chondroitin sulfates (as aggregates) in several types of connective tissue, including cartilage. Unlike the most of the proteoglycans (chondroitin sulfates, dermatan sulfate, heparin, and heparan sulfate), linkage region is different in keratan sulfate, which contains repeating units of N-acetyllactosamine that are O-sulfated.

2.3.6 *Starch* [277-291]

Starch is a high molecular weight polysaccharide, which occurs in plant tissue as a reserve energy resource. For instance, 75% and 65-80% of the dry weights of potato and corn is starch, respectively. Starch exists in plant tissues as granules, or crystalline beads of about 15 μm to 100 μm in diameter, in different crystalline forms. Strach is composed of the two major components, i.e., two biopolymers, amylose and amylopectin (Fig.9). The amylose is essentially a linear polymer of α-D-(1,4) glucan units, and makes up about 20% of the strach granule. The branch polymer, amylopectin, an α-D-(1,4) glucan, has α-D-(1,6) linkages at the branch point and exist in the strach granule up to 80%. The linear amylose chains have a molecular weight of 0.2-2 million, while the branched amylopectin molecules have molecular weights as high as 100-400 million. Due to linear structure and relatively shorter chains, amylose is crsytalline, but soluble in boiling water, and degraded faster than amylopectin.

Starch can be gelatinized using water and applying heat and pressure, which completely destroys the crystalline structure and results thermally processable amorphous starch, i.e., the so-called "thermoplastic starch". Products based on solely on starch are extremely water sensitive and of limited utility, therefore they should be modified by various means (see Chapter 6). The destructurised (thermoplastic) starch (90 to 99 percent starch, or very low (0.5) degree of substitution hydroxy propylated starch) can be processed into in soluble foams or other expanded items in a twin-screw extruder specially designed to hold the water (which functions as the plasticiser and blowing agent). Additives such as poly(vinyl alcohol) and other similar compounds are added to the starch to improve processibility. Modification of the starch hydroxyl groups by esterification chemistry to form starch esters of appropriate degree of substitution (ds) (1.5 to 3.0 ds) imparts thermoplasticity. Unmodified starch shows no thermal transitions except the onset of thermal degradation at around 260°C. Starch acetate of ds 1.5 shows a sharp glass transition at 155°C and starch propionate of same ds has T_g of 128°C. Plasticisers like glycerol triacetate and diethyl succinate are completely miscible with starch esters and can be used to improve processability. Water resistance of the starch esters is greatly improved over the unmodified starch. Appropriately formulated starch esters with plasticisers and other additives provide resin compositions that can be used to make injection-moulded products. Starch-cellulose blends are suitable for rigid and dimensionally stable injection molded items. Starch-poly(ε-caprolactone) blends can be produced by reactive extrusion within twin-screw extruders, in which starch phase is dispersed thoroughly within the polycaprolactone phase. Processability in film blowing and sealability, mechanical properties of these starch blends are equivalent to those of low-density polyethylene (LDPE). Therefore they can be processed by means of traditional blowing and casting equipment for LDPE with minor or no modifications.

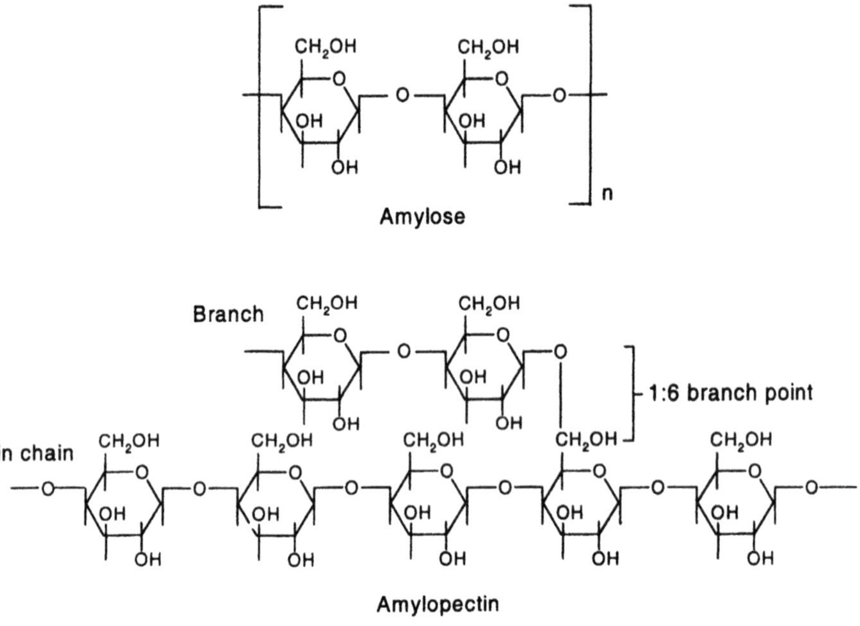

Fig. 9. Structure of starch.

In biomedical applications, starch has been mostly utilised in preparation of controlled drug delivery matrices. Polyacrylate modified starch microparticles have been proposed as carriers for proteins and drugs. Silicone-coated starch/albumin microparticles have been designed for protein release. Starch (amylose) has been proposed for surface modification of water insoluble drug particles. For drug delivery applications, starch and dextran microspheres have been prepared by using different techniques. Starch-based systems offer several advantages including ease of tablet preparation, potential of a constant release rate (zero-order) for an extended period of time and possibility to incorporate high percentages of drugs with different physicochemical properties. Hydroxyethyl starch is one of the most frequently used plasma substitutes. A variety of different hydroxyethyl starch solutions exist worldwide, which differ greatly in their pharmacological properties. Degradable starch microspheres carrying cytotoxic drug have been evaluated in liver cancer. Amylose, hydroxypropylated starch, and dextrin films and hydrogels have been also formed as edible coatings on foods and pharmaceutics to provide an oxygen or lipid barrier, and to improve appearance, texture and handling. The grafted starch solutions and the resulting hydrogels have been investigated as enzymatically degradable protective coatings in self-regulated drug delivery system applications. A biodegradable starch hydrogel matrix releasing basic fibroblast growth factor on the basis of protein metal coordination with the protein drug has been proposed.

Both amylose and amylopectin are readily hydrolyzed at the acetal link by enzymes. The α-1,4-link in both components is degraded by amylases and the α-1,6-link in amylopectin is attacked by glucosidases. With the aid of amylases, water molecules access the linkages, breaking the chains eventually to a mixture of glucose and maltose.

2.3.7 Alginates [292-310]

Alginates are a family of copolymers (linear heteropolysaccharides) composed of α-L-guluronic acid (G) and β-D-mannuronic acid (M) blocks with different sizes, therefore with different chain compositions (Fig. 10). Alginates selectively and cooperatively bind certain ions (e.g., Ca^{2+} ions) through the guluronate residues, and therefore form three-dimensional hydrogels. Alginates rich in G blocks bind more ions, and therefore gels with enhanced rigidity and strength are obtained. The copolymer chains with higher M contents are more flexible, and the gels, having this type of chains allows higher diffusion rates for the solutes through the gel matrix. Alginates can also form acid gels at pH below the pKa value of the uronic acid residues, in which M-blocks take also role in the gelation. Alginates (as polyanions) do form polyelectroylite gels with natural and synthetic polycations (e.g., poly(L-lysine), chitosan, DEAE-dextran, polyethyleneimine)

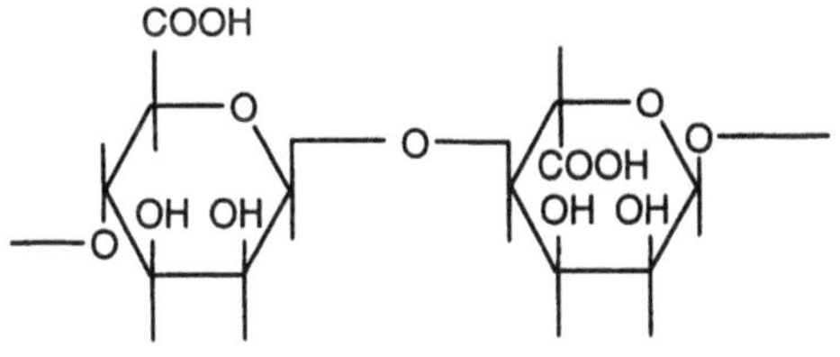

Fig.10. Structure of alginates.

Alginates can be produced from natural sources (like brown algae) by extraction. Alginate solutions purified by sterile filtration are commercially available. Alginates with desirable chain structures and therefore properties can also be produced by fermentation of several microorganisms (different strains, especially genetically modified ones). Chemically modified forms, for example those esterified with propylene oxide (in beers and salad dressings), alkylated with pharmacologically active alcohols (for drug delivery systems), or acrylated or allylated groups-attached (to obtain photo-crosslinkable gels) have also been synthesised.

Several methods can be applied in the manufacturing of alginate beads, micro-capsules, fibres and films/membranes. Aqueous solutions of alginates, containing also the components (drugs, enzymes, cells, etc.) to be immobilized/entrapped within the gel matrix, are introduced within the gelation medium (by extrusion or by simple injection), which is also aqueous and contains the gelling agent (i.e. the crosslinking cations, or polycations). Both alginate variables (e.g., chemical composition and molecular weight and distribution) and gel setting strategies (e.g., mixing of the phases, intruduction of the gelation agent into the alginate solution or the gelation medium) and gelation conditions (e.g., concentration of both alginate and gelation agent, temperature, pH, ionic strenght) may affect the final properties of the gel (e.g., strength, porosity/diffusion, homogenity of the crosslinking, swelling/shrinking, transparency, leaching of alginate from the gels).

Alginate is used in biomedical applications because of its non-toxicity and ability to form hydrogels under very mild conditions. High gel porosity of the alginate gels that allows for diffusion of macromolecules, made these gels an attractive carrier for the controlled delivery of macromolecular drugs (proteins, DNA drugs, etc.). Due to the very gentle, simple and rapid immobilisation procedure, alginate gels have been evaluated as carriers for biological entities (microorganisms, mammalian cells, and enzymes). In biological uses, it was proposed that alginates increase non-specific immunogenity that depends upon the content of mannurate residues, which may be even considered as a positive property of the alginate based materials in medical use.

The most important biomedical use of alginates is the microencapsulation of pancreatic islet cells as cellular transplants for the treatment of diabetes. Here, in order to control the pore structure and to improve the mechanical strength of the microcapsules, the outer layer has been coated with polycations, such as poly(L-lysine. Note that in order to increase the biocompatibility of the capsules (to eliminate the negative effect of the poly(L-lysine) layer) the outer layer of these microcapsules has been coated with a third layer of alginate. Calcium alginate has been extruded to make fibers, which are then woven into gauze-like products as dressings for the treatment of wounds and burns. Calcium alginate gels have been evaluated for sustained release of sensitive drugs in the gastro-intestinal tract. Alginate solutions containing chondrocytes have been investigated as slowly polymerizing injectable gels for tissue engineered cartilage repair. Alginate and alginate-chitosan beads have been utilized as carriers of several ligands (peptides and antibodies) for immunoaffinity purification and immunoassay.

Alginates are single stranded polysaccarides and are susceptible to degradation by cleavage of the glycosidic linkages by hydrolysis in acidic or alkaline media, or by oxidation with free radicals. In the living body enyzmatic degradation also plays an important role.

Alginate solutions can be sterilised by filtration. However, sterilisation of the solid alginate materials is not straight forward. Both heat and classical γ-irradiation are deleterious to the material, may cause significant breakage in the alginate chains. Sterilisation in electron accelerators for short times (e.g. 1 min) may cause less damage to the material.

2.3.8 Polyhydroxyalkanoates [311-322]

Polyhydroxyalkanoates (PHAs), also known as "microbial polyesters", or "bacterial plastics" are biosynthetic, biocomapatible, and biodegradable thermoplastics. They are bacterial storage polymers produced by various microorganisms (e.g., *A.eutrophus, P.oleovorans, R.rubrum, Rb.spaeroides*) in response to nutrient limitation occuring in the presence of an excess of carbon (see Chapters 9 and 10), and function as intra-cellular reserves of carbon and energy as well as ion sinks.

Poly-β-hydroxybutyrate (PHB) homopolymer and copolymers of hydroxybutyrate and hydroxyvalerate (PHB/HV) are the most studied members of the PHA family that are made by a controlled bacterial fermentation using different carbon sources (Fig. 11). The type of microorganism and carbon source (e.g., hydroxybutyric acid, butyric propionic or pentanoic acids), and also fermentation conditions determine the polymerisation yield, the chemical structure and the molecular weight/distribution of the resultant polyester. Today it is possible to reach very high yields (more than 95% polymer per unit dry weight of the cells) using the recombinant microorganisms (e.g., *E.coli* carrying the related genes from *A.eutrophus*). Copolymers composed of different amounts of HB and HV, or others, carrying different side groups with different functionalities can be produced. Very high molecular weight (usually >100,000, and up to 20 million) with a narrow polydispersity can be achieved.

PHB is a semicrystalline thermoplastic, and has properties similar to such commonly used plastics as polypropylene and polyethylene. Crystallinity of these polyesters is around 50%. Unlike other biological polymers, it is relatively hydrophobic, and being a thermoplastic, melt processable. The melting point depends on the polymer composition; PHB homopolymer melts at 177°C with a T_g at 9°C, while the 1:1 copolymer with HV melts at 91°C. PHB in itself is a stiff and relatively brittle thermoplastic. Its melting point is only slightly lower than the temperature at which it starts to degrade to crotonic acid, thus making processing by standard thermoplastics processing equipment difficult. The PHB/HV copolymers show a decreased stiffness and an increase in toughness and are thus easier to process and have properties more suitable for many commercial applications. Their physical properties, crystallinities and processibilities can also be improved by physical blending with various polymers, such as polyethylene oxide, ethylene-propylene rubber, polyvinyl acetate, polyvinyl chloride and polysaccharides.

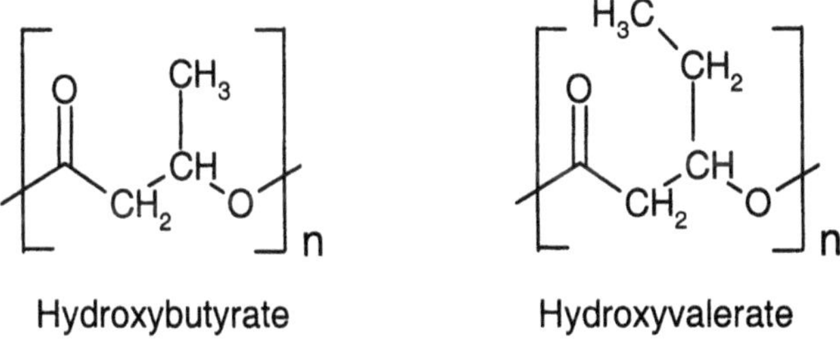

Fig. 11. Structure of polyhydroxyalkanoates.

When microbial polyesters are exposed to natural microorganisms in the ecosystem, degradation to carbon dioxide and water is achieved in about 1-6 months depending on composition. These polyesters can be degraded by enzymatic catalyzed hydrolysis (both intracellular and extracellular by endo- and exo-depolymerases) or simple chemical hydrolysis. The hydrolytic degradation is influenced by a number of factors, including chain composition, temperature, pH, content and sample structure and pro-duction. Studies of the hydrolytic degradation of HB polymers have shown that micro-spheres are degraded slowly in phosphate buffer at 85°C and after 5 months 20-40% of the polymer eroded under these conditions. Polymers having higher fractions of HV and low molecular weight polymers were more susceptible to hydrolysis.

Bacterial PHB and PHBV polyesters have been marketed and evaluated as bio-degradable, biocompatible and natural thermoplastics for medical and surgical devices. Drug delivery systems, sutures, wound dressings, bone plates and restorative absorb-able devices have been proposed using PHB homopolymers. The PHBV copolymers of various valerate contents containing several drugs (e.g., tetrascyclin, progestrone) in different forms (i.e., microsphres, microcapsules) have been studies in drug delivery systems and chemoembolisation agent. PHBV membranes have been applied for guided gingival tissue regeneration and studied *in vivo* for thoraco-abdominal tissue reconstruction in animals. The PHB homopolymer and the PHBV copolymers have been recently investigated for evaluations of biomedical implantable systems such as orthopaedic resorbable composites or tissue engineering substrates. In orthopaedic uses, piezoelectrical properties of the PHBV copolymers have been noted as an advantage. The use of PHB, a resorbable nerve conduit, as an alternative to primary nerve repair in reducing loss of neurons has been evaluated.

2.3.9 Other Natural Polymers [323-337]

Cellulose is the basic structural component of plant cell walls (90% of cotton and 50% of wood), and is the most abundant of all naturally occurring organic compounds. It is a complex carbohydrate consisting of 3,000 or more glucose units where two glucosyl-residues are linked with a *beta*-glycosidic bond (Fig. 12). It is degraded in nature, but it is non-degradable in the human body. The main biomedical use is as a dialyser membrane in haemodialysis, which is a routine clinical therapy. A wide variety of cellulose products are currently available in the medical field for use as wound-dressings, surgical wipe, treatment pad, burn bandage, etc. A multifunctional drug delivery system based on hydroxypropyl methylcellulose has been developed. Several membrane-affinity systems carrying a wide variety of affinity ligands (such as protein A, human immunoglobulin G, iminodiacetate-copper, and cibacron blue F3GA) have been based on cellulose membrane matrices. Several edible and biodegradable polymeric systems have been developed based on methyl cellulose, hydroxypropyl-methyl cellulose, hydroxypropyl cellulose, and carboxymethyl cellulose.

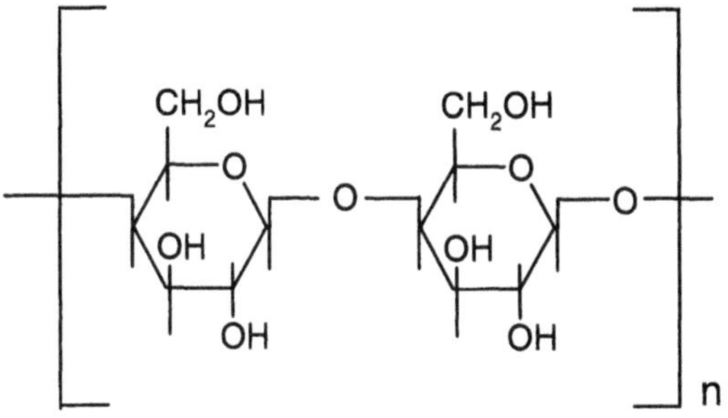

Fig. 12. *Structure of cellulose.*

Silk proteins have been used in textiles for thousands of years. Wounds have been sutured by using catgut and silk since ancient times. Many new applications of silk fibroin from various species of silkworm and spiders have been proposed as biomaterials. Plates coated with silk fibroin have been examined in animal cell cultures. One area of current research interest is the use of silk in immobilization technologies.

Other proteins related with biomedical and pharmaceutical coating that have been studied included corn zein, wheat gluten, soy protein isolate, and casein. Besides collagen and gelatin, corn zein is the only other protein that has been promoted commercially as an edible film or coating. The barrier, vitamin adhesion, and antimicrobial carrier properties of zein film coatings have been used on a variety of foods. Zein is also used on pharmaceuticals for coating capsules for protection, controlling release, and masking flavors and aromas.

Acknowledgement

Contributions of C.Babaç, D. Demirgöz, S. Dinçer, M.Duman, G.Güven, E.Kalaycı-oğlu, G.Özgöz, K. Tuzlakoğlu, M.Türk, D.Uzunok for the preparation of the text are kindly acknowledged. Special thanks to M.Duman for drawings.

References

1. Bronzino, J.D. (1995) *The Biomedical Engineering Handbook*, CRC Press, Boca Raton, FL.
2. Bruck, S.D. (1980) *Properties of Biomaterials in the Physiological Environment*, CRC Press, Boca Raton, FL.
3. Chu, C.C. (1983) Survey of clinically important wound closure biomaterials, in M. Szycher (ed.), *Biocompatible Polymers, Metals, and Composites*, Technomic Publ. Co., Lancaster, PA,,pp-477-523.
4. Davis, S.S., Illum, L., McVie, J.G. and Tomlinson, E. (1985) *Microspheres and Drug Therapy*, Elsevier, Amsterdam.
5. Hastings, G.W. (1992) *Cardiovascular Biomaterials*, Springer-Verlag, London.
6. Horncastle, J. (1995) Wound dressings. Past, present and future, *Med. Device Technol.* **6**, 30-36.
7. Kronenthal, R.L. , Oser, Z., Martin, E. (1975) *Polymers in Medicine and Surgery*, Plenum Press, New York.

8. Kroschwitz J.I. (1990) *Concise Encyclopedia of Polymer Science and Engineering,* A Wiley-Interscience Publ., New York, pp. 872-873.
9. Lanza, R.P., Langer, R. and Chick, W.L. (1997) *Principles of Tissue Engineering*, Acad. Press, San Diego, CA.
10. Liu, D.M. and Dixit, V. (1999) *Porous Materials for Tissue Engineering* Zurich, Trans Tech.Publications, Zurich.
11. Park, J.B. and Lakes, R.S. (1992) *Biomaterials: An Introduction*, Plenum Press, New York.
12. Patrick, C.W., Mikos, A.G. and McIntire, L.V. (1998) *Frontiers in Tissue Engineering*, Pergamon Press, New York.
13. Pişkin, E. and Chang, T.M.S. (1982) *The Past, Present and Future of Artificial Organs*, Meteksan, Ankara, Turkey.
14. Pişkin E. and Hoffman, A.S. (1986) *Polymeric Biomaterials*, Martin Nijhoff Publishers, Dordrecht.
15. Pişkin, E. (1992) *Biologically Modified Polymeric Biomaterial Surfaces,* Elsevier, Booking, UK.
16. Ratner, B.D., Hoffman, A.S., Schoen F.J. and Lemons J.E. (1996) *Biomaterials Science: An Introduction to Materials in Medicine*, Acad. Press, San Diego, CA.
17. Sayer, P.N. (1987) *Modern Vascular Grafts*, McGraw-Hill, New York.
18. Skalak, R. and Fox, C.F. (1998) *Tissue Engineering*, Riss AL, New York.
19. Szycher, M. (1991) *High Performance Biomaterials*, Technomic Publ., Lancaster PA.
20. Szycher, M. (1996) *"High Performance Biomaterials" A Comprehensive Guide to Mechanical Pharmaceutical* Applications, Technomic Publ. Co., Lancaster, Pennsylvania.
21. Vert, M., Cristel, P. and Chabot, F. (1984) *Macromolecular Biomaterials*, CRC Press, Baco Raton, FL.
22. Williams, D.F. and Lyman, D.J. (1982) *Blood Compatibility*, CRC Press, Boca Raton, FL.
23. Wise, D.L., Trantolo, D.J., Altobelli, D.E., Yaszemski, M.J., Greser, J.D. and Schwartz, E.D. (1995) *Encyclopedic Handbook of Biomaterials Bioengineering,* New York, Marcel Dekker.
24. Amass, W., Amass, A. and Tighe, B. (1998) A review of biodegradable polymers: uses, current developments in the synthesis and characterization of biodegradable polyesters, blends of biodegradable polymers and recent advances in biodegradation studies, *Polym. Internat.* **47**, 89-144.
25. Atala A., Mooney, D.J. and Arbor A. (1997) *Synthetic Biodegradable Polymer Scaffold: Tissue Engineering*, Birkhauser, Boston MA.
26. Barenberg, S.A., Brash, J.L., Narayan, R. and Redpath, A.E. (1990) *Degradable Materials. Perspectives, Issues and Opportunities,* CRC Press, Boca Raton, FL.
27. Barrows, T.H. (1986) Degradable implant materials: A review of synthetic absorbable polymers and their application, *Clin. Mater.* **1**, 233-257.
28. Barrows, T.H. (1991) Synthetic bioabsorbable polymers, in M. Szycher (ed.), *High Performance Biomaterials*, Technomic Publ., Lancaster PA, pp. 243-257.
29. Benicewicz, B.C. and Hopper, P.K. (1991) Polymers for absorbable surgical sutures-Part II. *J. Bioact.Comp. Polym.* **6**, 64-95.

30. Chandra, R. and Rustgi, R. (1998) Biodegradable polymers, *Progr. Polym. Sci.* **23**, 1273-1335.
31. Chasin, M. and Langer, R. (1990) *Biodegradable Polymers as Drug Delivery Systems*, Marcel Dekker, New York.
32. Chu, C.C. (1995) Biodegradable polymeric biomaterials: An overview, in J.D. Bronzino (ed.), *The Biomedical Engineering Handbook*, CRC Press, Boca Raton, FL, pp.611-626.
33. Coury, A.J. (1996) Chemical and biochemical degradation of polymers, in B.D. Ratner, A.S. Hoffman, F.J. Schoen and J.E. Lemons (eds.), *Biomaterials Science: An Introduction to Materials in Medicine*, Acad. Press, San Diego, CA, pp. 243-260.
34. Domb, A.J., Kost, J. and Wiseman, D. M. (1997) *Handbook for Biodegradable Polymers*, Harwood Acad. Publ., Singapore.
35. Engelberg, I. and Kohn, J. (1991) Physico-mechanical properties of degradable polymers used in medical applications: a comparative study, *Biomater.* **12**, 292-304.
36. Fambri, L., Migliaresi, C., Kesenci, K. and Pişkin, E. (2001) Biodegradable Polymers, in R. Barbucci (ed.), *Integrated Biomaterials Science*, Kluwer Acad. , Plenum Publ.., New York, pp. 119-187.
37. Feijen, J. (1986) Biodegradable polymers for medical purpose, in E. Pişkin and A.S. Hoffman, (eds.), *Polymeric Biomaterials*, Martinus Nijhoff Publ., Dordrecht, pp. 62-77.
38. Gilding, D.K. (1981) Biodegradable polymers, in D.F. Williams (ed.), *Biocompatibility of Clinical Implant Materials*, vol. 2, CRC Press, Boca Raton, FL, pp. 209-232.
39. Goupil, D. (1996) Sutures, in B.D. Ratner, A.S. Hoffman, F.J. Schoen and J.E. Lemons (eds.), *Biomaterials Science: An Introduction to Materials in Medicine*, Acad. Press, San Diego, CA, pp.356-360.
40. Hayashi, T. (1994) Biodegradable polymers for biomedical use, *Prog. Polym. Sci.*, **19,** 663-702.
41. Heller, J., Sparer R.V. and Zentner, G.M. (1990) *Biodegradable Polymers as Drug Delivery Systems,* Marcel Dekker, New York.
42. Hollinger, J.O. (1995) *Biomedical Applications of Synthetic Biodegradable Polymers*, CRC Press, Boca Raton, FL.
43. Kimura, Y. (1993) Biodegradable polymers, in T. Tsuruta, T. Hayashi, K. Kataoka, K. Ishihara and K. Kimura (eds.), *Biomedical Applications of Polymeric Materials*. Boca Raton FL, CRC Press, 1993.
44. Kopecek, J. and Ulbrich, K. (1983) Biodegradation of biomedical polymers, *Prog. Polym. Sci.* **9**, 1-58.
45. Leung, K.S., Hung, L.K. and Leung P.C.L. (1995) *Biodegradable Implants in Fracture Fixation*, World Scientific Publ. Co., Singapore.
46. Li, S. and Vert, M. (1999) Biodegradable polymers: Polyesters, in E. Mathiowitz (ed.), *Encyclopedia of Controlled Drug Delivery*, John Wiley, New York, Vol. I, pp.71-93.
47. MacGreger, E. A. and Greenwood, C. T. (1980) *Polymers in Nature*, John Wiley & Sons, New York, NY.

48. Middleton, J.C. and Tipton, A.J. (2000) Synthetic biodegradable polymers as orthopedic devices, *Biomater.* **21,** 2335-2346.
49. Ottenbrite, R.M., Huang, S.J. and Park K. (1996) *Hydrogels and Biodegradable Polymers for Bioapplications*, ACS Symp. Ser. 627, Am. Chem. Soc., Washington DC.
50. Parikh, M., Gross, R.A. and McCarthy, S.P. (1992) The effect of crystalline morphology on enzymatic degradation kinetics, Proc. ACS Div., *Polym. Mat. Sci. Eng.* **66**, 408-409.
51. Park, K., Shalaby, W.S.W. and Park, H. (1993) *Biodegradable Hydrogels for Drug Delivery*, Technomic Publ. Co., Lancaster PA.
52. Pietrzak, W.S., Verstynen, B.S. and Sarver, D.R. (1997) Bioabsorbable fixation devices: status for the craniomaxillo faxial surgeon. *J. Craniofaxial Surg.* **2,** 92-96.
53. Piskin, E. (1994) Biodegradable polymers as biomaterials, *J. Biomater. Sci. Polym. Ed.* **6**, 795-775.
54. Planck, H., Dauner, M. and Renaldy, M. (1990) *Medical Textiles for Implantation*, Springer-Verlag, Berlin, Germany.
55. Privalova, L.G. and Zaikov, G.E. (1990) Surgical sutures, *Polym. Plast. Technol. Eng.* **29**, 445-467.
56. Schacht, E.H. (1990) Using biodegradable polymers in advanced drug delivery systems, Med.Dev.Techn. **1**, 15-21.
57. Schindler, A., Jeffcoat, R., Kimmel, G.L., Pitt, C.G., Wall, M.E. and Zwiedinger, R. (1997) Biodegradable polymers for sustain drug delivery, *Contemp. Topics. Polym. Sci.,* **2,** 251-289.
58. Shalaby, S.W. (1988) Bioabsorbable polymers, in J. Swarbrick and J.C. Boylan (eds.*), Encyclopedia of Pharmaceutical Technology*, Marcel Dekker Inc., New York, pp. 465-476.
59. Shalaby, S.W. and Johnson, R.A. (1994) Synthetic absorbable polyesters, in S.W. Shalaby (ed.), *Biomedical Polymers. Designed-to-Degrade Systems*, Hanser Publ., Munich, Germany, pp. 1-34.
60. Sinha, V.R. and Khosla, L. (1998) Bioabsorbable polymers for implantable therapeutic systems, *Drug Dev. Ind. Pharm.* **24**, 1129-1138.
61. Smith, R., Oliver, C. and Williams, D.F. (1987a) The enzymatic degradation of polymers in vivo, *J. Biomed. Mater. Res.* **21**, 991-1003.
62. St. Pierre, T. and Chiellini, E. (1986) Biodegradability of synthetic polymers used for medical and pharmaceutical applications: Part I - Principles of hydrolysis, *J. Bioact. Compat. Polym.* **1**,467-497.
63. Tokiwa, Y. and Suzuki, T. (1977) Hydrolysis of polyesters by lipases, *Nature* **270**, 76-78.
64. Vainionpaa, S., Rokkanen and Tormala, P. (1989) Surgical applications of biodegradable polymers in human tissue, *Prog. Polym. Sci.* **14**, 679-716.
65. Vert, M. (1989) Bioresorbable polymers for temporary therapeutic applications. *Die Angewan Makromol. Chem.* **166/167,**155-168.
66. Williams, D.F. (1990a) Biodegradation of medical polymers, in D.F. Williams (ed.), *Concise Encyclopedia Medical and Dental Materials*, Pergamon Press, Oxford, pp. 69-74.
67. Zaikov, G.E. (1985) Quantitative aspects of polymer degradation in the living

body, *JMS-Rev. Macromol. Chem. Phys.* **25**, 551-597.

68. Zhang, X., Wyss, U.P., Pichora, D. and Goosen, M.F.A. (1993) Biodegradable polymers for orthopedic applications: synthesis and processability of poly(L-lactide) and poly(lactide-co-ε-caprolactone), *Pure Appl.Chem.* **A30**, 933-947.
69. Bergsma, J.E., de Bruijn, W.C., Rozema, F.R., Bos, R.R.M. and Boering, G. (1995) Late degradation tissue response to poly(L-lactide) bone plates and screws, *Biomater.* **16**, 25-31.
70. Britritto, M.M., Bell, J.P., Brenckle, S., Huang, S.J. and Knox, J.R. (1979) Synthesis and biodegradation of polymers derived from α–hydroxy acids, *J. Appl. Polym. Sci., Appl. Polym. Symp.* **35**, 405-414.
71. Carothers, W.H., Dorough, G.L. and Van Natta, F.J. (1932) Studies of polymerization and ring formation. X. The reversible polymerization of six-membered cyclic esters, *J. Am. Chem. Soc.* **54**, 761-772.
72. Christel, P., Chabot, F., Leray, J.L., Morin, C. and Vert, M. (1982) Biodegradable composites for internal fixation, in D.G. Winter, D.F. Gibbons and J. Plench Jr. (eds.), *Advances in Biomaterials*, Vol. 3, John Wiley and Sons, New York, pp.271-280.
73. Chu, C.C. (1981) Hydrolytic degradation of polyglycolic acid: Tensile strength and crystallinity study, *J. Appl. Polym.* Sci. **26**, 1727-1734.
74. Chu, C.C. and Williams, D.F. (1983) The effect of gamma irradiation on the enzymatic degradation of polyglycolic acid absorbable sutures, *J. Biomed. Mater. Res.* **17**, 1029-1040.
75. Chu, C.C., Zhang, L. and Coyne, L.D. (1995) Effect of gamma irradiation and irradiation temperature on hydrolytic degradation of synthetic absorbable sutures, *J. Appl. Polym. Sci.* **56**, 1275-1294.
76. Chujo, K., Kobayashi, H., Suzuki, J., Tokuhara, S. and Tanabe, M. (1967) Ring-opening polymerization of glycolide, *Die Makrom. Chemie* **100**, 262-266.
77. Cohn, D. and Younes, H. (1988) Biodegradable PEO/PELA block copolymers, *J. Biomed. Mater.Res.* **22**, 993-1009.
78. Ferguson, S., Wahl, D. and Gogolewski, S. (1996) Enhancement of the mechanical properties of polylactides by solid-state extrusion. II. Poly(L-lactide), poly(L/D-lactide), and poly(L/DL-lactide, *J. Biomed. Mater. Res.* **30**, 543-551.
79. Frazza, E.J. and Schmitt, E.E. (1971) A new absorbable suture, *J. Biomed. Mater. Res. Symp.* **1**, 43-58.
80. Furukawa, T., Matsusue, Y., Yasunaga, T., Shikinami, Y., Okuno, M. and Nakamura, T. (2000) Biodegradation behavior of ultra-high-strength hydroxyapatite/poly(L-lactide) composite rods for internal fixation of bone fractures, *Biomater.* **21**, 889-898.
81. Furukawa, T., Matsusue, Y., Yasunaga, T., Shikinami, Y., Okuno, M. and Nakamura, T. (2000) Histomorphometric study on high-strength hydroxyapatite/poly(L-lactide) composite rods for internal fixation of bone fractures, *J. Biomed. Mater. Res.* **50**, 410-419.
82. Fukuzaki, H., Yoshida, M., Asano, M. and Kumakura, M. (1989) Synthesis of copoly(D,L-lactic acid) with relatively low molecular weight and in vitro degradation, *Eur. Polym. J.* **25**, 1019-1026.
83. Gilding, D.K. and Reed, A.M. (1979) Biodegradable polymers for use in surgery-

polyglycolic/polylactic acid homo- and copolymers, *Polym.* **20**, 1459-1464.

84. Goodman, I. (1988) Polyesters, in H.F. Mark, N.M. Bikales, C.G. Overberger and G. Menges (eds.), *Encyclopedia of Polymers Science and Engineering*, 2nd Ed., John Wiley & Sons, New York, Vol. 12, pp. 1-75.
85. Grijpma, D.W. and Pennings, A.J. (1994) Copolymers of L-lactide: 2. Mechanical properties, *Macromol. Chem. Phys.* **195**, 1649-1663.
86. Hyon, S.H., Jamshidi, K. and Ikada Y. (1984) Melt spinning of poly-L-lactide and hydrolysis of the fiber in vitro, in S. Shalaby, A.S. Hoffmann, B.D. Ratner and T.A. Horbett (eds.), *Polymers as Biomaterials*, Plenum Press, New York, pp. 51-65.
87. Johns, D.B., Lenz, R.W. and Leucke, A. (1984) Lactones, in K.J. Ivin and T. Saegusa (eds.), *Ring-Opening Polymerisation*, Vol.I, Elsevier Appl. Sci. Publ. Ltd., pp. 461-521.
88. Grijpma, D.W., Zondervan, G.J. and Pennings, A.J. (1991) High-molecular-weight copolymers of L-lactide and epsilon-caprolactone as biodegradable elastomeric implant materials, *Polym. Bull.* **25**, 327-333.
89. Honda, M., Yada, T., Ueda, M. and Kimata, K. (2000) Cartilage formation by cultured chondrocytes in a new scaffold made of poly(L-lactide-epsilon-caprolactone) sponge, *J. Oral Maxil. Surg.* **58**, 767-775.
90. Karjalainen, T., Hiljanen, M., Malin, M. and Seppala, J. (1996) Biodegradable lactone copolymers. III. Mechanical properties of ε–caprolactone and lactide copolymers after hydrolysis in vitro, *J. Appl. Polym. Sci.* **59**, 1299-1304.
91. Kricheldorf, H.R., Jonte, J.M. and Berl, M. (1985) Polylactones.3.Copolymerization of glycolide with L,L-Lactide and other lactones, *Makromol. Chem.* **12**, 25-38.
92. Kulkarni R.K., Pani K.C., Neuman C. and Leonard F. (1966) Polylactic acid for surgical implants, *Arch. Surg.* **93**, 839-843.
93. Kulkarni, R.K., Moore, E.G., Hegyeli, A.F. and Leonard, F. (1971) Biodegradable poly(lactic acid) polymers, *J. Biomed. Mater. Res.* **5**, 169-181.
94. Leenslag, J.W., Kroes, M.T., Pennings, A.J. and Van der Lei, B. (1988) A compliant, biodegradable vascular graft: Basic aspects of its construction and biological performance, *New Polym. Mater.* **1**, 111-126.
95. Lewandrowski, K.U., Gresser, J.D., Wise, D.L., Trantolo, D.J. and Hasirci, V. (2000) Tissue responses to molecularly reinforced polylactide-co-glycolide implants, *J. Biomater. Sci. Polym. Ed.* **11**, 401-414.
96. Lewis, D. H. (1990) Controlled release of bioactive agents from lactide/glycolide polymers, in M. Chasin and R. Langer (eds.), *Biodegradable Polymers as Drug Delivery Systems,* Marcel Dekker, New York, pp.1-41.
97. Maquet V., Martin D., Malgrange B., Frazen R., Schoenen J., Moonen G. and Jerome R. (2000) Peripheral nerve regeneration using bioresorbable macroporous polylactide scaffolds, *J. Biomed. Mater. Res.* **52**, 639-651.
98. Marra, K.G., Szem, J.W., Kumta, P.N., DiMilla, P.A. and Weiss, L.E. (1999) In vitro analysis of biodegradable polymer blend/hydroxyapatite composites for bone tissue engineering, *J. Biomed Mater. Res.* **47**, 324-335
99. Matsusue, Y., Yamamuro, T., Oka, M., Shikinami, Y., Hyon, S.H. and Ikada, Y. (1992) In vitro and in vivo studies on bioabsorbable ultra-high-strength poly(L-

lactide) rods, *J. Biomed. Mater. Res.* **26**, 1553-1567.

100. Migliaresi, C., Cohn, D., De Lollis, A. and Fambri, L. (1991a) Dynamic mechanical and calorimetric analysis of compression molded PLLA of different molecular weights, *J. Appl. Polym. Sci.* **43**, 83-95.
101. Migliaresi, C., De Lollis, A., Fambri, L. and Cohn, D. (1991b) The effect of the thermal treament on the crystallinity of different molecular weight PLLA biodegradable polymers, *Clinical Mater.* **8**, 111-118.
102. Migliaresi, C., Fambri, L. and Cohn, D. (1994) A study on the in vitro degradation of poly(lactic acid), *J. Biomater. Sci. Polym. Edn.* **5**, 591-606.
103. Nakamura, T., Shimizu, Y., Matsui, T., Okumura, N., Hyon, S.H. and Nishiya, A. (1992) A novel bioabsorbable monofilament surgical suture made from (ε–caprolactone, L-lactide) copolymer, in H. Planck, M. Dauner and M. Renardy (eds.), *Degradation Phenomena on Polymeric Biomaterials,* Springer-Verlag, Berlin-Heidelberg, pp. 153-162.
104. Pegoretti, A., Fambri, L. and Migliaresi, C. (1997) In vitro degradation of poly(L-lactic acid) fibers produced by melt spinning, *J. Appl. Polym. Sci.* **64**, 213-223.
105. Pietrzak, W.S., Sarver, D.R. and Verstynen, M.L. (1997) Bioabsorbable polymer science for the practicing surgeon, *J. Craniofacial Surg.* **8,** 87-91.
106. Pitt, C.G., Jeffcoat, A.R., Zweidinger, R.A. and Schindler, A. (1979) Sustained drug-delivery systems. I. The permeability of poly(ε-caprolactone), poly(DL-lactic acid) and their copolymers, *J. Biomed. Mat. Res.* **13**, 497-507.
107. Pitt, C.G., Gratzl, M.M., Kimmel G.L., Surles, J. and Schindler, A. (1981) Aliphatic polyesters II. The degradation of poly(DL-lactide), poly(ε-caprolactone), and their copolymers in vivo, *Biomater.* **2**, 215-220.
108. Pitt, C.G. and Schindler, A. (1984) Long-Acting Contraceptive Delivery Systems, Harper and Row, Philadelphia.
109. Pitt, C.G. (1990) Poly-ε-caprolactone and its copolymers, in M. Chasin and R. Langer (eds.), *Biodegradable Polymers as Drug Delivery Systems,* Marcel Dekker Inc., New York, pp. 71-120.
110. Sinclair, R.G. (1977) Copolymers of L-lactide and epsilon caprolactone, *US Patent*, 3 057 537.
111. Sanz, L.E., Patterson, J.A., Kamath, R., Willett, G., Ahmed, S.W. and Butterfield, A.B. (1988) Comparison of Maxon suture with Vicryl, chromic catgut, and PDS sutures in fascial closure in rats, *Obstet. Gynecol.* **71**, 418-422.
112. Schmitt, E.E. and Polistina, R.A. (1967) Surgical sutures, *US Patent*, 3,297,033.
113. Ural, E., Kesenci, K., Fambri, L., Migliaresi, C. and Pişkin, E. (2000) Poly(D,L-cactide/epsilon-caprolactone)/hydroxyapatite composites, *Biomater.* **21**, 2147-2154.
114. Tormala, P., Vasenius, J., Vainionpaa, S., Laiho, J., Pohjonen, T. and Rokkanen, P. (1991) Ultra-high-strength absorbable self-reinforced polyglycolide (SR-PGA) composite rods for internal fixation of bone fractures: in vitro and in vivo study, *J. Biomed. Mater. Res.* **25**, 1-22.
115. Tunç, D.C. (1995) Orientruded polylactide based body-absorbable osteosynthesis devices: a short review, *J. Biomater. Sci. Polym. Ed.* **7**, 375-380.
116. Vanhoorne, P., Dubois, P., Jerome, R. and Teyssie Ph. (1992) Macromolecular engineering of polylactones and polylactides.7. structural-analysis of copolyesters of epsilon-caprolactone and L-caprolactone or D,L-lactide initiated by AL(OIPR)3,

Macromol. **25**, 37-44.

117. Wang, N., Wu, X.S., Lujan-Upton, H., Donahue, E. and Siddiqui, A. (1997) Synthesis, characterization, biodegradation and drug delivery application of biodegradable lactic/glycolic acid oligomers: I. Synthesis and characterization, *J.Biomat. Sci., Polym.Ed.,* **8,** 905-917.
118. Verheyen, C.C.P.M., de Wijn, J.R., van Blitterswijk, C.A. and de Groot K. (1992) Evaluation of hydroxylapatite/poly-(L-lactide) composites: Mechanical behavoiour, *J. Biomed. Mater. Res.* **26**, 1277-1296.
119. Vert, M. and Guerin, P. (1991) Biodegradable aliphatic polyesters of the poly(hydroxy acid)-type for temporary therapeutic applications, in M.A. Barbosa (ed.), *Biomaterial Degradation: Fundamental Aspects and Related Clinical Phenomena*, Elsevier, Amsterdam, pp. 35-51.
120. Vion, J.M., Jerome, R. and Teyssie, P. (1986) Synthesis, characterization and miscibility of caprolactone random copolymers, *Macromol.* **19**, 1828-1838.
121. Wehrenberg, R.H. (1981) Polylactic acid polymers: strong, degradable thermoplastics, *Mater. Eng.* **94**, 63-66.
122. Williams, D.F. (1981) Enzymatic hydrolysis of polylactic acid, *Eng. in Med.* **10**, 5-7.
123. Zhang, X., Wyss, U.P., Pichora, D. and Goosen, M.F.A. (1994) An investigation of poly(lactic acid) degradation, *J. Bioact. Compat. Mater.* **9**, 80-100.
124. Zhu, K.J., Xiangzhou, L. and Shilin, Y. (1990) Preparation, characterization and properties of polylactide(PLA)-poly(ethylene glycol) (PEG) copolymers: a potential drug carrier, *J. Appl. Polym. Sci.* **39**, 1-9.
125. Benoit, M.A., Gillard, B.B. (1999) Preparation and characterization of protein-loaded poly(epsilon-caprolactone) microparticles for oral vaccine delivery, *J. Int J. Pharm.* **184**, 73-84.
126. Cai, Q., Bei, J., Wang, S.G. (2000) Synthesis and degradation of a tri-component copolymer derived from glycolide, L-lactide and epsilon-caprolactone, *J. Biomater. Sci. Polym. Ed.* **11**, 273-288.
127. Das, G.S., Rao, G.H.R., Wilson, R.F. and Chandy, T. (2000) Colchicine encapsulation within poly(ethylene glycol)-coated poly(lactic acid)/poly(epsilon-caprolactone) microspheres-controlled release studies, *Drug Del.* **7**, 129-38.
128. Dendunnen, W.F.A., Vanderlei, B., Robinson, P.H., Holwerda, A., Pennings, A.J. and Schakenraad, J.M. (1995) Biological performance of a degradable poly(lactic acid-epsilon-caprolactone) nerve guide – influence of tube dimensions, *J. Biomed. Mater. Res.* **29**, 757-766.
129. Eliaz, R.E. and Kost, J. (2000) Characterization of a polymeric PLGA-injectable implant delivery system for the controlled release of proteins, *J. Biomed. Mater. Res.* **50**, 388-396.
130. Gogolewski, S., Pineda, L. and Busing, C.M. (2000) Bone regeneration in segmental defects with resorbable polymeric membranes: IV. Does the polymer chemical composition affect the healing process?, *Biomater.* **21**, 2513-2520.
131. Perez, M.H., Zinutti, C. and Lamprecht, A. (2000) The preparation and evaluation of poly(epsilon-caprolactone) microparticles containing both a lipophilic and a hydrophilic drug, *J. Control Rel.* **65**, 429-438.

132. Aime, S., Botta, M., Garino, E., Crich, S.G., Giovenzana, G., Pagliarin, R., Palmisano, G. and Sisti, M. (2000) Non-covalent conjugates between cationic polyamino acids and Gd-III chelates: A route for seeking accumulation of MRI-contrast agents at tumor targeting sites, *Chem-Eur J.* **6,** 2609-2617.
133. Aprahamian, M., Lambert, A. and Balboni, G. (1987) A new reconstituted connective-tissue matrix preparation, biochemical, structural and mechanical studies, *J. Biomed. Mater. Res.* **21**, 965-977.
134. Arem, A. (1985) Collagen modifications, *Clin. Plast. Surg.* **12**, 209-220.
135. Arnold, L. J., Dagan, A. and Kaplan, N.O. (1983) *Targeted Drugs*, John Wiley and Sons Ltd., New York.
136. Averbatch, B.L. (1978) in: Proc. 1st Int. Conf. *Chitin/Chitosan,* MIT, Cambrigde, pp. 199.
137. Banaszczyk, M.G., Lollo, C.P., Kwoh, D.Y., Phillips, A.T., Amini, A., Wu, D.P., Mullen, P.M., Coffin, C.C., Brostoff, S.W. and Carlo, D.J. (1999) Poly-L-lysine-graft-PEG-comb-type polycation copolymers for gene delivery, *J. Macromol. Sci. Pure* **36**, 1061-1084.
138. Birchall, A.C. and North, M. (1998) Synthesis of highly branched block copolymers of enantiomerically pure amino acids, *Chem. Commun.* 1335-1336.
139. Caponetti, G., Hrkach, J.S., Kriwet, B., Poh, M., Lotan, N., Colombo, P., Langer, R. (1999) Micro particles of novel branched copolymers of lactic acid and amino acids: Preparation and characterization, *J. Pharma Sci.* **88**, 136-141.
140. Constancis, A., Meyrueix, R., Bryson, N., Huille, S., Grosselin, J., Gulik-Krzywicki, T. and Soula, G. (1999) Macromolecular colloids of diblock poly(amino acids) that bind insulin, *J. Colloid and Inter. Sci.* **217**, 357-368.
141. Hayashi, T., Nakanishi, E., Iizuka, Y., Oya, M. and Iwatsuki, M. (1995) Preparation and properties of copoly(N-hydroxyalkyl-D,L-glutamine) membranes, *Eur. Polym. J.* **31,** 453-458.
142. Hoes, C.J.T. and Feijen, J. (1989) *Drug Carrier Systems*, John Wiley and Sons Ltd., London, pp. 57-109.
143. Huang, S.J. and Leong, K.W. (1989) Biodegradable polymers. Polymers derived from gelatin and lysin esters, *Polym. Preprints* **20**, 552-554.
144. Kohn, J. (1990) Pseudo poly(amino acids) in M. Chasin and R Langer (eds.), *Biodegradable Polymers as Drug Delivery Systems*, Marcell Dekker, New York, Chapter 6, pp.195-229.
145. Marck, K.W., Wildevuur C.H., Sederel W.L., Bantjes, A. and Fejen, J. (1977) Biodegradability and tissue reaction of random copolymers of L-leucine, L-aspartic acid and L-aspartic acid esters, *J. Biomed. Mater. Res.* **11**, 405-422.
146. Miller, A.G. (1964) Degradation of synthetic polypeptides. III., Degradation of polylysine by proteolytic enzymes in 0.20 M sodium chloride, *J. Am. Chem. Soc.* **86**, 3818-3822.
147. Nikiforov, T.T. and Jeong, S. (1999) Detection of hybrid formation between peptide nucleic acids and DNA by fluorescence polarization in the presence of polylysine, *Anal. Biochem.* **275**, 248-253.
148. Nukui, M., Hoes, K., Vandenberg, H. and Feijen, J. (1991) Association of macromolecular prodrugs consisting of adriamycin bound to poly(L-glutamic acid), Macromol. Chem. **192**, 2925-2942.

149. Pouton, C.W., Lucas, P., Thomas, B.J., Uduehi, A.N., Milroy, D.A. and Moss, S.H. (1998) Polycation-DNA complexes for gene delivery: a comparison of the biopharmaceutical properties of cationic polypeptides and cationic lipids, *J.Control. Rel.* **53**, 289-299.
150. Sidman, K.R., Schwope, A.D., Steber, W.D., Rudolph, S.E. and Poulin, S.B. (1980) Biodegradable, implantable sustained release systems based on glutamic acid copolymers, *J. Mem. Sci.* **7**, 277-291.
151. Akbari, H., Attwood, D. and D'Emanuele, A. (1998) Effect of fabrication technique on the characteristics of polyanhydride matrices, *Pharma.Dev.Technol.* **3**, 251-259.
152. Chasin, M., Lewis, D. and Langer, R. (1988) Polyanhydrides for controlled drug delivery, *Biopharm. Manufac.* **1**, 33-46.
153. Chasin, M., Domb, A., Ron, E., Mathiowitz, E., Langer, R., Leong, K., Laurencin, C., Brem, H. and Grossman, S. (1990) Polyanhydrides as drug delivery systems, in M. Chasin and R. Langer (eds.), *Biodegradable Polymers as Drug Delivery Systems*, Marcel Dekker, New York, pp. 43-70.
154. Domb, A.J., Amselem, S., Langer, R. and Maniar, M. (1994) Polyanhydrides as carriers of drugs, in S. W. Shalaby (ed.), *Biomedical Polymers. Designed-to-Degrade Systems,* Hanser Publ., Munich, Germany, pp. 69-96.
155. Erdmen, L., Macedo, B. and Uhrich, KE. (2000) Degradable poly(anhydride ester) implants: effects of localized SA release on bone, *Biomater.* **21**, 2507-2512.
156. Gopferich A. (1999) Biodegradable polymers: Polyanhydrides, in E. Mathiowitz (ed.), *Encyclopedia of Controlled Drug Delivery*, John Wiley, New York, Vol.I, pp.60-71.
157. Laurencin, C., Gerhart, T., Witschger, P., Satcher, R., Domb, A., Hanff, P., Edsberg, L., Hayes, W. and Langer, R. (1989) Bioerodible Polyanhydrides for Antibiotics Drug Delivery: In Vivo Osteomyelitis Treatment Studies, *Proc. Int. Symp. Rel.Bioact. Mater.*, Controlled Release Society, Inc.
158. Leong, K.W., Brott, B.C. and Langer, R. (1985) Bioerodible polyanhydrires as drug carrier matrices I: characterization, degradation and release characteristics, *J. Biomed. Mater. Res.* **19**, 941-955.
159. Ron, E., Turek, T., Mathiowitz, E., Cahsin, M. and Langer R. (1989) Release of Polypeptides From Polyanhydride Implants, *Proceed. Intern. Symp. Rel. Bioact. Mater.* 16, Controlled Release Society, Inc.
160. Choi, N.S. and Heller, J. (1978) Drug delivery devices manufactured from polyorthoesters and polyorthocarbonates, *US Patent,* 4 093 709.
161. Choi, N.S. and Heller, J. (1979) Erodible agent releasing device comprising polyorthoesters and polyorthocarbonates, *US Patent*, 4 138 344, Feb 6.
162. Daniels, A.U., Chang, M.K.O., Andriano, K.P. and Heller, J. (1990) Mechanical properties of biodegradable polymers and composites proposed for internal fixation of bone, *J. Appl. Biomater.* **1,** 57-78.
163. Davies, M.C., Khan, M.A., Lynn, R.A., Heller, J. and Watts, J.F. (1991) X-ray photoelectron spectroscopy analysis of the surface chemical structure of some biodegradable polyorthoesters, *Biomater.* **12**, 305-308.
164. Heller, J. (1983) Use of polymers in controlled drug release, in M. Szycher (ed.), *Biocompatible Polymers, Metals, and Composites*, Technomic Publ. Co.,

Lancaster, PA, Chapter 24, pp. 551-584.
165. Heller, J. (1990) Development of polyorthoesters: A historical overview, *Biomater.* **11,** 659-665.
166. Heller, J., Sparer, R.V. and Zentner, G.M. (1990a) Polyorthoesters, in M. Chasin and R Langer, (eds.), *Biodegradable Polymers as Drug Delivery Systems,* Marcel Dekker, New York, pp. 121-161.
167. Heller, J., Ng, S.Y., Fritzinger, B.K. and Roskov, K.V. (1990b) Controlled drug release from bioerodible hydrophobic ointments, *Biomater.* **11,** 235-237.
168. Heller, J., Ng, S.Y. and Fritzinger, B.K. (1992) Synthesis and characterization of a new family of polyorthoesters, *Macromol.* **25,** 3362-3364.
169. Heller, J. and Daniels, A.U. (1994) Polyorthoesters, in S.W. Shalaby (ed.), *Biomedical Polymers. Designed-to-Degrade Systems,* Hanser Publ., Munich, Germany, pp. 35-67.
170. Heller, J. (1996) Drug delivery systems, in B.D. Ratner, A.S. Hoffman, F.J. Schoen and J.E. Lemons (eds.), *Biomaterials Science: An Introduction to Materials in Medicine*, Acad. Press, San Diego, CA, pp. 346-356.
171. Heller, J. and Gurny, R. (1999) Polyorthoesters, in E. Mathiowitz (ed.), *Encyclopedia of Controlled Drug Delivery*, John Wiley, New York, Vol.II, pp. 852-874.
172. Leadley, S.R., Shakesheff, K.M., Davies, M.C., Heller, J., Franson, N.M., Paul, A.J., Brown, A.M. and Watts, J.F. (1998) The use of SIMS, XPS and AFM to probe the acid catalysed hydrolysis of polyorthoesters, *Biomater.* **19,** 1353-1360.
173. Mao, H.Q., Kadiyala, I., Leong, K.W., Zhao, Z. and Dang, W. (1999) Biodegradable polymers: Polyphosphoesters, in E. Mathiowitz (ed.), *Encyclopedia of Controlled Drug Delivery*, John Wiley, New York, Vol.I, pp. 45-60.
174. Allcock, H.R. (1972) *Phosphorus-Nitrogen compounds. Cyclic, Linear, and High Polymeric Systems*, Acad. Press, New York, USA.
175. Allcock, H.R., Fuller, T.J., Mack, D.P., Matsumura, K. and Smeltz, K.M. (1977) Phosphazene compounds. Synthesis of poly[(amino acid alkylester)phosphazenes], *Macromol.* **10,** 824-830.
176. Allcock, H.R., Fuller, T.J. and Matsumura, K. (1982) Hydrolysis pathways for aminophosphazenes, *Inorg. Chem.* **21,** 515-521.
177. Allcock, H.R. (1990) Polyphosphazene as new biomedical and bioactive materials, in M. Chasin and R. Langer (eds.), *Biodegradable Polymers as Drug Delivery Systems,* Marcel Dekker, New York, pp.163-193.
178. Allcock, H.R., Pucher, S.R. and Scopelianos, A.G. (1994) Poly(amino acid ester)phosphazenes as substrates for the controlled release of small molecules, *Biomater.* **15,** 563-569.
179. Allcock, H.R., Pucher, S.R. and Scopelianos, A.G. (1994) Synthesis of poly(organophosphazenes) with glycolic acid ester and lactic acid ester side groups - Prototypes for new bioerodible polymers, *Macromol.* **27,** 1-4.
180. Allcock H.R. (1998) Functional polyphosphazenes, in A.O. Patil, D.N. Schulz and B.N. Novak (eds.), *Functional Polymers. Modern Synthetic Methods and Novel Structures*, ACS Symp. Ser., Washington, pp. 261-275.
181. Andrianov, A.K., Cohen, S., Visscher, K.B., Payne, L.G., Allcock, H.R. and Langer, R. (1993) Controlled release using ionotropic polyphosphazene hydrogels,

J. Control. Rel. **27,** 69-77.

182. Crommen, J., Vandorpe, J. and Schacht, E. (1993) Degradable polyphosphazenes for biomedical applications, *J. Control. Rel.* **24,** 167-180.
183. De Jaeger, R. and Gleria, M. (1998) Polyorganophosphazenes and related compounds: synthesis, properties and applications, *Progr. Poym. Sci.* **23,** 179-276.
184. Goedemoed, J.H. and De Goot, K. (1988) Development of implantable antitumor devices based on polyphosphazene, *Macromol. Chem. Macromol. Symp.* **19,** 341-365.
185. Grolleman, C.W.J, De Visser, A.C., Wolke, J.G.C., Klein, C.P.A.T., Van der Goot, H. and Timmerman, H. (1986) Studies on a bioerodible drug carrier system based on a polyphosphazene, *J. Control. Rel.* **3,** 143-154.
186. Ibim, S.M., Ambrosio, A.A., Larrier, D., Allcock, H.R. and Laurencin, C.T. (1996) Controlled macromolecule release from polyphosphazene matrices, *J. Control. Rel.* **40,** 31-39.
187. Ibim, S.M., Ambrosio, A.A., Kwon, M.S., El-Amin, S.F., Allcock, H.R. and Laurencin, C.T. (1997) Novel polyphosphazene/poly(lactide-co-glycolide) blends: miscibility and degradation studies, *Biomater.* **18,** 1565-1569.
188. Langone, F., Lora, S., Veronese, F.M., Caliceti, P., Parnigotto, P.P., Valenti, F. and Palma, G. (1995) Peripheral Nerve Repair Using a Poly(organo)phosphazene Tubular Prosthesis, *Biomater.***16,** 347-353.
189. Laurencin, C.T., Koh, H.J., Neenan, T.X., Allcock, H.R. and Langer, R. (1987) Controlled release using a new bioerodible polyphosphazene matrix system, *J. Biomed. Mater. Res.* **21,** 1231-1246.
190. Scopelianus, A.G. (1994) Polyphospazenes as new biomaterials, in S.W. Shalaby (ed.), *Biomedical Polymers. Designed-to-Degrade Systems,* Hanser Publ., Munich, Germany, pp. 153-172.
191. Singler, R.E., Sennett, M.S. and Willingham, R. A. (1988) Phosphazene polymers: synthesis, structure, and properties, in M. Zeldin, K.J. Wynne and H.R. Allcock (eds.), *Inorganic & Organometallic Polymers*, ACS Symp. Ser. Vol. 360, 360.
192. Vandorpe, J., Schacht, E., Dunn, E., Hawley, A., Stolnik, S., Davis, S.S., Garnett, M.C., Davies, M.C. and Illum, L. (1997a) Long circulating biodegradable poly(phosphazene) nanoparticle surface modified with poly(phosphazene)-poly(ethylene oxide) copolymer, *Biomater.***18,** 1147-1142.
193. Veronese, F.M., Marsilio, F., Lora, S., Caliceti, P., Passi, P. and Orsolini, P. (1999) Polyphosphazene Membranes and Microspheres in Periodontal Diseases and Implant Surgery, *Biomater.* **20,** 91-98.
194. Verweire, I., Schacht, E., Qiang, B.P., Wang, K. and De Scheerder, I. (2000) Evaluation of fluorinated polymers as coronary stent coating, *J. Mater. Sci.: Mater. in Med.* **11,** 207-212.
195. Amiel, G.E., Sukhotnik. I., Kawar, B. and Siplovich, L. (1999) Use of N-butyl-2-cyanoacrylate in elective surgical incisions-longterm outcomes, *J. Am. Coll. Surg.* **189,** 21-25.
196. Beattie, G.C., Kumar, S. and Nixon, S.J. (2000) Laparoscopic total extraperitoneal hernia repair: Mesh fixation is unnecessary, *J. Laparoendosc. Adv.* **10,** 71-73.
197. Ciapetti, G., Stea, S., Cenni, E., Sudanese, A., Marraro, D., Toni, A. and Pizzoferrato, A. (1994) Cytotoxicity testing of cyanoacrylates using direct contact

assay on cell cultures, *Biomater.* **15,** 63-67.

198. Courtney, P.J. and Sorenson, J. (1996) Adhesive bonding of medical plastics, *Med. Plast. Biomater.***3,** 1-20.
199. Couvreur, P. Kante, B. and Roland, M. (1979) Polycyanoacrylate nanocapsules as potential lysosomotropic carriers- preparation, morphological and sorptive properties, *J. Pharm. Pharmacol.* **31,** 331-332.
200. Damage, C., Vranckx, H., Balschmidt and Couvreur, P. (1997) Poly(alkyl cyanoacrylate) nanospheres for oral administration of insulin, *J. Pharm. Sci.* **86,** 1403-1409.
201. Fattal, E., Blanco-Prieto, M. J., Leo, E., Puisieux, F. and Couvreur, P. (1997) Design of nanoparticles for vaccine delivery in antigen delivery systems, Hardwood Acad. Publ., pp.139-157.
202. Fattal,E., Peracchia, M.T. and Couvreur, P. (1997) Polyalkylcyanoacrylates in A.J. Domb, J. Kost and D.M. Wiseman (eds.), *Handbook of Polymer,* Hardwood Acad. Publ., pp.183-202.
203. Jaffe, R., Wade, C.W.R., Hegyeli, A.F., Rice, R. and Hodge, J. (1986) Synthesis and bioevaluation of alkyl 2-cyanoacryloyl glycolates as potential soft tissue adhesives, *J. Biomed. Mater.Res.* **20,** 205-212.
204. Lenaerts, V., Couvreur, P., Christiansen-Leyh, D., Joiris, E., Roland, M., Rollman, B. and Speiser, P. (1984) Degradation of poly(isobutyl cyanoacrylate) nanoparticles, *Biomater.* **5,** 65-68.
205. O'Connor, J. and Dreifus, D. (1986) *Cyanoacrylates,* Loctite Corp. Internal Technical Memorandum.
206. Peracchia, M.T., Harnisch, S., Pinto-Alphandary, H., Gulik, A., Dedien, J.C., Desmaële, D., d'Angelo, J., Müller, R.H. and Couvreur, P. (1999) Visualization of in vitro protein-rejecting properties of PEGylated stealth ® polycyanoacrylate nanoparticles, *Biomater.* **20,** 1269-1275.
207. Ronis, M.L., Harwick, J.D., Fung, R. and Dellavecchia, M. (1984) Review of cyanocrylate tissue glues with emphasis on their otorhinolaryngological applications, *Laryngoscope* **94,** 210-213.
208. Tseng, Y., Tabata, Y., Hyon, S. and Ikada, Y. (1990) In vitro toxicity of 2-cyanoacrylate polymers by cell culture method, *J. Biomed. Mater. Res.* **24,**1355-1367.
209. Tuncel, A., Çiçek, H. and Pişkin, E. (1995) Degradation and drug-release characteristics of monosize polyethylcyanoacrylate microspheres, *J. Biomat.. Sci-Polym. Ed.* **6,** 845-856.
210. Wang, M.Y., Levy, M.L., Mittler, M.A., Liu, C.Y., Johnston, S. and Mc Comb, J.G. (1999) A prospective analysis of the use of octyl acrylate tissue adhesive for wound closure in pediatric neurosurgery, *Pediatr. Neurosurg.* **30,** 186-188.
211. Wohlgemuth, M., MacHtle, W. and Mayer, C. (2000) Improved preparation and physical studies of polybutylcyano acrylate nanocapsules, *J. Microencapsul.* **17,** 437-448.
212. Borgdorff, P., Van den Berg, R.H., Vis, M.A., Van den Bos, G.C. and Tangelder, G.J. (1999) Pump-induced platelet aggregation in albumin-coated extracorporeal systems, *J. Thorac. Cardiovasc. Surg.* **118,** 946-952.
213. Bos, G.W., Scharenborg, N.M., Poot, A.A., Engbers, G.H., Beugeling, T., Van

Aken W.G. and Feijen, J. (1999) Blood compatibility of surfaces with immobilized albumin-heparin conjugate and effect of endothelial cell seeding on platelet adhesion, *J. Biomed. Mater. Res.* **47,** 279-291.

214. Bouillot, P., Ubrich, N., Sommer, F., Duc, T.M., Loeffler, J.P. and Dellacherie, E. (1999) Protein encapsulation in biodegradable amphiphilic microspheres, *Int. J. Pharm.* **181,** 159-172.
215. Guillaume, Y.C., Peyrin, E. and Berthelot, A. (1999) Chromatographic study of magnesium and calcium binding to immobilized human serum albumin, *J. Chromatogr. B: Biomed. Sci. Appl.* **728,** 167-174.
216. Kocisova, E., Jancura, D., Sanchez-Cortes, S., Miskovsky, P., Chinsky, L. and Garcia-Ramos, J.V. (1999) Interaction of antiviral and antitumor photoactive drug hypocrellin A with human serum albumin, *J. Biomol. Struct. Dyn.* **17,** 111-120.
217. Kramer, P. A. (1974) Albumin microspheres as vehicles for achieving specificity in drug delivery, *J. Pharm. Sci.* **63,** 1646-1647.
218. Lee, T.K., Sokoloski, T.D. and Royer, G.P. (1981) Serum albumin beads; an injectable, biodegradable system for the sustained release of drugs, *Science* **213,** 230-235.
219. Arem, A. (1985) Collagen modifications, *Clin. Plast. Surg.* **12,** 209-220.
220. Auger, F.A., Rouabhia, M., Goulet, F., Berthod, F., Moulin, V. and Germain, L. (1998) Tissue-engineered human skin substitutes developed from collagen-populated hydrated gels: clinical and fundamental applications, *Med. Biol. Eng. Comput.* **3,** 801-812.
221. Bartone, F.F., Shervey, P.D. and Gardner, P.J. (1976) Long term tissue responses to catgut and collagen sutures, *Invest. Urol.* **13,** 390-394.
222. Brumback, G.F. and McPherson, S.D Jr. (1967) Reconstituted collagen sutures in corneal surgery. An experimental and clinical evaluation, *J. Ophthalmol.* **64,** 222-227.
223. Choi, Y.S., Hong, S.R, Lee, Y.M., Song, K.W., Park, M.H. and Nam, Y.S. (1999) Studies on gelatin-containing artificial skin: II. Preparation and characterization of cross-linked gelatin-hyaluronate sponge, *J. Biomed. Mater. Res.* **48,** 631-639.
224. Chvapil, M., Kronenthal, R.L. and van Winkle, W. (1973) Medical and surgical applications of collagen, *Int. Rev. Connect. Tissue Res.* **6**, 1-61.
225. Coleman, W.P. (1996) Assessment of a new device for injecting bovine collagen-The ADG needle, *Dermat. Surg.* **22,** 175-176.
226. Engler, R.J., Weber, C.B. and Turnicky, R. (1986) Hypersensitivity to chromated catgut sutures: a case report and review of the literature, *Ann. Allergy* **5,** 317-320.
227. Friess, W. (1998) Collagen-biomaterial for drug delivery, *Eur. J. Pharm. Biopharm.* **45,** 113-136.
228. Friess, W., Uludağ, H., Foskett, S., Biron, R. and Sargeant, C. (1999) Characterization of absorbable collagen sponges as rhBMP-2 carriers, *Int. J. Pharm.* **187,** 91-99.
229. Goissis, G., Marcantonio, E. Jr., Marcantonio, R.A., Lia, R.C., Cancian, D.C., and de Carvalho, W. M. (1999) Biocompatibility studies of anionic collagen membranes with different degree of glutaraldehyde cross-linking, *Biomater.* **20,** 27-34.
230. Gorham, S.D. (1991) Collagen, in D. Byron (ed.), *Biomaterials, Novel Materials*

From Biological Sources, Stockton Press, New York, Chapter 2.

231. Huc, A. (1985) Collagen biomaterials characteristics and applications, *J. Amer. Leather Chem. Ass.* **80,** 195-212.
232. Lou, X. and Chirila, T.V. (1999) Swelling behavior and mechanical properties of chemically cross-linked gelatin gels for biomedical use, *J. Biomater. Appl.* **14,** 184-191.
233. Nimni, M.E. (1983) Collagen: structure, function and metabolism in normal and fibrotic tissues, *Semin. Arthritis Rheum.,* XIII, pp. 1-86.
234. Okada, T., Hayashi, T. and Ikada Y. (1992) Degradation of collagen suture in vitro and in vivo, *Biomater.* **13,** 448-454.
235. Pachence, J.M., Berg, R.A. and Silver, F.H. (1987) Collagen: Its place in the medical device industry, *Med. Device Diagn. Ind.* **9,** 49-55.
236. Rogalla, C.J. (1997) Autologous collagen: a new treatment for dermal defects, *Minim. Invasive Surg. Nurs.* **11,** 67-69.
237. Schlegel, A.K., Möhler, H., Busch, F. and Mehl, A. (1997) Preclinical and clinical studies of a collagen membrane (Bio-Gide), *Biomater.* **18,** 535-538.
238. Silver, F.H., Pins, G.D., Wang, M.C. and Christiansen, D. (1995) Collagenous biomaterials as models for tissue inducing implants, in D.L. Wise (ed.), *Encyclopaedic Handbook of Biomaterials and Bioengineering,* Part A: Materials, Marcel Dekker Inc., New York, pp. 63-70.
239. Stone, K.R., Webber, R.J., Rodkey, W.G. and Steadman, J.R. (1989) Prosthetic meniscal replacement: In vitro studies of meniscal regenaration using copolymeric collagen prostheses, *Arthroscopy* **5,** 152-158.
240. Sung, H.W., Huang, D.M., Chang, W.H., Huang, L.L., Tsai, C.C. and Liang, I.L. (1999) Gelatin-derived bioadhesives for closing skin wounds: An in vivo study, *J. Biomater. Sci., Polym. Ed.* **10,** 751-771.
241. Ulrich, S., Kuntz, G. and Anita, R. (1992) Haemostyptic preparations on the basis of collagen alone and as fixed combination with fibrin glue, *Clinical Mater.* **9,** 169-177.
242. Yannas, I.V. (1972) Collagen and gelatin in solid-state, *J. Macromol. Sci. Revs. Macromol. Chem.* **C7**, 49-57.
243. Yannas, I.V. and Burke, J.F. (1980) Design of an artificial skin: I. Design principles, J. Biomed. Mater. Res. **14,** 65-81.
244. Averbatch, B.L. (1978) *Proc. 1st Int. Conf. Chitin/Chitosan,* MIT, Cambrigde, pp. 199.
245. Chandy, T., Das, G.S. and Rao, G. (2000) 5-Fluorouracil-loaded chitosan coated polylactic acid microspheres as biodegradable drug carriers for cerebral tumors, *J. Microencap.* **17,** 625-38.
246. Chung, L.Y., Shmidt, R.J., Hamlyn, P.F., Sagar, B.F. and Andrews, A.M. (1994) Biocompatibility of potential wound management products: Fungal mycelia as a source of chitin/chitosan and their effect on the proliferation of human F1000 fibroblasts in culture, *J. Biomed. Mat. Res.* **28,** 463-469.
247. Erbacher, P., Zou, S., Bettinger, T., Steffan, A.M. and Remy, J.S. (1998) Chitosan-based vector/DNA complexes for gene delivery: biophysical characteristic and transfection, *Pharm. Res.* **15,** 1332-1339.

248. Gallaher, C., Munion, J., Hesslink, R.Jr., Wise, J. and Gallher, D.D. (2000) Cholesterol reduction by glucomannan and chitosan is mediated by changes in cholesterol absorption and bile acid and fat excretion in rats, *J. Nutr.* **130,** 2753-2759.
249. Hudson, S.M. (1994) Review of chitin and chitosan as fiber and film formers, *J.Mater. Sci.* **34,** 375-437.
250. Kuen, Y.L., Wan, S.H. and Won, H.P. (1995) Blood compatibility and biodegradibility of partially N-acylated chitosan derivatives, *Biomat.* **16,** 1211-1216.
251. Kurita, K. (1998) Chemistry and application of chitin and chitosan, *Polym. Degrad. Stabil.* **59,** 117-120.
252. Leong, K.W., Mao, H.Q., Truong-Le V.L., Roy, K., Walsh, S.M. and August, J.T. (1998) DNA-polycation nanospheres as non-viral gene delivery vehicles, *J.Control. Rel.* **53**, 183-193.
253. Muzzarelli, R.A. (1993) Biochemical significance of exogenous chitins and chitosans in animals and patients, *Carbohydrate Polym.* **20,** 7-15.
254. Muzzarelli, R.A., Jeuniaux, C. and Gooday, G.W. (1986) Evaluation of Chitosan as a New Hemostatic Agent: Vitro and In Vivo Experiments, in G. Fradet, S. Brister, D. Mulder, J. Lough, and B.L. Averbach (eds.), *Chitin in Nature and Technology*, Plenum Press, New York.
255. Nagai, T., Sawayanagi, Y. And Nambu, N. (1984) *Chitin, Chitosan and Related Enzymes,* Acad. Press, Orlando, FL
256. Sandford, P.A. (1989) Chitosan: Commercial uses and potential applications, in T. Anthonsen and P. Sandford (eds.), *Chitin and Chitosan: Sources, Chemistry, Biochemistry Physical Properties and Applications.* Elsevier Applied Science, N.Y. , pp: 51-69.
257. Stone, C.A., Wright, H., Clarke, T., Powell, R. and Devaraj, V.S. (2000) Healing at skin graft donor sites dressed with chitosan, *J. Plast. Surg.* **53,** 601-606.
258. Suh, J.K. and Matthew, H.W. (2000) Aplication of chitosan-based polysaccharide biomaterials in cartilage tissue engineering, *Biomat.* **21,** 2589-2598.
259. Varum, K.M., Myhr, M.M., Hjerde, R.J.N. and Smidsrud, O. (1997) In vitro degradation rates of partially N-acetylated chitosans in human serum, *Carbohydrate Res.* **299**, 99-101.
260. Zizokis, J.P. (1984) *Chitin, Chitosan and Related Enzymes,* Acad. Press, Orlando, FL.
261. Abatangelo, G., Barbucci, R., Brun, P. and Lamponi, S. (1997) Biocompatibility and enzymatic degradation studies on sulphated hyaluronic acid derivatives, Biomaterials **18,** 1411–1415.
262. Anderson, A.B. and Clapper, D.L. (1998) Coatings for blood-contacting devices, *Med. Plast. Biomat.* **3,**16-20.
263. Balazs, E.A. (1995) Hyaluronan biomaterials: Medical applications, in D.L. Wise (ed.), Handbook of Biomaterials and Applications, New York, Marcel Dekker, pp. 2719–2741.
264. Baumann, H., Mueller, U. and Keller, R. (1997) Which glycosaminoglycans are suitable for antithrombogenic or athrombogenic coatings of biomaterials? Part I: Basic concepts of immobilized GAGs on partially cationized cellulose membrane,

Sem. in Thrombosis and Hemostasis **23,** 203-213.

265. Butler, C.E., Navarro, F.A., and Orgill, D.P. (2001) Reduction of abdominal adhesions using composite collagen-GAG implants for ventral hernia repair, *J.Biomed.Mater.Res.* **58,** 75-80.
266. Davis, W.M., (1998) The role of glucosamine and chondroitin sulfate in the management of arthritis, *Drug Topics* 3S-13S.
267. Denuzière, A., Ferrier, D. and Domard, A. (2000) Interactions between chitosan and glycosaminoglycans (chondroitin sulfate and hyaluronic acid): physicochemical and biological studies, *Ann. Pharmaceut. Francaises* **58,** 47-53.
268. Gogly, B., Dridi, M., Hornebeck W., Bonnefoix, M., Godeau, G. and Pellat, B. (1999) Effect of heparin on the production of matrix metalloproteinases and tissue inhibitors of metalloproteinases by human dermal fibroblasts, *Cell. Biol. Int.* **23,** 203-209.
269. Hoekstra, D. (1999) Hyaluronan-modified surfaces for medical devices, *Med. Dev. Diagn. Ind. Mag.* **2,** 48-52.
270. Laurent, T.C. (1970) Structure of hyaluronic acid, in E.A Balazs (ed.), *Chemistry and Molecular Biology of the Intercellular matrix,* Acad. Press, London, pp. 703–732.
271. Oerther, S., Le Gall, H., Payan, E., Lapicque, F., Presle, N., Hubert, P., Dexheimer, J. and Netter, P. (1999) Hyaluronate-alginate gel as a novel biomaterial. *Biotechnol. Bioeng.* **63,** 206-215.
272. Oerther, S., Maurin, A.C., Payan, E., Hubert, P., Lapicque, F., Presle, N., Dexheimer, J., Netter, P. and Lapicque, F. (2000) High interaction alginate-hyaluronate associations by hyaluronate deacetylation for the preparation of efficient biomaterials, *Biopolym.* **54,** 273-281.
273. Paulsson, M., Gouda, I., Larm, O. and Ljungh, A. (1994) Adherence of coagulase-negative staphylococci to heparin and other glycosaminoglycans immobilized on polymer surfaces, *J.Biomed.Mater.Res.* **28,** 311-317.
274. Rastelli, A., Beccaro, M., Biviano, F., Calderini, G. and Pastorello, A. (1990) Hyaluronic acid esters, a new class of semisynthetic biopolymers: chemical and physico-chemical properties, in G. Heimke, U. Soltesz and A.J.C. Lee (eds.), *Clinical Implant Materials-Advances in Biomaterials,* Vol. 9, Elsevier, Amsterdam, pp.199-206.
275. Roche, S., Ronzière, M.C., Herbage, D., and Freyria, A.M. (2001) Native and DPPA cross-linked collagen sponges seeded with fetal bovine epiphyseal chondrocytes used for cartilage tissue engineering, *Biomat.* **22,** 9-18.
276. Tomihata, K. and Ikada, Y. (1997) Preparation of cross-linked hyaluronic acid films of low water content, *Biomater.* **18,** 189-195.
277. Artursson, P., Edman, P., Laakso, T., and Sjöholm, I. (1984) Characterization of polyacryl starch microparticles as carriers for proteins and drugs, *J. Pharma Sci.* **73**, 1507-1513.
278. Brook, M.A., Jiang, J.X., Heritage, P., Underdown, B. and Mc Dermott, M.R. (1997) Silicone-modified strach/protein particles: protecting biopolymers with a hydrophobic coating, *Coll. Surf. B: Interfaces* **9,** 285-295.
279. Brookfield, P., Murphy, P., Harker, R. and MacRae, E. (1997) Starch degradation and starch pattern indices; interpretation and relationship to maturity, *Posharvest*

Biol. Tec. **11,** 23-30.

280. Franssen, O., Stenekes, R. J. and Hennink, W. E. (1999) Controlled release of a model protein from enzymatically degrading dextran microspheres, *J. Control. Rel.* **59,** 219-228.
281. Franssen, O., Vandervennet, L., Roders, P. and Hennink, W. E. (1999) Degradable dextran hydrogels: controlled release of a model protein from cylinders and microspheres, *J. Control. Rel.* **60,** 211-221.
282. Heller, J., Pangburn, S.H., Roskos, K.V. (1990) Development of enzymatically degradable protective coatings for use in triggered drug delivery systems: derivatized starch hydrogels, ***Biomater.*** **11,** 345-350.
283. Huijun, L., Ramsden, L. and Corke, H. (1998) Physical properties and enzymatic digestibility of acetylated and normal maize starch, *Carbohydr. Polym.* **34,** 283-289.
284. Kumada, T., Nakano, S., Sone, Y., Kiriyama, S., Hisanaga, Y., Rikitoku, T., Tamoto, A. and Honda, T. (1999) Clinical effectiveness of degradable starch microspheres in patients with liver cancer, *Gan To Kagaku Ryoho* **26,** 1678-1683.
285. Lawton, J.W. (1996) Effect of starch type on the properties of starch containing films, *Carbohydr. Polym.* **29,** 203-208.
286. Ratto, J.A., Stenhouse, P.J., Auerbach, M., Mitchell, J. and Farrell, R. (1999) Processing, performance and biodegradability of a thermoplastic aliphatic polyester/starch system, *Polymer* **40,** 6777-6788.
287. Rein, H., and Steffens, K.J. (1997) Surface modification of water-insoluble drug particles with starch, *Starch-Starke* **49,** 364-371.
288. Rindlav-Westling, A., Stading, M., Hermansson, A.M. and Gatenholm, P. (1998) Structure, barrier and mechanical properties of amylose and amylopectin films, *Carbohydr. Polym.* **36,** 217-224.
289. Tabata, Y. and Ikada, Y. (1999) Vascularization effect of basic fibroblast growth factor released from gelatin hydrogels with different biodegradabilities, *Biomater.* **20,** 2169-2175.
290. Treib, J., Baron, J.F., Grauer, M.T. and Strauss, R.G. (1999) An international view of hydroxyethyl starches, Intensive Care Med. **25,** 258-268.
291. Wurzburg, O.B. (1986), *Modified Starches: Properties and Uses*, CRC Press, Boca Raton, FL.
292. Dennis, J.M. (1987) Production,properties and uses of alginates, in J.M. Dennis, (ed.), *Production and Utilization of Products from Commercial Seaweeds,* Food and Agriculture Organization of United Nation, US, pp. 58-115.
293. Douxian, S., Yan, Z., Anlie, D., Goosen, M.F.A. and Sun, A.M. (1991) Studies on the degradation of chitosan and preparation of alginate-chitosan microcapsules, *Polym. Biomater.* **3,** 295-300.
294. Edwards-Levy, F. and Levy, M.C. (1999) Serum albumin-alginate coated beads: mechanical properties and stability, *Biomater.* **20,** 2069-2084.
295. Ertesvag, H. and Valla, S. (1998) Biosynthesis and applications of alginates, *Polym Degrad Stabil.* **59,** 85-89.
296. Gaserod, O., Smidsrod, O. and Skjak-Braek., G. (1998) Microcapsules of alginate-chitosan-I.A quantative study of the interaction between alginate and chitosan, *Biomater.* **19,** 1815-1825.

297. Gaserod, O., Sannes, A. and Skjak-Braek, G. (1999) Microcapsules of alginate-chitosan – II. A study of capsule stability and permeability, *Biomater.* **20,** 773-783.
298. Gutowska, A., Jeong, B. and Jasionowski, M. (2001) Injectable gels for tissue engineering, *The Anatomical Record.* **263,** 342-349.
299. Horncastle, J. (1995) Wound dressings. Past, present, and future, *Med. Device Technol.* **6,** 30-36.
300. Kurt, I.D., Skjak- Braek, G. and Olav, S. (1997) Alginate based new materials, *Int. J. Biol. Macromol..* **21,** 47-55
301. Lee, O.S., Ha, B.J., Park, S.N. and Lee, Y.S. (1997) Studies on on the pH-dependent swelling properties and morphologies of chitosan/calcium-alginate complexed beads, *Macromol. Chem. Phys.* **198,** 2971-2976.
302. Leo, W.J., Mc Loughlin, A.J. and Malone, D.M. (1990) Effects of sterilization treatments on some properties of alginate solutions and gels, *Biotechnol Progr.* **6,** 51-53.
303. Martinsen, A. and Smidsrod, O. (1989) Alginate as immobilization material: Correlation between chemical and physical properties of alginate gel beads, *Biotechnol Bioeng.* **33,** 79-89.
304. Murata, Y., Maeda, T., Miyamoto, E. and Kawashima, S. (1993) Preparation of chitosan-reinforced alginate gel beads – effects of chitosan on gel matrix erosion, *Int.J Pharm.* **96,** 139-145.
305. Ribeiro, A.J., Neufeld, R.J., Arnaud, P. and Chaumeil, J.C. (1999) Microencapsulation of lipophilic drugs in chitosan-coated alginate microspheres, *Int. J. Pharm.* **187,** 115-123.
306. Smidsrod, O. and Skjak-Braek G. (1990) Alginate as immobilization matrix for cells, *Trends Biotechnol.* **8,** 71-78.
307. Strand, B.L., Morch, V.A. and Skjak-Braek, G. (2000) Alginate as immobilization matrix for cells, *Minevra Biotechnol.* **12,** 223-233.
308. Thu, B., Bruheim, P., Espevik, T., Smidsrod, O. and Skjak-Braek, G. (1996) Alginate polycation microcapsules - I. Interaction between alginate and poycation, *Biomater.***17,** 1031-1040.
309. Thu, B., Bruheim, P., Espevik, T., Smidsrod, O. and Skjak-Braek, G. (1996) Alginate polycation microcapsules – II. Some functional properties, *Biomater.***17,** 1069-1079.
310. Vacanti, C., Langer, R., Schloo, B. and Vacanti, J.P. (1991) Synthetic polymers seeded with chondrocytes provide a template for new cartilage formation, *Plast Recon Surg.* **88,** 753-758.
311. Atkins, T.W. and Peacock, S.J. (1996) In vitro biodegradation of poly(beta-hydroxybutyrate-hydroxyvalerate) microspheres exposed to Hanks' buffer, newborn calf serum, pancreatin and synthetic gastric juice, *J. Biomater. Sci. Polym.* **7,** 1075-1084.
312. Boeree, N.R., Dove, J., Cooper, J.J., Knowles, J. and Hastings, G.W. (1993) Development of a degradable composite for orthopaedic use: mechanical evaluation of a hydroxyapatite-polyhydroxybutyrate composite material, *Biomater.* **14,** 793-796.
313. Doi, Y. (1990) *Microbial Polyesters*, Carl Hanser Verlag, New York.
314. Gangrade, N. and Price, J.C. (1991) Poly(hydroxybutyrate-hydroxyvalerate)

microspheres containing progesterone: preparation, morphology and release properties, *J. Microencapsul.* **8,** 185-202.

315. Kassab, A.C., Xu, K., Denkbas, E.B., Dou, Y., Zhao, S. and Piskin, E. (1997) Rifampicin carrying polyhydroxybutyrate microspheres as a potential chemoembolization agent, *J. Biomater. Sci., Polym. Ed.* **8,** 947-961.
316. Kemnitzer, J.E., Gross, R.A. and McCarthy, S.P. (1992) Stereochemical and morphoplogical effects on th degradation kinetics of polyhydroxybutyrate: a model study, Proc. ACS Div., *Polym. Mat. Sci. Eng.* **66,** 405-407.
317. Kostopoulos, L. and Karring, T. (1994) Guided bone regeneration in mandibular defects in rats using a bioresorbable polymer, Clin. Oral Implants Res. **5,** 66-74.
318. Ljungberg, C., Johansson-Ruden, G., Bostrom, K.J., Novikov, L. and Wiberg, M. (1999) Neuronal survival using a resorbable synthetic conduit as an alternative to primary nerve repair, *Microsurg.* **19,** 259-264.
319. Ottenbrite,R.M., Huang, S.J. and Park, K. (1996) *Hydrogel and Biodegradable Polymers for Bioapplications,* ACS Symp. Ser. 627, Am. Chem. Soc. Washington DC, Maryland, pp. 69-92.
320. Rouxhet, L., Duhoux, F., Borecky, O., Legras, R. and Schneider, Y.J. (1998) Adsorption of albumin, collagen, and fibronectin on the surface of poly(hydroxybutyrate-hydroxyvalerate) (PHB/HV) and of poly (epsilon-caprolactone) (PCL) films modified by an alkaline hydrolysis and of poly(ethylene terephtalate) (PET) track-etched membranes, *J. Biomater. Sci., Polym. Ed.* **9,** 1279-1304.
321. Sendil, D., Gursel, I., Wise, D.L. and Hasirci, V., 1999, Antibiotic release from biodegradable PHBV microparticles. *J. Control. Rel.* **59,** 207-217.
322. Timminis, M.R., Gilmore, D.F., Fuller, R.C. and Lenz, R.W. (1993) Bacterial polyesters and their biodegradation, in C. Ching, D. Kaplan and E. Thomas (eds.), *Biodegradable Polymers and Packaging.* Technomic Publ. Co., Lancaster, PA, pp. 119-130.
323. Hagenmaier, R.D. and Shaw, P.E. (1990) Moisture permeability of edible films made with fatty acid and hydroxypropyl methylcellulose, *J. Agric. Food Chem.* **38,** 1799-1803.
324. Hoenich, N.A. and Stamp, S. (2000) Clinical investigation of the role of membrane structure on blood contact and solute transport characteristics of a cellulose membrane, *Biomater.* **21,** 317-324.
325. Kester, J.J. and Fennema, O. (1986) Edible films and coatings: A review, *J. Food Sci.* **40,** 47-59.
326. Krogel, I. and Bodmeier, R. (1999) Development of a multifunctional matrix drug delivery system surrounded by an impermeable cylinder, *J. Control. Rel.* **61,** 43-50.
327. Nose, Y. (1990) Artificial kidney, is it really not necessary? *Artif. Organs,* **14,** 245-251.
328. Shi, Y., Ploof, J. and Correia, A. (1999) Increasing antibody production with hollow-fiber bioreactors, *IVD Technol. Mag.* **5,** 37-40.
329. Grasset, L., Cordier, D. and Ville A. (1977) Woven silk as a carrier for the immobilization of enzymes, *Biotechnol. Bioeng.* **19,** 611-618.
330. Heslot, H. (1998) Artificial fibrous proteins: A review, *Biochim.* **80,** 19-31.

331. Kaplan, D., Adams, W., Farmer, B., and Viney, C. (1993) *Silk Polymers: Materials Science and Biotechnology,* ACS Symp. Ser.
332. Liu, Y., Chen, X., Qian, J., Liu, H., Shao, Z., Deng, J. and Yu, T. (1997) Immobilization of glucose oxidase with the blend of regenerated silk fibroin and poly(vinyl alcohol) and its application to a 1,1'-dimethylferrocene-mediating glucose sensor, *Appl. Biochem. Biotechnol.* **62,** 105-117.
333. Minoura, N., Tsukada, M. and Nagura, M. (1990) Physico-chemical properties of silk fibroin membrane as a biomaterial, *Biomater.* **11,** 430-434.
334. Santin, M., Motta, A., Freddi, G. and Cannas, M. (1999) In vitro evaluation of the inflammatory potential of the silk fibroin, *J. Biomed. Mater. Res.* **46,** 382-389.
335. Seves, A., Romano, M., Maifreni, T., Sora, S. and Ciferri, O. (1998) The microbial degradation of silk: a laboratory investigation, Internat. *Biodeterior. Biodegrad.* **42,** 203-211.
336. Vollrath, F. (2000) Sternght and structure of spider's silk, *Rev. Mol. Biotechnol.* **74,** 67-83.
337. Zhang Y.O., Zhu, J. and Gu R.A. (1998) Improved biosensor for glucose based on glucose oxidase-immobilized silk fibroin membrane, *Appl. Biochem. Biotechnol.* **75,** 215-233.

11

ENVIRONMENTALLY BIODEGRADABLE WATER-SOLUBLE POLYMERS

GRAHAM SWIFT
GS Polymer Consultants
10378 Eastchurch, Chapel Hill,
North Carolina 27517.

1 Introduction

Environmentally biodegradable water-soluble polymers represent a very important goal in the drive towards responsible waste management and environmental protection in a major segment of the polymer industry. The water-soluble polymer segment is often forgotten or neglected when environmental issues associated with the disposal of polymers and plastics in general are evaluated. Water-solubility is sometimes magically considered to confer environmental acceptance and biodegradabilityon such polymers: this is categorically untrue as indicated in this chapter. The scope of this chapter is broad to include and address the key issues of definitions and testing protocols, that need to be addressed before a polymer may be labelled as biodegradable. Currently acknowledged biodegradable water-soluble polymers and synthetic approaches to new polymers are also covered. Structurally similar water-soluble biodegradable polymers designed for biomedical and controlled drug delivery applications rather than environmental degradation are not included in this chapter. However, necessary comparisons and correlations with the extensive work and literature on biodegradable plastics are included where essential for the understanding of biodegradable water-soluble polymers and their development.

Interest in environmentally biodegradable polymers began over forty years ago with the recognition that common commodity packaging plastics including polyethylene (PE), polypropylene (PP), polystyrene (PS), poly(vinyl chloride) (PVC), and poly(ethylene terephthalate)(PET) were accumulating in the environment. Their carefully designed and well established resistance to environmental degradation was observed to be

G. Scott (ed.), Degradable Polymers, 2nd Edition, 379-412.

contributing to landfill depletion and litter problems due to careless disposal after use. At that time, the major focus was on replacing these synthetic packaging plastics with environmentally biodegradable substitutes. Subsequently, the more reasonable goal of selective replacement with biodegradables has been the focus with recycle and incineration for example, as other viable options.

In the last 25 years, it has become increasingly apparent that, in addition to the major commodity synthetic plastics, water-soluble commodity and specialty polymers and plastics, such as poly(acrylic acid)s, polyacrylamide, poly(vinyl alcohol) and poly(alkylene oxide)s and even some modified natural polymers, for example cellulosics and starch, may potentially contribute to environmental problems and should also be targets for biodegradable substitutes. Water-soluble polymers are widely used in a myriad of valuable commercial applications such as coatings additives, temporary packaging, temporary coatings, pigment dispersants, mining, water treatment, detergents, adhesives, etc. The fact that they are water-soluble and pass unseen into the environment is not an indicator that they are environmentally acceptable or biodegradable, which for so long seemed to be the accepted reasoning. Just the opposite pertains, water-soluble polymers are capable of moving freely and rapidly through the aqueous environment with the potential for many deleterious environmental effects on plant and animal life. Demonstrated biodegradation or proof of no significant (i.e. acceptable) riskduring disposal are essential for these polymers to be considered environmentally benign. The surest way of confirming environmental acceptability is to unequivocally establish biodegradability on disposal within a specified time in any aqueous compartment that the polymers are likely to contact and preferably in the primary disposal site.

There are many well documented similarities in the chemistries, biodegradation requirements, and synthetic routes evaluated for the development of biodegradable plastics and water-soluble polymers [1-35]. All are recommended reading even though this chapter focuses only on synthesis approaches and characteristics of biodegradable water-soluble polymers. The design and synthesis of biodegradable water-soluble polymers have been widely based on natural renewable resources, non-renewable petrochemical resources, and combinations of the two. They are, primarily, with few exceptions such as temporary coatings and water-soluble packaging polymers and plastics, designed to be biodegradable on disposal in aqueous environments after use, such as municipal wastewater treatment plants, on-site treatment facilities, lakes, rivers, etc. Ideally, as most of these polymers are discarded in sewage streams, they should biodegrade completely in industrial or municipal wastewater treatment plants before entering the natural environment at large. This would ensure that they do not enter rivers and streams where they can readily move around the environment with unknown fate and effects, unless biodegradation can be established any or all of these subsequent compartments. Composting is a disposal option for water-soluble packaging plastics, however care must be excercised to ensure compliance with biodegradation or environmental acceptability, as with biodegradable plastics in general.

Water-soluble biodegradable polymers may degrade in the environment by any of the accepted degradation pathways well established for biodegradable plastics. These include photodegradation, biodegradation, and chemical degradation, which includes hydrolytic and oxidative degradation. It should be recognized that, of these environmental degradation pathways, only biodegradation (biotic degradation) can lead to complete removal from the environment and accordingly it has justifiably received the highest attention. The other degradation pathways are appropriately described as abiotic environmental degradation, deterioration or disintegration to produce low molecular weight fragments and / or other chemical compounds that may or may not biodegrade and may or may not be harmful to the environment. Clearly, biodegradation and abiotic environmental degradation of polymers are inter-related and biodegradation is often dependent on prior abiotic environmental degradation.

There are well known alternative disposal technologies competing with biodegradation for the waste-management of plastics such as recycling. Recycling of plastics has many facets including; recycling of the plastic as is or in blends with other plastics, recycling of plastics to monomers and subsequent repolymerization to the same or new polymers; recycling to olefinic feedstocks by pyrolysis; recycling by biodegradation to renewable resources; and recycling to the energy stream for incineration. In contrast, disposal options for water-soluble polymers is very limited, with the exception of incineration, none of these re-use alternatives is readily applicable due to the difficulty of recovering them from a dilute aqueous disposal stream after use. Incineration of water-soluble polymers that are adsorbed on wastewater treatment plant sludge is practiced in some locations.

Acceptable definitions for environmentally degradable and biodegradable polymers and meaningful laboratory testing protocols for quantitatively measuring biodegradation and environmental fate and effects are essential for progress in the design and development of environmentally biodegradable polymers. It is critically important to establish that laboratory biodegradation test protocols correlate closely with real world exposures on disposal in order to have confidence in their predictability for new polymer developments. Consequently, definitions and test methods will be addressed early in the chapter, followed by a discussion pertaining to recognized biodegradable water-soluble polymers and many of the important approaches evaluated for the synthesis of biodegradable water-soluble polymers. The synthesis approaches fall into the two broad categories indicated below:

- Synthetic polymers
  - Carbon-carbon chain polymers
  - Heteroatom chain polymers
- Modified natural polymers
  - Graft polymers
  - Chemically modified polymers

The chapter concludes with some projections for future directions for research and commercial opportunities for environmentally biodegradable water-soluble polymers.

2. Definitions

There have been numerous definitions proposed for environmentally degradable and biodegradable water-soluble polymers and plastics, almost every article written on the subject includes the author's favorite definitions. All tend to encompass the same broad concepts but are slightly different in phraseology depending on the author's perspective and discipline, chemist, biochemist, layman, lawyer, legislator, etc. This is both fortunate and unfortunate since, on the one hand, it ensures that there is broad understanding of the problem but, on the other hand, it means that it is exceedingly difficult to reduce to words an acceptable definition which has a world-wide consensus. Indeed, consensus may be more easily reached by establishing specifications for (bio)degradable polymers based on the testing protocols for various degradation pathways to be discussed later. However, definitions are important because they are indicative of the mechanisms involved in environmentally degradable and biodegradable polymers as well as the types of testing protocols that are needed to establish the acceptability of polymers in a particular disposal environment. At this time, the definitions developed by the American Society for Testing and Materials (ASTM) [1] for degradable, biodegradable, photodegradable, hydrolytically degradable, and oxidatively degradable plastics, indicated below, are probably the most widely accepted as written or in some slightly modified form. They are equally applicable to water-soluble polymers and are used as such in this chapter.

Degradable plastic, a plastic designed to undergo a significant change in its chemical structure under specific environmental conditions resulting in a loss of some properties as measured by standard test methods appropriate to the polymer.

Biodegradable plastic, a degradable plastic in which the degradation results from the action of naturally-occurring micro-organisms such as bacteria, fungi, and algae.

Hydrolytically degradable plastic, a degradable plastic in which the degradation results from hydrolysis promoted either by chemical or enzymatic processes.

Oxidatively degradable plastic, a degradable plastic in which the degradation results from oxidation.

Photodegradable plastic, a degradable plastic in which the degradation results from the action of natural sunlight (oxygen is also required).

Note, these definitions do not quantify the extent of degradation by any of the pathways; they only indicate of the mechanism that is operating to promote degradation. While this is acceptable in a scientific sense to define a process or mechanism, the definitions do not meet or satisfy the requirement for environmentally acceptable degradable and biodegradable polymers, which, in the minds of legislators and lay people, is the key issue. To satisfy this very practical requirement, specifications for acceptability have to be set and monitored by standard testing protocols, based on the above definitions, which will be discussed later.

Environmental degradation processes are interrelated as shown schematically in Fig. 1. All four environmental degradation pathways for polymers, biodegradation, oxidation, hydrolysis and photodegradation initially give intermediate fragments. These initial fragments may degrade further by biotic or abiotic processes to some other residue or biodegrade completely and be removed from the environment entirely, and ultimately mineralized as indicated. Alternatively, the fragments may remain unchanged in the environment.

It should be noted that mineralization is correctly defined as a slow process that refers to the complete bioconversion of an organic compound to minerals such as carbon dioxide and / or methane (depending on aerobic or anaerobic environment), water, and mineral salts. Here, in accord with many generalizations in the polymer biodegradation arena, mineralization is used loosely to indicate complete or total removal from the environment to carbon dioxide or methane, water, and biomass. In other words, complete conversion to inorganics (mineralization) is not necessarily observed or needed for the acceptance of the biodegradation of a polymer.

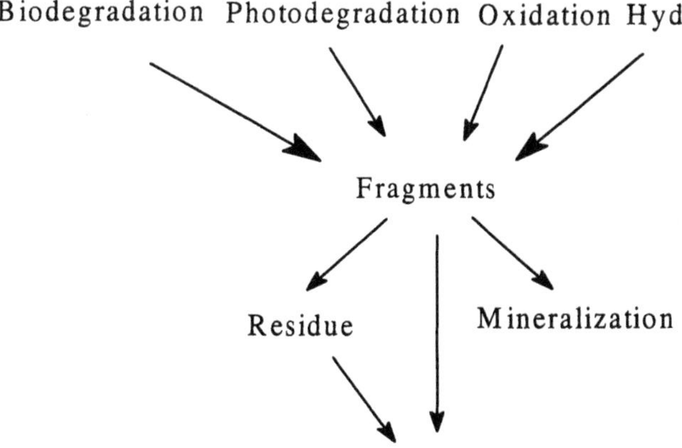

Fig. 1. Interrelationships of Environmental Degradation Pathways for Polymers

Clearly, only biodegradation has the potential to remove polymers completely from the environment and this should be recognized when developing and designing polymers for degradation by any of the other pathways. The final degradation stage should preferably be complete biodegradation and removal from the environment with ultimate mineralization. In this way, the polymers are recycled through nature into microbial cells, plants and higher animals [2], and finally back into renewable resource chemical feedstocks. Where degradation residues do remain, however, it is essential that these must be established as an acceptable low risk in the environment by rigorous fate and effect evaluations.

Based on the above discussion, an acceptable environmentally degradable polymer may be defined as one, which degrades by any or a combination of several, defined mechanisms such as, biodegradation, photodegradation, oxidation, and hydrolysis, to leave no harmful residues in the environment. This definition has the advantage of being independent of the rate or degree of degradation for a particular polymer but requiring sufficient testing of fragments and degradation products that are incompletely removed from the environment to ensure no long-term damage or adverse affects to the ecological system. Polymers meeting this definition should be acceptable for disposal in the appropriate designated environment anywhere in the world. Nevertheless, it must be emphasized that the most significant goal for plastics and water-soluble polymers in the environment should ultimately be biodegradation as this categorically precludes any unwanted adverse environmental impacts and the reqirement for prolonged fate and effects evaluations.

3. Disposal Options for Biodegradable Water-Soluble Polymers

Waste-management requirements and disposal options are the drivers for research to develop environmentally biodegradable polymers and to establish laboratory testing protocols. It is commercial opportunities that indicate polymer types, structures, properties and uses that dictate waste management and disposal requirements. Disposal methods and locations identify the testing protocols that must be established for polymers in order to evaluate their environmental degradation under laboratory simulated environmental exposure conditions. Major options for the disposal and waste-management of water-soluble polymers are indicated in Fig.2.

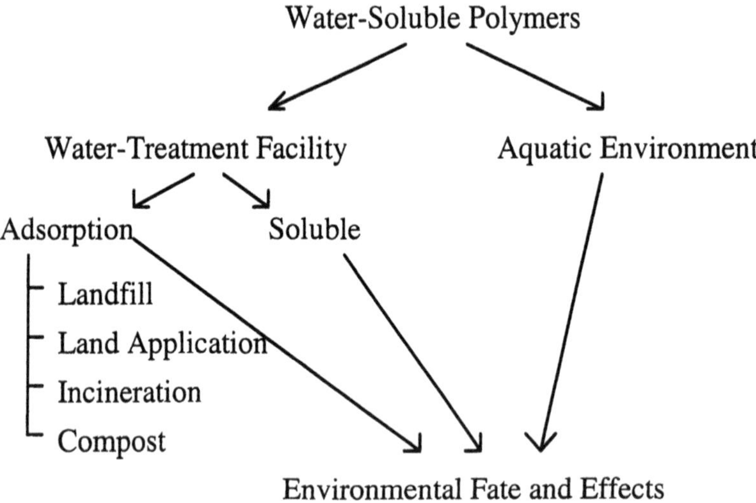

Fig. 2. Major Disposal Options for Water-Soluble Polymers

After use, water-soluble polymers are usually very dilute aqueous solutions and are preferably disposed of through wastewater treatment facilities though sometimes they are directly dumped into the aquatic environment, as shown above. On entering a wastewater treatment plant, a water-soluble polymer may remain soluble and pass straight through, into streams, rivers, lakes and other aquatic environments, or it may be adsorbed onto the suspended solids in the treatment plant. If it passes straight through, it is no different from direct disposal into those aqueous environments, and both raise similar questions as to the environmental fate and effects of the polymers involved. On the other hand, adsorption of a polymer onto sewage sludge results in other disposal options such as land filling, incineration, composting, and land application as fertilizer or for soil amendment. The chosen option will likely depend on local customs and needs. However, whichever disposal option is chosen, environmental fate and effects need to be addressed. It must be established how and where these polymers move in their new environmental compartments, and what is evolved from incineration and where byproducts are ultimately deposited.

Clearly, for water-soluble polymers there is a preference for complete biodegradation in wastewater treatment facilities, usually the first environmental exposure. Biodegradable water-soluble polymers, where possible, should be designed to be completely biodegradable and removed in the immediate disposal compartment, generally the sewage treatment facility. If biodegradation is not complete in this disposal environment, it must be assessed, often with increasing difficulty, in subsequent compartments. Once the biodegradation of a polymer has been confirmed in any environmental compartment, no further uncertainty remains as to its fate and effects. The advantage of complete biodegradation is that it is a positive result that can be categorically established; whereas, with less or slowly biodegradable polymers, the assessment of environmental fate and effects are risk assessments and are negative results, based on limited studies with only a few aquatic species. Such fate and effects results may not be representative of all the species that a water soluble polymer, or its partial degradation fragments, may contact as it moves through many aqueous environment compartments.

A major difference between the wastemanagement of biodegradable water-soluble polymers and biodegrdable plastics is that there is a well established widely available disposal infrastructure of wastewater treatment plants for the former; whereas plastics, being developed largely for composting, will require the large scale implementation of a composting infrastructure and this will slow their acceptance. Hence, as biodegradable water-soluble polymers are developed to protect the aqueous environment they may be immediately implemented as disposal infrastructures are in place in almost every country in the world.

4. Test Methods for Biodegradable Water-Soluble Polymers

Even though environmentally biodegradable polymers may degrade by a combination of several inter-related environmental degradation mechanisms, as discussed earlier, standard test method development for water-soluble polymers has focused almost entirely on measuring degree of biodegradation. A few reported studies indicate that some water-soluble polymers degrade by hydrolysis, ozonolysis and free radical oxidation to low molecular weight fragments, which may or may not subsequently biodegrade. However, there are no standard test methods to measure intermediate degradation steps and correlate them with ultimate biodegradation.

The goals of standard test method development are to define and develop laboratory simulations of real world conditions to which a particular polymer will be exposed on disposal or use. Results are expected to indicate the rate and extent of biodegradation under test conditions, which should correlate with real world environmental exposure. Standard laboratory test methods should, therefore, be able to predict the response of polymers to the environment. This prediction is not simple or straightforward since natural environments differ widely in microbial composition, pH, temperature, moisture, etc and are not easily reproduced in the laboratory. Once an environment has been sampled and placed within the confinements of a laboratory vessel it can no longer dynamically interact with the greater environment in response to an added xenobiotic, such as a polymer, and results may not always be representative of real world exposures. Consequently, care should be exercised in the interpretation of biodegradation results obtained in standard laboratory test methods. Biodegradation observed in a given environment is generally consistent with biodegradation in a natural environment, but lack of biodegradation in the laboratory is not conclusive of recalcitrance in a natural environment. This should be clearly understood by all in the field of designing and developing test methods for biodegradable polymers. Repeated laboratory testing for biodegradation in several environmental samples is a recommended practice before deciding a particular polymer is not biodegradable.

Biodegradation may occur aerobically, in the presence of oxygen, or anaerobically, in the absence of oxygen. Test methodology development for the former has received the most attention, as oxygen is usually present in disposal environments for water-soluble polymers. However, anaerobic conditions do occur, for example, in anaerobic digesters in sewage treatment plants. The basic chemistry of the two different biodegradation pathways are expressed by the equations below:

Aerobic Environments:

$$\text{Polymer} + O_2 \rightarrow CO_2 + H_2O + \text{Biomass} + \text{Residue}$$

AnaerobicEnvironments:

$$\text{Polymer} \rightarrow CO_2 + CH_4 + \text{Biomass} + \text{Residue}$$

Analytical techniques are needed for all the reactants, polymers and oxygen, and products in order to quantitatively assess the degree of biodegradation in either environment. Usually, test methods measure oxygen uptake, biochemical oxygen demand (BOD), or CO_2 production for aerobic degradation and CO_2 and CH_4 production for anaerobic degradation. Analysis for residue should also be done where possible as this is also a good indicator of biodegradation and sometimes the mechanism operating. It is usual to run blank tests, with killed inocula, in concert with the authentic tests with a living environment to differentiate abiotic and biotic degradation. Established biodegradable standard polymers or organic compounds are simultaneously tested to confirm the viability of the test environment.

Water-soluble polymers are initially tested for biodegradation in simple tests, but results may sometimes be difficult to interpret because of variability due to issues mentioned earlier. Hence, several tests of increasing complexity are often coupled to determine whether a polymer is biodegradable. Simple tests such as biochemical oxygen demand (BOD) as a fraction of theoretical oxygen demand (ThOD) or chemical oxygen demand (COD) in aerobic environments may be misleading as the result may be due to competing abiotic oxidation or impurities in the polymer sample. Advanced confirmatory tests such as the measurement of CO_2 (CO_2 and CH_4 for anaerobic environments) and removal of soluble residues or total organic content (TOC) in laboratory simulated sewage plants are more definitive. However, in some tests where solids are present from sewage, it must be ascertained whether removal of TOC is by biodegradation or adsorption on solids. Ultimately, the total fate of a polymer needs to be established to claim biodegradability or environmental degradability or acceptability. To do this categorically, isotopic labeling of the polymer with, for example, radiocarbon or tritium, is often used and strongly recommended for complete and thorough confirmative testing. A very comprehensive and authoritative account of biodegradation testing for water-soluble polymers and organic chemicals is given by Swisher [7]. Additionally, work by Larson and Swift and their coworkers, especially for water-soluble polycarboxylates [36, 39], and others [37, 38, 40], are recommended reading as they are aimed at improving test methodologies.

In addition to the above test methods, microbial activity in polymer solutions has been used to assess biodegradation by measuring population growth using optical density as an indicator of activity. This approach is amenable to the quantification and identification of metabolites, which is useful to establish biodegradation mechanisms. Analytical tools such as mass spectroscopy and gas chromatography are widely used for metabolite analysis. Gel permeation chromatography is used to monitor changes in polymer molecular weight and molecular weight distribution as an indicator of biodegradation. Careful attention to molecular weight and distribution changes may be used to establish endo-biodegradation or exo-biodegradation mechanisms.

As a precautionary note, when working with biodegradable water-soluble polymers, as with any biodegradable polymers, it is important to recognize that low molecular weight fractions and impurities are usually present. These may be more biodegradable than higher molecular weight fractions and hence give rise to false positive results, particularly in short term tests such as 5 day BOD tests sometimes run as rapid screens as indicators of biodegradation potential. The reader is also warned that often results claimed by the original authors are not always substantiated due to difficulty in estimating test reliability. However, references to these studies are included as the polymers are considered to be biodegradable to some extent and hence are a good indicator for future research directions and opportunities.

5. Biodegradable Water-Soluble Polymers

Water-soluble polymers may be anionic, cationic, or nonionic in character, which influences both their applications and their biodegradability. They may be totally synthetic, totally natural, or synthetically modified natural polymers.

The synthetics may be sub-divided into addition polymers and condensation polymers, the former being mostly carbon-chain backbone polymers and the latter being chiefly heteroatom chain backbone polymers. Poly(acrylic acid) and poly(vinyl alcohol) are representative of carbon-chain polymers, which may be considered functional polyolefins. Polyesters, polyamides, and polyethers are examples of condensation polymers, which are structurally similar to many naturally occurring polymers and hence good structural models for the development of synthetic water-soluble biodegradable polymers. The predominant commercially successful synthetic water-soluble polymers are largely carbon-chain polymers and these are usually resistant to fast easily measurable biodegradation as are their counter parts in nature such as lignins, rubbers, etc. In fact, they are a good example of the future requirements for test method development in which the needs of slowly biodegradable polymers are addressed (see Chapter 14). Nevertheless, cost drives commercial success and much effort has been expended, as we shall see, on trying to develop carbon-chain biodegradable water-soluble polymers, especially those with carboxylic acid functionality that meet current expectations as defined by test methodology that depends on fast biodegradation..

All natural polymers are considered to be biodegradable or at least not harmful to the environment. However, modification of natural polymers either by grafting synthetic polymers or by chemical conversions such as oxidation and esterification, changes their structure, properties and biodegradation characteristics significantly. Therefore, polymers produced by any of these modifications must be evaluated for biodegradability in the same manner as purely synthetic polymers by the tests indicated.

5.1 CARBON-CHAIN POLYMERS

Functional water-soluble carbon-chain polymers show limited propensity to biodegrade at molecular weight (Mn) useful for commercial properties which is usually > 1000. Suzuki [41] demonstrated this limitation very elegantly by ozonizing higher molecular weight polymers such as poly(acrylic acid), polyvinylpyrrolidone, and polyacrylamide to oligomers of molecular weight < 1000 which were generally demonstrably much more biodegradable, though polyacrylamide was slower to biodegrade than the other two polymers. He, similarly, showed that poly(ethylene oxide) and poly(vinyl alcohol), both of which are established as biodegradable over low to high molecular weight ranges, biodegraded more rapidly at lower molecular weight.

The biodegradability of functional derivatives of polyethylene, particularly poly(vinyl alcohol) and poly(acrylic acid) and derivatives have received attention. Both have high water-solubility, high commercial value and high volume use, and frequent disposal into the aqueous environment through wastewater treatment plants. Poly(vinyl alcohol) is used in a wide variety of applications, including textiles, paper, plastic films, temporary packaging, etc.; and poly(acrylic acid) is widely used in detergents as a builder, super absorbent for diapers and feminine hygiene products, water treatment, thickeners, pigment dispersant, etc.

5.1.1 Poly(vinyl alcohol)

Poly(vinyl alcohol), commercially available by the hydrolysis of poly(vinyl acetate), is probably the only carbon chain polymer widely accepted to be fully biodegradable in the environment, particularly in waste-water treatment facilities, and confirmed in current standard tests. But, even in these laboratory tests, acclimation is usually essential to effect rapid biodegradation, indicating a further complexity in the laboratory testing procedures mentioned earlier. The biodegradation mechanism is considered to be a random chain cleavage of 1,3 diketones formed by an enzyme-catalyzed oxidation of the polymeric secondary alcohol functional groups. Biodegradation was first observed by Yamamoto [42] as a reduction in aqueous viscosity of the polymer in the presence of soil bacteria. Subsequently, T. Suzuki [43] identified a *Pseudomonas* species as the soil bacteria responsible for the degradation of poly(vinyl alcohol) over a degree of polymerization range of 500 to 2000 utilizing the polymer as a sole carbon source. An aqueous polymer solution at a concentration of 2700 ppm was reduced to 250-300 ppm concentration in 7-10 days at pH 7.5-8.5 and 35-45°C.

An oxidative endo mechanism was proposed based on the data shown in Table 1 and later substantiated [44] by quantifying the oxygen uptake as one mole for every mole of hydrogen peroxide produced and identifying the degradation products as ketones and carboxylic acids, as shown in Scheme 1, below.

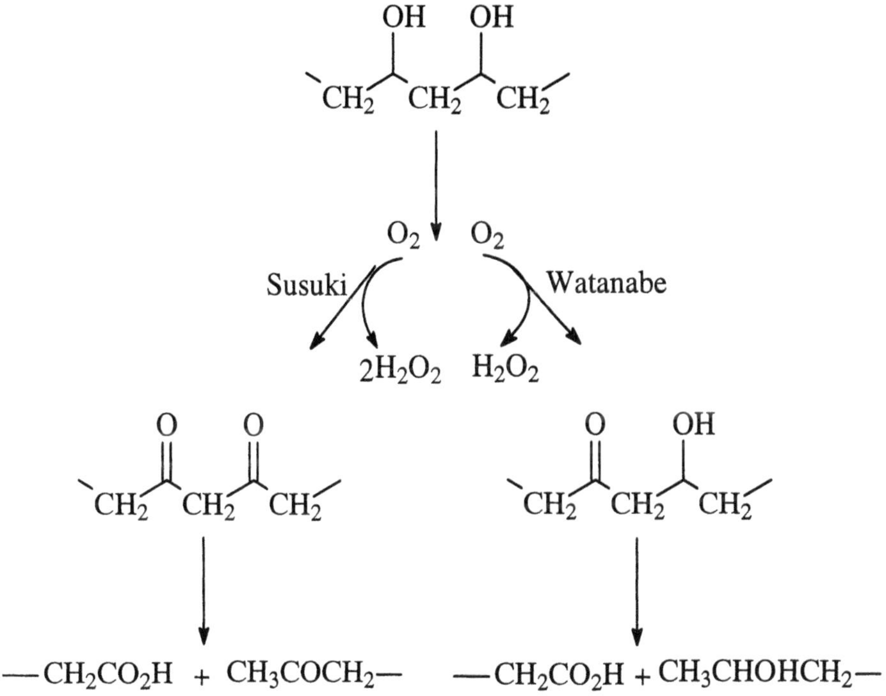

OH OH

CH_2 CH_2 CH_2

O_2 O_2

Susuki Watanabe

$2H_2O_2$ H_2O_2

O O O OH

CH_2 CH_2 CH_2 CH_2 CH_2 CH_2

$—CH_2CO_2H + CH_3COCH_2—$ $—CH_2CO_2H + CH_3CHOHCH_2—$

Scheme 1: Biodegradation pathways for poly(vinyl alcohol)

Included in the diagram is the alternative mechanism proposed by Watanabe [45, 46] in which the products were identified as alcohols and carboxylic acids. This was subsequently proved in error and the Suzuki mechanism is now widely accepted for the degrdation of poly(vinyl alcohol) by the microbes evaluated in his work.

Other bacterial strains identified as biodegrading poly(vinyl alcohol) include *Flavobacterium* [50]and *Acinetobacter* [51] , and many others as well as fungi, molds, and yeasts [52]. Industrial evaluations at DuPont [53] and Air Products [54] indicate that over 90% of poly(vinyl alcohol) entering wastewater treatment plants is removed in those locations and hence environmental pollution is unlikely.

Table 1. Consumption of oxygen and stoichiometry of metabolites in the biodegradation of poly(vinyl alcohol)

Reaction time (hr)	Oxygen consumed (μM)	Hydrogen peroxide (μM)	Ketones (μM)	Carboxylic acids (μM)
1.0	2.92	2.94	ND	ND
2.0	5.40	5.15	ND	ND
6.0	ND	4.0	1.8	1.8
24.0	ND	7.2	3.5	3.1

The rapid biodegradation of the polyketone obtained by the chemical oxidation of poly(vinyl alcohol) [47, 48, 49] shown in Scheme 2 supports the proposed mechanism of Suzuki. The byproduct is hydrogen peroxide.

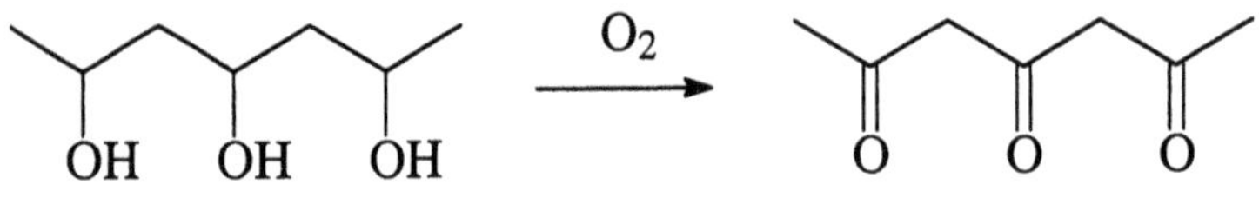

OH OH OH $\xrightarrow{O_2}$ O O O

Scheme 2. Oxidation of poly(vinyl alcohol)

Poly(vinyl acetate), the precursor for poly(vinyl alcohol), hydrolyzed to less than 70%, is claimed to be non-biodegradable under conditions similar to those that biodegrade the fully hydrolyzed polymer [55] There are several packaging applications of partially hydrolyzed poly(vinyl acetate) whereby the rate of biodegradation is retarded deliberately to give acceptable use life for the package. Hot water disposable films and fabrics [56], body wrappings for mortuary use [57], moldings [58], and microfibrils for synthetic paper pulp [59] are reported examples.

Blends of poly(vinyl alcohol) with polycaprolactone [60] and ethylene-maleic anhydride copolymers [61] are used in biodegradable packaging, though the latter blend is unlikely to be fully biodegradable.

Other interesting articles and reviews desribe poly(vinyl alcohol) composites, generally [62], with proteins [63, 64, 65], starch [66], chitosan [67], simple sugars [68], and a Baeyer-Villiger oxidation [95], Scheme 3, of a an ethylene / vinyl alcohol copolymer to produce a biodegradable polyhydroxypolyester.

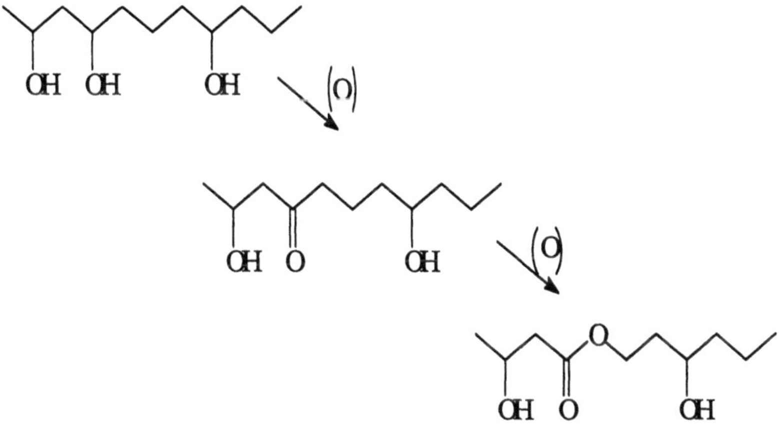

Scheme 3. Baeyer-Villiger oxidation of ethylene / vinyl alcohol copolymer

5.1.2 Polycarboxylates

Carboxylate derivatives of poly(vinyl alcohol) are biodegradable and functional in detergents as co-builders, although too costly to be practical replacements for poly(acrylic acid), at this time. Matsumura polymerized vinyloxyacetic acid [69, 70] and Lever has patented polymers based on vinyl carbamates obtained from the reaction of vinyl chloroformates and amino acids such as aspartic and glutamic acids [71]. Both hydrolyze, Scheme 4, to poly(vinyl alcohol) which is biodegradable.

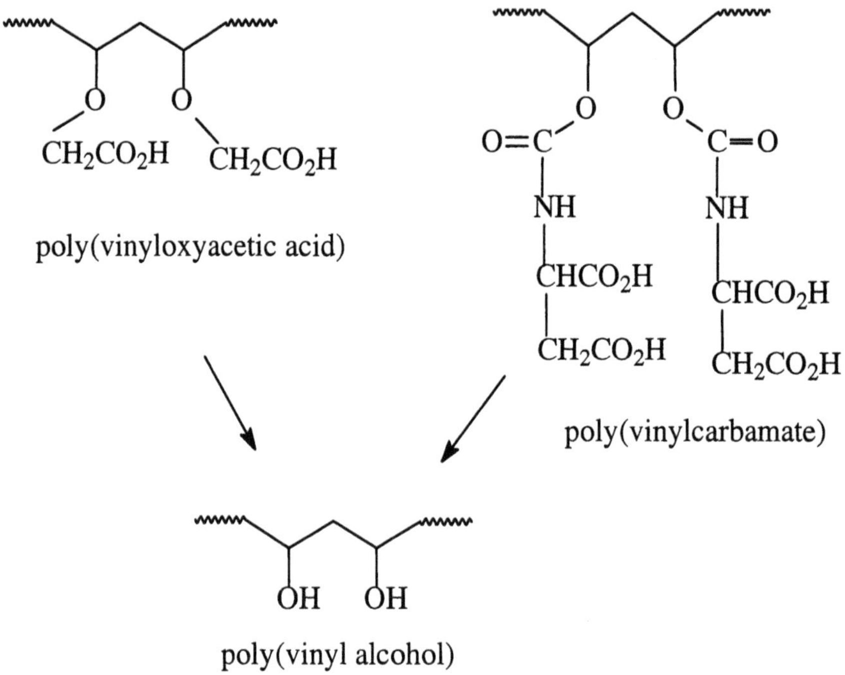

O O
CH_2CO_2H CH_2CO_2H

poly(vinyloxyacetic acid)

O O
O=C C=O
NH NH
$CHCO_2H$ $CHCO_2H$
CH_2CO_2H CH_2CO_2H

poly(vinylcarbamate)

OH OH

poly(vinyl alcohol)

Scheme 4. Hydrolysis of carboxylated poly(vinyl alcohol) derivatives

Copolymers of vinyl alcohol with acrylic and / or maleic acid have been evaluated in detergents as potentially biodegradable co-builders in a number of laboratories [72, 73, 74], but the results were not encouraging for balanced biodegradation and performance. Higher than 80 mole % of vinyl alcohol is required for biodegradation, and less than 20 mole % for acceptable performance.

The use of poly(carboxylic acids) in detergents, particularly poly(acrylic acid)s, and the search for biodegradable alternatives is well established and has been well reviewed in many articles mentioned earlier [12, 18, 19, 20, 21] and by Hunter [75]. The current polymers have little biodegradability at preferred performance molecular weights, ca. 5000 Dalton for poly(acrylic acid) and 70000 Daltons for copoly(acrylic/maleic acids). Even though there are no data to indicate harmful environmental effects, negative results do not categorically indicate complete environmental acceptability and this has resulted in a massive search for cost effective biodegradable replacements.

Many efforts to radically copolymerize acrylic and maleic acids with a whole range of vinyl monomers to produce biodegradable polymers [76-83], and graft substrates, including polysaccharides [84], have failed. Therefore, it must be considered that the pioneering biodegradation work of Suzuki [41] with carbon-chain backbone functional polymers is probably correct and only low molecular weight oligomeric carbon chain polymers are likely to be biodegradable, at least at any reasonable rate. Recent confirmation of Suzuki's work has come from Kawai [85] and research in NSKK and Idemitsu laboratories in Japan [86, 87].

As indicated earlier, Suzuki [41] ozonized high molecular weight poly(acrylic acid), polyacrylamide, and poly(vinyl pyrrolidone) to oligomers with molecular weights less than 1000 Daltons and observed a marked increase in their biodegradability. The work of Kawai and Tani (NSKK) was based on carefully synthesized oligomers. The results all indicate that poly(acrylic acids) are not completely, or at least rapidly, biodegradable above about a degree of polymerization of about 6-8 (400-600 Daltons). The work reported by Larson and Swift in their methodology development [39] also supports this result.

Other efforts to use radical polymerization to synthesize carbon chain biodegradable carboxylated polymers have been based on combining low molecular weight oligomers through degradable linkages and by introducing weak links into the polymer backbone. BASF [88] and NSKK [89] have patented acrylic oligomers chain branched with degradable linkages (X) as in Scheme 5, below. More recently, Idemitsu has patented variations on these themes [90]. Grillo Werke has patented copolymers of acrylic acid and enol sugars [91]. The degradability of these polymers has not been categorically established, the partial biodegradation observed is probably related to the lack of molecular weight control of the fragmented acrylic oligomer portion.

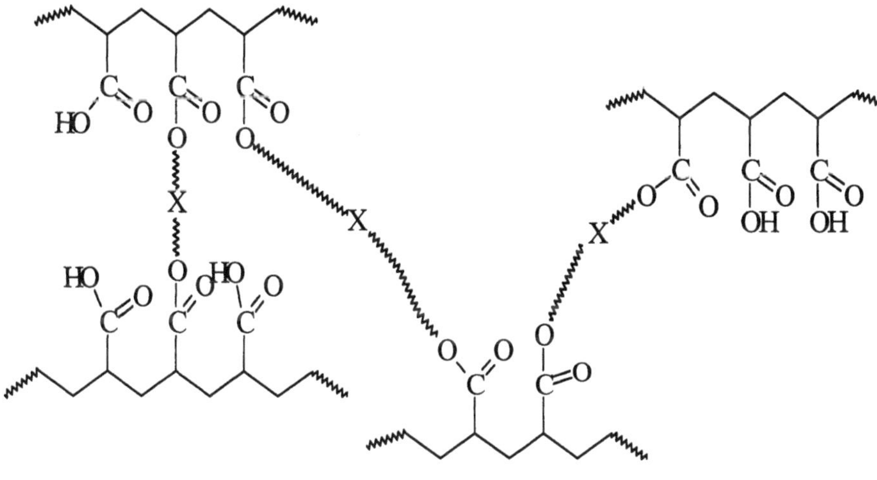

Scheme 5. Branched Acrylic Oligomers

Several miscellaneous carbon chain polymers have been claimed as biodegradable without clear supporting evidence, these include copolymers of methyl methacrylate and vinyl pyridinium salts [92, 93, 94], where the pyridinium salt is hypothesized as a 'magnet' for bacteria which then cleave the chain into small fragments which ultimately biodegrade completely. Careful analysis indicates absorption on sewage sludge present for the laboratory test and, since biodegradation was measured by TOC loss, the results were both misleading and misinterpreted. The work at Deutsche Gold-und Silber [96] with copolymers of acrolein and acrylic acid post-converted by a Cannizzaro reaction to copolymers containing hydroxyl and carboxyl functionality, Scheme 6, indicated high levels of biodegradability by BOD. However, low molecular weight fragments and other organics present in the Cannizzaro product lend uncertainly and speculation to the results and no further work has been reported.

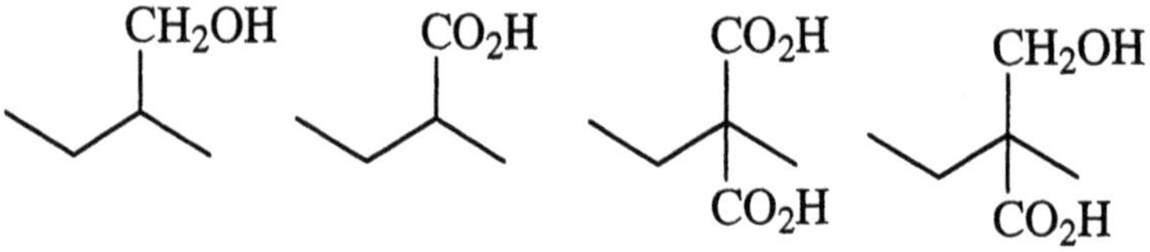

CH_2OH CO_2H CO_2H CH_2OH

CO_2H CO_2H

Scheme 6 Cannizzaro Functional Products from Acrylic / Acrolein Polymers

Polymers of α-hydroxyacrylic acid [97] are prepared as indicated in Scheme 7 from the polymerization of alpha chloroacrylic acid and subsequent hydrolysis. The claims of biodegradability for these polymers [98, 99, 100] were not substantiated by Mulders and Gilain [97] using radiolabelled carbon polymers. There was no clear differentiation between adsorption on sludge and bacterial cellular adsorption in the sewage testing protocol.

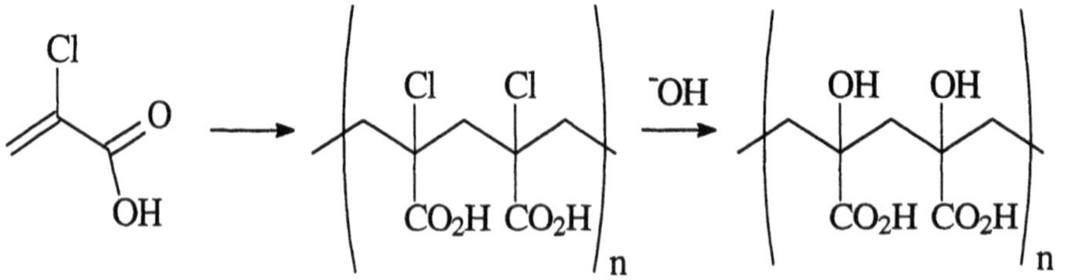

Cl O OH → Cl Cl CO_2H CO_2H n $\xrightarrow{^-OH}$ OH OH CO_2H CO_2H n

Scheme 7. Synthesis of • hydroxyacrylic acid polymers

Mitsubishi [101] claimed a unique biodegradable random copolymeric polycarboxylate containing ethylene, carbon monoxide, and maleic anhydride monomers in a variety of ratioins as seen in Scheme 8.

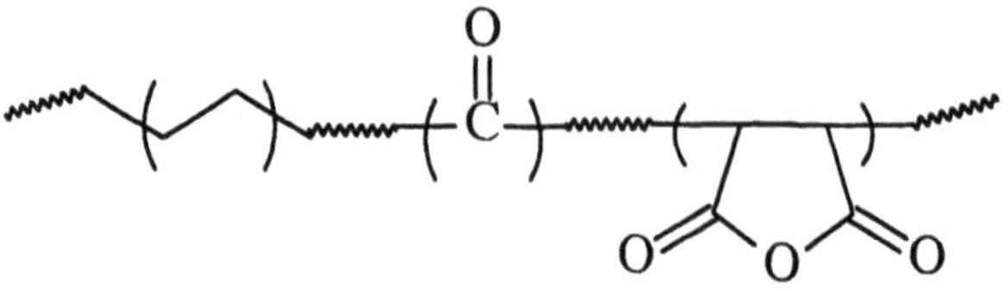

O C O O O

Scheme 8. Photo / Biodegradable Copolymeric Carboxylates

The initial degradation step for these polymers is photodegradation to yield low molecular weight fragments as indicated in Table 2. Unfortunately, no biodegradation data were reported on these low moleculare fragments, it was assumed that biodegrdation would follow because of the low molecular weight achieved.

Table 2: Photodegradation of photo / biodegradable Carboxylates

CO (mole %)	MW Initial	MW 2 hrs	MW 4 hrs	MW 5 hrs
0	94,000	83,100	73,500	57,200
8	120,500	7,200	4,500	2,300
12	202,400	4,600	2,300	7,800

5.2 HETEROATOM CHAIN POLYMERS

This class of polymers includes polyesters, polyamides, polyethers, and other condensation polymers have been widely studied from the initiation of research on biodegradable polymers. Their chemical linkages are quite frequently found in nature and these polymers are considered, and usually found to be more likely to biodegrade than the carbon chain hydrocarbon-based polymers discussed in the previous section.

5.2.1 Polyesters and Polyamides

Bailey developed a very clever free radical route introduce weak ester linkages into the backbones of hydrocarbon polymers and render them susceptible to biodegradation [101-106]. Free radical copolymerization of ketene acetals with vinyl monomers incorporates an ester linkage into the polymer backbone by rearrangement of the intermediate chain end ketene acetal radical as it incorporates into the polymer chain. The ester bond inserted is a potential site for biotic (enzyme hydrolysis) and abiotic (chemical hydrolysis) attack to yield low molecular weight fragments likely to biodegrade . Bailey demonstrated the chemistry with ethylene, see Scheme 9 below, and it has been extended to copolymers of acrylic acid [107, 108]. The biodegradation of the resulting copolymers has not been demonstrated, just claimed.

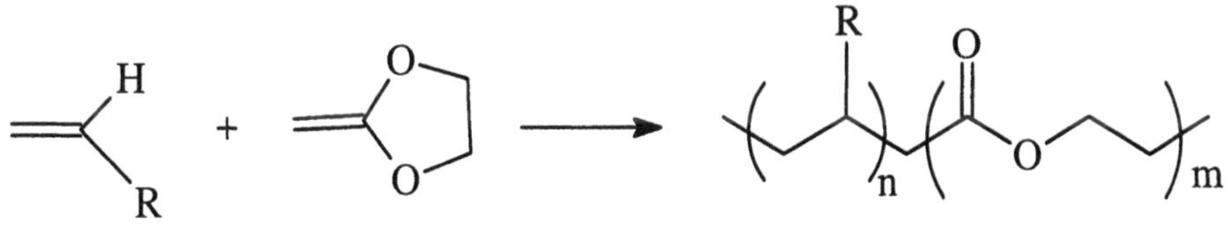

H R + O O → R O O n m

Scheme 9. Copolymerization with ketene acetals

Water-soluble polyesters and polyamides containing carboxyl functionality are reported by BASF to be biodegradable detergent polymers by BASF and may be obtained by condensation polymerization of monomeric polycarboxylic acids such as citric acid, butane-1, 2, 3, 4-tetracarboxylic acid, tartaric acid, and malic acid with polyols [109], amino compounds, including amino acids [110], and polysaccharides [111]. Earlier work by Matsumura [112] and Lenz and Vert [113] demonstrated the self-condensation of malic acid to biodegradable carboxylated polyesters regardless of the ester linkage, alpha or beta, formed. While Procter and Gamble has patented succinylated poly(vinyl alcohol) [114] as a detergent polymer.

Nonionic water-soluble biodegradable polyamides are reported by Bailey [115], Chiellini [116], and Ahmed for disposable fibers and webs [117].

Polyanionic polyamides are available by the condensation of polycarboxyamino acids such as glutamic acid and aspartic acid. Though both homopolymers are known and claimed as biodegradable, L-aspartic acid is more amenable to a practical industrial thermal polymerization since it has no tendency to form an internal N-anhydride as is the case with glutamic acid. An alternative synthesis for poly(aspartic acid) is from ammonia and maleic acid via an unisolated D,L-aspartic acid intermediate.

Acid catalyzed condensation of L-aspartic acid yields an authenticated biodegradable polymer [118] by standard biodegradation test methodology in wasterwater treatment plant water. The non-catalyzed polymerization and the ammonia/maleic acid processes give partially (ca 30 wt % residue remains in the Sturm test) biodegradable polymers after the 30 day stipulated period due to the molecules being highly branched and more resistant to enzymatic attack. Prolonged testing indicates that complete biodegrdation will occur.

The two polymeric structures obtained from L-aspartic acid are shown below in Scheme 10. Pathway 'A' is the acid catalyzed thermal condensation, and 'B' is the non-catalyzed thermal condensation. The polysuccinimide intermediates hydrolyze at the points indicated to give mixtures of α and β poly(D,L- aspartic acid) salts. Regardless of the stereochemistry of the starting aspartic acid, D or L, the final polymeric product is always the DL racemate.

There are many patents and publications related to poly(aspartic acid)s from aspartic acid [119-124] and ammonia/maleic [125, 126] processes indicating that the products should find use in many applications including dispersants [127,128] and detergents [129, 130]. In a structural variation BASF has an aspartic acid copolymer patent with carbohydrates and polyols [131], and Procter and Gamble [132] has a patent for poly(glutamic acid), both for biodegradable detergent cobuilders.

This exciting field of biodegradable poly(aspartic acid)s has generated many new opportunities and ideas for anionic polymers, though no large commercial success is at hand. Some examples of an ever-expanding list include dispersants for oil production [133], tobacco filters [134], cosmetics [135], hydrophobic associating thickeners [136], adhesives [137], crosslinked superabsorbents [138, 139, 140]. A recent review on poly(aspartic acid)s up to 1997 includes a large body of information [141] for medical and industrial applications.

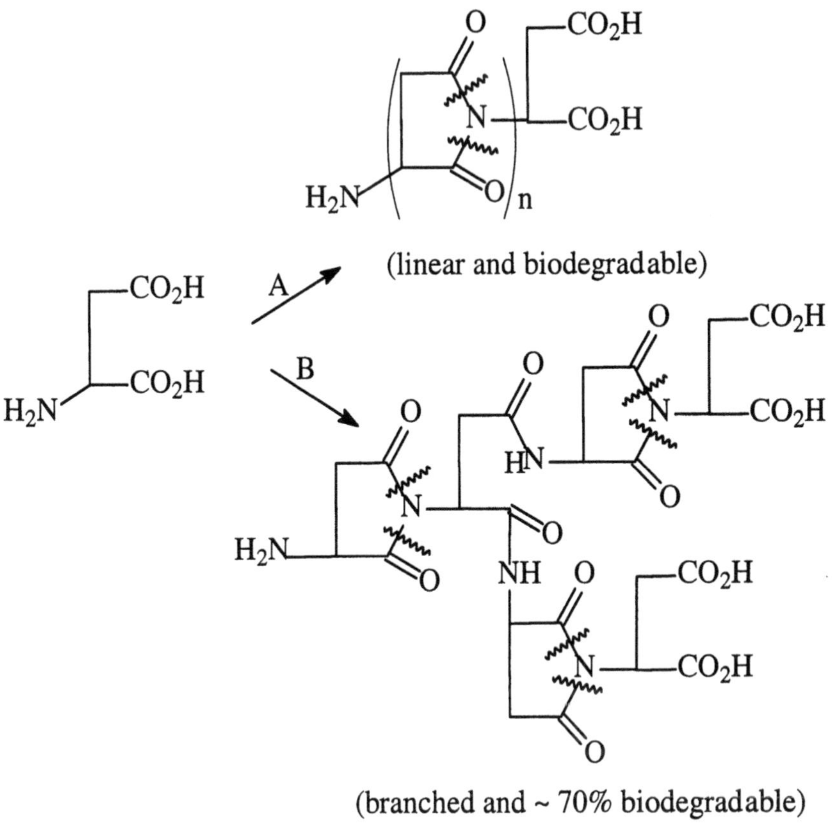

(linear and biodegradable)

(branched and ~ 70% biodegradable)

Scheme 10. Linear and branched Poly(aspartic acid)s

A recent huge advance in this chemistry is reported by Sikes [142], and shown in Scheme 11. This significant technology advance allows the synthesis of water-soluble reactive intermediates by copolymerizing monosodium aspartate with aspartic acid in a pre-selected ratio. The intractable polysuccinimide intermediate obtained in other polymerizations of aspartic acid and ammonia / maleic anhydride mentioned earlier in alternative approaches is avoided and the copolymer is soluble in water for further functionalization. Easier handling is very important for future characterization and development of applications.

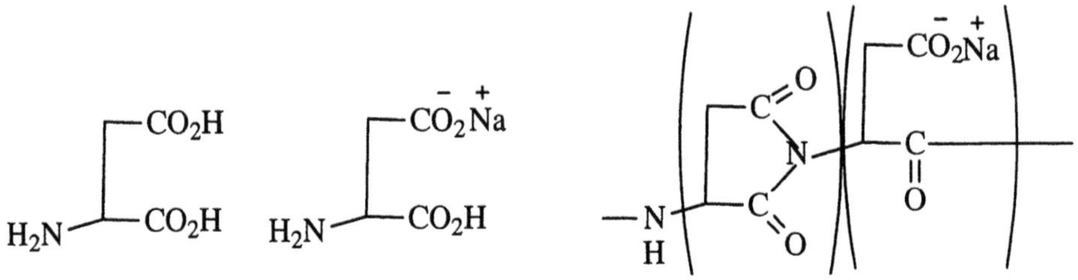

Scheme 11: Copolymerization of aspartic acid and monosodium aspartate

5.2.2 *Polyethers*

The biodegradation of polyethers has been investigated since about 1962, especially poly(ethylene oxide)s which are water-soluble and widely used in detergents and shampoos, and as a synthesis intermediate for polyurethanes, for example. This whole field has been very comprehensively review by Kawai [142] who established a symbiotic nature for two organisms for the degradation of poly(ethylene oxide)s with molecular weights higher than 6000 Daltons, while below a molecular weight of 1000 Daltons the polymer is biodegraded by many individual bacteria. The enzymatic exobiodegradation pathway described by Kawai is shown in Scheme 12 below, the first stage is promoted by a dehydrogenaase, the second stage is promoted by an oxidase, the third stage is an oxidation followed by a hydrolysis to remove a two-carbon fragment as glyoxylic acid. Biodegradation of poly(ethylene oxide)s at molecular weights as high as 20,000 Daltons has been reported. The biodegradation of higher poly(alkylene oxide)s is hindered by their lack of water solubility. Only the low oligomers of poly(propylene oxide)s are biodegradable with any certainty [146, 147, 148], as are those of poly(tetramethylene oxide)s [149]. A similar exo oxidation mechanism to that described for poly(ethylene oxide)s is proposed.

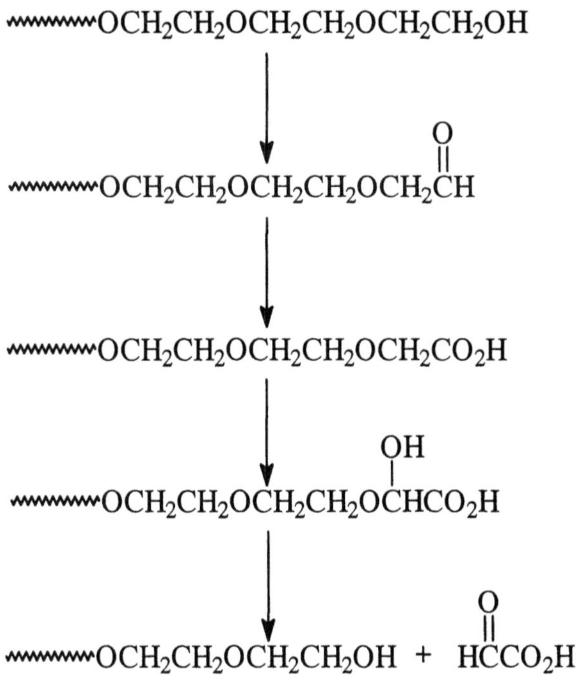

$$\sim\sim OCH_2CH_2OCH_2CH_2OCH_2CH_2OH$$

$$\downarrow$$

$$\sim\sim OCH_2CH_2OCH_2CH_2OCH_2\overset{O}{\overset{\|}{C}}H$$

$$\downarrow$$

$$\sim\sim OCH_2CH_2OCH_2CH_2OCH_2CO_2H$$

$$\downarrow$$

$$\sim\sim OCH_2CH_2OCH_2CH_2O\overset{OH}{\overset{|}{C}}HCO_2H$$

$$\downarrow$$

$$\sim\sim OCH_2CH_2OCH_2CH_2OH \; + \; H\overset{O}{\overset{\|}{C}}CO_2H$$

Scheme 11. Exobiodegradation of poly(ethylene oxide)s

Anaerobically, poly(ethylene oxide)s degrade slowly although molecular weights up to 2000 Daltons have been reported to completely biodegrade [143, 144].

Polyether carboxylates have been evaluated as biodegradable detergent polymer, initially by Crutchfield [150], and later by Procter and Gamble [151], and Matsumura [152]. All these polymers fit the general structure of the series made in Matsumura's extensive evaluation of anionic and cationic polymerized epoxy compounds shown in Scheme 13. Polymers with a molecular weight range of several hundreds to a few thousands, where X or Y may be carboxyl functionality and X or Y may be hydrogen or a substituent bearing a carboxyl functionality were evaluated. The extent of biodegradability, based on biochemical oxygen demand (BOD), is structure dependent.

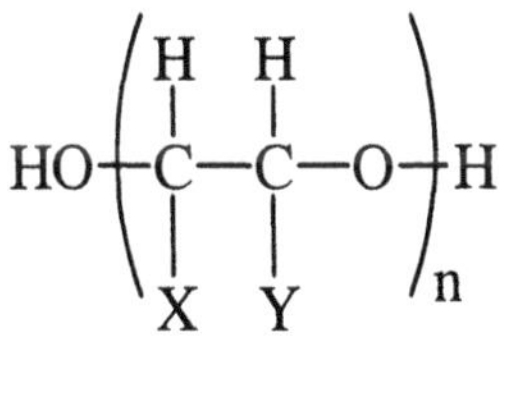

$$HO\left(\begin{array}{c}H\\|\\C\\|\\X\end{array}-\begin{array}{c}H\\|\\C\\|\\Y\end{array}-O\right)_nH$$

Scheme 13. Functional polyethers

Water-soluble biodegradable polycarboxylates containing an acetal weak link were the clever invention of Monsanto scientists [153] in their search for biodegradable detergent polymers. Clever science, however, is no match for economics and the polymers never reached commercial activity due to cost / performance limitations. The polymers are based on the anionic or cationic polymerization of glyoxylic esters at low temperature (molecular weight is inversely proportional to the polymerization temperature) and subsequent hydrolysis to the salt form of the polyacids. The salt is reasonably stable under basic conditions being a hemi acetal, or ketal when methylglyoxylic acid is used. Hydrolytic instability results from the pH drop a detergent polymer solution experiences as it leaves the alkaline laundry environment, pH ca 10 and enters the sewage or ground water environment where pH is closer to neutral. The polymer is unstable at the lower pH and hydrolyzes to monomer which rapidly biodegrades. The chemistry is outlined in scheme 14, below and is reported in many patents [153] and several publications [154,155].

Other copolymers based on polyacetal structures were prepared by BASF scientists from ω-dialdehydes and aliphatic hydroxycarboxylic acids derived from sugars [156,157], shown below in scheme 15. Alternatively, carboxypolyacetals are available by the addition of polyhydroxy carboxylic acids, tartaric acid for example, to divinyl ethers [158].

Miscellaneous recent acetal polymer chemistry to produce biodegradable water-soluble polymers includes superabsorbents [159], the use of hydroxycarboxylic acid to produce lubricants [160], scale inhibitors [161], and water-soluble packaging materials [162].

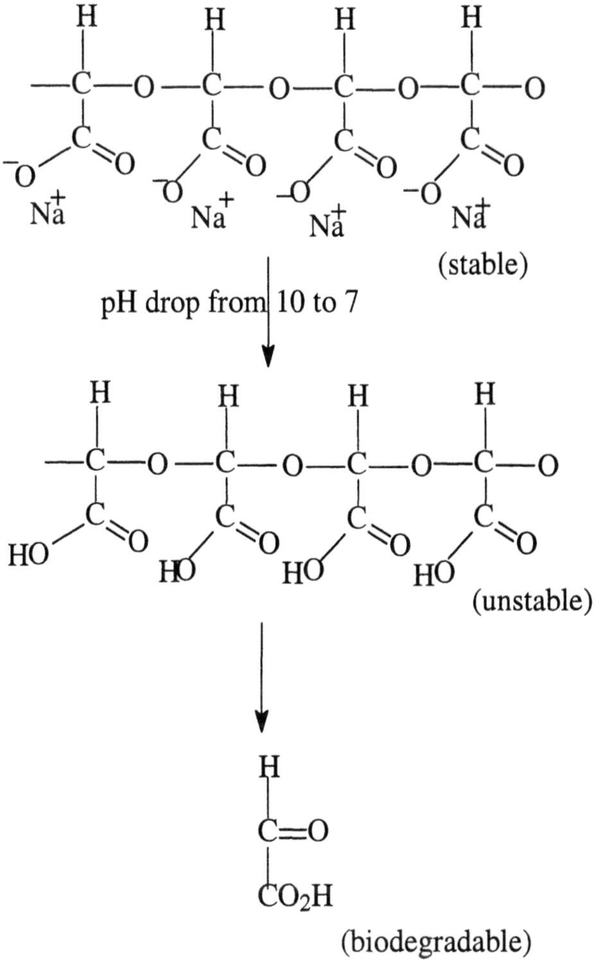

H H H H
—C—O—C—O—C—O—C—O
C=O C=O C=O C=O
$^-$O Na$^+$ $^-$O Na$^+$ $^-$O Na$^+$ $^-$O Na$^+$
(stable)

pH drop from 10 to 7

H H H H
—C—O—C—O—C—O—C—O
C=O C=O C=O C=O
HO HO HO HO
(unstable)

H
C=O
CO_2H
(biodegradable)

Scheme 14. Hydrolysis / biodegrdation of polyacetal carboxylates

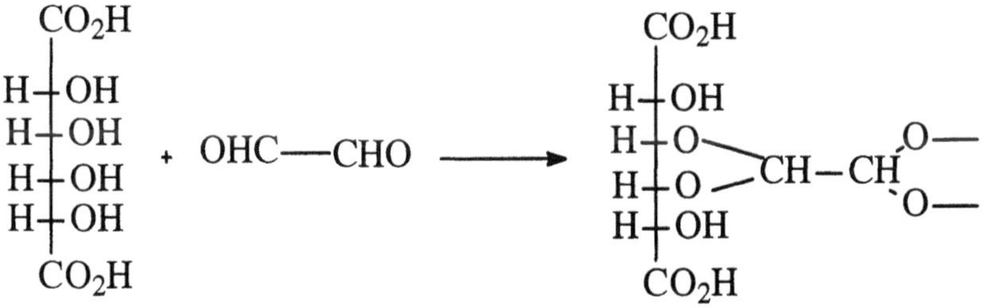

CO_2H
H—OH
H—OH
H—OH
H—OH
CO_2H

+ OHC—CHO ⟶

CO_2H
H—OH
H—O
H—O
H—OH
CO_2H
CH—CH
O—
O—

Scheme 15. Polyacetal formation from carboxysugars and dialdehydes

6. Modified Natural Polymers:

Modification of natural polymers such as starch, cellulose, proteins, etc. is a way of capitalizing on the well-accepted biodegradability of the base material with the intention of developing biodegradable or environmentally acceptable polymers that function in variety of commercial applications. It is essential, of course, to demonstrate that the modification does not to interfere with the biodegradation process and the product meets the guide listed earlier for environmental acceptability, namely:

- either it must be demonstrated to be totally biodegraded and to be removed from the environment or biodegradable to the extent that no environmentally harmful residues remain.

With this in mind, the approaches that have received the major attention are either grafting of another polymeric composition and chemical modification to introduce some desirable functional group by oxidation or by some other simple chemical reaction such as esterification or etherification.

6.1 GRAFT POLYMERS

Starch, as mentioned earlier in the work of Sanyo for detergent additives, has been a substrate of choice for biodegradable water-soluble polymers by grafting with synthetic polymers. The intention is to achieve property improvement and new properties such as carboxyl functionality not available in starch, whiel retaining as much biodegradability as possible. As noted earlier this was not successful due to the high molecular weight of the carbon chain graft. Further attempts have been made to meet the low molecular weight limitations mentioned earlier (DP less than ca 6-8). Acrylic acid grafts onto polysaccharides in the presence of alcohol chain transfer agent [163] were not completely biodegradable, nor were those based on the use of metal initiators [164] such as Ce^{4+} and mercaptans as chain transfer agents [165, 166]. Other efforts include methanol chain transferred graft of methyl acrylate onto starch [167], sodium acrylate grafts [168], maleic anhydride grafts [169], poly(vinyl alcohol) grafts [170], and polysuccinimide grafts [171].

Protein substrates[172] for grafting are expected to be similar to the starch grafts, the fundamental problem remains the control acrylic acid polymerization to the oligomers range, in order to have complete biodegradability.

Other grafts to natural materials are exemplified by Dordick's work [173] in which he produced polyesters from sugars and monomeric polycarboxylates by enzyme catalysed condensation polymerization. These polymers and the method of synthesis may well be the future of renewable resource chemistry. .

6.2 CHEMICAL MODIFICATION OF POLYSACCHARIDES

Simple chemical reactions on natural polymers are well known to produce water-soluble polymers such as hydroxyethyl cellulose, hydroxypropyl cellulose, carboxymethyl cellulose, cellulose acetates and propionates, and many others that have been in commerce for many years. Their biodegradability is often taken for granted, but, in many cases, is not at all well established. Carboxymethyl cellulose, for example, has been claimed as biodegradable below a degree of substitution of about 2, which is similar to the claim for cellulose acetate. More recently, there has been attempts to more rigorously quantify the biodegradation of the cellulose acetates [174,175], and to establish a property/biodegradation relationship. Rhone-Poulenc indicates that cellulose acetate with a degree of substitution of about 2 is biodegradable, in agreement with their earlier conclusion [176]. Cellulose has been discussed as a renewable resource [177]. A recent publication [178] on chitosan reacted with citric acid indicates that the ampholytic product is biodegradable. Chitosan acetate liquid crystals [179], and hydrophobic amide derivatives [180], and crosslinked chitosan [181] are also claimed to be biodegradable.

Carboxylated natural polymers have been known for many years with the introduction of carboxymethyl cellulose, as noted above. This product has wide use in detergents and household cleaning formulations, even though of questionable biodegradability at the level of substitution required for performance. Nevertheless, carboxylated polysaccharides are a desirable goal for many application and the balance of biodegradation with performance has been recognized as an attractive target with a high probability of success by many people. Three approaches have been employed, esterification, oxidation and Michael addition of one or more of the hydroxyl groups with a suitable vinyl carboxylic acid receptor. Attempts have also been made to react specifically at the primary hydoxyl, the 6 position, or secondary sites the 2,3 positions of polysaccharides.

Esterification with polycarboxylic anhydrides can be controlled to minimize diesterification and crosslinking to produce carboxylated cellulosic esters. Eastman Kodak in a recent patent claimed the succinylation of cellulose to different degrees, 1 succinyl group per three anhydroglucose rings [182] and 1 per 2 rings [183] . Henkel [184] also has a patent for a surfactant by the esterification of cellulose with alkenylsuccinic anhydride, presumably substitution degree governs the hydrophile/hydrophobe balance of the product and its surfactant properties. No correlation of structure with biodegrdation was noted.

Oxidation of polysaccharides is a far more attractive route to polycarboxylates, potentially cheaper and cleaner than esterification. Selectivity at the 2,3-secondary hydroxyls and the 6-primary hydroxyl is claimed possible. Total biodegradation with acceptable property balance has not yet been achieved, though. For the most part, oxidations have been with hypochlorite/periodate under alkaline conditions, but more recently catalytic oxidation has appeared as a more viable possibility, and chemical oxidations have also been developed that are specific for the 6-hydroxyl group.

Matsumura [185-188] has oxidized a wide range of polysaccharides, starch, xyloses, amyloses, pectins, etc. with hypochlorite/periodate. The products are either biodegradable at low oxidation levels or functional at high oxidation levels, the balance has not yet been established for commercial success. Other than Matsumura, van Bekkum, at Delft University, has been a major player in the search to control the hypochlorite/periodate liquid phase oxidations of starches [189-191]. He has been searching for catalytic processes to speed up the oxidation with hypochlorite. Hypobromite is a more active oxidant than hypochlorite but more expensive, however it may be generated in situ from the cheap hypochlorite and bromide ion in one solution [191, 192]. This is shown in Scheme 16

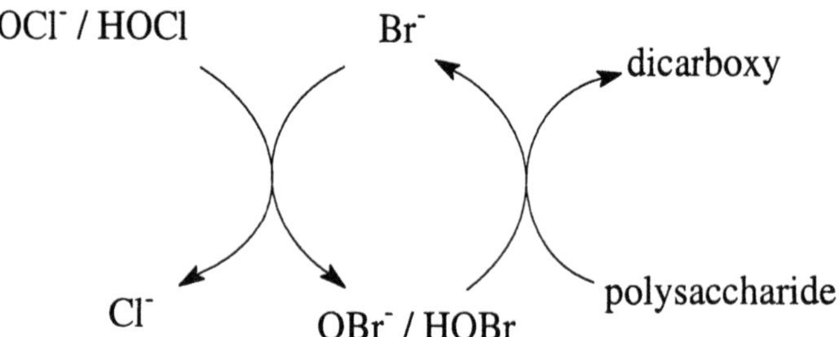

Scheme 16. Catalytic Oxidation of polysaccharides with Br^- / ^-OCl

van Bekkum [193] has also published a catalytic method for oxidizing specifically the 6-hydroxyl group (primary) of starch by using TEMPO and bromide / hypochlorite as shown below inn Scheme 17.

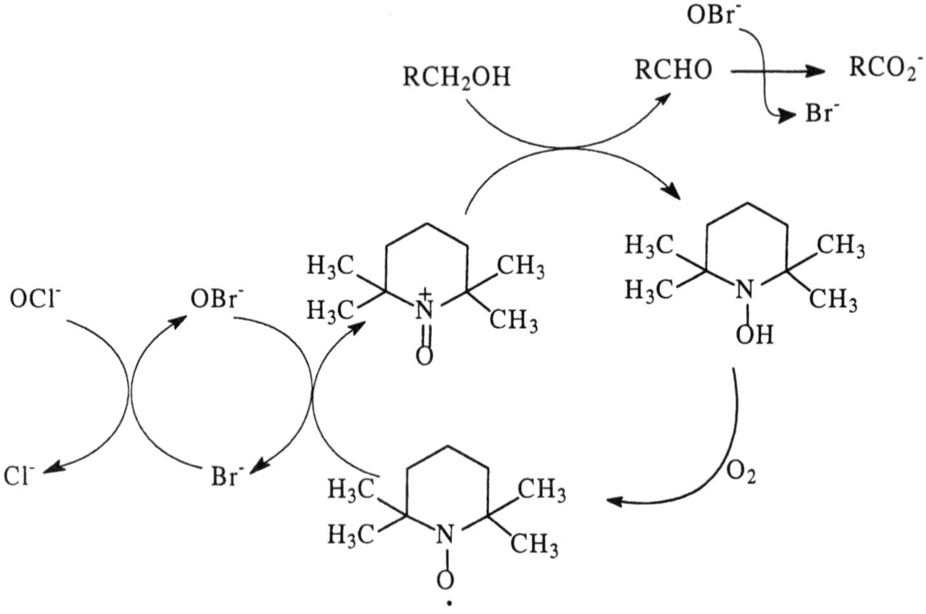

Scheme 17. Catalytic oxidation of C-6 hydroxyl of polysaccharides

Chemical oxidation of polysaccharides with strong acid is reportedly selective at the 6-hydroxyl, with a mixture of nitric acid / sulfuric acid / vanadium salts [194] which is claimed as specific for up to 40% conversion. Alternatively, dinitrogen tetroxide in carbon tetrachloride has similar specificity up to 25% conversion [195].

Catalytic oxidation in the presence of metals and oxygen is claimed as both non-specific and specific for the 6-hydoxyl oxidation depending on the metals used and the conditions employed for the oxidation. Non-specific oxidation is achieved with silver or copper and oxygen [196] and with noble metals in combination with bismuth and oxygen [197)]. Specificity results with platinum catalyst at pH 6-10 in water in the presence of oxygen [198]. A related patent to Hoechst on the oxidation of ethoxylated starch produces a water-soluble carboxylated starch and another on the oxidation of sucrose to a tricarboxylic acid are all specific for primary hydroxyls and use a platinum catalyst at a pH near neutrality in the presence of oxygen [199, 200].

Further reading on polysaccharides as raw materials in the detergent industry may be found in articles by Swift et al.[201, 202].

7. Conclusions

Although there has been much research activity to produce totally biodegradable water-soluble polymers for a variety of applications, especially for detergents, few commercial successes have been registered beyond poly(vinyl alcohol), poly(ethylene oxide)s and, in a limited way, poly(aspartic acid)s. Almost all efforts to obtain new carbon chain polymers that are completely biodegradable have failed, at least in the short term testing protocols currently in use. Some promising leads are noted in condensation polymers, poly(aspartic acid)s, acetals, etc. and in the modification of natural polysaccharides.

This review suggests that, if current test methodology is unchanged, future research directions for biodegradable water-soluble polymers should be based condensation polymers and modified natural polymers, especially starch as a very cheap and widely available natural resource. The research will require close attention to cost / performance characteristics to develop competitive products since current non-biodegradable products are not apparently creating environmental problems. Although there are nagging doubts about this, all evidence to date is based on the absence of negative responses.

A large caveat, though, is that current test methodology is designed for organic compounds and polymers that undergo rapid biodegradation. This limitation and concern, which mirrors that of the test methodology for plastics, needs attention, if the most cost effective polymers are to be developed. It is essential to extend current methods and develop new methods that are capable of measuring longer term biodegradation. Without this advance, progress to develop new cost effective biodegradable water-soluble polymers is likely to be severely hindered. The current tests are more aligned with the acceptability of condensation polymers which are similar to many natural polymers and are more likely to biodegrade rapidly. However, such polymers are also more costly to produce. The most cost effective water-soluble polymers now in commercial use are based on hydrocarbon backbones which, even in nature are well known to be slow to biodegrade and fail current biodegradation tests. Rather than neglect cost effective opportunities, the development of new test methodology needs to be seriously considered.

References

1. ASTM D 883 - 93, *Terminology Relating to Plastics.*
2. Narayan, R. (1990)Annual Meeting of the Air and Waste Management Assoc., June 24- 29-40 (1990).
3. ASTM, (1993*) Standards on Environmentally Degradable Plastics*, ASTM Publication Code Number (PCN): 03-420093-19.
4. Potts, J. E., et al., (1973) *Polymers and Ecological Problems*, Plenum Press(1973); EPA Contract, CPE-70-124 (1972).
5. Clendinning, R. A., Cohen, S. and Potts, J. E. (1974) *Great Plains Agric.Council Pub.*, 68, 244.
6. Fields, R. D., Rodriquez, F. and Finn, R. K. J. (1973) *Amer. Chem. Soc., Divn. Poly., Chem.,*14, 2411.
7. Swisher, R. D. (1987) *Surfactant Biodegradation*, Marcel Dekker, 2nd. Edition,.
8. OECD Guidelines for Testing Methods. (1981) *Degradation and Accumulation Section*; OECD: Washington, DC,; Nos. 301A-E, 302A-C, 303A, and 304A.
9. Tanna, R. J., Gross, R. and McCarthy, S. P. (1992) *Poly. Mater. Sci. Eng.,* **67,** 230-231.
10. J. D. Gu, S. P. McCarthy, G. P. Smith, D. Eberiel, R. Gross, *Polym. Mater. Sci. Eng*., **67,** 294-295 (1992).
11. Palmisano, A. C. and Pettigrew, C. A. (1992) *Bioscience*, 42(9), 680-685.
12. Swift, G. (1994) *Polymer News,* 19, 102-106.
13. Swift, G. (1993) *International Biodegradable Polymer Workshop*, Osaka, Japan, November.
14. Andrady, A. L. (1994) *JMS-Rev. Macromol. Chem. Phys*. C34(1), 25-76.
15. Swift, G. (1994) *Polymer Degradation and Stability*, 45, 215.
16. Swift, G. (1998) *Polymer Degradation and Stability*, 59,19.
17. Swift, G. (1996) *Environmentally Degradable Polymers*, Kirk-Othmer Encyc. of Chemical
Technology, Fourth Edition, Volume 19, ISBN 0-471-52688-6,
18. Swift, G. (1990) *Degradability of Commodity Plastics and Specialty Polymers*, Agricultural and Synthetic Polymers, ACS Symposium Series Number No. 433, J. E. Glass and G. Swift, Editors.
19. Swift, G. (1996) *A Review of Synthetic Approaches to Biodegradable Polymeric Carboxylic* Acids *for Detergent Applications,* ACS Advances in Chemistry series No. 248, J. E. Glass, Editor.
20. Swift, G. (1993) *Accounts of Chemical Research*, 26, 105.
21. Swift, G. (1998) *Macromol Symposia*, 130, 379.
22. Lo, F., Petchonka, J. and Hanly, J. (1993) *Chem. Eng. Prog.,* 89(7), 55.
23 Chiellini, E. and Solari, R. (1996) *Adv. Mater.,* 8(4), 305.
24. Tian, Y., Hou, L. Yu, X. and X. Tang, X. (1999) *Shanghai Huagong*, 24(6), 28.
25. Matsumura, S. (1995) *Yukagaku,* 44(2), 97.

26. Matsumura, S. (1995) *Petrotech,* (Tokyo), 18(3), 207.
27. Matsumura, S. (1990) *Zairyo Gijutsu*, 8(5), 164.
28. Swift, G. (1997) *213th ACS National Meeting, San Francisco*, PMSE Divn., 283.
29. Swift, G. (1990) *Polym. Mater. Sci. Eng.,* 63, 846.
30. Kawai, F. (1993) *Kobunshi Kako*, 42(4), 176.
31. Kawai, F. Macromol. Symp., (1996) 123 (*37th Microsymposium on Macromolecules, Biodegradable Polymers Chemical, Biological and Environmental Aspects (1996)* 177.
32. Matsumura, S. (1996) *Sen'I Gakkaishi*, 52(5), 209.
33. Kawai, F. (1995) *Advances in Biochemical Engineering/Biotechnology*, 52, 151.
34. Smith E. A. and Oehme F. W. (1991) *Reviews on Environmental Health*, 9(4), 215.
35. Takahashi, S. (1995) *Idemitsu Giho*, 38(5), 480.
36. Larson, R. J., Williams, R. and Swift, G. (1992) *Polymer Materials Science and Eng.* 67, 348
37. Doi, J. Ecol. (1997) *Assess. Polym.*, 33.
38. Suzuki, T. (1975) *Kobunshi*, 24(6), 384.
39. Larson, R. J., Bookland, E. A., Williams, R. T., Yocom, K. M., Saucy, D. A., Freeman, M. B. and Swift, G. (1997) *Environ. Polym. Degrad.*, 5(1), 41.
40. Scholz, N. (1991) *Tenside, Surfactants, Detergents*, **28,** 277-281.
41. Suzuki, T. Hukushima, K. and Suzuki, S. (1978). *Environ. Sci. Technol.*, 12(10), 1180.
42. Yamamoto, T., Inagaki, H., Yagu, J. and Osumi, T. (1966) *Abst. Annual Meeting Agric. Chem.Soc.* Japan, 133.

43. Suzuki, T., Ichihara,Y., Yamada, M. and Tonomura, K. (1973) *Agric Biol. Chem.*, 34(4), 747.
44. Suzuki, T., Ichihara, Y., Yamada, M. and Tonomura, K. J. (1979) *Appl. Poly. Sci., Appl. Poly. Symp.*, 35, 431.
45. Watanabe, Y., Morita, M., Hamada, N. and Tsujisaka, Y. (1975) *Agric. Biol.Chem.*, 39(12), 2448.
46. Watanabe, Y., Morita, M., Hamada, N. and Tsujisaka, Y. (1976) *Arch. Biochem.Biophys.*, 174, 575.
47. Huang, S. J., Quingua, E. and Wang, I. F. (1982) *Org. Coat. Appl. Polym.Sci. Proc.*, **46,** 345.
48. Huang, J. C., Shetty, A. S. and Wang, M.S. (1990) *Adv. Polym. Tech.*, 10, 23.
49. Kuraray. *JP* 03263406 and *JP* 03263407.
50. Fukanaga F. et al. (1977) Japan Kokai 7794471.
51. Fukanaga F. et al. (1976) *Japan. Kokai*, 76125786.
52. Shimao, M. and Kato, N. (1990) *International Symp. on Biodegradable Polymers Abstr.,* 80.
53. Casey, J. P. and Manley, D. G. (1976) *Proc. 3rd. International Biodeg. Symp. Appl. Sci. Publ.*, 731-741.
54. Wheatley, O. D. and Baines, C. F. (1976) *Text. Chem. Color*, 8(2), 28-33.
55. Matsumura, S., Maeda, S., Takahashi, J. and Yoshikawa, S. (1988) *Kobunshi Robunshu,* 45(4), 317.

56. Planet Polymers. (1996) *US Patent* 5658977.
57. Negoce et Distribution S.a.r. (NEDI) (1985) *French Patent* 2553423.
58. Unitika Chem Kk,. (1993) Japan Kokai, 05295210.
59. Robeson, L. M., Axelrod, R. J., Kittek, M.R. and Pickering, T. L. (1994) *Polym. News,* 19,(6), 167.
60. Haschke, H., Tomk I. and Keilbach, A. *Monatsh. Chem.*, (1998) 129(4), 365.
61. Coombs, S., Christie, Y. B. G. and Yu, L. (1999) *PCT Int. Applic. WO* 2000036006.
62. Chiellini, E., Cinelli, P., Corti, A., Kenawy, E. R., Fernandez, E. G. and Solaro, R. (2000) *Macromol. Symp.*152, 83.
63. Breitenbach, A., Nykamp, G. and Kissel, T. (1999) *Proc. Int. Symp. Controlled Release Bioact. Mater.*, 150.
64. Breitenbach A. and Kissel, T. (1998) *Polymer,* 39 (14), 3261.
65. Alexy, P., Bakos, D., Kolomaznik, K., Sedlak, M. and Sedlakova, E. (2000) *Int. Pat., PCT Int. Appl.* WO 20001019.
66. Bastioli, C., Belloti, V., Camia, M., Giudice, D. L. and Rallis, A. (1994) *Stud. Polym. Sci.* 12, 200.
67. Kweon, D.K., Kang, W. D. and Chung, J. N. (1998) *Polymer(Korea),* 22(5), 786.
68. Haschke, R. H., Rauch, R., Reiterer, F. and Wehrmann, F. (1996) *PCT Int. Appl.* 9603443.
69. Matsumura, S., Takahashi, J., Maeda, S. and Yoshikawa, S. (1988) *Macromol. Chem. Rapid Commun.,* 9(1), 1-5.
70. Matsumura, S., Takahashi, J., Maeda, S. and Yoshikawa, S. (1988) *Kobunshi Ronbunshi,* 45(4), 325.
71. Lever, *US Patent* 5062995.
72. Matsumura, S., et al., (1993) *J. Am. Oil. Chem. Soc.*, 70, 659-665; and ACS Symposium Series (1996) 627 *Hydrogels and Biodegradble Polymers for Bioapplications*, 137.
73. Matsumura, S., Ii, S., Shigeno, H., Tanaka, T., Okuda, F., Shimura, Y. and Toshima, T. (1993) *Makromol. Chem.* 194(12), 3237.
74. Rohm and Haas *US Patent* 5191048.
75. Hunter, M., daMotta Marques D. M. L., Lester, J. N. and Perry, (1988) *Environ. Technol. Lett.* 9, 1-22.
76. Matsumura, S., et al. (1984) *Yukagaku* 33, 211, 228.
77. Matsumura, S., et al. (1985) *Yukagaku* 34, 202, 456.
78. Matsumura, S., et al. (1985) *Yukagaku* 35, 167.
79. Matsumura, S., et al. (1986) *Yukagaku* 35, 937.
80. Matsumura, S., et al. (1980) *Yukagaku* 30, 31.
82 Matsumura, S., et al. (1981) *Yukagaku,*30, 757.
83. T. Yamamoto and K. Itakura, JP Kokai 04304216 (1992).
84. Sanyo *JP* 6131498.
85. Kawai, F. *JP* 05237200-A, and (1993) *Appl. Microbiol. and Biotech.* 39(3), 382-385.
86. Tani, Y. et al. (1993) *Applied and Environ. Microbiol.* 1555-1559.
87. Matsuo, S. (1999) *JP Kokai* 11172039.
88. BASF (1988) *DE* 373348A, *EP* 289827A; (1988) *DE* 3716543A,

EP 291808A; (1988) *DE* 3716544, *EP* 292766A; (1988) *EP* 289787A *EP* 289788A; (1989) *DE* 381426); (1990) *US* 4952655; (1994) *DE* 4319934.

89. NSKK *EP* 529910A.
90. Matsuo S. (Idemitsu), (2000) *JP Kokai* 2000281801.
91. Grillo Werke, *EP* 289895A (1988).
92. Kawabata, N. (1992) *Prog. Polym. Sci.* 17(1), 1; (1993) *Nippon Gomu Kyokaishi* 66(2), 80-87.
93. Peng X. and Shen J. (1999) *J. Appl. Polym. Sci.* 71(12), 1953.
94. Yamamoto, T., Kakajima O. and Katsuhiko K. (Mitsubishi Petro.) (1993) K. *JP Kokai* 05059130
95. Quantum Chemicals *US Patent*, 5219930.
96. Deutsche Golde-und Silber, US Patent 3686145 (1972); US Patent 3896086 (1975); US Patent 3923742 (1975).
97. Mulders, J. and Gilain, J. (1977) *Water Res.* 11(7), 571-574.
98. Henkel et Cie. (1972) *DE* 2061584
99. Solvey et Cie. (1972); *Belg. Patent* 786464; (1978) *US Patent* 4107411 and (1980) *US Patent* 4182806.
100. Hoechst, A. G. (1975) *US Patent* 3890288.
101. Mitsubishi Petrochemicals (1988) *EP* 0281139 and (1989) *JP Kokai* 63284296.
102. Bailey, W. J. et al. (1979) *Contemp. Topics Polym. Sci.* 3, 29.
103. Bailey, W. J. et al. (1975) *J. Polym. Sci., Polym. Lett. Edn.*13, 193.
104. Bailey, W. J., Gu, W. J., Lin, Y. and Zheng, Z. (1991) *Makromol. Chem. Makromol. Symp.* 42/43, 195.
105. Bailey, W. J. and Gapud, B. (1985) *Polym. Stab. And Degrad.*, 280, 423.
106. Bailey, W. J. and Zhou, L. L. (1992) *Macromolecules*, 25(1), 3.
107. American Cyanamid (1990) *US Patent* 4923941.
108. Toa Gosei Chemical Industry Co. (2000) *JP Kokai* 2000038595.
109. BASF *EP* 4223807; *US* Patent 5217642; *WO* 9216493-A1; *DE* 4108626-A1.
110. BASF *DE* 4225620-A1, *DE* 4213282-A1.
111. BASF *DE* 4108626-A1, *DE* 4034334-A1.
112. Abe, Y., Matsumura, S. and Imai, K. (1986)*Yukagaku* 35(11), 937.
113. Lenz, R. W. and Vert, M. (1978) ACS Polym. Preprints, 20, 608.
114. Procter and Gamble *US Patent* 5093170.
115. Bailey, W. J., Okamoto, Y., Kuo, W_C. and Narita, T. (1976) *Proc. Int. Biodegradation Symp.* 765.
116. Chiellini, E., Bizzarri R. and Solari R. (1999) *J. Bioact. Compat. Polym.* 14(6), 504.
117. Ahmed, S. U. (1999) *US Patents* 5869596, 5869597.
118. Swift, G., Freeman, M. B., Paik, Y. H., Wolk, S.and Yocom, K. M. (1994) *ACS Biotech. Sectetariat Abstr., San Diego, Spring*, and (1994) *6th Internatonal Conference on Polymer Supported Reactions in Organic Chemistry(POC), Venice*, June 19-23, Abstr. page21.13., and (1997) *35th, IUPAC International Symposium on Macromolecules, Akron,* Ohio, July 11-15, 1994, Abstr. 0-4.4-13th, page 615; and (1997) *Macromol. Symp.* 123, 195.
119. Cygnus *US Patent* 5219952.
120. Rohm and Haas *EP* 578448-A1.

121. Rhone-Poulenc *EP* 511037-A1.
122. Donlar Corporation *US Patent* 5221733.
123. Rao, V. S. (1993) *Makromol Chem.*194, 1095.
124. Donlar Corporation *WO* 9214753.
125. SR Chem. *US Patent* 5288783.
126. SR Chem. *WO* 9323452.
127. Sikes, S. University of South Alabama, *US Patent* 5260272.
128. Donlar Corporation *US Patent* 5116513.
129. Montedipe *EP* 454125.
130. Lever *EP* 561464, and *EP* 561452.
131. BASF *DE* 4221875-A1.
132. Procter and Gamble *WO* 93 06202.
133. Fan, G., Kokan, L. P. and Ross, R. J. (1998*) 215th ACS Abstracts. National Meeting, ENVR-* 029.
134. Taniguchi, H. and Nishimura, K. (1995) *JP Kokai* 95-200764.
135. BASF (1998) *DE* 19631380.
136. Nakato, T., Tomida, M., Suwa, M., Morishima, Y., Kusuno A. and Kakuchi,T. (2000) *Polym. Bull. (Berlin),* 44(4), 385.
137. Bayer Corp. (2001) *EP Appl.* 1085072.
138. Mitsui Chemicals (1998) *EP Appl.* 856539.
139. Mitsui Chemicals (1998) *JP Kokai* 11217436.
140. NSKK (1996) JP Kokai 08059820.
141. Roweton, S., Huang S. J. and Swift,G. (1997) *J. Environ. Polym. Degrad.* 5(3), 175.
142. Sikes, S. (1999) *US Patent* 5981691.
143. Kawai, F. (1987) *CRC Critical Reviews in Biotechnology,* C. R. C. Press, 6, 273.
144. Schink, B. and Strab, H. (1983) *Appl. Environ. Microbiol.* 45, 1905.
145. Schink, B. and Strab, H. (1986) *Appl. Microbiol. Biotech.* 25, 37.
146. Kawai, F., Hanada, K., Tani, Y. and Ogata, K. (1977) *J. Ferment. Technol.* 55, 89.
147. Kawai, F., Okamoto, T. and Suzuki T (1985) *J. Ferment. Technol.* 63, 239.
148. Kawai, F. (1982) *J. Kobe Univ. Commerce*, 18(1-2), 23.
149. Kawai, F. and Yamanaka, H. (1986) *Ann. Meeting Agric. Chem. Sic. Japan, Kyoto.*
150. Crutchfield, M. M.(1978) *J. Amer. Oil Chem. Soc.*,**55,** 58.
151. Procter and Gamble, *US Patents* (1987) 4654159, 4663071, 4689167, and *EP Patents* (1986) 192441), 192442, and (1987) 236007, 264977.
152. Matsumura, S., Hashimoto, K. and Yashikawa, S. (1987) *Yukagaku* **36(110),** 874-881.
153. Monsanto *US Patents*, 4144226, 4146495, 4204052, 4233422, 4233423.
154.. Gledhill, W. E. and. Saeger, V. W, (1987) *J. Ind. Microbiol.* **2(2),** 97.
155. Gledhill, W. E. (1978) *Appl. Environ. Microbiol.***12,** 591.
156. BASF *DE* 4204808-A1.
157. BASF *DE* 4106354-A!, WO 9215629-A1.
158. BASF *DE* 4142130-A1.
159. NSKK (1997) *JP Kokai* 09124754.
160. Clariant G.m.b.h.(1998*) DE* 19636688.
161. Mitsubishi Gas Chemical Company (1997) *EP Appl.* 764675.

162. NSKK. (1998) *JP Kokai* 10081817.
163. Taechang Moolsan Co. Ltd. *WO* 9302118-A1.
164. Rhone-Poulenc *EPs* 465286, and 465287.
165. BASF *DE* 4003172.
166. Stockhausen *WO* 9401476-A1.
167. Xu, X. Q., Duan, M.L., Dong, X. H. and Feng, J. X. Huaxue (2000) *Gongye Gongcheng (Tianjin),* 17(5), 307.
168. Mitsubishi (1992) *JP Kokai* 04055411, 04055412.
169. NSKK (1994) *JP Kokai* 06298866.
170. Tokiwa, Y. and Kitagawa, M. (1998) *Sci. Technol. Polym. Adv. Mater.* 447.
171. Dainippon Ink and Chemicals (2000) *JP Kokai* 2000290502.
172. BASF *US Patent* 5027941, and *DE* 4029348.
173. Dordick, J. Univeristy Iowa, State Res. Found. *WO* 92221765.
174. Kim, S., Stannett, V. T. and Gilbert, R. D. (1973) *J. Polym. Sci.Lett. Edn.* 11(12), 731.
175. Buchanan. C. M., Komanek, R., Dorschel, D., Boggs, C. and White, A. W.(1994) *J. Polym. Sci.* 52(10), 1477, and Gu, J. D., T.Ebereiel, D., McCarthy, S. P. and Gross, R. A. (1994) *J. Environ. Polym. Degrad.* 1(2), 143.
176. Rhone-Poulenc (1993) *Announcement Eur. Plastics News* #20, 16.
177. Arch, A. (1993) *J. Macromol. Sci.* A30(9/10), 733.
178. Monal W. A. and Covac, C. P. (1993) *Macomol. Chem. Rapid Commun.* 14, 735.
179. Canon KK (1994) *JP Kokai* 06329966.
180. Yu, C. L., Kumar, R., Pu J. and Mccarthy, S. (1998) *216th ACS Book of Abstracts, Poly.* 251.
181. Mayer J. M. and Kaplan, D. L. (1991) *US Patent* 5015293.
182. Eastman Kodak *WO* 92210521-A1.
183. Eastman Chemical *EP* 560891-A1.
184. Henkel *EP* 254025-B.
185. Masumura S., et al. (1993) *Angew. Makromol. Chem.* 205, 117.
186. Matsumura, S., Maeda, S. and Yoshikawa, S. (1993) *Macromol. Chem.* 191(6), 1269.
187. Matsumura, S., et al.(1992) *Poly Preprints Japan* 41(7), 2394.
188. Matsumura, S., Aoki, K. andd Toshima, K. (1994) *J. Amer. Oil. Chem. Soc.* 71(7), 755.
189. van Bekkum, H., et al. (1987) *Prog. Biotech.* 3, 157.
190. H. van Bekkum, (1988) *Starch Starke* 192.
191. A. C. Besemer (1993) *Thesis (Delft)*, and *EP* 4273459-A2, *WO* 9117189.
192. Bessemer, A. C. and van Bekkum, H. (1994) *Starch-Starke* 46, 95-100; 46, 101-106.
193. deNooy, A. E. J., Bessemer, A. C. and van Bekkum, H. (1994) Rec. Trav. Chim. 113(3),165-166.
194. Procter and Gamble *EP* 542496-A1.
195. Henkel *DE* 4203923-A1; *WO* 9308251-A1.
196. Novamont *EP* 548399-A1, *WO* 9218542-A1, *WO* 9238205.
197. Roquette Freres *EP* 455522-A, *US Patent* 4985553.

198. Mercian Corp. *JP* 05017502.
199. Hoechst *US Patent* 5223642, *WO* 9102712..
200. Hoechst *US Patent* 5238597.
201. Swift, G., Paik, Y. H., Simon, E. S. (1993) *ACS Polym. Mater. Sci. and Eng. Abstr.* **69,** 496; and (1995) *Chemistry and Industry*, Jan 16th, 55.
202. Swift, G. (1998) *Book of Abstracts, 216th ACS National Meeting, Poly*-002

12

PLASTICS AND THE ENVIRONMENT

JAMES GUILLET
Department of Chemistry
University of Toronto
Canada

1 Introduction

Over the past half-century, synthetic plastics have become the major new materials for everything from replacements for human body parts to the construction of supersonic aircraft and spacecraft. Much of this growth has taken place at the expense of more traditional materials, such as steel, aluminum, paper and glass. Quite understandably, the industries associated with their manufacture have fought back through public relations campaigns intended to protect their own specific markets. Unfortunately, advertising agencies are not the most reliable sources of scientific information, and as a result, the public perception of the role of plastics in society is based more on what can only be described as "mythology" than on demonstrable facts. It is the purpose of this chapter to deal with a number of the misconceptions about plastics generally and about the role of degradable plastics in particular. The first of these will involve resource considerations.

The first North American scientific meeting on this subject was held in New York City, August 27-September 1, 1972, sponsored by three Divisions of the American Chemical Society and the National Academy of Science. The proceedings were published in 1973 under the title *Polymers and Ecological Problems* [1]. The work in this current chapter was initiated after an invited lecture given by the author to a plenary session of the Society of Plastics Engineers in San Francisco, May 14 1974, entitled "Plastics, Energy and Ecology", much of which was published in *Plastics Engineering* in 1974 [2] and later in *Degradable Plastics*, (1st Edition) [3]. The present chapter contains more recent information and précis written with the kind permission of Dr.George Harlan of some material on ethylene-carbon monoxide (E-CO) polymers from his chapter in the 1st edition of the latter publication. Since EC-O is by far the largest volume degradable plastic now in production, it was felt that this material should be summarized in the present volume.

G. Scott (ed.), Degradable Polymers, 2nd Edition, 413-448.

2 Resource considerations for plastics

One of the common mythologies about plastics is that because they are made from oil or natural gas, which are "non-renewable" resources, their use should be discouraged to extend the useful lifetime of these raw materials for other purposes. Apart from the fact that chemical raw materials represent less than 5% of the total oil and gas production world-wide, plastics can be made from a wide variety of carbon sources, and in fact, the restriction of automobile speeds to 55 miles per hour in the US in the 1970s and more recent legislation on automobile fuel economy have saved more oil than was necessary to produce all of the world's plastics.

Figure 1 shows a prediction made in 1972 of the rate of plastics production in the US, based on an annual increase of 6%. This is also the actual average recorded for the most recent 10 year period (1989-1999). The experimental points added later show that in spite of three oil crises and several recessions, production is still on target. At some point within the range of the actual data, the total volume of plastics exceeded the total volume of steel. It has been traditional to name the ages of human society on the basis of the materials which are used in making tools and in construction. The Stone age, the Bronze age, the Iron age and the Steel age have passed and it is now clearly the Plastics Age.

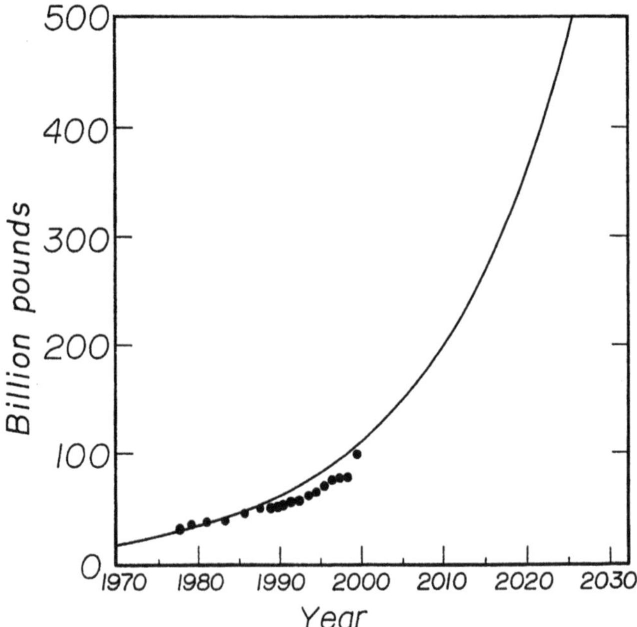

Figure 1. Computer estimate of plastics production in the U.S (solid curve). Data points are actual production.

The changes from the Stone age to the Bronze age and the Bronze age to the Iron age took hundreds – maybe thousands – of years, whereas the change from the Steel age to the Plastics Age has taken only a single lifetime. It is quite understandable that these rapid changes are difficult to come to grips with, especially for people who are not associated with the technology. As a result, there is much public concern about the long-term effects of this new technology. For example, Barry Commoner [4] in the 1960s, suggested that the earth could be wrapped in the current production of nylon film, and that because nylon was a new material and there were no biological organisms which were capable of destroying it, soon the earth's surface would be buried ten feet deep in plastic waste. This was one of the first instances where concern about the biodegradability of plastics was raised.

However, not many people realize that polyethylene, another product that Commoner was very worried about, was originally produced in Britain by the fermentation of grain to produce alcohol and dehydration of alcohol to produce ethylene. It and many other common plastics including polystyrene and polyester can be produced by known chemical processes from "renewable resources" such as *wheat*, cellulose, starch, biomass, etc. through the following chemical reactions.

$$\text{Starch} \xrightarrow{\text{fermentation}} CH_3CH_2OH \tag{1}$$

$$CH_3CH_2OH \xrightarrow{\text{dehydration}} CH_2{=}CH_2 + H_2O \tag{2}$$

$$CH_2{=}CH_2 \xrightarrow[\text{pressure}]{\text{heat}} \text{polyethylene} \tag{3}$$

$$CH_2{=}CH_2 \longrightarrow \text{benzene} \xrightarrow{+\ CH_2{=}CH_2} CH_3CH_2\text{–}C_6H_5 \tag{4}$$

$$CH_3CH_2\text{–}C_6H_5 \longrightarrow CH_2{=}CH\text{–}C_6H_5 \text{ (styrene)} \xrightarrow{\text{heat}} \text{polystyrene} \tag{5}$$

However, as pointed out in earlier publications [1-2] the use of renewable resources such as biomass, to replace synthetic plastics or fibres, requires the utilization of vast areas of agricultural land to provide the raw materials – a scenario which is unacceptable in a world in which millions of people starve every year. Furthermore, even when grain production is highly subsidized in most developed countries, the cost of production would be many times greater than processes using gas, oil or coal as raw materials.

For example, it has been suggested that plastics should be replaced by wood, a natural renewable resource. It takes a lot of land to grow wood. The maximum production in Canada is about 5000 pounds per acre per annum for virgin timber that has been growing for approximately 500 years on the west coast. The average annual production rate for natural fir forests in the U.S. is about 2,000 lbs per acre per year [5]. If the U.S. plastics production in the year 2000 of 103 billion pounds were somehow replaced by wood and paper from natural forests, it would require harvesting at least an additional 52 million acres of land. To put this in scale, it is equivalent to the area of all the agricultural land in five states the size of Michigan, Louisiana and Virginia. Any strategy that requires a return to the use of so-called renewable resources like wood, cotton, leather, and wool results in a depletion of the biosphere, the thin green layer that covers our earth. In order to replace synthetic materials, more trees will need to be cut down and more land devoted to growing trees, grain, and other sources of cellulose. This is a strategy that is not sustainable over the long term and would cause untold economic and physical hardships, even in the short term, particularly among the less developed countries of the world. In fact, what are called renewable resources are almost all renewed using the energy of the sun. If there is enough energy, all resources are in principle renewable, as will be explained in the next section.

Plastics are based on the common elements: carbon, hydrogen, nitrogen and oxygen, and smaller amounts of chlorine and sulphur. These are among the most common elements on the surface of the earth and will never be depleted. Their cost will vary more or less, depending on the energy required to convert them into chemical intermediates.

Strategies for resource allocation can best be evaluated on an energy cost basis, since calories (or joules) do not change over time, unlike the value of the currencies used to acquire them. A preliminary attempt to establish a basis for such analyses was given in 1974 [2].

2.1 ENERGY AND RESOURCE ANALYSIS

It is commonly believed that the natural resources of the earth are being used up, and that our fate as a species on this planet depends in a critical way on the conservation of these resources. Shortly after Commoner [4] and Meadows and Meadows [6] came up with proposals for "zero growth", many countries in the developed world experienced cycles of less than zero growth and found it very unpleasant. The new watchword then became "sustainable growth", but no one determined how to measure it. Furthermore, there is no consensus on what is sustainable, or indeed how to define "growth".

Theologians have stepped in where economists fear to tread and a common view was expressed in a recent encyclical. It is stated that the advanced countries have used up so much of the world's resources that there is not enough left to bring the underdeveloped countries up to the economic levels of the developed countries. At the Environment Summit in Rio de Janeiro in 1993, it was proposed that a supernational body should be set up to ration the resource expenditures of the developed countries, particularly hydrocarbons used for fuel and energy.

This idea that the control of resources is a critical matter for the survival of a nation-state was the source of most of the wars of the last two centuries. For example, both Japan and Germany wished to gain more territory in order to control resources, because they believed if they did not own key resources they would not be able to achieve their full industrial potential. Yet these nations who lost the wars and lost control of occupied or colonial territory are in fact among the most powerful economic countries in the world. They do not territorially control their resources. Neither Japan nor Germany have very extensive natural resources, and yet they have been very productive economically. It is not physical resources that make a major industrial nation. It is the resource of an educated population with the imagination, the scientific insight and well functioning communities that make resources useful. The Japanese, for example, are quite happy to buy oil at any price because it represents such a small percentage of the total value of the products that they produce. In 1992, Hong Kong, which has a total area not much greater than the city of Toronto and has no natural resources, was the tenth largest exporter of manufactured goods in the world.

2.2 THERMODYNAMIC CONSIDERATIONS

In order to consider resources in scientific terms it is necessary to make concise definitions of certain well known phrases or terms. What, for example, is meant by growth? This cannot easily be expressed in currencies, because their value changes with timc. Even adjustments to "constant dollars" to account for inflation or deflation is not enough since many of the most important characteristics of society cannot be given a monetary value and are seldom included in common economic terms such as gross national product (GNP). Most environmentalists equate growth with "consumption" of resources. How then are resources "consumed"?

A scientific approach can help to clarify our understanding of these terms. Whereas monetary values change, energy units do not. In thermodynamic terms, what is measured is the value in energy units, such as calories or joules. Planet Earth is called, in chemical thermodynamics, an isolated system. It is isolated in the sense that very little mass enters or leaves the earth. Nearly everything that is here has always been here, except for a few meteorites that have come in from outer space and a few satellites that have been sent out and are now lost in space. To a very good first approximation the only thing which enters or leaves our terrestrial sphere is radiation. The radiation coming in is primarily that of the sun, but small amounts also arrive from outer space. In return, the earth radiates energy back to space.

The word "consumption" is intimately related to chemistry and energy. When a piece of bread is eaten (consumed), its mass is not destroyed, nor is its overall atomic composition changed. Some chemical components of the bread are "metabolized", that is, they are absorbed by the body and converted to other chemicals through a complex series of chemical reactions. The food components of the bread are mostly carbohydrates and these are used primarily to provide energy by a reaction (respiration) very similar to combustion.

$$(CH_2O)_n + nO_2 \longrightarrow nCO_2 + nH_2O + \text{energy} \tag{6}$$

Some components of the bread may also be used in the construction of body parts, and the remainder is excreted. In all of these chemical and physical processes there is no change in total mass, i.e., the sum of the masses of the reagents equals the sum of the masses of the products. This is a simple statement of the First Law of Thermodynamics (the law of conservation of matter) and it applies to nearly all chemical transformations except for those involving nuclear reactions in which mass is converted to energy.

For example, the grain from which the bread was made was once stored in a grain elevator near the farms where the grain was produced. As such it represents a valuable resource to the farming community which produced it. Its monetary value can be evaluated easily by determining its mass (weight) and the current price farmers can get for grain of that quality, less the cost of shipping if that is not paid by the customer. Suppose that a tornado destroys the grain elevator and spreads the grain around the surrounding fields. The weight of the wheat is unchanged but the resource is in a much higher state of "entropy". Entropy is the measure of the "mixed upness" of a system. In order to be sold it must be collected, cleaned, dried and replaced in another storage area. Its new value as a resource can be evaluated exactly, using an equation analogous to that used originally to evaluate the energy changes in chemical reactions:

$$G = H - TS \tag{7}$$

where G is the free energy (or useful work), T is the temperature, H is the enthalpy and S is the entropy. The resource analogy can be written

$$V = V^0 - V_R \tag{8}$$

where V is the value of the resource in its present form, V^0 is the value of the resource in its pure concentrated state, and V_R is the cost or value of the work required to convert the resource back into its pure concentrated state. In the grain elevator example it seems likely that V_R will be greater than V^0, leading to a negative value for V, and the farmers will most likely apply for a government grant to replace the loss if they are not covered by crop insurance. In principle, the value of all resources can be estimated in this way. Canadians are particularly familiar with this arithmetic in the estimation of the value of mineral resources (but not in the stocks of the companies that mine them!). The value of a gold deposit is given by

$$\text{Value} = nV = n(V^0 - V_R) \tag{9}$$

Where n is the estimated number of ounces of gold in the deposit, V^0 is the price of an ounce of gold as quoted daily on the gold exchange and V_R is the estimated cost of extraction of an ounce of gold from the ore.

It is clear from this analysis that high entropy is associated with high recovery costs in mineral refining. The same principle applies to recycling strategies where, to become economic, it is necessary to reduce the entropy of a resource like municipal waste by presorting the more valuable components. Furthermore, it seems likely that in most instances, the major component of what is meant by "consumption" is the increase in entropy which occurs when a resource is consumed.

What is a "non-renewable resource"? The models produced by Meadows and Meadows [6] and Commoner [4] suggested that the earth's population would suffer a catastrophic decline around the year 2030 because the so-called non-renewable resources would have been used up. A typical example of a non-renewable resource is an oil well. The oil at the bottom has been created over millions of years essentially by trapping the energy of sunlight. It is stored solar energy which has become converted into oil. When a pipe is put down, the pressure of the rock above forces the oil up. About 95% of the world's oil is burned with oxygen to give carbon dioxide, water and energy.

$$(CH_2)_n + 3n/2\ O_2 \rightleftarrows nCO_2 + nH_2O + \text{energy} \qquad (10)$$

However, if energy is available, this process can be reversed. For example, it is possible, in principle, to use the energy of a nuclear reactor to collect CO_2 from the atmosphere and convert it to hydrocarbon fuels which could be pumped to underground caverns for storage, thus renewing the resource.

Photosynthesis in green plants, which is the major process for providing energy for living systems, uses the sun's energy to renew that resource, i.e.;

$$\text{Solar energy} + nCO_2 + nH_2O \longrightarrow \underset{\textit{carbohydrate}}{(CH_2O)n} + nO_2 \qquad (11)$$

This process produces both carbohydrate fuel from CO_2 and water, and also renews the supply of oxygen, which is essential to life on earth. A careful analysis of what are called renewable resources reveals that all the important ones such as hydroelectricity and biomass are renewed by solar energy. This suggests that further scientific attention should be paid to other types of solar processes.

A further example of consumption of resources is illustrated by the production of iron and steel. Iron is one of the most common elements in the world, and it is proposed that it is the major component of the earth's molten centre. On the earth's surface, it occurs mainly as iron oxide, often in large concentrated deposits. It is converted to iron by heating with carbon (usually in the form of coke) i.e.;

$$\text{Energy} + 2Fe_2O_3 + 3C \longrightarrow 4Fe + 3CO_2 \qquad (12)$$

The iron is used as such, or converted into steel by the addition of more carbon. Iron and steel are used to make everything from cars and bridges to metal cans. Left alone in the environment it usually returns (oxidizes) back to Fe_2O_3, but in a state of much higher entropy. Iron from ships and cars is effectively recycled by a melting process. Smaller articles are less likely to be recycled and their value as resources is reduced substantially by the high entropy factor.

If the resources are not actually consumed, how can one best describe what ultimately happens as resources are depleted? Here is a restated form of the First Law of Thermodynamics:

$$\text{Production} = \text{garbage} + \text{litter} \qquad (13)$$

Everything that is made (produced) is ultimately going to be disposed of. Garbage is that part of the total production that the community actually takes charge of and does something with. It is burnt, buried in the ground or recycled. Litter is that portion that escapes the intentional disposal system. Everything that is made, whether it be a building, a chair, a car, a can, a bottle, or a roll of plastic film, will obey this law. Even the pyramids will eventually disappear through various forms of erosion. It is just a question of time!

It is now generally accepted that a modern economy must show some growth (sustainable growth) to provide an equitable society. More consumer goods must be produced. If more goods are produced it is necessary to think about the disposal of those goods after they have served their useful function, whether this be in 100 years, or five years, or two weeks. It seems incredible that governments believe that they can continue to increase industrial production and at the same time reduce the disposal costs associated with it. The real problem with science and technology in this area is that almost all of the scientific work has been devoted to the production side of the equation. Huge sums of money are spent every year in industrial countries developing new products and production processes. Comparatively little has been done to improve or expand disposal procedures. Most disposal of waste is still by burial or combustion techniques which are more than 5000 years old. With the huge expansion in production capacity which is expected to occur during the present century this subject must be approached with more urgency.

3. Environmental considerations for packaging materials

3.1 ENERGY AND RESOUCES

A major concern of the environmental movement world-wide, has been the increasing use of disposable packaging (frequently plastics) which is considered to waste non-renewable resources and impose unacceptable burdens on municipal disposal facilities. In concurrence with susceptible politicians, laws and regulations have been promulgated in many jurisdictions which have increased both the cost to the taxpayer of garbage disposal and the loss of energy resources.

Using the energy analysis outlined in the previous sections, a rationale can be established for evaluating the potential benefits (if any) of various legislative proposals. Energy costs for the production of various packaging materials (Table 1) and products were first discussed by Guillet [2], based on the data of Makhijani and Lichtenberg [7]. More detailed calculations were published by Boustead and Hancock [8] and these incorporated some of the improvements in energy recovery in more modern manufacturing plants, but do not differ much from Guillet's earlier calculations.

Table 1. Energy requirements for the production of material used in packaging applications

Material	Energy requirements (kWh, thermal per pound)
Aluminum	33.6
Steel	6.3
Glass	3.6
Paper	3.2
Plastic	1.4

There is, however, one fundamental difference between the two sets of calculations in that, for plastics, Boustead and Hancock added the energy value of the petroleum raw material used for the manufacture of plastics, but not that for wood, paper, iron or aluminum, all of which could be burned to supply energy. This unfairly distorts the balance of the calculations against the use of plastics. For this reason, the data used in the following tables are based on the earlier studies by Makhijani and Lichtenberg [7].

The energy requirements for selected beverage containers are shown in Table 2. It is obvious that the energy cost of plastics per pound is substantially less than that of all its major competitors. The saving of energy is even more obvious when one includes the weight factor, as is done in Table 2. The energy cost of glass bottles and aluminum cans, for example, is of the order of 20 to 30 times that of comparable plastic containers; therefore many returns per container are necessary to compete with disposable plastics.

It is not possible to make a general rule about the savings of energy with different packaging systems since it depends to a large extent on the number of returns that one can expect with a returnable glass bottle, and the amount of energy required to return the bottle and wash it, as compared with the disposable system. What is clear from these tables is that returnable systems do not automatically save either energy or raw materials. In a society such as the United States in which a large part of the electrical energy is obtained from burning hydrocarbons, it is clear that a considerably larger amount of petroleum or natural gas is consumed in synthesizing a glass bottle than in making a plastic container to hold the same amount of liquid.

Table 2. Energy requirement per beverage container

Container	Weight (ounces)	Energy used per container (kWh)
Aluminum can	1.41	3.00
Returnable soft drink bottle	10.6	2.40
Returnable glass beer bottle	8.83	2.00
Steel can	1.76	0.70
Paper milk carton (1 pint)	0.92	0.18
Plastic beverage container	1.23	0.11

It is often assumed that the use of reusable container such as glass bottles saves energy and resources. This is clearly not the case, as is demonstrated by the data on milk containers in Table 3.

*Table 3. Comparison of energy cost of manufacture of disposable and returnable milk containers **

Container	Weight (ounces)	Energy used in manufacture (kWh)	Energy Ratio	Heat Content (kcal) †
Two-quart glass milk bottle	37.1	8.36	99.5	0
Two-quart plastic Pouch (PE plastic bag)	0.97	0.84	1.0	317

*Complete energy cost estimates of using the system would involve additional costs for the returnable system in transporting the extra weight of the glass bottle (2.3 lb) and in washing and sterilizing it for reuse. An allowance for the energy cost of garbage collection must also be made for the disposable system.
† This amount of heat energy would be obtained by burning the plastic in air; it is sufficient to raise the temperature of about one gallon of water to boiling point.

When the comparison is made between a glass milk bottle distribution system and plastic pouches, there is a clear energy advantage to the use of plastic disposable pouches and probably advantages in sanitation as well. The energy cost of the bottle is one hundred times that of a plastic pouch. A bottle would have to be used more than 100 times before it would be a lower energy cost system of packaging. Furthermore, since the primary raw material for glass is sand, a cheap and abundant raw material, the primary non-renewable resource used is energy. Fifty times more oil or natural gas would be needed to make a glass bottle than a plastic package for milk.

A further consideration, now that extensive recycling systems are in place, is to re-evaluate the use of aluminum cans that are promoted because they are easy to recycle and provide a large savings of resources when recycled. However, even after extensive publicity, and much popular support, the maximum recovery of aluminum beverage containers in the US and Canada seldom exceeds 50%. A country like Canada might be expected to use about 10 billion beverage containers per annum. At 50% recovery, the energy wastage of the five billion not recovered represents a total of 15×10^9 kilowatt hours (kWh) as compared with 0.3×10^9 kWh for the plastic bottle, a factor of 30 times less. At an energy cost of 10 cents per kWh, the cost of energy lost in the aluminum containers would be about $1.5 billion dollars per annum.

Clearly, the proper way to dispose of plastic materials in garbage is to burn them in an incineration system equipped to use the heat of combustion either to generate electricity or to provide steam for municipal heating. In this way, one can consider that the use of a plastic container is simply borrowing a barrel of petroleum to make the plastic, using it once or twice as a container, and then recovering about the same amount of heat from it as if the barrel of petroleum had been burned in the first place. Such applications obviously represent a considerable conservation of energy and raw materials.

3.2 PAPER VERSUS PLASTICS – AN ENVIRONMENTAL ASSESSMENT

In spite of much scientific evidence to the contrary, the public's perception of plastics is that they are potentially toxic and incompatible with nature. For this reason, many consumers demanded that paper packaging be used, for example, in supermarkets and fast food chains. Some municipalities in the US and Canada have attempted to ban the use of plastic packaging for environmental reasons.

A possible substitute for plastic packaging is paper. A typical grocery bag made from high-density polyethylene weighs approximately 5 g, whereas a heavy kraft paper grocery bag weights 35 about g. Replacement of the plastic grocery sack by paper would thus increase the actual weight of garbage from this source by a factor of about seven! Low-density polyethylene grocery and shopping bags are somewhat heavier (about 10 g) but are still about one-third that of paper. There are similar advantages to the use of foamed polystyrene fast food containers as compared to paper products.

The resource consequences of a shift from plastic back to paper are enormous. Assuming that the lower figure is taken (i.e., a factor of one-third the weight of a paper package), the replacement of plastic by paper would increase the total weight of garbage by 55 billion pounds per year, and would require the additional production of 82 billion pounds of paper per annum. Packaging grade paper requires about 1.1 to 1.2 pounds of wood per pound of paper. To produce the required 82 billion pounds of paper per annum would require tens of millions of acres of forest land devoted to paper production. When the environmental damage already inflicted on the earth by cutting down trees in countries such as Brazil and China is considered, a solid waste strategy which might result in harvesting many millions more in North America does not appear to be beneficial to the environment, particularly when it is known that forests are important in reducing the CO_2 concentration in the atmosphere.

Finally, there is the question of the energy cost of a conversion from plastic to paper. Makhijani and Lichtenberg [7] estimate the energy requirements for the production of paper and plastic (polyethylene) at 3.2 and 1.4 kWh per pound, respectively. A simple calculation indicates that the extra energy required to produce the paper would cost about 224 billion kilowatt hours, or 25.5 million kilowatt years. It would be unwise to generate this extra energy using coal, oil, or gas because of the greenhouse gases produced. A possible solution would be the construction of about fifty 500-megawatt nuclear power reactors. The issue of materials for packaging is a very complex one, and simple solutions such as banning plastics can result in very undesirable ecological consequences.

3.3 AIR AND WATER POLLUTION IN PAPER PRODUCTION

A further consideration is the air and water pollution associated with the manufacture of paper and plastics packages. Table 4 shows data produced by the West German Federal Office of the Environment, Berlin [9] on the air and water pollution associated with the production of 50,000 carrier bags of polyethylene, unbleached kraft paper and "paper combinations". The latter is the formulation used for most paper carrier bags approved for use in Germany. The production of plastic carrier bags causes significantly less air pollution, and as much as 200 times less water pollution compared to that of paper carrier bags.

Table 4. Air and water pollution associated with the production of 50,000 carrier bags

Environmental Burden	Polyethylene	Unbleached kraft paper	Paper combinations
Energy (GJ) for production process	29	67	69
Air pollution (kg)			
SO_2	9.9	19.4	28.1
NO_x	6.8	10.2	10.8
CH_3	3.8	1.2	1.5
CO	1.0	3.0	6.4
Dust	0.5	3.2	3.8
Waste water burdens (kg)			
COD	0.5	16.4	107.8
BOD_5	0.02	9.2	43.1

The report concludes with the statement: "The replacement of polyethylene by paper carrier bags makes no sense ecologically. The production of polyethylene carrier bags requires less energy, and in the process results in less burden to the environment. There is no significant difference in the disposal of polyethylene and paper bags at landfill sites or in incineration plants"

3.4 THE ROLE OF PHTODEGRADABLE PLASTICS IN PACKAGING

Environmental groups and the general public have often failed to distinguish between two distinct problems relating to solid-waste management of packaging materials. One is the disposal or recycling of waste packages included in the municipal garbage stream. The other, which is more difficult, is the problem of litter, i.e., that part of solid waste that escapes collection and contaminates beaches, forests, and other natural regions which can be hundreds of miles away from the site where the litter was discarded.

"Garbage" can be defined as the discarded solid-waste products of household or industry which are collected and disposed of in some central facility such as a dump, landfill, or incinerator. Litter, on the other hand, may be defined as a synthetic object in a place where it should not be. For example, a fallen tree in the forest is not litter, but a discarded wooden box made from the same material in the same place is. Paper in a rubbish bin is not litter. The same piece of paper blowing along the side of a road definitely is. Surveys of litter show that by far the greatest proportion consists of containers or packages of various kinds used for food, beverages or tobacco.

Plastics have one major advantage over glass and metal in packaging applications in that they are inherently organic materials, just like banana skins and coconut shells, and it is therefore possible in principle to make them degrade by natural mechanisms once they have performed their primary function as a temporary container. As has been shown in the previous sections, the proper use of disposable packages can result in savings of both energy and resources.

Considering these principles, it is possible to draw up a list of the desirable characteristics for a packaging material [10].

1. It must be resistant to the material which it is to contain and not contribute to the taste, odor, or toxicity, particularly if it is a food product.
2. It must be light in weight and easily formable into an attractive package.
3. It must be cheap and represent a minimal expenditure of natural resources in its manufacture.
4. It must be resistant to microorganisms which might otherwise attack the materials which it contains.
5. It must be stable and maintain its desirable physical properties for at least the lifetime of the product which it contains.
6. It must be disposable or recyclable by conventional disposal technology.
7. It should degrade by some natural mechanism if it becomes litter.

Plastics, as currently manufactured, fulfil all of these characteristics except the last, and until recently point 7 has been considered to be inconsistent with point 4 since it was felt that if the plastic were biologically degradable, it would no longer afford adequate protection against the attack of microorganisms on the product which the plastic is intended to contain. Now, however, it is clear that these two requirements need not be mutually exclusive.

It has been found that the resistance of conventional plastics to microorganisms is primarily due to two factors: (1) the low surface area and relative impermeability of plastic films and moulded objects and (2) the very high molecular weight of the plastic material. Microorganisms tend to attack the ends of large carbon-chain molecules and the number of ends is inversely proportional to the molecular weight. In order to make plastics degradable, it is necessary first to break them down into very small particles with large surface area, and secondly to reduce their molecular weight.

Although the merits of "degradable plastics" as a means of solving some of the problems associated with the *disposal* of packaging materials in the solid-waste stream remain to be demonstrated conclusively, their effective use in litter control is now well established. One of the most successful examples is the Hi-Cone™ beverage carrier developed by the Illinois Tool Works Inc. (Chicago, Illinois, USA), a rectangular sheet of about 1.5 mm polyethylene into which are punched six holes the size of conventional beverage containers. Weighing only a few grams, the manufacture of these packages makes few demands on our non-renewable resources, and they replace plastic and paper carriers which are much more demanding in their resource and energy content. This minimalist design is a triumph of the concept of "under packaging" and reduces both the litter and wildlife problems caused by careless discarding of the carriers by the public.

Since the 1980s, a number of US states, including California, required this package to be photodegradable because it is highly litter-prone. It also endangered the lives of birds who often

became entangled in the rings. After a number of years of development, an ethylene copolymer containing *c.* 1% carbon monoxide was selected for the manufacture of the degradable variety of this six-pack carrier. The resin was available from major plastics producers in the US, and it is estimated that the total volume may now exceed 100 million pounds per annum for this one application [11]. Furthermore, beach and other surveys of plastic litter show at least a ten-fold reduction in the number of these carriers is observed. Its success confirms the validity of the computer models of anti-litter strategies published by Guillet [2].

An amendment to the Plastics Pollution Control Act of 1987 was enacted by the US Government in 1990 which requires all ring carriers in the US to be degradable. Currently, all such carriers, as defined by US Public Law 100-556, are manufactured from a photodegradable ethylene-carbon monoxide (E-CO) resin. This is the largest application of degradable plastics for packaging of any kind in the world to date.

3.5 COMPARATIVE STRATEGIES FOR LITTER ABATEMENT

Photodegradable plastics are designed to address the litter problem, not necessarily the problem of garbage disposal or landfill capabilities. Computer simulations are a useful way of assessing the value of various litter abatement strategies. Guillet and Ainscough [12] developed programs based on US plastics production in 1970, which over the previous ten years had grown at an annual rate of 6%. It was shown that the rate of litter accumulation was almost independent of the rate laws assumed for litter degradation, and depended primarily on the expected lifetime of the littered object, as shown in Figure 2.

The assumptions behind this model are as follows. (1) The production of plastics in 1970 in the United States was 19,600 million pounds. (2) The production of plastics will increase annually by 6%. (3) The proportion of plastics production which will be used in packaging applications will be constant at 20% per annum. (4) Two per cent of plastics packaging will become litter.

The model was then used to predict the results of various strategies for litter abatement. The results are shown in Figure 3. Curve (a) shows the accumulation of litter if the average lifetime is 10 years. Curve (b) shows the accumulation if, as a result of education or fines, the amount of litter is reduced by a factor of five. Very little improvement occurs and the exponential increase in accumulated litter starts again 10 to20 years later. Curve (c) shows the effect of reducing the lifetime of litter to 0.2 years (approximately two months), a target well within the range of modern plastics technology. An immediate improvement in litter accumulation occurs and after 20 years, litter has almost disappeared. Curve (d) shows that nearly the same thing occurs even if people throw away twice as much litter because they think it is degradable. This confirms that the most effective way to deal with the litter problem is by reducing the "lifetime" of the littered object. Photodegradable plastics technology provides the best way to accomplish this objective. These conclusions have been confirmed by actual studies of the accumulation of degradable six-pack plastic rings on Pacific beaches. Those states whose legislation required the use of photodegradable rings showed a substantial decrease in the number of such packages accumulated on their beaches. The remarkable success of the photodegradable products substantiates the value of computer modelling in establishing anti-litter strategies.

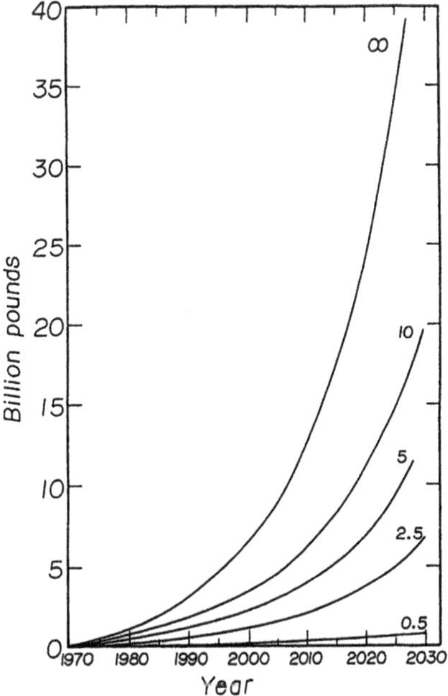

Figure 2. Computer estimates of the effect of half-life on litter accumulation. The curves show half-lives of 0.5, 2.5, 5, 10, and an infinite number of years.

4. Technology of photodegradable plastics

Degradable plastics can be manufactured by a variety of processes [13]. An early approach to the problem was to add oxidation accelerators such as benzophenone to make "unstable" plastics. As the technology developed, more sophisticated systems were devised since packaging materials must not only degrade, but must do so at a controlled and predictable rate. A number of these systems have been described in other chapters of this book.

An alternative approach was adopted in the author's laboratories at the University of Toronto. In this technology, small quantities of a sensitizing group are chemically attached to the macromolecular chains. When a plastic containing this sensitizing group is exposed to natural sunlight, the sensitizing group absorbs radiation, which causes the chain to break at that point and thus form smaller segments. Since the physical properties of a plastic depend on the length of the chain, if the chain is broken, the plastic will become very fragile. If the chain is broken in enough places, it becomes biologically degradable.

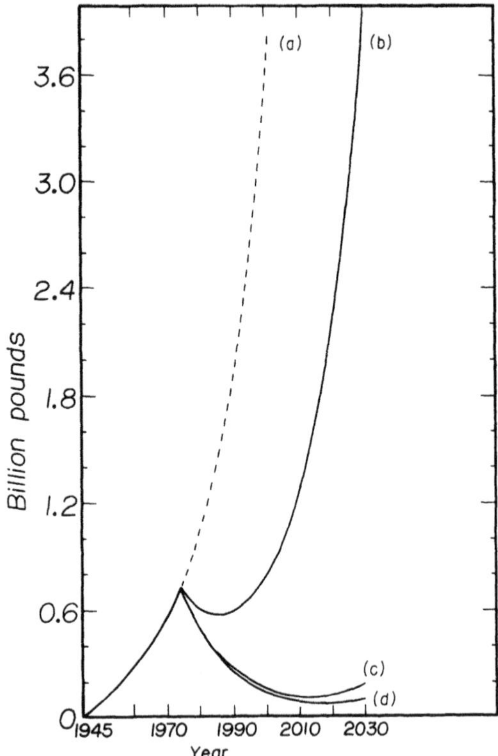

Figure 3. Computer estimates of the effect of various strategies for litter abatement (see text).

An alternative approach was adopted in the author's laboratories at the University of Toronto. In this technology, small quantities of a sensitizing group are chemically attached to the macromolecular chains. When a plastic containing this sensitizing group is exposed to natural sunlight, the sensitizing group absorbs radiation, which causes the chain to break at that point and thus form smaller segments. Since the physical properties of a plastic depend on the length of the chain, if the chain is broken, the plastic will become very fragile. If the chain is broken in enough places, it becomes biologically degradable.

The process was first disclosed in an article in Time Magazine on May 11, 1970, and was awarded Canadian Patent 1,000,000 and a Gold Medal for creative invention by the Canadian Government in 1976.

4.1 THE ECOLYTE PROCESS

The Ecolyte^{TM} process [1] developed at the University of Toronto involves inclusion in the backbone of the chain a polymer of a group of the general structure where R and R′ are various alkyl and aryl substituents. By changing the nature of these two groups, one can control the rate of the degradation process. When the carbonyl group absorbs a quantum of ultraviolet light, the classical photochemical reactions which occur are (a) the Norrish type I reaction, a free-radical split occurring at the carbonyl group to give two free radicals, and (b) the type II process, an intermolecular rearrangement resulting in a scission of the main chain to give a methyl ketone and a terminal double bond.

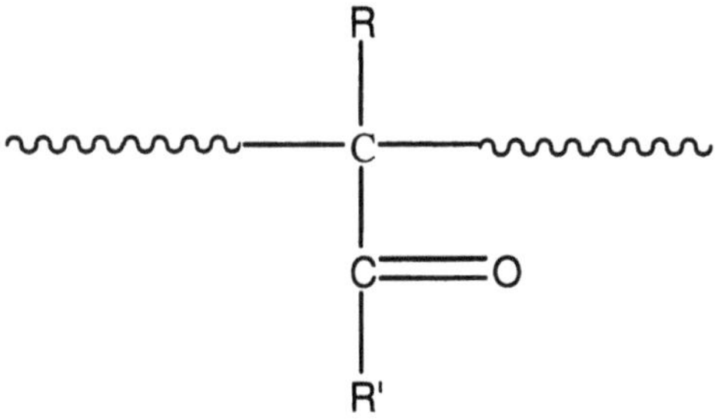

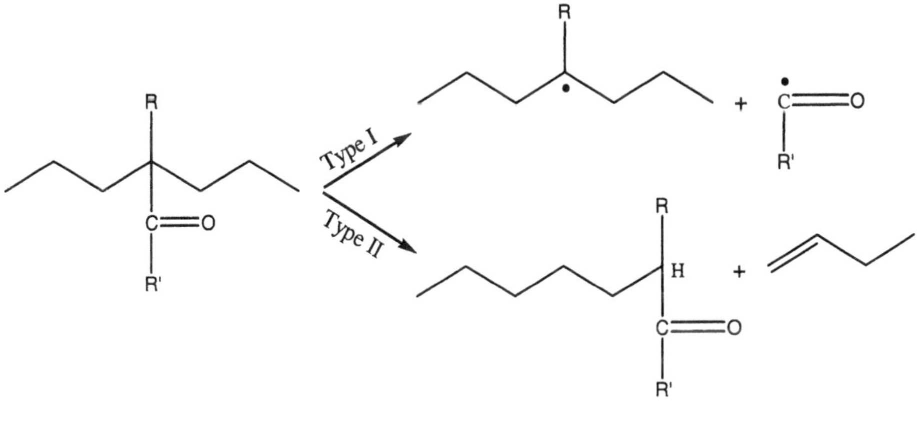

Copolymers of this type were first described by Guillet and Norrish [14,15] in 1955 and an ethylene copolymer suitable for packaging applications was disclosed by Hartley and Guillet in 1968 [16].

The type I process in ketone-containing polymers gives two free radicals, one polymeric and one small acyl radical. The polymeric radical can undergo a rearrangement known as β-scission which results in a break of the C—C bond in the backbone of the polymer and a consequent reduction in molecular weight. The type II reaction, however, is the major photodecomposition process that causes chains to break. In the presence of oxygen, both radical sites can induce photooxidation processes which cause chain degradation over a longer time scale.

Alkyl ketones have absorbances with a maximum at around 280-290 nm and they cut off rather sharply at about 330 nm. This is important since most synthetic polymers do not absorb light in the region above 300 nm. Figure 4 shows a solar spectrum along with that of other light sources. The emission spectrum of the sun has an approximate Boltzmann distribution of radiation which cuts off rather sharply at 300 nm because of the absorption of the ozone layer in the upper atmosphere. On the other hand, the emission of typical fluorescent and incandescent lamps cuts off around 330 nm. As a result, ketones are stable compounds photochemically as far as visible light is concerned and undergo photochemical reactions only if they are irradiated with light of wavelength less than 330 nm, such as occurs in natural sunlight. This gives the possibility of producing a packaging material which is stable in visible light, but which degrades when thrown away in the outdoor environment. The critical portion of the sun's spectrum, between 290 and 330 nm is called the "erythmal region" and is the radiation responsible for tanning and sunburn of the human skin.

Table 5 shows the relative intensity of sunlight in various regions of the spectrum. For the erythmal radiation, noon sunlight in Arizona is about 300 or 400 times the intensity of an ordinary fluorescent light. In the near UV range there is more radiation in artificial sources although it is still rather small compared to solar radiation. It is quite obvious that there will be a much larger effect in sunlight than there is under any of these normal lighting conditions. The other factor which is important is that ordinary window glass filters out the erythmal radiation of the sun, so that packages will not degrade if exposed to the sun behind store windows.

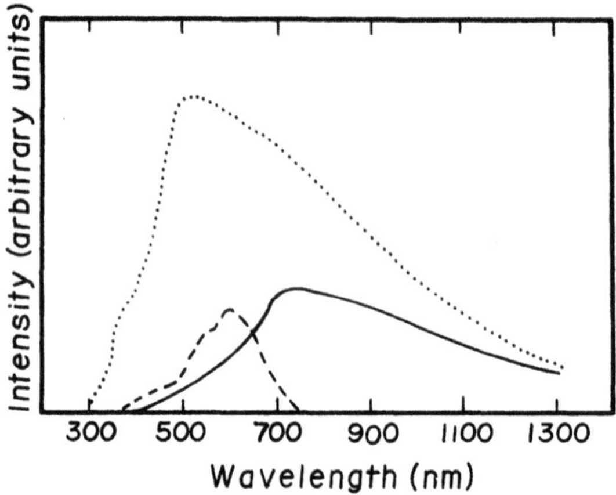

Figure 4. Emission spectra of the sun and artificial light sources: • • • sunlight; ——incandescent lamp; - - - fluorescent lamp.

Table 5. Output of artificial lighting compared to solar radiation ($\mu W\ cm^{-2}$)

Type of lamp (W)	Erythmal (λ=280-320 nm)	Near UV (λ=320-400 nm)	Visible and IR (λ>400 nm)
Incandescent			
40	0	0.21	21
100	0	0.89	71
500	0	6.55	409
Fluorescent			
40	0.8		
Noon sunlight, Arizona	259	4640	88,000

It is not necessary for the plastic to be in direct sunlight in order for the degradation process to occur. Over 50% of the total amount of ultraviolet radiation comes from the sky rather than from the sun itself. Consequently, even if the plastic is in the shade it will still be receiving skylight and hence will degrade. In fact, as long as the plastic can be seen outdoors, it will be undergoing degradation. The rate at which the chains will be broken depends only upon the intensity and duration of the UV light absorbed by the sample. In northern latitudes such as in Canada, the intensity of this UV light will vary with the time of year. This means that in the winter the rate of degradation will be rather slow, while during the summer, the rate will be considerably more rapid. In equatorial regions the intensity of UV radiation does not vary appreciably throughout the year.

Surprisingly, the total amount of UV radiation does not vary much over the surface of the globe. During the Arctic summer, for example, the amount of UV radiation is comparable to that of more temperate regions simply because of the longer daylight hours. This then provides a mechanism for degradation in very cold Arctic regions where biological processes are either very slow or non-existent.

Although the chain-breaking process begins as soon as the plastic is exposed to solar radiation, there is a certain time necessary before an appreciable change in the physical properties occurs. The reason for this is that above a certain molecular weight, which is sometimes called the critical molecular weight, there is only a small change in the physical properties of the polymer as the molecular weight decreases. Once the critical molecular weight is reached however, any subsequent decrease in molecular weight will cause a drastic change in the properties of the material. This means that even after exposure to solar radiation, the plastic material will still retain its useful properties for a certain period of time, and this time can be controlled at will in the manufacturing process.

Ecolyte plastics are made by copolymerizing ketone-containing comonomers in small amounts with ethylene, styrene, or other monomers used in the manufacture of commercial plastics. The process is covered by a number of patents [17]. Condensation polymers such as nylon and polyesters can also be made photodegradable by this method [18]. It is even possible with polymers such as poly(vinyl chloride) [19] and poly(acrylonitrile) [20] which normally do not degrade by chain scission. Poly(ethylene terephthalate) (PET) can also be made photodegradable by copolymerization with glycols or diacids containing ketone groups. This can be done either in the initial synthesis of the polymer, or in a later extrusion process using recycled PET as the raw material [21]. It was found that in order to provide an acceptable rate of degradation for polystyrene, for example, it is only necessary to include about 1% of these carbonyl groups in the polystyrene molecule.

In many commercial applications, where rapid degradation is not required, it is convenient to prepare a concentrate containing 2 to 5% of the ketone monomer. This concentrate, which may also be colored with selected pigments, can then be blended with natural resin in ratios of 1:9 to 1:20 and extruded or molded to provide products with the desired rate of degradation. Control of the rate is provided by changes in concentrate and/or pigment concentration. In some cases, the ketone groups can be introduced by a chemical post-treatment. Because of the minor amount of modification required, the physical properties of the photosensitive resin are almost identical with those of the untreated plastic.

Since photodegradable plastics may be used for food packaging, the question of food approval for packages made from them is important. In the Ecolyte process, essentially all of the ketone has been introduced into the polymer chain and is chemically attached to it. For this reason, the ketone groups cannot be extracted from the polymer film or package and hence can have no effect on the toxicity or the taste of the packaged product. This represents a particular advantage of the Ecolyte system in that most other processes make use of additives which are merely dissolved and are not chemically bonded into the plastic and therefore may migrate from the plastic into any food packaged in it.

The first industrial application of Ecolyte technology was in the manufacture of photodegradable polystyrene. It was found that a copolymer of c. 1% methyl isopropenyl ketone (MIPK) with styrene photodegraded in a few weeks when exposed to sunlight in foamed formulations for such applications as coffee cups or fruit basket separators. The patents on the Ecolyte process had been assigned to the University of Toronto by the inventors, and the rights were licensed to a Toronto Company named EcoPlastics Limited. The industrial work was supported financially by a joint venture with a Dutch company, Van Leer, of Amstelveen whose president, Oscar van Leer, was a staunch environmentalist. A commercial process was developed and used both in Holland, Denmark and Canada to produce photodegradable coffee cups for fast-food restaurants like MacDonalds. Because the ketone groups were chemically bonded to the polymer chain there was no migration into the coffee and hence no possible toxicity or taste. The formulation later passed all US. FDA requirements but was not finally approved because of the opposition of a senior administrator of the US Environmental Protection Agency (EPA) who opposed it on the basis that "if people know it is degradable, they will throw much more of it away". She did not accept that the amount of litter depends not on how much is thrown away, but on the lifetime of the litter in the environment, as shown by our computer calculations and confirmed by our experience with degradable E-CO six-pack rings. Van Leer abandoned the project three years later due to the unavailability of some reagents during the first oil crisis and the patents were licensed to Canada's largest plastics producer (Polysar) of Sarnia, Ontario who later were acquired by Nova Corporation of Calgary, Alberta, one of the world's largest producers of polystyrene resin.

Ecolyte later developed a masterbatch process for polyethylene in which MIPK was grafted to a low-density polyethylene substrate. This was produced for agricultural applications and the film was produced by Atlantic Packaging Limited of Toronto. The plastic film was sold mainly for farms in Florida growing melons and tomatoes. Atlantic found it difficult to compete with local producers who used recycled material. The masterbatch was later also marketed by Eco Atlantic in Baltimore, Maryland.

4.2 ETHYLENE CARBON MONOXIDE (E-CO) COPOLYMERS

The copolymerization of carbon monoxide with ethylene in high pressure polyethylene reactors was first discovered by Imperial Chemical Industries (ICI) in England during the late 1930s and early 1940s. Ethylene gas made from the dehydration of ethanol frequently contained CO as an impurity and it was discovered that the presence of the CO group in polyethylene destroyed its unique properties as an insulator for high frequency electric currents, such as found in long-distance telephone circuits and military radar. One of the first commercial applications of partition gas chromatography was its use by ICI to analyze ethylene to control the CO concentration in ethylene to negligible values. A US patent on the synthesis of E-CO copolymers was issued to M. M. Brubaker of the Du Pont Company in 1950. [22].

The development of ethylene-carbon monoxide copolymers as environmentally friendly plastics began in 1963 as part of a fundamental study of the photosensitivity of organic polymers. Hartley and Guillet, at the University of Toronto, studied polymers of ethylene and carbon monoxide donated by Tennessee Eastman who had made them as part of an experimental program in their polyethylene reactors. The authors were surprised to find that even in the solid phase the efficiency of photochemical reaction was relatively high and almost independent of polymer molecular weight, a result which most scientists would not have predicted. This made it possible to investigate a very wide range of polymer reactions of commercial importance, including photoresists for the manufacture of computer chips. Although such copolymers had been known for many years, their UV photochemistry (and hence usefulness in this application) was first reported by Hartley and Guillet [16] in 1968.

The photochemistry of E-CO resins is similar to that of the Ecolyte vinyl ketone copolymers except that the ketone carbonyl group is in the backbone of the polymer chain. As a result, the yield of free radicals by the type I process is very low; hence the polymer does not photooxidize very rapidly and is less useful in a "masterbatch" application.

During the late 1970s, the Hi-Cone division of Illinois Tool Works, Inc. Chicago, IL USA, inventors and producers of the polyethylene loop carriers for beverage-can 6-packs, known as Hi-Cone™ carriers produced a degradable version of their product based on E-CO copolymers. Degradable carriers reduced both the litter and wildlife entanglement problems caused by careless discarding of the carriers by the public.

An E-CO formulation was developed that conformed to government statutes in 1977 and since that time photodegradable can-carriers have been the major use of E-CO as well as being the world's largest application of degradable plastics for packaging of any kind to date [11]. The majority of E-CO carriers are made in the United States. Until recently, there were three manufacturers of E-CO polymers, namely Dow, DuPont and

Union Carbide. Due to mergers in the plastics business, Dow is now the major producer of E-CO.

Ethylene-carbon monoxide is produced commercially by the high pressure co-polymerization of ethylene and carbon monoxide using techniques similar to those used to make high-pressure, low-density polyethylene homopolymer (LDPE). The monomers undergo random copolymerization under well controlled temperatures and high pressures in either tubular or stirred autoclave reactors:

$$CH_2{=}CH_2 + CO \xrightarrow[\textit{catalyst}]{\textit{Heat/pressure}} \text{E-CO copolymer} \tag{14}$$

Wide ranges of average molecular weight and of CO incorporation can be achieved by varying the reaction conditions and monomer concentrations. Typical E-CO resins for extrusion contain 0.5-4.0 weight per cent CO and are 0.5 to 1.5 g/10 minutes in melt index. Densities from 0.928 to 0.936 g ml^{-1} can result for a CO range of 0.5 to 1.6 weight per cent.

Typical data shown in Tables 6 and 7 for an E-CO and a corresponding LDPE homo-polymer, demonstrates the similarities between the two resins. In appearance, as film or extruded items, they are also seemingly identical. It is now known that the C=O groups fit into the PE crystal lattice and hence have little effect on the overall crystallinity. However, E-CO density increases with increases in CO content. It has been noted that at around 16 % CO, E-CO will not float in fresh water, and at about 20% CO, the copolymer will sink in salt water. For rapid photodegradability of marine litter, those CO levels are the upper limitations to assure exposure to the light necessary for photodegradation.

Resin melting points and solidification points decline slightly as CO content increases, but most other physical properties of unexposed E-CO change little as the CO content is varied, at least up to around 13% [23]. The rheological behaviors of E-CO and LDPE are much alike as evidenced by their similar extrusion characteristics.

Due to similarities in their rheological and thermal properties, E-CO resins (with CO levels up to at least 5%) and LDPE of comparable melt index (MI) and molecular weight distribution (MWD) behave in sheet extrusion and blown film processing very much alike. Both E-CO and LDPE, when similarly stabilized with typical polyolefin antioxidants, exhibit very good processing thermal stability, even with high levels of reprocessing, and at temperatures of up to at least 260°C. E-CO copolymers of appropriate melt index can be injection molded and blow molded similarly to LDPE. Mold release, shrinkage, cycle times, clamping pressures, and other factors in the molding of E-CO should be much like those encountered with LDPE of comparable MI and MWD.

*Table 6. General Properties of E-CO and LDPE **

	E-CO	LDPE	Test method
CO content, %	0.9	0	Non-standard
Melt index, g/10 min	0.8	0.8	ASTM D-1238
Specific gravity	0.930	0.927	ASTM D-1505
Tensile strength, Pa	17	18	ASTM D-638
Ult, elongation, %	580	570	ASTM D-638
Secant modulus of elasticity, Pa	190	200	ASTM D-638
Vicat softening temperature, °C	104	102	ASTM D-1525
Brittleness temperature, °C	< -100	< -100	ASTM D-746
Melting point, °C	113.0	115.4	

*Typical values of 2 mm extruded sheet.

It has been shown that ethylene-carbon monoxide copolymers, because of their internal ketone group, degrade at ambient temperatures mainly by Norrish Type II chain scission

*Table 7. Film properties of E-CO and LDPE**

	E-CO	LDPE	Test method
CO content, %	0.9	0	Non-standard
Tensile strength, MD/TD,† Pa	20/17	22/19	ASTM D-882
Ult, elongation, MD/TD, %	330/490	300/490	ASTM D-882
Secant modulus of elasticity, MD/TD, Pa	228/228	234/241	ASTM D-882
Tear resistance, g mm^{-1}	5100	5600	ASTM D-1004
Oxygen transmission, ml m^{-2} (h)	220	228	ASTM D-1434
Moisture vapor transmission, g m^{-2} (day)	1.1	0.9	ASTM E-96/C-355

*Typical values of 1.7 mm blown film.
†MD = machine direction; TD = transverse direction.

reactions when subjected to ultraviolet radiation in the 290 nm region. The reaction has been found to be generally independent of temperature [16].

$$—CH_2—CH_2—\overset{\overset{\large O}{\|}}{C}—CH_2—CH_2—CH_2— \xrightarrow{h\nu}$$

$$—CH_2—CH_2—\overset{\overset{\large O}{\|}}{C}—CH_3 \quad + \quad CH_2{=}CH— \tag{15}$$

A Norrish type I free radical reaction is also known to occur with E-CO polymers, at elevated temperatures. At room temperatures, about 20% of E-CO chain scission has been estimated to be via Norrish I reaction, while at 120°C, it becomes closer to 50%. The Norrish I reaction for E-CO is

$$—CH_2—CH_2—\overset{\overset{\large O}{\|}}{C}—CH_2—CH_2—CH_2— \xrightarrow{h\nu}$$

$$—CH_2—CH_2—\overset{\overset{\large O}{\|}}{C}\bullet \quad + \quad \bullet CH_2—CH_2—CH_2— \tag{16}$$

The radical species so formed rapidly react with oxygen to initiate an oxidative chain reaction which usually results in embrittlement and further loss of tensile strength.

E-CO products exposed to ultraviolet undergo losses in physical properties that reflect the effects of the on-going chain scissions. For instance, tensile strength and ultimate elongation rapidly diminish to the point of product embrittlement, typically within two to four months after being placed outdoors, nearly anywhere.

The effects of UV exposure upon the ultimate elongation shown in Table 8 are for a 0.9% CO copolymer and a control homopolymer LDPE using a Q-U-V Panel accelerated weathering tester. The E-CO is shown to have degraded (lost elongation) more in 16 hours than did the homopolymer in 380 hours. Concurrently with the indicated loss of elongation, the E-CO copolymer lost 75% of its weight average molecular weight (to 55,000) during the first 72 hours of UV exposure, while the homopolymer's loss was only about 3%.

The volatile products of irradiation of a 1% CO copolymer were reported in a study by Li and Guillet [24] to be mostly CO (65.5 mol %) when exposed to a full mercury arc in an inert atmosphere. The other major component was acetone (13.8 mol %). The rest were mostly low MW hydrocarbons.

In a litter simulation test in New Jersey, USA, the wintertime outdoor disintegration of E-CO sheeting and other common multi-can packaging materials were compared by Harlan [25]. Both plastic and paper materials were included. The E-CO (2.7% CO) lost nearly all elongation within a month,.embrittled within two months and broke up into small pieces and blew away in five months.

Table 8. Effects of UV exposure on tensile properties of a 0.9% CO copolymer and a LDPE homopolymer control

	Tensile strength, Pa		Ultimate elongation, %	
UV exposure (h)	E-CO	LDPE	E-CO	LDPE
0	17.9	17.4	570	580
16	10.7	18.5	30	620
24	8.7	17.4	15	620
48	2.8	16.7	<10	640
72	3.2	14.9	<10	580
384	1.9	13.8	<10	40

Homopolymer LDPE sheet and shrink film maintained their elongations relatively well over that period. A polypropylene thermoformed snap-on cover for a six-pack of cans did lose all elongation with five months, but remained intact and non-brittle to the touch. The paper products (light-weight fruit juice carton, beer carton, and beer corrugated box) generally lost some tensile strength and their printed surfaces had faded over the five-month exposure period, but maintained their physical integrity. This study demonstrated that E-CO copolymers do disintegrate more rapidly than other beverage-can packaging materials when discarded as litter, even in colder weather.

Recycling of degradable plastics is a topic of debate within the plastics industry. Studies have shown that no deleterious effects on tensile strength or ultimate elongation were observed from the addition of 20 wt % embrittled E-CO to a typical LLDPE blow molding resin. Subsequent UV exposure of that mixture showed little difference in degradation characteristics between it and the polyethylene controls [26]. Thus, the presence of E-CO copolymers in the recycle stream should not be a deterrent to recycling of plastics.

5 Biodegradation studies on photodegraded plastics

Public concern about the biodeterioration of synthetic materials is a relatively recent phenomenon. In the early 1950s, serious environmental problems were identified with the discovery of the persistence of chlorinated pesticides in the ecosystem, ably publicized by Rachel Carson in her book, *The Silent Spring*. About the same time, the widespread use of automatic dishwashers and laundry machines required the use of synthetic detergents in place of conventional soaps. These early detergents were not completely degraded in municipal sewage plants, and caused foaming and other related problems in the downstream water effluent. As a result, soap companies quickly developed so-called "biodegradable" detergents in which the branched alkyl chains used in detergent synthesis were replaced by linear alkanes, thus accelerating the rate of biodegradation of these compounds in the biologically active sewage treatment process. The success of this new chemistry gave the general public the impression that something that was biodegradable was inherently good for the natural environment.

This simplistic notion gained further momentum as a result of an advertising campaign initiated by the paper industry at about the same time, in a desperate attempt to retain their markets in the packaging industry against the expanding use of plastics. After only a very cursory examination of the properties of the two materials, the paper industry claimed that paper packages should be used because paper was biodegradable and plastic was not. No scientific evidence was presented to substantiate either the rate or extent of the biodegradabilty of paper or its supposed ecological advantages. The plastics industry agreed that plastics were not very biodegradable, but proposed that this was an advantage, citing the problems arising from the byproducts of biodegradation, i.e., methane and carbon dioxide, which are now known to be "greenhouse" gases. In fact, neither of these extreme positions is scientifically correct. While certain types of paper products (e.g., toilet paper, facial tissues, etc.) do indeed degrade rather rapidly in sewage disposal systems, newspaper and coated paper packages may take many decades to completely biodegrade in sanitary landfills. On the other hand, all synthetic plastics degrade extensively after outdoor exposure over a period of time and hence become more susceptible to attack by biological organisms. Some plastics can be made so that they biodegrade at predetermined rates in living systems [27]. Further public confusion on these issues resulted from the failure to distinguish between the problems of garbage disposal and litter in the natural environment.

The term biodegradability has no exact scientific meaning. Several recent international conferences of scientists working in the field have failed to come up with a mutually acceptable definition. Some experts claim that "complete biodegradation" of organic compounds containing carbon, hydrogen and oxygen occurs when all of the compound is converted to carbon dioxide and water. However, there is little information as to how long this process might take, even with the simplest of natural materials. Jansson [28] reported that complete biodegradation (which he called mineralization) of straw would take about ten years, and take place in three stages. This would be true only if the straw were decomposed by living organisms with relatively short lifetimes. If, however, the straw were eaten by an animal with a long lifetime, this period could be much longer because some of the carbon becomes a part of the animal's body, and does not convert to carbon dioxide until the animal itself dies and is consumed. Even then, not all of the carbon would be released as carbon dioxide, since some of this would be incorporated in bone, shells and other solid parts of living organisms. Most museums can demonstrate that human beings are not completely biodegradable, since human skeletons are often on display which have been buried for tens of thousands of years and show little structural degradation. The White Cliffs of Dover and European Dolomite Mountains are monuments to the non-biodegradability of the bodies of sea creatures who lived millions of years ago. They also demonstrate that materials which are not biodegradable will accumulate in the global environment. In the case of shell fish, this process of calcification (i.e., making a non-biodegradable body part) is essential in maintaining life on earth, because it removes carbon dioxide from the biosphere thus reducing global warming due to the greenhouse effect.

In view of these complexities, an exact measure of the rate and extent of biodegradation of polymeric materials like plastics is difficult to define. The following definitions are proposed [29].

- **Biodegradable**. Capable of being chemically transformed by the action of biological enzymes or microorganisms into products which themselves are capable of further biodegradation.
- **Rate of biodegradation**. The rate of attack of microorganisms or enzymes measured by CO_2 production relative to that of compounds of similar chemical structure produced in nature which are known not to accumulate in the environment.
- **Complete biodegradability**. If a material contains more than one chemical species, then it must be fractionated into its components and each fraction must show a rate of biodegradation comparable to that of a compound known to be non-accumulative in the environment.
- **Per cent biodegradable**. Percent by weight of the fractions of the product which show biodegradability.

5.1 EXPERIMENTAL STUDIES OF BIODEGRADATION

The biodegradation of synthetic plastics has been studied extensively in a number of laboratories. In this chapter only the recent studies on Ecolyte photodegradable polymers will be discussed. These plastics are designed for use in packaging or other products which tend to be highly litter-prone. Since it is well established that these plastics will break up into small particles at rates controlled by the duration and intensity of their exposure to ultraviolet light from the sun, it is important to determine if the remaining particles will undergo long-term biodegradation so that there is no accumulation of the products in the environment.

Early studies of the biodegradability of photodegraded Ecolyte polystyrene and polyethylene used classical oxygen uptake procedures using high-activity media such as sewage sludge and also a variety of natural soil samples [30,31]. Typical results for Ecolyte polyethylene and polypropylene are shown in Figure 5. The apparent levelling off of the uptake is typical of a static test, since addition of more bacteria increases the rate again.

By using repetitive transfer to minimal media in which the only carbon source is degraded polymer, it was possible to isolate bacteria which are capable of attacking degraded polyethylene and polypropylene [32].

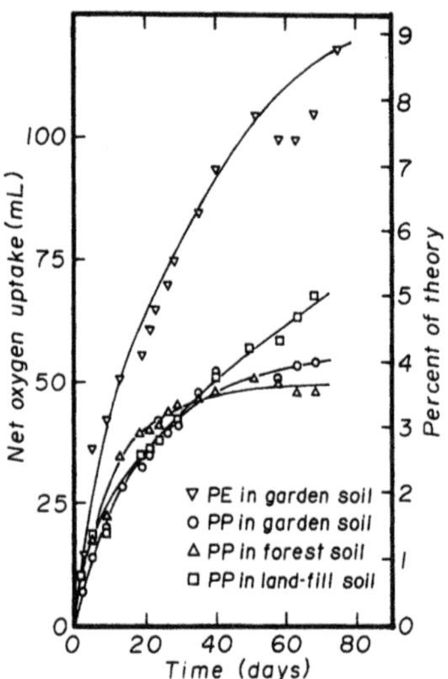

Figure 5. Biodegradability of photodegraded Ecolyte polystyrene and polyethylene in various soil environments

The organisms were separated into two groups by their Gram-staining characteristics. Each organism in each group was then subjected to a set of standard microbiological identification tests to determine the genera. In certain cases, specific tests were employed where the standard tests could not easily distinguish between two genera. When tentative identification was made, the genera were subjected to any further tests which would confirm their identification. The following genera were identified: *Pseudomonas*, *Alcaligenes*, *Achromobacter, Flavobacterium*, and *Gamella* (all Gram-negative), and *Arthrobacter*, *Aerococcus, Cellulomonas* and an *Asporogenous* bacillus (all Gram-positive). Most of these bacteria are common in soils and *Pseudomonas* and *Achromobacter* are known to attack both aliphatic and aromatic hydrocarbons.

A further conclusion of this work is that those bacteria which have been demonstrated to attack plastic residues are of relatively common varieties which would be expected to be widely distributed in soils in most terrestrial environments. No fungi capable of utilizing the polymers were isolated by this technique. Furthermore, none of the fungi isolated from the soil were capable of utilizing these polymers as the sole source of carbon. Further details of static tests developed for testing the biodegradability of plastics have been published elsewhere [27,29,30].

However, the most definitive tests for biodegradability of plastics involves the use of ^{14}C-labelled plastics in long-term simulated-soil-burial procedures. The ^{14}C procedure has been used in medical research for many years to decipher the intricate pathways of important biological reactions involving enzymes and microorganisms, and its practice is based on thoroughly rigorous scientific procedures. All of these tests confirm that photodegraded plastics are attacked by soil microorganisms much more rapidly than undegraded plastic and at rates approaching that of many natural biopolymers. A typical enclosed terrarium for soil studies in shown in Figure 6.

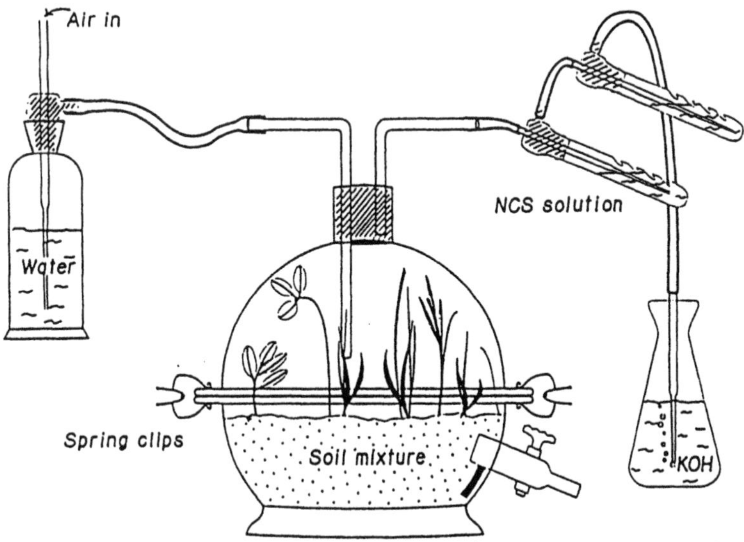

Figure 6. Terrarium for biodegradation of ^{14}C-labelled polymers

5.2 STUDIES OF ECOLYTE POLYSTYRENE

Studies of the biodegradation of ^{14}C-labelled Ecolyte polystyrene [33] have confirmed the presence of the radio label in the growing hyphae of bacterial species as well as in proteins, nucleic acid derivatives, and the CO_2 developed during the culture. In experiments with a closed terrarium, it was demonstrated that up to 5% of the carbon in a photodegraded ^{14}C-labelled Ecolyte polystyrene in the soil was converted into CO_2 or directly into the growing plants during a five-month period.

Data on the radiocarbon tracer detected in a mixture of plant species from the terrarium is shown in Table 9. These results confirm that polymer fragments produced by the photo-degradation of certain plastic molecules are indeed attacked and metabolized by soil micro-organisms, and are assimilated into the natural carbon cycle. Because radiocarbon assay of the CO_2 produced by plant and animal respiration does not reflect the total carbon assimilated in the bodies of plants and microorganisms, this procedure will tend to underestimate the actual rate of biodegradation of the plastic fragments in a natural environment. In these tests, almost half of the carbon metabolized was found in the bodies of the growing plants.

There is no indication of a slowing of the rate of biodegradation with time because of the living plants in the terrarium test. Extrapolation of the data would indicate that substantially all of the polystyrene would be biodegraded in about ten years. This is about the same time as has been estimated for the complete biodegradation of straw [28], well known as a biodegradable material.

Table 9. Total ^{14}C derived from ^{14}C-labelled Ecolyte polystyrene after five months in terrarium

	dpm			
	^{14}C from plant leaves	^{14}C-labelled CO_2 from traps	Total ^{14}C	% of initial polymer ^{14}C added to soil
Mean value	126,000	130,000	256,000	4.4

5.3 STUDIES ON ECOLYTE POLYETHYLENE

Guidelines provided by some government agencies have required that plastics used for packaging should be 90% biodegradable. An appropriate test of 90% biodegradability would be to fractionate the plastic by well established methods into at least ten fractions. These would be tested for biodegradability using the rapid procedure described elsewhere [29]. If 90 wt % of the fractionated materials is attacked by bioactive agents, this would demonstrate 90% biodegradability.

If a plastic composition contains more than one chemical component, it must be separated into its components and each component tested for biodegradability by the same procedures. In the case of synthetic polymers, which were often mixtures of components of similar chemical composition but differing in molecular weight, this may require separation on the basis of molecular weight.(see Section 5.1) In the case of plastics containing additives such as starch, which is known to be biodegradable, it is necessary to remove this component to test the degradability of the mixture, or alternatively to show that the biological degradation of the starch actually increases the rate at which the remaining polymeric components (usually polyethylene) degrade under bioactive conditions.

Experiments on photodegradable Ecolyte polyethylene were carried out and the results are shown in Table 10. The biodegradation index (BDRI) records the rate of biodegradation of the fraction relative to lauric acid, a compound known to biodegrade in the environment. The determination of BDRI is carried out by a rapid procedure described by Guillet [29] using sewage sludge as the biologically active medium. In this case only the highest molecular weight fraction showed no biodegradability by this test, and it represented less than 10% of the total polymer.

5.4 RADIOTRACER STUDIES ON THE BIODEGRADATION OF POLYETHYLENE IN SEWAGE SLUDGE AND ACTIVE SOIL

It has frequently been claimed that biodegradable film can be made by blending starch into polyethylene. On the other hand, films produced by the Ecolyte process are claimed to be biodegradable only after they have been photodegraded. In the author's opinion, the only unequivocal evidence of the biodegradation of the plastic components of such films is the production of carbon dioxide from biological attack on the carbon atoms in the polymer molecules. The rate of such processes can be determined accurately and unambiguously by the preparation of polymer films in which some of the ^{12}C atoms have been replaced by ^{14}C. The weak radiation given off by ^{14}C can be readily detected by scintillation counting.

Table 10. Biodegradation index (BDRI) for fractions of photodegraded Ecolyte polyethylene in sewage sludge

Sample	BDRI
Control	0
Lauric acid	100
PE fraction 2	18
PE fraction 3	21
PE fraction 4	16
PE fraction 5	14
PE fraction 6	21
PE fraction 7	20
PE fraction 8	20
PE fraction 9	18
PE fraction 10	2

Three film samples were prepared, each of which contained 25% ^{14}C labelled polyethylene: **Blend A**, Ecolyte low-density polyethylene (photodegraded); **Blend B**, low density polyethylene (not photodegraded); and **Blend C**, low density polyethylene with 16% starch.

Blends A and B were tested in powder form. Blend A only was photodegraded by exposing it to a commercial GE sunlamp (the equivalent of about two months of summer sunlight). Blend C was tested in film form since it is claimed that films containing starch are biodegradable without exposure to sunlight.

The results of sewage sludge tests (Figure 7) and the soil terrarium (Figure 8) were remarkably similar in form, but the sludge degradation is much more rapid. The starch composition showed very little degradation to give ^{14}C-labelled CO_2. Undegraded labelled high-density polyethylene was very slow, but measurably faster than the starch composition. The photodegraded Ecolyte composition biodegraded at a rate more than an order of magnitude faster than the starch composition. The decrease in rate after three days is due to the fact that the sewage sludge rapidly loses its activity after about three days, so this test cannot be used directly to determine the total amount of degradable material present.

Long-term tests in natural soil were run in an enclosed terrarium containing living plants, fungi and bacteria in a standard soil mixture. The radio-labelled CO_2 produced by all of the living organisms in the terrarium is trapped and analysed by a scintillation counting procedure. The results (Figure 8) are consistent with the accelerated tests and show no detectable biodegradation for the starch sample, but relatively rapid and consistent biodegradation of photodegraded Ecolyte. The combination of evidence from these tests is that photodegraded polyethylene is indeed biologically degradable over a time scale of about 20 to 30 years, depending on the activity of the soil microorganisms.

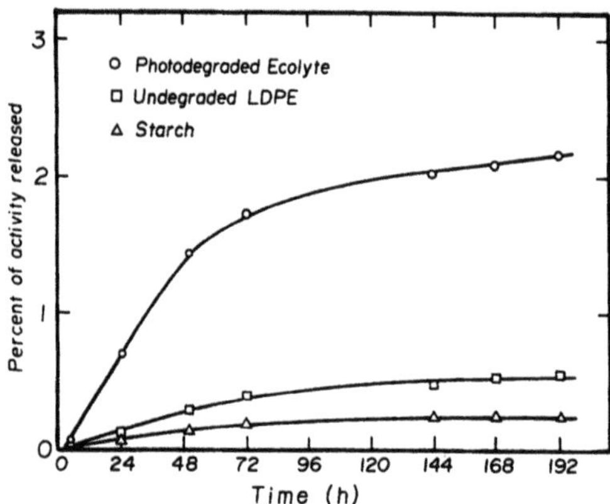

Figure 7. Biodegradation of various polyethylenes as measured by release of $^{14}CO_2$ in sewage sludge test

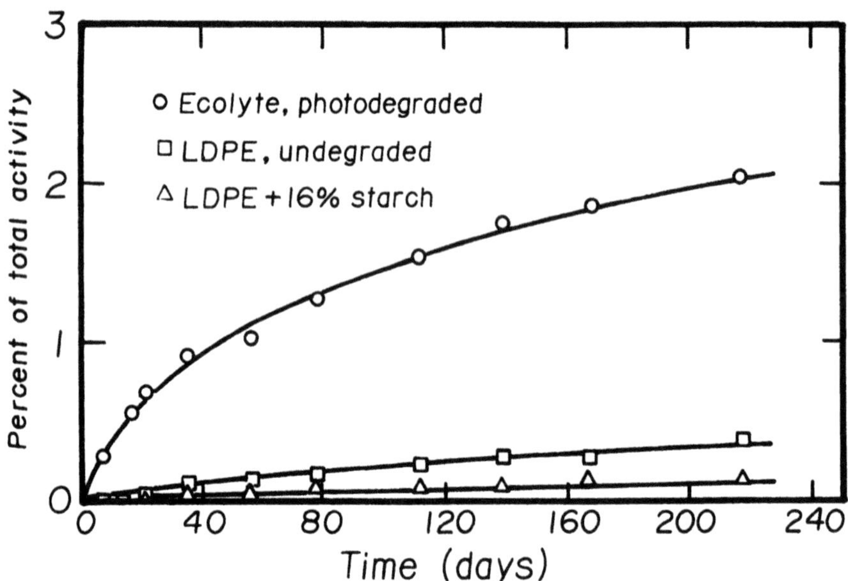

Figure 8. Biodegradation of various polyethylenes as measured by loss of $^{14}CO_2$ in a terrarium.

5.5 ECOLYTE POLY(ETHYLENE TEREPHTHALATE)

Similar testing procedures on ^{14}C-labelled Ecolyte poly(ethylene terephthalate) (PET) [34] showed that photodegraded PET initially degraded much more rapidly in the soil terrarium than the undegraded polymer (Figure 9) , but after about two months began to degrade rapidly, eventually catching up with the photodegraded material (Figure 10). Preliminary results on plant and soil assays indicate that the plants contain nearly as much ^{14}C as the CO_2 released. A similar result was observed in earlier studies with ^{14}C-labelled Ecolyte polystyrene [33] . Taken together with the CO_2 results, this suggests that PET is more than 30% biodegraded after burial for two years in an active soil environment, a result comparable to those reported for straw by Jansson [28] .

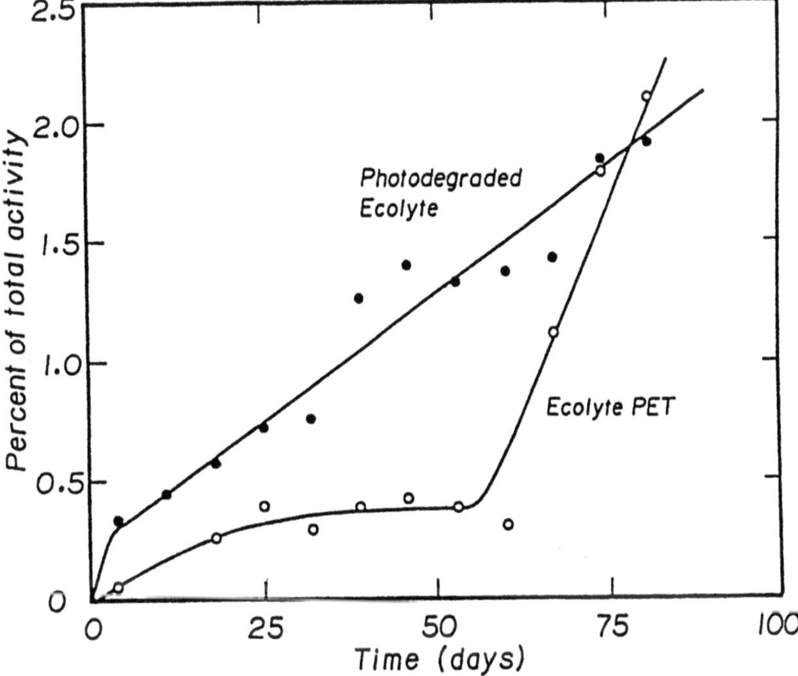

Figure 9. Terrarium biodegradation of photodegraded and non-photodegraded Ecolyte PET.

Due to the fact that PET contains hydrolysable linkages in the backbone of the polymer chain, these results suggest that PET may be an inherently biodegradable polymer when exposed to appropriate conditions. This should lead to a much expanded use of the polymer, particularly for beverage containers and packaging films.

6. Future directions

Potential uses for photodegradable polymers, such as ethylene-carbon monoxide and Ecolyte polystyrene and poly(ethylene terephthalate), will exist wherever plastics littering occurs. It has been estimated that almost a billion pounds of plastics find their way into the world's waterways annually [35] . Because E-CO and foamed plastics float and are photodegradable when in water, there should be commercial opportunities for them in marine packaging, fishing gear, and similar applications.

Many food contact applications are expected to be developed if the US Food and Drug Administration sanctions the use of E-CO films for food packaging. E-CO resins of up to 30% carbon monoxide do comply with regulation 21 CFR Part 175.105(c)(5) which covers the use of adhesives as components in articles for use in food contact applications

[36] . Food approval for foamed polystyrene beverage containers made from photodegradable polystyrene would provide a much more satisfactory product than the current plastic coated paper cups which have little heat insulation and cause serious litter problems.

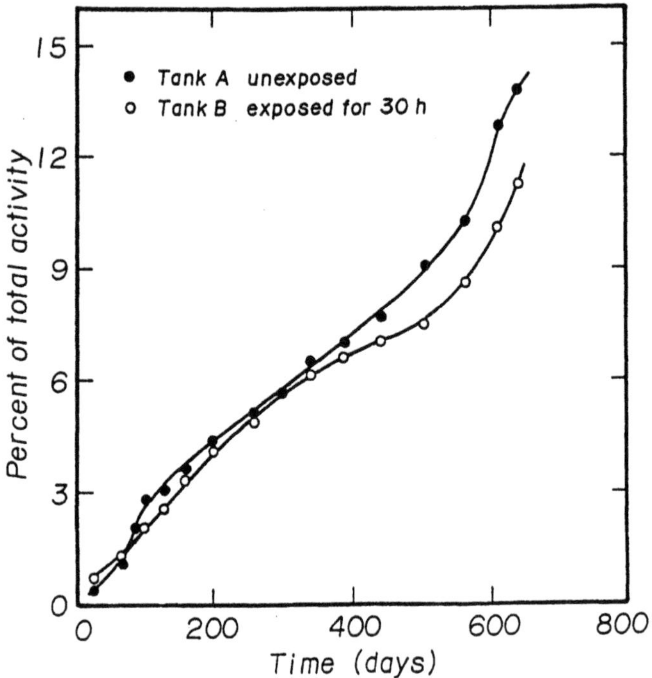

Figure 10. Terrarium biodegradation of photodegraded and non-photodegraded Ecolyte PET (at longer times).

Solutions to the solid waste disposal problems facing the world will include recycling, incineration and land fill where feasible. However, in those instances where collection is prohibitive such as roadside and beach littering, and on the oceans and other large bodies of water, photodegradable plastics will remain a viable solution and will find additional applications, repeating the success of E-CO as the material of choice for the Hi-Cone carrier for beverage cans.

Conclusion

During the latter half of the 20th Century, the production of synthetic plastics and fibres had grown so that the total volume of plastics produced worldwide now exceeds that of steel. This chapter has been concerned with theoretical and experimental studies relating to the environmental consequences of such a rapid shift from a technology based primarily on agriculture, forestry and metallurgy to one based on chemical raw materials such as oil, coal and natural gas. It is shown that plastics and synthetic fibres have the lowest energy costs of nearly all comparable materials and cause less environmental pollution in their production and fabrication. They are easily recycled when not contaminated with other materials and can be manufactured in photo- or biodegradable modifications tailored to highly litter-prone applications.

The technology and role of photodegradable plastics is considered and computer models are described to evaluate strategies for litter abatement, and experimental studies of the synthesis and biodegradation of conventional and photodegradable polyethylene, polypropylene, poly(ethylene terephthalate), and polystyrene.

References

1. Guillet, J. E. (1973) in *Polymers and Ecological Problems*, (ed. J. E. Guillet), Plenum Press, New York.
2. Guillet, J. E. (1974) *Plastics Engineering*, August, 47-56.
3. Scott, G. and Gilead, D. (1995) *Degradable Polymers*, Chapman & Hall, London.
4. Commoner, B. (1972) "The environmental cost of economic growth", *Chemistry in Britain*, 8(2), 52-66.
5. Tuskan, G. "Bioenergy Feedstock Program", Oak Ridge National Laboratory, P. O. Box 2008, Oak Ridge, TN 37831-6422.
6. Meadows, D. H., Meadows, D. L., Randers, J. and Behrens, W. W. III (1972) *Limits to Growth*, New American Library, NY.
7. Makhijani, A. B. and Lichtenberg, A. J. (1971) "An assessment of energy and materials utilization in the USA", Memorandum no. ERL-M310 (Revised), Electronics Research Laboratory, College of Engineering, University of California, Berkeley.
8. Boustead, I. and Hancock, G. F. (1981) *Energy and Packaging*, Ellis Horwood Publishers, Chichester.
9. West Germany Federal Office of the Environment (1988) Vergleich der Umweltauswirkungen von Polyethylene- und Papiertragetaschen, Umwelt Bundes Amt.
10. Guillet, J. E. (1990), "Photodegradable Plastics", in *Degradable Materials: Perspectives, Issues and Opportunities*, (eds. S. A. Barenberg, J. L. Brash, R. Narayan and A. E. Redpath), CRC Press, Boca Ratan, Florida, 55-97.
11. Heppenheimer, T. A. (1988) "Plastics makers clean up from litter", *High Technology Business*, Aug. 30.
12. Guillet, J. E. and Ainscough, A. N. "Studies of the accumulation of plastic litter by computer simulation". Internal report available from the author.
13. Heskins, M. and Guillet, J. E. (1976) "Photodegradation, Controlled", in *Encyclopedia of Polymer Science and Technology*, Suppl. vol. 1, Wiley-Interscience, New York.
14. Guillet, J. E. and Norrish, R. G. W. (1954) *Nature* **173**, 625-627.
15. Guillet, J. E. and Norrish, R. G. W. (1955) *Proc. Roy Soc. A*. **233**, 153-172.
16. Hartley, G. H. and Guillet, J. E. (1968) *Macromolecules*, **1**, 165-169.
17. U. S. Patents 3,753,952, 3,811,931, 3,853,814, and 3,860,538.
18. U. S. Patent 3,878,167.
19. Heskins, M., Reid, W. J., Pinchin, D. J. and Guillet, J. E. (1976) *ACS Symp. Ser.* No. **25**, 272.

20. Alexandru, L. and Guillet, J. E. (1975) J. *Polym. Sci., Polym. Chem. Ed.*, **13**, 483.
21. Guillet, J. E., Treurnicht, I. and Li, R. S., US Patent 4,833,857.
22. Brubaker, M. M. (1950), US Patent 2,495,286 (to DuPont).
23. Statz, R. J. and Dorris, M. C. (1987) "Photodegradable polyethylene", *Proceedings of Symposium on Degradable Plastics*, SPI, Washington, D.C., June 10, 51-55.
24. Li, S. K. L. and Guillet, J. E. (1980), *J. Polym. Sci., Polym. Chem. Ed.*, **18**, 2221-2238.
25. Harlan, G. M. and Nicholas, A. (1987) "Degradable ethylene-carbon monoxide copolymer", *Proceedings of Symposium on Degradable Plastics*, SPI, Washington, D. C. June 10, 14-17.
26. Kmiec, C. (1990) "Ethylene-carbon monoxide copolymer: The established degradable plastic", *RECYCLE '90, Forum and Exposition*, Davos, Switzerland, May 29-31.
27. Barenberg, S. A., Brash, J. L., Narayan, R. and Redpath, A. E. (eds) (1990) *Degradable Materials: Perspectives, Issues and Opportunities*, CRC Press, Boca Ratan, Florida.
28. Jansson, S. L. (1963) "Nitrogen transformation in soil organic matter", in *The Use of Isotopes in Soil Organic Matter Studies*, Report of the FAO/IAEA Technical Meeting, 9-14 Sept., Pergamon Press, Oxford.
29. Guillet, J. E., Huber, H. X. and Scott, J. (1992) in *Biodegradable Polymers and Plastics* (eds M. Vert *et al.*), The Royal Society of Chemistry, Cambridge, 55-70.
30. Jones, P. H., Prasad, D. and Heskins, M. (1974) *Environ. Sci. Technol.*, **8**, 929.
31. Guillet, J. E., Regulski, T. W. and McAneney, T. B. (1974) *Environ. Sci. Technol.*, **8**, 923.
32. Spencer, L. R., Heskins, M. and Guillet, J. E. (1976) *Proceedings of the Third International Biodegradation Symposium* (eds J. M. Sharpley and A. M. Kaplan), Applied Science Publishers, London.
33. Guillet, J. E. (1990) *Polym. Mat. Sci. Eng.*, **63**, 946.
34. Guillet, J. E., Huber, H. X. and Scott, J. A. (1995) *J. Macromol. Sci., Pure Appl. Chem.* A, **32**, 823.
35. Smock, D. (1987) "Are shipboard plastics all washed up?", *Plastics World*, Sept., 75-79
36. Ward, R. M. and Kelley, D. C. (1988), *TAPPI J.*, 140-144.

13

DEGRADABLE HYDROCARBON POLYMERS IN WASTE AND LITTER CONTROL

GERALD SCOTT[a] AND DAVID M.WILES[b]
a *Aston University* b *Plastichem Consulting*
Birmingham, UK *Victoria, Canada*

1. What are wastes?

Wastes are by-products of nature's productive activities including human activity. Most naturally occurring wastes are not normally perceived to cause environmental problems; for example, even when natural polymers become durable litter, as in the case of trees, branches etc., they are eventually incorporated into the biological carbon cycle [1]. By contrast the by-products of human activity are not seen in this way although they rarely remain in the outdoor environment as long as fallen trees [2]. Firstly, synthetic polymers look different from nature's wastes and many man-made products, particularly those manufactured from non-renewable resources are not considered to be bioassimilable into the natural cycle. The latter view, which is popular among environmentalists, is in fact a misunderstanding since there are very few man-made carbon-based polymers that are not ultimately bioassimilated and those that are not degraded are so stable that they cause no environmental hazard. The second problem with man-made wastes is that they are produced mainly in cities and towns where acceptable disposal becomes a logistical challenge.

Litter differs from waste only in that it is discarded on the surface of ground or water where it is both an aesthetic nuisance and, on occasions, a danger to animals and fish through accidental ingestion [3]. Conventional plastics used in agriculture, if left in the soil as litter, may have an adverse effect on plant growth, unlike traditional biological materials such as straw, which do slowly biodegrade [4].

2. The Management of Wastes

The management of municipal and industrial wastes is now one of the most pressing preoccupations of local and district authorities. Not many years ago it was sufficient to identify 'holes in the ground', often resulting from previous mineral extraction, and to use collected municipal wastes to remedy previous environmental damage. Since the

G. Scott (ed.), Degradable Polymers, 2nd Edition, 449-479.

1980s the situation has changed dramatically. It is now recognised by governments that natural resources for the production of materials are limited. Although it is not now believed, as was predicted by some environmentalists in the 1960s, that oil and natural gas resources will dry up in the early decades of the present century, it is nevertheless apparent that they are not infinite and at the present rate of depletion as fuels they will be too expensive to simply burn in the foreseeable future.

Half of the waste produced in the industrialised countries is generated by households [5] and has to be transported, sometimes for considerable distances, as available landfill sites become scarcer. Domestic waste has become an important resource from which to 'recover' potentially valuable materials by recycling to useful products. Initially the emphasis was on mechanical recycling, that is the reprocessing of the recovered material to the same or similar applications (see Chapter 1). Collection and recycling has long been practiced for traditional materials such as metals, glass, paper and textiles, particularly as wastes from industrial operations, where they are segregated at source. It was relatively simple to apply municipal recycling to these materials by the 'bring' collection system where the onus is on the householder to segregate the waste before bringing it to central waste collection systems.

However, the 'bring' system does not work well for plastics, since unlike traditional materials, they are very difficult to identify and segregate into generic types. Plastics packaging present in the domestic waste stream is composed of variety of different polymers that are visually similar but chemically and physically different. When processed together they are incompatible, leading to mechanically weak secondary products [6,7] (see Chapter 1). Furthermore, the major carbon-chain plastics, namely polyethylene (PE), polypropylene (PP), polystyrene (PS) and polyvinyl chloride (PVC) are chemically changed by the reprocessing operation with loss of mechanical properties and decrease in durability.

Many mistakes were made in early attempts to make quality products from recycled plastics until it was eventually recognised that to recycle polymers to the original products, the polymer itself must be formulated for recycling [5]. Furthermore, the product itself must be physically designed for easy recovery. This process, which has been given the name 'closed-loop' recycling is finding application in quality products particularly in the automotive industry where up to 90% of the plastics used may be reused, generally by blending with virgin materials.

By contrast, mixed (co-mingled) plastics wastes have a more limited second-life potential and apart from relatively minor uses as timber substitutes where their ecological value has yet to be demonstrated (Chapter 1, Section 2.1), very few waste-utilising applications have so far been found in spite of the development of modified processing equipment to handle mixed plastics [5]. It is labour-intensive to segregate most items of packaging from the domestic waste stream, although there are exceptions to this generalisation. For example, poly (ethylene terephthalate), PET, is an exception to this generalisation, since it can be readily identified in the domestic waste stream by non-specialists and it can be both mechanically recycled and readily hydrolysed to its component monomers.

PE, PP and PS, often as copolymers or blends with other carbon-chain polymers, are the major polymer components of packaging waste and unlike the traditional packaging materials, they have a calorific value when incinerated similar to that of the oil from which they were originally manufactured [5]. They therefore have a potential second life as fuels and because they replace an equivalent weight of fuel oil, they do not add to the 'greenhouse effect'. Energy recycling must then be considered in principle to be an ecologically acceptable practice. However, there is deep public distrust of incineration, particularly of PVC, when carried out near conurbations and indeed, there is some justification for this since a number of studies have shown increased ill-health risks down-wind from municipal incinerators [5]. An alternative to incineration is the recovery of fuels or in some cases (e.g. polystyrene) of vinyl monomers by pyrolysis [8]

3. Ecological aspects of the recycling of post-consumer wastes

For the reasons outlined above, materials recycling is not always a satisfactory solution to the beneficial utilisation of domestic wastes. It is often over-looked that the collection, cleansing and reprocessing of plastics wastes involves an energy input. About one third of the energy used in the manufacture of polyethylene from crude oil is used in the conversion process and this is essentially the same as that used in the reprocessing operation. When all the energy input components of the transport segregation and mechanical recycling of polymer wastes are added together, the ecological viability of materials recycling as a general solution to reclaiming the energy value of polymer wastes is very much called into question [5]. Since the energy produced by incineration of PE approaches the total used in the manufacture of PE from oil, incineration with energy recovery is in principle a more ecologically favourable process (Chapter 1}.

The energy balance for materials recycling becomes even less favourable when the use of plastics in agriculture is considered. Large amounts of PE and PP are now used in agricultural feedbags, fertiliser sacks, silage and hay wrap and binder twines. Very little of this discarded packaging is recycled for logistical reasons and because it is contaminated by 'farmyard' materials and most of it remains as litter in the fields and on riverbanks, caught on walls and shrubs. Although the percentage of plastics used in agricultural packaging is relatively small compared with that used in food distribution [5], the litter it produces is a serious environmental nuisance because it is very visible in recreational areas where it is most often found. Very little of it is collected for treatment in the recycling regimes discussed in the previous section. Organised collection of hay and silage wrap from farms has been found to be uneconomic due to the distances involved in retrieval of the waste.

It should be noted that there are other uses of plastics in agriculture and horticulture where the above problems do not arise. The first is in 'durable' greenhouse films where the polymers are formulated to give as long a service life as possible with available antioxidants and light stabilisers. Since these are well-stabilised materials and can be collected easily by the farmer, they can often be recycled to secondary products at the end of their first life. The second is in mulching films and tunnels used for one or at the most two seasons and which generally end up as litter on the soil. It will be seen in Section 7.1 that mulching films bring considerable economic benefit to the farmer and the problem of litter is now solved by the used of programmed-life plastics (mainly polyethylene) that disintegrate and biodegrade at the end of their useful life. Experience gained from the use of economically beneficial degradable polymers can now be applied to other aspects of litter in the countryside.

4. Biocycling

A substantial proportion of domestic waste collected by local authorities is biodegradable in the form of foodstuffs and garden wastes. At present much of this still ends up in landfill in the USA and UK but on the European continent and in Scandinavia, there is an increasing trend to 'recover' domestic waste materials by composting for the benefit of agriculture and horticulture [9].

Environmental surveys suggest that there is a very substantial demand from the farming and horticultural industries, market gardens and municipal parks and gardens for good quality compost to replace fertilisers and peat. Demand far exceeds current availability and biological cycling of carbon-based wastes appears to have the greatest potential for recovering value from domestic wastes for the foreseeable future.

Fig. 1 summarises the currently available options for the recycling of urban and rural wastes. They are listed in approximate order of decreasing ecological priority but it must be recognised that, because the effects of these operations will impact on the public, they may not appear in this order in practice.

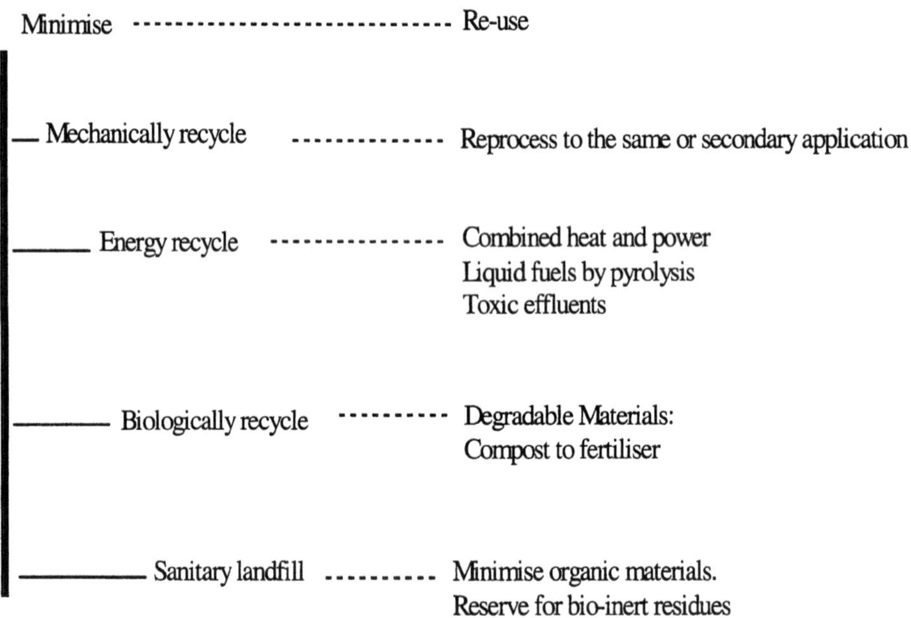

Fig. 1. Waste management options[1]

It seems inevitable, however, that recycling practice will involve a combination of all options. This is categorised as the Best Practical Environmental Option (BPEO) by waste management specialists and is defined as;

"The option that has the best combination of environmental characteristics whilst being practical in terms of economic viability"
The BPEO will of course vary from one country to another and even within communities depending on the availability of landfill sites, commercial composting facilities and on the structure of the waste recovery services ('bring', curbside or none). Ideologically, landfill is looked upon as a last resort to be used only for non-biodegradable materials and residues and some European counties have already banned landfilling of putrescible household wastes. Elsewhere, in countries with large land areas (USA and Canada are typical), landfill will probably continue to be used for some time to come.

BPEO means that in practice the same disposable product may end up in any one of the alternative options discussed above. Consequently the material used should ideally be accommodated in any of the procedures used. Thus for example, if a biodegradable product is to be mechanically recycled, it should be capable of being reprocessed at the same temperature as the rest of the polymeric waste. This has proved to be difficult in the case of many bio-based materials. Degradable polyethylene can be recycled normally at polyolefin processing temperatures [10] whereas most hydro-biodegradable polymers depolymerise or 'scorch' at these temperatures and cannot be recycled with commercial synthetic polymers in standard reprocessing equipment.

5. The biodegration environment

The potential for utilising degradable polymers in waste disposal is substantial and varied. Applications range from landfill covers through compostable garden waste bags to compostable disposable plates, cups and cutlery and personal hygiene products such as diapers. All degradable materials do not end up in the same biotic environment and it is not possible to devise a single biodegradability standard that will satisfy all biodegradable products (see Section 9). Fig. 2 shows typical time-scales for biodegradation in different environments [11].

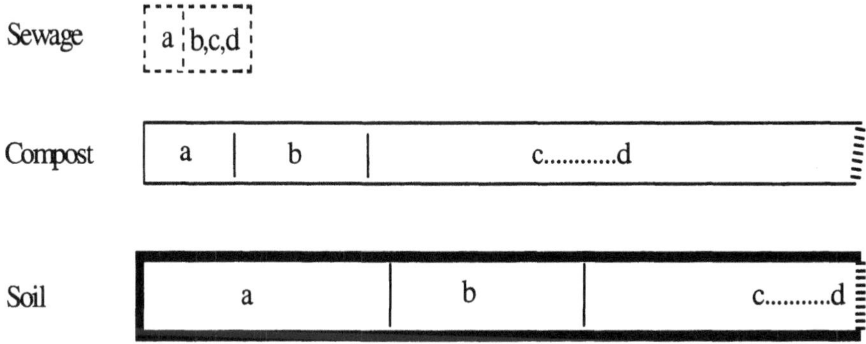

a No change in chemical or mechanical properties; a requirement of all polymers
b Chemical and physical degradation; loss of mechanical properties
c... d Formation of cell biomass and carbon dioxide, leading ultimately to complete mineralisation

Fig. 2. Typical time scales for biodegradation in different environments [11]

It is clear from fig. 2 that, in view of the very short residence time in wastewater and sewage systems, biodegradation to carbon dioxide and water should ideally occur in days rather than weeks. For this purpose cellulose can be looked upon as nature's model for bioassimilation in the aqueous waste environment. If the materials are biodigested to give methane (biogas) for use as fuel then the time scale must be similarly be short for economic reasons. Thus polymers derived from cellulose or starch would appear to be suitable for a number of aqueous disposal applications. Similarly, some hydro-biodegradable aliphatic polyesters such as polyvinyl alcohol (PVOH), polycaprolactone (PCL) or polylactic acid (PLA) can fill a role as a flushable waste where traditional materials would biodegrade too slowly. Hydro-biodegradable polymers and their applications in disposable packaging are discussed in other Chapters.

The compost environment is quite different. Degradable polyethylene, as a result of its hydrophobic nature, is not suitable for disposal in aqueous waste systems but is well adapted for recovery by composting. The normal input into compost (green waste) contains a considerable amount of ligneous materials in the form of twigs, straw, roots and organic fibrous materials. These are not considered to be deleterious in commercial compost because they facilitate aeration and conditioning of the soil. Furthermore, as seen in Chapter 3, these materials are a major source of humus by oxo-biodegradation which, rather than converting them immediately to carbon dioxide and water, conserves the carbon nutrients in the soil. Unnecessarily rapid conversion of synthetic carbon-chain polymers to carbon dioxide is also an additional burden on the environment.

Degradable polymers destined for composting should, therefore, be modelled on the behaviour of lignin as far as possible [12]. Oxo-biodegradable polyolefins, particularly the polyolefins and polystyrene, have been shown to be particularly suitable for this application. They cannot, because of their chemical structures, ultimately produce any products other than carbon dioxide and water. The practical applications of these materials are the main focus of this Chapter and composting standards for degradable polyolefins will be discussed in Section 9.

The third environment of interest to farmers and horticulturists is the land itself. The primary criterion here is that particulate materials, whether of biological origin or not, should be small enough not to interfere with root growth. Undue amounts of woody materials can reduce compost quality, as can large pieces of plastic film (> 50 x 50 mm) that may interfere with root penetration. Other criteria of quality are toxicity to macroorganisms in the soil (worms, daphnia, etc) and possible toxicity to plants and the animals that eat them from transition metal ions (see Section 9).

6. Degradable polymers in municipal waste management

A primary target for degradable plastics is in waste and litter control and most manufacturers of such materials have made claim to the environmental acceptability of their products as replacements for the commodity packaging polymers. A primary criterion of acceptability is cost and few hydro-biodegradable polymers can at the moment approach the hydrocarbon polymers in this respect. Consequently, in the following Sections the emphasis will be on synthetic commodity polymers with enhanced biodegradability in the natural environment. By way of clarification, hydrocarbon polymers include not only the polyolefins and polystyrene and their copolymers and blends, but also the carbon-chain rubbers (polydienes) that are used in some degradable polyolefins to enhance the rate of abiotic peroxidation. Some synthetic rubbers contain other elements besides carbon and hydrogen and caution is always essential when there is the possibility of environmental persistence of breakdown products. This would certainly apply to any polymers containing halogens because, although halocarbons are quite safe when they form part of a high molecular weight polymer, the toxicity of their lower molar mass degradation products is generally unknown and this requires a much more rigorous investigation than is normally required for hydrocarbons where the ultimate degradation products can only be carbon dioxide and water. Nitrogen-containing polymers, such as nitrile-butadiene rubber (NBR), are more tolerable in the small concentrations normally used as additives for the polyolefins but their eco-toxicity must be evaluated if they are to end up in the environment.

6.1 DEGRADABLE POLYETHYLENE PACKAGING IN THE LANDFILL ENVIRONMENT

In spite of widespread enthusiasm for the concept of materials recycling of post-consumer plastics, it has not become as popular as was predicted 10 years ago, for a variety of practical reasons. Notwithstanding the significance of the energy "stored" in, and recoverable from plastics, incineration with heat recovery is not a widely used or growing technology for municipal solid waste (MSW) streams containing used plastics. The reasons for this were discussed in Section 2.2. Thus, a significant quantity of packaging plastics is disposed of in municipal landfills after use, and this situation is expected to continue. Furthermore, it is predicted [13] that the market for food and beverage packaging will increase substantially, in part because of increased demand for "eco-friendly" packaging, new technologies, and continuing replacement of traditional packaging. A large fraction of synthetic plastics in MSW are virtually inert in a landfill environment, but it is worthwhile to consider why this can be a problem, and how "eco-friendliness" can be incorporated in the array of properties characteristic of good packaging plastics.

The commodity of value in a landfill is space and, when it is all used up, the expense and environmental trauma of locating and engineering a new landfill site are formidable obstacles. Let us consider how the use of degradable polymers in many kinds of plastics packaging can prolong the useful life of landfills. It is known from archeological excavations [14] that, for example, polyethylene bags, films and containers persist for 2 or more decades in landfills. If polyethylene and other commonly-used packaging plastics were modified so that they would degrade in a landfill environment the volume (space) saved by the "disappearance" of the plastics themselves would be minimal. It is commonly the case, however, that food waste and paper, for instance, are enclosed in polyethylene bags - the ubiquitous trash bags, grocery bags that are used to collect food waste in the kitchen, bags from litter bins, etc. - prior to being delivered to a landfill. This means that biodegradable wastes in MSW are prevented from biodegrading in landfills as a result of being trapped in impenetrable bags and films that are inherently bioinert. Indeed, food items and newspapers have been identified after incarceration in a landfill for many years.

Owing to differences in climate and regulations, it is not realistic to attempt to define a typical landfill, and yet there are characteristics that are common to a good many waste disposal facilities that are subject to some kind of monitoring and control. The so-called sanitary landfill normally has an impervious layer at the bottom, and is operated with a daily cover so as to reduce odour and visual impact, prevent scavenging by birds and rodents, and to prevent windblown litter. Microbial activity is common in landfills and, at greater depths and longer periods of use, activity of anaerobic bacteria to produce methane from bioassimilable carbon in the buried waste is usually observed. It is impractical and usually illegal to use a "capped" landfill site for any other purpose until microbial activity has ceased and the terrain is stable. This can take many, many years and is a problem in any situation where land is scarce and/or valuable.

It follows that there are three reasons why the use of degradable plastics in place of conventional packaging plastics in landfill disposal is encouraged by both financial and environmental considerations.

(i) Relatively rapid embrittlement and fragmentation of trash bags, shopping bags and other film-based products can speed up the biodegradation of food waste, paper and the like in which these readily biodegradable materials are enclosed. This will prolong the useful life of the landfill facility owing to the reduction in volume occupied by these wastes;

(ii) Although it is not known in general how long aerobic conditions persist in a given space as the landfill is filling up, the more aerobic biodegradation that occurs the better since the carbon dioxide produced under these conditions is much less serious a greenhouse gas than the methane produced by anaerobic microbes. Fragmentation of the polyethylene bags and films will prolong aerobic conditions by allowing the vertical flow of water and gases.

(iii) The more rapidly the biodeterioration of susceptible materials occurs in a landfill before it is capped, the sooner the land will stabilize subsequently and can be used for other purposes.

There are no suitable lab-scale tests to evaluate the aerobic biodegradation of plastics in a landfill environment and, those which are supposed to approximate anaerobic conditions, are focused on the wrong criterion. The practical requirement is that packaging plastics should undergo aerobic degradation, initiated by moderate heat, to reach embrittlement in a matter of a year or so. Bags and films that fragment under the movement characteristic of landfill operations (e.g., from spreading and compaction equipment deployed above) will allow the aerobic biodegradation of all types of organic waste. It is not important that microbial attack on the oxidised fragments of plastic may take several more years.

6.2 DEGRADABLE POLYETHYLENE IN LANDFILL COVERS

Regulations governing the operation of a sanitary landfill include the use of a daily cover applied to the active face. This is for visual and hygienic reasons. It is common to use several inches of soil as the cover, but this is likely to be expensive and it is certainly wasteful of space. A thin film of polyethylene would serve the purpose well, but the stability (including bioinertness) of this plastic would be a problem by restricting the vertical flow of water and air and therefore retarding the biodegradation of organic waste material for many years. A degradable polyethylene film would, however, serve as an excellent daily cover by becoming brittle and losing its physical integrity within the first year or so. Such a product has been developed by EPI, and it is sold in many countries as Enviro®Cover using EPI prodegradant TDPA™ additive formulations. The blended polyethylene films, whose oxidative degradation in landfills in Canada and China is illustrated in Table 1, were used daily as landfill cover films.

Evidence has been obtained [14] that microbially-generated warmth (30 to 54^{0}C [15]) in landfills is sufficient to cause significant oxidative degradation of polyethylene containing prodegradant within a few months. The data shown in Table 1 involved samples inserted in operating landfills for relatively short periods during the winter. The decrease in tensile properties with time in both landfills indicates degradation. The increase in carbonyl absorption (compared to the controls) as recorded with FTIR spectroscopy shows that it is oxidative degradation, and the increases at the 2 m level over the 1 m location indicate that the oxidation is initiated thermally. The prodegradant additive in the blended polyethylene films imparts photosensitivity as well although this is a factor only in the sample exposed on the landfill surface in China. The programmed, early onset of the fragmentation of degradable films occurred during a relatively short exposure in landfills in a cool time of year. It is reasonable to conclude that the increasing use of degradable polyolefins in bags and other film products could engender more aerobic biodegradation of food waste, paper and the like in landfills, and that this would have noticeable environmental benefits, for the reasons given above.

Table 1. Oxidation of degradable polyethylene films (LLDPE + LDPE) as a result of burial in a landfill

Sample	Location	TBS[a] (MPa)	EaB[b] (%)	Absorbance at 1715 cm^{-1}
Control	Canada[c]	24.5	550	0.18
1 m below surface		14.0	450	0.26
2 m below surface		8.0	130	0.42
Control	China[d]	24.0	480	0.24
On the surface		12.7	90	0.45
20 cm below surface		22.6	450	0.26
2 m below surface		10.1	40	0.59

a. tensile breaking strength
b. elongation at break
c. landfill at Chilliwack, British Columbia, Dec. 1995 to March 1996
d. Shenzehn Xiapin landfill, October to December 1998.

6.3 DEGRADABLE POLYETHYLENE GARDEN WASTE BAGS

In view of the importance of composting in the spectrum of methods for recovering value from significant portions of domestic and industrial waste, there is an obvious requirement for inexpensive, one-way containers for collecting and transporting compostable organic materials. Such containers need to be strong, light, flexible and stretchable, and have high wet strength, and these characteristics together with low cost favour oxo-biodegradable polyolefins. Since compostable plastic bags will, after shredding, become part of the input to composting operations, these plastics must not interfere with the biodegradation of the normal input (kitchen and garden wastes, for example) and they must themselves undergo biodegradation. Most importantly, the resulting compost must be of the highest quality, i.e., no defects in appearance or texture, and no toxicity effects during the growth of anything in the soil to which the compost is subsequently applied (see Section 9). Demonstrating that compost bags made from oxo-biodegradable polyethylene can meet all these requirements has been a challenge. As a result, the oxo-biodegradable polyethylene developed by EPI Environmental Plastics Inc. was evaluated in a number of commercial composting facilities [17], rather than in a laboratory environment. The results are summarized in Table 2. Although all of the results indicate that the two-stage degradation of these plastics occurs as expected, some of the data were not quantitative, and the quality of the compost was not always tested, largely owing to the formidable experimental difficulties of sampling and analysis. Accordingly, EPI's product was also evaluated [17] in the highly instrumented municipal composting plant of Vienna Neustadt in Austria, under the direction of Professor Bernhard Raninger. In 1998, this facility treated about 10,000 tons of mixed household and garden waste, using a standardized two-stage procedure: a forced aeration "tunnel" process, followed by 10 weeks in an outdoor windrow with weekly watering and turning, and then a further 3 months of outdoor storage, all according to Austrian National Standard ON S 2200. The compost product is used for fertilization, mainly in landscaping and gardening. In the EPI trial in 1999, the input to one of the 3 automated tunnels contained in the 60 m^3 (60 tons) of organic waste just over 1% by weight of polyethylene bags incorporating EPI's TDPA™ additive package. This involved 10,000 bags weighing about 660 kg. After 12 weeks of bioprocessing, the compost met all of the Austrian Standard requirements for Compost Quality A. After a further 26 weeks at the Vienna Neustadt facility, the end product compost met fully the standards of the Austrian Compost Quality Seal.

Table 2. The biodegradability of EPI TDPA™ plastics in composting environments

Date/ Country Quality	Plant operator	Process type	Biodegradability	Compost
Oct./95 USA	Nelson Composting Services	Open windrow facility	95-97% degraded after 4 months	Not investigated
Feb./96 USA	Creative Landscaping & Compost facility	Open windrow	10% biodegraded after 10 days, fully after 30 days	Not investigated
May/97 Canada	Pictou County District Commission	28 days in closed vessel	non-detectable level achieved	Not investigated
July/97 Canada	Fraser Richmond Biocycle	Open windrow facility	Ultimate biodegradation	Passed toxicity tests; earthworm & lettuce seed
Aug./99 Germany	Bio-Energie-Consult. BEC	Small container with forced aeration. 20 days intensive, 30 days post maturation	3 of 4 tests with total degradation, 1 with remaining matter	Accord. to AGA M10: VS 31.6%, C/N 15, maturing stage II
Dec./99 Germany	GOA Gesellschaft Ostalbkreises für maturation Abfallwirt-with forced schaft aeration	14 days container, 10-12 stage, weeks post	Visible after 1st fully degraded after 12 weeks approp. for fertilization	Achieved RAL compost Quality Seal,

As mentioned above, compost that includes degradable plastic bags must be top quality so it will have commercial value as a fertilizer/soil conditioner for agricultural and gardening purposes. The product must therefore contain substantial amounts of biomass, be free of visible contamination, and be free of toxic components. The compost from the Vienna Neustadt trial showed minimum or no trace of heavy metals, and passed the plant tolerance and seeds and propagules tests, according to Austrian National Standard On S 2023 (as required by ON S 2200). Furthermore, in additional testing at an independent laboratory in Belgium, the compost had no negative effects in the cress test, the summer barley plant growth test, the daphnia test, and the earthworm test.

It is reasonable to conclude that the heat generated by microorganisms in a commercial composting operation triggers the oxidative degradation of degradable polyethylene to the extent that film embrittlement and fragmentation occur. Bioassimilation of oxidized molecular fragments will generate biomass, some production of carbon dioxide and, of course, more heat. The combination of these abiotic and biotic processes need have progressed only so far by the end of six months as to preclude any deleterious visual or textural effects of residual plastic on the value of the compost, and any "ecotoxicity" effects on its use. There is unequivocal scientific evidence that oxidized polyethylene is biodegradable [18,19] and that all the carbon will eventually be converted to carbon dioxide [20]. No requirement exists that this conversion should be complete in 6 months or less.

6.4 DEGRADABLE POLYETHYLENE IN DISPOSABLE CONSUMER PRODUCTS

Some items of packaging are potentially dangerous to animals and birds when discarded. Six-pack beer can collars have been especially indicted and photo-degradable polyethylene has been found to be particularly useful in causing rapid fragmentation This can be achieved either by chemical modification of the polymer back-bone (E-CO polymers) [21] or by photosensitised peroxidation by transition metal ions (Chapter 3). The chemistry of the former was discussed in Chapter 12.

There is another specific type of disposal consumer product for which degradability can be especially useful. Disposable diapers are sold in the tens of billions annually and most of them, in all probability, are disposed of in landfills. Great improvements in design and component materials have drastically reduced the bulk and increased the efficiency of these products since they were first introduced and, as a result, they still probably represent less than 2% by volume of collected solid waste in spite of increasingly widespread usage. Many of the components of most modern disposable diapers are biodegradable in a landfill, but the impervious polyolefin backsheet and topsheet in most products are not. Thus the biodegradation of the components - and the contents – of used disposable diapers is significantly impaired because they are wrapped or enclosed in non-biodegradable plastic films and fabrics. In a detailed joint development program, a disposable diaper company (Absormex S.A. de C.V, Mexico) and a degradable plastics additive company (EPI Environmental Products Inc., USA) has developed polyolefin components that enable the degradation of disposable diapers in a simulated active landfill test (ASTM D-5525). Under the same conditions the same diapers with conventional polyolefin components do not degrade. The widespread use of truly degradable disposable diapers should help to reduce a potential health hazard in landfills, and could significantly prolong the useful lifetime of landfills.

7 Degradable polymers in agriculture

Plastics have achieved a dominant position in agriculture during the past 20 years. The earliest applications were in greenhouse films. Since the relatively heavy gauge films have to replace glass, they must remain tough and strong for several years in sunlight. At the opposite end of the stability spectrum are the degradable polymers. The applications of degradable polymers fall into two distinct categories. The first, which was discussed in Section 6, makes use of degradable materials in the disposal or recovery of polymers in order to improve the environment at the end of the useful life of domestic and industrial products. The second utilises degradability to enhance the technical function of products. There are two main area where this as been pursued with considerable success.

(a) Biomedical applications such as sutures and structural nets that are required to disintegrate and to be harmlessly bioassimilated into the body in a controlled way. Here, cost is much less important than the required performance. The medical industry can afford to pay many times more for polymers used in surgery than can the packaging industry. The science and technology of biomedical degradable polymers is discussed in Chapter 11.

(b) For protective films in agriculture and horticulture where variable "programmed" environmental life is essential. Although cost is more important than in medicine, the economic benefits of using environmentally degradable polymers in agriculture and horticulture far outweigh the extra cost of the polymers. The following are the main areas of application or potential application [22-24].

7.1 PROTECTIVE FILMS

The use of degradable plastics in tunnels and mulching films for the growing of soft fruits and vegetables has become an important economic tool in commercial horticulture. Scott-Gilead (S-G) time-controlled degradable films originally developed by G.Scott of Aston University and D.Gilead of Plastopil has been utilised in soft fruit growing in Israel and Southern Europe since the early 1980s and the technology has been discussed by Gilead [23] and Fabbri [24] in the first edition of this book. Major advantages in early cropping, coupled with water and fertiliser conservation were discussed [24]. In addition the desalination of soils by the refluxing of water, taking the salts out of reach of the plant roots was reported [23].

Protective films make possible the growth of crops such as chilli, sweetcorn and sweet potatoes in many parts of the world that until recently could only be grown in warmer climates. Degradable mulch is also used in cereal growing, notably maize and sweet corn and in forestry and environmental improvement schemes (e.g. growing of shrubs on road embankments). Mulching films are used in a variety of different ways and when used to their maximum potential, they greatly increase the value of commercial crops as a result of earlier and heavier cropping (Table 3).

Mulching films and tunnels made from conventional plastics have to be removed from the fields before the next planting season since otherwise they interfere with root growth and reduce crop yields [11,23]. This is a labour intensive process since it is particularly difficult to remove all plastics from the soil due to the poor mechanical properties of the residual plastic. With degradable films disintegration is programmed to commence at the time of cropping. It must not occur too early or the micro-environment at the roots of the plants is destroyed and much reduced yields result The use of mulching films reduces water and fertiliser usage to less than half that on open ground [24]. The same is of course true for degradable plastics mulch, but to optimise the benefits of this technology, they must remain intact until just before harvesting. This is particularly important in the case of soft fruits such as melons, bell peppers and sensitive vegetables that are normally irrigated with aqueous nutrients. Since mulching films create a microenvironment at the roots of the plant and the plants take up only the water and nutrients that they need, excess water from heavy rainfall can be just as damaging to sensitive crops as too little water. If then the films degrade prematurely, much of the benefit of the protective mulch may be lost with consequent loss of income to the farmer. In automated film laying, which is normally accomplished by turning under the 'tuck' to eradicate wind-lift, the young plants are rooted through pre-punched holes in the film in a single operation. Consequently, the film then has to be tough enough to resist this mechanical process.

Table 3 Ratio of increased income to cost of mulching film [11].

Crop	Increased income: cost
Melons	13.0
Vegetables	5.0
Peanuts	3.9
Sugar cane	3.6
Cotton	3.0
Maize	2.5

With normal plastics, after cropping, the tough plastic residues clog the cropping machinery and manual removal is essential [11,23]. Modern mulching film technology uses low-micron films (8-10 μm) that are so fragile after environmental expose that it is virtually impossible to manually remove them from the soil [23]. They are, nevertheless, an impediment to plant growth if left on the soil and are also a potential hazard to animals when the land is subsequently put down to grass. Films made from regular polymers such as polyethylene and polypropylene tend to accumulate from year to year due to the durability of commercial products. Programmed-life polyolefins on the other hand degrade sharply at the end of their service life and do not accumulate in the soil. S-G additives have been used for 15 years on the same fields in Israel and the USA and apart from "the tuck", which is not subjected to direct sunlight exposure during the first season, the photo-oxidised polyethylene can be ploughed into the soil, becoming part of the soil structure. This is followed by slower bioassimilation, leaving no visible residues at the beginning of the next planting season. When the buried "tuck" is exposed by ploughing, it in turn photo-oxidises and the process of boassimilation begins in the following season (see Chapter 3).

Reduced water and fertiliser usage are particularly important in arid areas and degradable mulch has considerable potential in the recovery of desert land to agricultural use. The above requirements demand accurate time-control in the service-life of the plastic films and it is crucial that they remain tough just as long as is required but then rapidly fragment (lose elongation, E_b) just before cropping. This is achieved in the S-G system by the use of effective but photo-transient light stabilisers based on the transition metal complexes (see Chapter 3, Section 4).

S-G technology for degradable polyethylene in agricultural applications provides the film manufacturer with a range of time-controlled concentrates. These are added to regular commercial PE generally at a standard addition rate. However, the additives give very different use lifetimes before physical disintegration (Table 4) . In spite of the differences in user lifetime, once fragmentation of the films has started, bioassimilation commences and the bioassimilation time is not very different for all the grades.

Table 4 Standard S-G grades [25].

S-G additive grade	Time to embrittlement*
1. #221	6 weeks
2. #131	3 months
3. #19	4 months
4. #12	6 months
5. #112	12 months

* Average times for mid-Europe or mid-USA for spring planting

It has been found by experience that the S-G grades listed above cover requirements in every situation encountered where mulching films and tunnels are used. Although there is a shift in fragmentation (embrittlement) time between the hot sunny southern climates to the more northerly cooler climates, it can normally be predicted fairly accurately which additive grade will give the desired result in each climate, based on UV incidence and temperature. However, the preferred protocol with time-controlled materials is that an initial small-scale trial is carried out in the field so that the results of field tests can be compared with the behaviour of the same films in laboratory accelerated weathering tests. S-G time-controlled degradable plastics are widely used in many counties throughout the world under the brand name of the film manufacturer.

The concentrate additive approach to degradable agricultural films provides the most cost-effective technology for protective films in agriculture. In northern climes where silage film waste is a major problem, the longest S-G formulation provides a user lifetime of 18 months before disintegration of film integrity occurs (see Section 7.2). The commercial success of the additive concentrate technology is based on the fact that it does not require a separate and relatively small-scale manufacture for each grade of degradable polymer since the additive concentrate replaces the additive package that is added to conventional polyolefins by film manufacturers to provide the environmental durability.

An interesting comparison was recently made among three commercial degradable mulch films in 1999 at the SAC Crichton Royal Farm in Dumfries, Scotland with forage maize as the crop. The maturity of this crop (evaluated as dry matter and starch content) and thus its commercial value is restricted by the amount of heat it receives and this, of course, is dependant on weather conditions. Although a major concern is the degradability both above and below ground of the mulch film, it is necessary to keep the soil covered for approximately 6 – 8 weeks. The greatest benefit to the maize plant is likely to occur in the first 4 to 6 weeks after sowing, first by encouraging germination, and then by helping to overcome the "stagnant growth phase" which often occurs in May and early June.

A relatively late, high-yielding variety of seed ("Wallis") was sown using a Jeantil drill, which also laid the plastic, at a target rate of 85,000 plants/ha. Normal fertilisation, soil preparation and weed control techniques were used. Plastic film covers were applied in a randomised block design so that statistically significant results could be identified. Dry matter (DM) values were measured gravimetrically; starch and in vitro digestibility (DOMD) were determined by infrared spectroscopy; and metabolisable energy (ME) was calculated from DM and DOMD values. Meteorological data were collected at the farm, and subsurface temperatures were measured hourly. The degradable film products evaluated in this trial were: a PE film based on TDPA additive technology developed by EPI, IP plastic (LLDPE); and Degradyl (PVC).

Absolute values of costs, crop yields and the like are a function of the weather, the variety of seed, sowing and harvesting dates, etc. Comparison among the mulch film candidates is nevertheless interesting. As might be expected all three types of plastic cover increased DM and starch content values although TDPA™ showed significantly superior performance as measured by other criteria. For example, TDPA™ had the lowest costs per unit weight for DM, for ME and for starch, and the lowest total costs for the trial. At harvest, visible residues of plastic were least for TDPA™, which began to break down after approximately 6 weeks of cover, as intended. The underground residue from TDPA™ became embrittled and was partly broken up by discing. In contrast, that part of the Degradyl and IP plastic that had survived wind damage was still intact on the surface at harvest, but began to be broken up by harvest traffic and subsequent cultivation by discs. This resulted in numerous small pieces being blown about the fields, causing concern for environmental acceptability. In addition, of course, perceived environmental problems from the use of a chlorine-containing plastic would be a concern with the use of the Degradyl product.

The conventional way of using plastic mulching films is in a single cropping regime in which the film disintegrates at the time of cropping (see Fig. 3, S-G #221) Recently, time-controlled biodegradable S-G films have been evaluated in Taiwan in 'dual crop' cultivation of fast-growing vegetables. The objective is to produce two crops in quick succession on the same degradable film. In this regime the film is applied late in the year and the second crop is planted in the holes left after the harvest of the first crop [12,26]. The film is timed to degrade as the **second** crop is being harvested (see Fig. 3, S-G #131). This procedure offers considerable advantages in irrigated systems since quite apart from the lower costs, the irrigation tubes are not disturbed, ensuring a fast changeover. Trials in Taiwan have shown [26] that a late crop planted in September can harvested in December and this may be followed by a second very early cash crop in the next season by planting on the same film in December or January. For longer cropping times #19, #12# and #112 may be used.

Another procedure that is also gaining acceptance is to sow seed directly into the soil under a complete plastic cover. This has the advantage of avoiding the 'shock' of transplanting with consequent earlier maturity of the crop. 'Mid-bed trenching' as this process is called [23,24] involves sowing the seed in a trench and laying the plastic over the growing plants (Fig. 4). The plastic film must be timed to break under the pressure of wind and weather (i.e. $E_b < 10\%$) when the leaves of the plant contact the cover. If the film breaks too early, the greenhouse effect will be lost and if it breaks too late the plant will be misshapen.

October 27 - 28 days after laying

December 30 – 92 days after laying

Fig. 3 Single (#221) and double (#131) cropping of fast-growing vegetables with Scott-Gilead formulations over the winter period in Tainan, Taiwan [26].

Programmed-life films have also been evaluated as solar sterilisation films in tropical climates [23]. The principle involved is that photodegradable films are laid before the crops are planted. This results in the destruction of pathogenic bacteria, which tend to accumulate in intensively cultivated land. The high temperatures achieved under transparent films destroy pathogenic bacteria but leave untouched beneficial microorganisms. Normally, the films are allowed to fragment before the crop is grown on a new photodegradable film but if the initial film is designed to remain strong, as described above, the plants could be grown through holes punched in the original film.

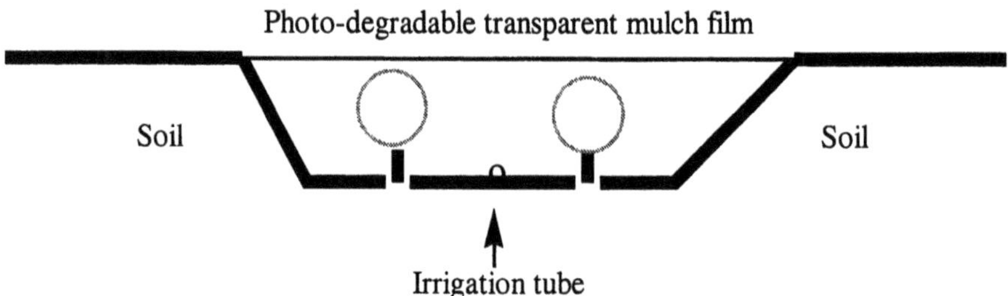

Fig. 4. Mid-bed trenching using photo-biodegradable polyethylene

7.2 AUXILIARY PRODUCTS

Biodegradable plastics are being increasingly used in auxiliary products for agriculture and horticulture that frequently end up in the environment as litter. These include irrigation tubing, plant pots, soil sterilisation films, polyolefin baler twines, fruit protection bags and crop-protection netting [23]. In some of these applications, for example plant pots which are intended to be used only once, there is a useful application for mechanically recyclable polymers, particularly in polypropylene.

Some auxiliary plastics are not at present degradable. Typically silage and hay wrap films and fertiliser and animal feed sacks persist in the environment for many years and the wind-blown plastics detritus is a particularly serious nuisance in areas of environmental sensitivity. There is little economic incentive to the plastics manufact-urer to make use of biodegradability when there is no cost-benefit other than a cleaner environment. Some auxiliary products such as packaging materials can be recovered in a relatively clean form for recycling but the subsequent performance of the secondary products produced from them [5] is generally inferior to that of virgin materials. Most plastics detritus recovered from farms is not worth recycling at all because it is seriously contaminated by biological matter and by transition metal ions picked up from the soil. Mechanical recycling of farmyard plastics is in most cases less ecologically viable than making the same products from virgin polymers [5,12,28] and biocycling offers a better environmental solution to this problem.

7.2.1 Twines and protective netting

Polyolefin baler twines, which have largely replaced sisal due to lower production costs, become heavily contaminated during service. These hold the bales together during storage and are split open when the fodder is required. Consequently they end up as litter and are trampled underfoot by cattle. In the USA, biodegradable polyolefin twines are now being manufactured by Ambraco [25] using S-G technology because of the environmental benefits they bring. In the case of polypropylene twines, a lifetime of one year has been found to be adequate, followed by rapid disintegration and subsequent bioassimilation. The same technology has been applied to protective netting and fastenings for fruit bushes and vines, which are difficult to remove manually after use [25].

7.2.2 Stretch-wrap films for silage and hay-storage

It is now common practice to store hay for use as silage in an airtight bag so that the nitrogen produced by fermentation is contained within the forage. This involves completely sealing the hay after harvesting and the seal is not normally broken until the contents are fed to animals the following winter. The films have to remain tough and strong during the fermentation period but after the silage is fed to animals, the residual plastic is carried by the wind, sometimes for many miles, and becomes an environmental nuisance after being deposited in trees and hedges along river banks. This waste material is again highly contaminated through contact with the soil and is expensive to collect for disposal.

The cost of a clean environment is not easy to calculate but in a recent survey in a UK National Park [29] it was estimated that about 500 tonnes of film are used in just one area of outstanding natural beauty annually (about 65,000 tonnes nationally) and this accumulates from year to year since almost none is routinely collected for disposal. The cost of landfill disposal is at present £30/tonne and is increasing year-by-year. Controlled biodegradability is an obvious solution to this problem and the technology is now available to allow even black pigmented polyolefins to be made photo-biodegradable with a time delay of one year (or more if required). So far, the agricultural film industry has shown little enthusiasm for this environmentally acceptable technology [27].

7.2.3 Controlled release systems for fertilisers

An important development in Japan is the use of biodegradable polyolefins in controlled release of fertilisers by encapsulation [30]. This results in controlled release in leaching environments over an extended period of time compared with direct application. This in turn effectively reduces the pollution of streams and the eutrophication of lakes and watercourses. Controlled release of pesticides by encapsulation has also considerable potential by matching the application time to the life cycle of the pest [11].

8 Control of oxo-biodegradation in the environment

It was seen in Chapter 3 that hydrocarbon polymers, of which natural rubber is a naturally occurring example, degrade both abiotically and biotically by a peroxidation chain mechanism. This process is accelerated by the introduction of photosensitive groups into the polymer chain (e.g. E-CO polymers, Chapter 13) or by transition metal ions that accelerate peroxidation both in the absence and presence of light. Antioxidants and light stabilisers retard these processes (Chapter 3). Phenolic (chain-breaking) antioxidants control melt degradation of polyolefins during processing and also control metal-catalysed peroxidation in the outdoor environment for short periods. However, polyolefin films that must remain intact out-of-doors for 3 months or more in sunny climates (Section 7) require a different solution. The Scott-Gilead process utilises peroxidolytic antioxidants that catalytically destroy hydroperoxides as they are formed in the polymer in a process not involving free radical formation (Chapter 3, Section 4). Combinations of peroxidolytic antioxidants with varying photo-stability permit the design of polyolefin products with a wide range of environmental stabilities (Section 8.1). Once the antioxidant ligand is destroyed either by heat or light, the prooxidant transition metal ions are released and catalyse the peroxidation of hydrocarbon polymers to low molar mass products even in the absence of light.

9 Standards for biodegradable polymers

During the 1980s some packaging manufacturers exploited the excess of corn starch on the American continent and promised the 'green' consumer 'environmentally friendly' biodegradable packaging materials by the incorporation of starch into polyethylene. At that time, the principles underlying environmental degradation were not fully understood by manufacturers of biodegradable packaging and it was thought that by simply adding a hydro-biodegradable filler such as starch to a commodity polymer, biodegradation would be induced in the hydrophobic matrix. This approach was subsequently shown to be a serious misunderstanding (at least at the concentrations then used) and the claims made for them were stigmatised as 'deceptive' by USA legislative authorities. This set back by many years the acceptance of the premise that polyolefin-based plastics with controlled biodegradability may play a part in the control of waste and litter. An investigation by the Association of Attorneys General of the USA led to the publication of the 'The Green Report' in 1990 [31] which drew together the criticisms of scientifically unjustified claims about the fate of some of the 'biodegradable' plastics used in packaging. The conclusions of this report and their implications for the manufacturers of degradable polymers were discussed in Chapter 1.

Although this report was a constructive contribution to the public debate about the value of degradable polymers to industry and society at large, it was not a scientific review of the technologies available at that time and was concerned almost entirely with starch-filled polyolefins and largely ignored parallel developments on time-controlled oxo-biodegradation of the polyolefins. This had the unfortunate consequence that many scientists subsequently concentrated their research efforts on modifying natural hydro-biodegradable polymers such as starch in order to regulate their biodegradability to the total exclusion of the ecologically viable polyolefins [12,32].

The main achievement of the 'Green Report' was the recognition of the need for internationally recognised standards. International Standards Organisation (ISO), the American Society for Testing and Materials (ASTM) and the Comité Européen de Normalisation (CEN) have attempted to address the problem of the characterisation of biodegradable materials. The starting point for these investigations were the biometric tests developed in the 1950s to combat the foam pollution of rivers and waterways.

Typical is the Sturm test, which depends on the measurement of carbon dioxide to determine the extent of degradation of detergents in the presence of typical river microorganisms. This test and the related oxygen absorption test are required to show that the detergents are substantially converted to carbon dioxide in the time taken for a volume of water to pass down a typical river; generally days rather than weeks. The above biometric tests were then applied by ISO and ASTM to biodegradable polymers in quite different environments and the following standards resulted;

ISO/DIS 14852 *Plastics – Evaluation of ultimate aerobic biodegradation of plastics materials in an aqueous medium – Method of analysis of released carbon dioxide.*

This was then adopted by ASTM in the following standards
D 5271 *Activated-sludge-wastewater-treatment system* and D 5320 *Municipal sewage sludge system*

The biodegradability of plastics in compost is characterised in the following ASTM Standard.
D5338-98 *Determining aerobic biodegradation of plastic materials under controlled composting conditions*
This standard describes the biodegradation, in a laboratory-scale composter, of cellulose at 58°±2° C. No extent or time-scale are quoted for pass or fail but cellulose, which gives 75.3% CO_2 formation at 50°C is quoted at a "positive reference" from which it must be assumed that the polymer should biodegrade at a similar rate to cellulose.

It is unfortunate that the first standards for biodegradable polymers were proposed without reference to degradable materials already available. Consequently, existing biodegradable packaging materials (including cardboard and the polyolefins) and all new materials had to satisfy standards intended for quite a different purpose. This skewed tests for biodegradation toward very short mineralization times and discriminated against carbon-chain polymers that biodegrade by a different mechanism.

9.1 STANDARDS FOR THE BIODEGRADABILITY OF PLASTICS PACKAGING IN COMPOST

It was recognised by the EU in the 1980s that composting of plastics is a valuable alternative to mechanical recycling [9] and the following variation on D5338-98 has been proposed by CEN TC 261 and accepted by the EU Commission.

EN 13432:2000 *Packaging – Requirements for packaging recoverable through composting and biodegradation – Test scheme and evaluation criteria for final acceptance of packaging* [33].

In this standard, the compostability of biodegradable polymers is assessed by the following criteria, all of which must be satisfied.

1. **Characterisation**: identification of packaging constituents, dry solid content, ignition residues, hazardous metal residues.
2. **Biodegradability**: 90% of the total theoretical CO_2 evolution in compost or

simulated compost in 6 months.

3. **Disintegration:** Not more than 10% shall fail to pass through a >2mm fraction sieve.
4. **Compost quality**: No negative effects on density, total dry solids, volatile solids, salt content, pH, total nitrogen, ammonium nitrogen, phosphorus, magnesium and potassium eco-toxicity effects on 2 crop plants.
5. **Recognisability**: "must be recognisable as compostable or biodegradable by the end user by appropriate means"

The methods of measuring **Disintegration** and **Biodegradability** are elaborated in the following additional draft standards currently in preparation;

PrEN 14045 *Packaging – Evaluation of the disintegration of packaging materials in practical oriented tests under defined composting conditions*

PrEN 14046 *Packaging – Evaluation of the ultimate aerobic biodegradability and disintegration of packaging materials under controlled composting conditions – Method by analysis of released carbon dioxide.*

The reference material for the biometric test in PrEN 14046, as in EN 13432 and ASTM D5338-98, is pure crystalline cellulose since it can be shown to be 68% converted to carbon dioxide at 65°C in 32 days but all other constituents of natural origin are excluded from the biodegradability test because they are considered to be biodegradable "by definition". CEN TC 249 explains the distinction between natural and synthetic polymers as follows [34];

"Natural products (leaves, wood, small stones are….generally known to be non-toxic. They are universally .recognised as biodegradable. On the other hand, residues of synthetic polymers would be ***perceived by the general public*** as being contamination of the compost…. The accumulation of lignin in the environment is a natural event which is beneficial for the fertility of the soil. On the other hand, the accumulation of other foreign materials cannot be encouraged because, while it is well known that lignin is ultimately degradable and helps environment and soil structure, this cannot be claimed for synthetic products whose behaviour in the environment is not known ….Therefore, the CEN scheme considers lignin and natural non-chemically modified materials as biodegradable ***by definition"*** (The italic emphasis is the authors')

However, public perception is a very doubtful basis upon which to judge the environmental acceptability of materials. Scientific evidence must be the keystone in the development of objective standards if safety is not to be compromised

All the standards based on biometric evolution of carbon dioxide have valid application to polymers that are intended to end up in waste waters or sewage where the time-scale for biodegradation has to be short (see Fig. 2) but EN 13432 and prEN 14046 have little relevance to packaging materials or agricultural films that are intended to have a service life of weeks or months before disintegration and biodegradation. Furthermore, it clear from the data presented in Chapter 3, that the CEN 'biodegradability test' does not comply with the EU 'Waste Framework Directive" 1991, [9] which requires that the "reclamation" of organic substances by composting should result in "benefit to agriculture or ecological improvement" (see Chapter 1, Section 5). EN 13432:2000 does not fulfil this requirement. Rather than producing biomass and humus, which are valuable as the seed-bed for future biological growth [1], the waste is transformed unnecessarily rapidly to carbon dioxide and thus plays no part in the remediation of soil. Instead it adds to the CO_2 burden in the environment.

Rapid mineralization is favoured by parts of the composting industry because it is a convenient way of disposing of packaging wastes rapidly and economically in the environment. However, nature does not dispose of the enormous quantities of lingocellulosic wastes in this way. As discussed in Chapter 3, natural biodegradation serves as a model for the disposal of synthetic as well as natural wastes. The principles of 'natural' bioassimilation have been outlined in the following statement prepared for CEN by a number of leading scientists working in the field of polymer degradation [32];

"Biodegradability tests that have been developed (by the Standards Committees) largely reflect the behaviour of hydro-biodegradable polymers (e.g. aliphatic polyester, modified starch). These materials are ideal for rapid biodegradation in sewage sludge where a maximum rate and extent of mineralization is required. The fundamental characteristic and most positive value of compost or mulches is the presence of biomass. Without biomass, there simply would be no product. Therefore rapid mineralzation is not ideal for polymers in compost where the carbon in the original plastic should be converted over a longer period of time to biomass and only slowly to carbon dioxide. The oxo-biodegradable polymers (e.g. the polyolefins) are ideal for this purpose since controlled peroxidation is the <u>rate-determining</u> step in the overall process. Furthermore they cannot give toxic or otherwise objectionable by-products during bio-assimilation".

In contrast to the rapid biometric tests, PrEN 14045 provides a more realistic basis for the establishment of a test method for degradable plastics packaging in compost. It requires fragmentation of the degradable plastic during the composting process. The emphasis in this test is on the quality of the compost obtained in pilot scale composting tests. It is by no means clearly established that composting carried out on a pilot scale provides the same added value to the soil as full scale composting and alternative full scale composting procedures are also currently in preparation by CEN. Fig. 5 [35] outlines the tests that should be carried out in full scale composting tests in order to fulfil the requirement to add value to the soil. The mandatory requirements are shown in Fig. 5 in heavy type and are discussed in the following Section.

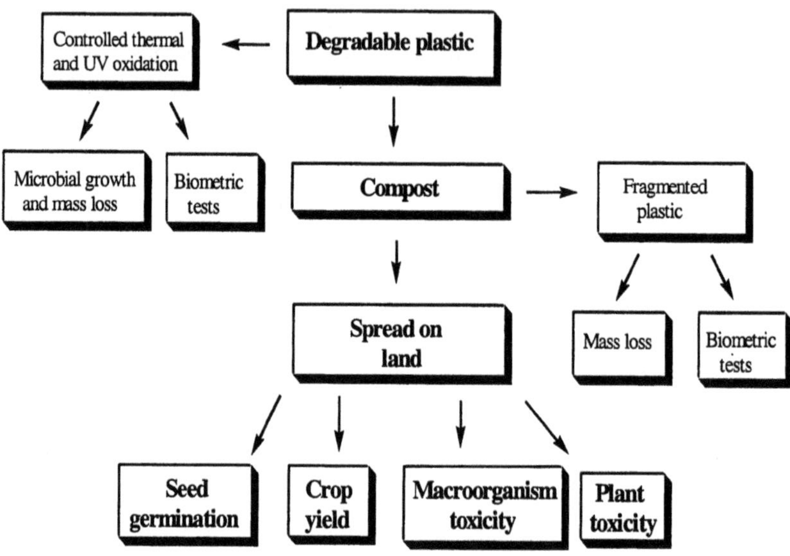

Fig. 5. Testing protocol for oxo-bidegradable materials in compost.

9.1.1 *Eco-toxicity requirements of degradable plastics in soil*

Seed germination is compared with that of soil that contains no plastic particles in Table 5 (from unpublished work by Raninger, Section 6.3) for a typical agricultural degradable plastic (EPI TDPA™). It is evident that the plastics detritus has no adverse effect on the germination rate of either seeds.

Table 5. Germination rates of typical plants in soil containing TDPA™ plastics

Compost formulation	Species	Germination %	Plant yield g
Control	Cress	32.3	1.42
TDPA™	Cress	33.3	1.68
Control	Barley	92	14.0
TDPA™	Barley	94	14.2

The growth rate of typical crop plants in compost to which partially degraded plastics have been added is shown in Table 6 [36] for the same plastics. Again there is no evidence that fragmented plastics have any effect on plant growth and within the limits anticipated from year to year, the loading of plastics detritus appears to make little difference to crop yields.

Table 6. Effect of degraded plastics mulch debris on yields [36]

Planting date	Yields (kg/15.6m^2) Without debris	With debris
January 1992	37.8	39.3
January 1994	35.2	38.2
February1995	32.4	34.5
December 1995	52.0	55.4
October 1996	32.5	38.7
October 1997	40.1	40

The effect of the plastics component in the compost on the population of beneficial macroorganisms (worms, daphnia, etc) has been evaluated. The results of such a test are given in Table 7A and 7B (from unpublished work by Raninger, Section 6.3). Again no detrimental effects were observed.

Table 7A. Effects of degraded plastics in compost on survival of daphnia

Medium water/compost Dilution factor[a]	Survival (%) Blank compost[b]	EPI compost[c]
26.2/1	100	100
16.4/1	100	100
10.2/1	100	97
6.4/1	60	83
4.0/1	12	40

a. survival rate is 100% in standard fresh water (no compost)
b. compost, from Vienna Neustadt facility, that included no degradable plastic
c. compost, from Vienna Neustadt facility, in which 1.1% of PE containing TDPA™ had been included

The accumulation of toxic transition metal ions from the plastics in the stems, leaves, fruit and tubers from the growing of soft fruits and vegetables. Table 8 [24] shows that even if the soil is loaded with much higher concentrations of Ni salts than can ever be obtained from degraded plastic films, the plants take up only the amount of metal ion they require. The concentration in the plant remains the same whatever the application rate.

Table 7B. Effects of degraded plastics in compost on survival of earthworms(14 days)

Medium compost in artificial soil mix[a]	Survival % Blank compost[b]	[Live weight] g/worm EPI compost[c]
65% compost	88 [0.36]	100 [0.43]
80% compost	10 [0.26]	68 [0.39]
100% compost	0 [-]	28 [0.27]

a. survival rate in 100% artificial soil is 100%, live weight is 0.56 g/worm
b compost, from Vienna Neustadt facility, that included no degradable plastic
c compost, from Vienna Neustadt facility, in which 1.1% of PE containing TDPA™ had been included

Table 8. Effects of Ni treatment of soils on concentrations in melons [24].

	Control	60 years*	120 years*	180 years*
Leaves	17.3	15.2	13.5	13.7
Stems	5.0	4.5	5.2	5.0
Flesh	2.7	2.0	3.0	3.2
Skin	3.0	3.5	3.2	3.0

The soil was sprayed with $NiSO_4$ to give nickel concentrations in the topsoil equivalent to the accumulation from S-G mulching films used for the number of years indicated.

The eco-toxicity results reported above are practical tests that can be quickly and easily carried out for any new degradable plastic that is likely to persist in compost or remain in the soil. Results demonstrated above refer to two different kinds of degradable plastic but it does not follow that all degradable plastic will be so benign. Each new formulation must be assessed in order to assure compliance.

9.1.2 Scientific verification of biodegradation

The evaluations indicated in the side-branches of Fig. 5 have a different purpose. They are intended to show from a scientific standpoint that the oxidation products formed from degradable polyolefins in the natural environment are bioassimilated by soil microorganisms. It would be ideal to take samples of plastics from compost at intervals to carry out biometric (carbon dioxide formation) tests and to measure mass-loss due to bioerosion of the plastic. However, this is a difficult procedure since CO_2 formation is concomitant with mass loss during composting and it is difficult to achieve even an approximate mass balance.

It is easier to achieve a carbon mass balance by temporally separating the peroxidation process from the biodegradation process. As discussed in Chapter 3, several workers have successfully applied this technique to degradable rubbers and polyolefins. CO_2 formation begins abiotically during thermal (and photo-) oxidation and continues during the bioassimilation of the polymer. In the case of rubbers it has been found possible to correlate mass-loss with the mass of the protein produced by the polymer in soil.

There is not a 1:1 correlation of polymer mass lost and microbial weight gained, since the microorganisms utilise other elements in the soil besides carbon. These assessments are confirmatory rather than diagnostic but since the rate of abiotic peroxidation can be measured at different temperatures in the laboratory, it is possible in principle to predict the lifetime of any polymer in the environment if the temperature history is known (see Chapter 3). It has also been shown that abiotic and biotic peroxidation of carbon-chain polymers occur together during the biodegradation process and the effect of this synergism is being studied.

9.2 STANDARDS FOR THE BIODEGRADABILITY OF PLASTICS LITTER

There are at present no standard tests for the biodegradation of litter in the environment. However, scientific studies described in the last Section show that substantial mass-loss occurs for polyolefins from which antioxidants have been removed (see Chapter 3). Moreover, it is not clear that such tests, if they were to be produced would serve any useful purpose or be helpful to environmental protection since the policing of packaging litter is notoriously difficult. Much of the litter ending up on the shores of island countries like the United Kingdom are sea-borne [37] after being discarded by international shipping and the MARPOL protocol, which was introduced in the 1980s to eliminate this practice has had almost no effect in the reduction of packaging litter [3] since heavily stabilised polymers continue to be used.

Agricultural plastics litter is rather different because it is used with a positive purpose and, as discussed in Section 7, it is designed to disintegrate and biodegrade as part of its essential function. The fate of the plastics residues is therefore followed very closely by the farmer and there is already a good deal of field evidence showing that degradable polyolefin mulching films disintegrate to particles within one season (see Section 7.1) and become an inseparable part of the structure of the soil within two seasons. No accumulation of plastic has ever been observed although mulching films have been used in the same fields in Florida and Israel for up to 15 years. Consequently it would be virtually impossible to devise a standard test to encompass all plastics in all soils. Instead, in considering agricultural wastes, the emphasis must be on the identification of potential hazards from the disintegration product, which includes, not only the effects of particulate materials on crop yields and macroorganisms, but also on the potential release of products other than carbon dioxide and water from the degraded polymers (Section 8).

10. The present position and the future

Research over the past ten years has taught us a great deal about how materials are recycled in the biosphere. The most important conclusion is that nature does not depend on just one degradation mechanism. In 1992 at the Second International Scientific Workshop on Biodegradable Plastics and Polymers, after a very intensive discussion, the following definition of a biodegradable polymer was agreed [38].

"**A degradable polymer** is one in which the degradation is mediated at least partially by a biological system"

In the opinion of the present authors, no better definition has been proposed. It embraces the ISO definition [39] which defines the external influences on the polymer as "chemical, physical and biological interactions" It is apparent from earlier chapters in this book that both hydro-biodegradable and oxo-biodegradable polymers frequently involve an abiotic component (i.e. hydrolysis or oxidation). Abiotic peroxidation is generally a necessary precursor to biological attack in order to induce colonisation of microorganisms on hydrophobic polymer surfaces. During bioassimilation both abiotic and biotic processes occur together synergistically.

A good deal is now known about how hydrocarbon polymers degrade, both abiotically and biotically and, as indicated above, by measuring the kinetics of these processes by known techniques, it is becoming increasingly possible to control the rates at which the polyolefins in particular are assimilated into the biosystem. There are many potentially important applications for programmed-life degradable plastics in the processes of waste disposal and in agricultural technologies. Degradable polyolefins are already widely used in waste sacks that are likely to end up in landfill or in compost. There is an equally varied range of potential uses for degradable polyolefins in agriculture, where the plastics remain on or in the soil. In general, the service life of agricultural plastics is longer than for domestic packaging, but all have at some stage to become part of the soil structure and must be capable ultimately of being transformed, after performing a nutritional function for growing plants or as soil conditioner, to carbon dioxide and water. Although environmental protection must be the primary purpose in the recovery of wastes, this must be based on overall eco-assessment of products. In this connection, an understanding of natural biocycling mechanisms provides a blueprint for the recovery of man-made polymers of value to the environment. Additional Standards now in preparation encompass slowly biodegrading oxo-biodegradable plastics and wood products.

Standards protect the public from unscrupulous overclaiming by manufacturers but they also provide a "level playing field" for industry (see Chapter 1). However, they are not intended for the promotion of the products of any one company. This latter proscription has not always been evident in the past when convenient but inappropriate biometric tests have been made mandatory without due consideration of the scientific principles underlying environmental biodegradation. Consequently standards developed for the 'recovery' of polymers as compost should not require short-term mineralization, since this test is an artificial barrier to the commercial development of products that biodegrade more slowly by an oxidative mechanism. Scientific evidence must be the only reliable basis for the development of standards for degradable polymers.

Acknowledgements

We are grateful to Professor Jacques Lemaire, Anne-Marie Delort and their colleagues of Clermont-Ferrand University and Professor Emo Chiellini of Pisa University for permission to use information from their laboratories, which is currently in the process of publication. We also thank Mr. Joseph Gho of EPI Environmental Products Inc. for permission to discuss previously unpublished information on TDPA™ products. We are also grateful to Dr. Shaw-rong Yang for unpublished field information on field trials of S-G degradable polyethylene in Taiwan and Professor Bernhard Raninger for permission to publish the work on composting of TDPA™.

References

1. Scott, G. and Gilead, D. (1995) *Degradable Polymers, Principles and Applications*, 1st Edition, eds.
   G.Scott and D.Gilead, Chapman and Hall (Kluwer), 250-253.
2. Scott, G, (1999) *Polymers and the Environment*, Royal Society of Chemistry, p.97.
3. *Proceedings of the Second International Conference on Marine* Debris, Shomura, R.S. and Godfrey,
   M.L., Eds. US Department of Commerce (1990).
4. Janssen, S.L (1963) in *The use of isotopes in soil organic matter studies, Report of theFAO/IAEA Technical Meeting,* 9-14 Sept, Pergamon Press.
5. Scott, G, (1999) *Polymers and the Environment*, Royal Society of Chemistry, Chapter 4.
6. Sadrmohaghegh, C., Scott, G. and Setudeh, E. (1985) *Polym. Plast. Technol. Eng.*, **24**, 149-185.
7. Scott, G. (1990}in *International Conference on Advances in the Stabilisation and Controlled Degradation of Polymers*, Ed. A.Patsis, Technomic Pub. Co., 215.
8. Kaminsky, W., Menzel, J. and Sinn, H. (1976) *Conservation Recycling*, **1**, 91-110
9. European Union, *Waste Framework Directive*75/442/EEC and Amendment 91/155/EEC.(see also ref. 5).
10. Al-Malaika, S, Chohan, M, Coker, M. Scott, G., Arnaud, R, Dabin, P., Fauve, A and Lemaire, J. (1995), *J. Macromol. Sci., Pure App. Chem.* A32 (4) 731.
11. Scott,G, (1999) *Polymers and the Environment*, Royal Society of Chemistry, Chapter 5.
12. Scott, G and Wiles, D.M. (2001) *Biomacromolecules*, **2**, 615-622.
13. Anon (2001*) Chemistry in Britain*, July, 16.
14. Rathje, W. L. (1991) *National Geographic*, 116.
15. Tung, J-F., Wiles, D. M., Cermak, B. E., Gho, J. G., and Hare, C. W. J. (1999) *Proceedings of the Fifth International Plastics Additive sand Modifiers*

Conference, Prague, Oct.27&28, paper 17.

16. Gho, Joseph, G., EPI Environmental Products Inc., personal communication.
17. Wiles, D. M., Tung, J-F., Cermak, B. E., Hare, C. W. J., Gho, J. G. (2000) *Proceedings of the Biodegradable Plastics 2000 Conference*, Frankfurt, June 6 & 7.
18. Arnaud, R., Dabin, P., Lemaire, J., Al-Malaika, S., Chohan, S., Coker, M., Scott, G. and Fauve, A. (1994) Polym. Deg. Stab., **46**, 211.
19. Scott, G. (1997) *Trends in Polymer Science*, **5**, 361.
20. Chiellini, E (2001), personal communication.
21. Harlan,G and Kmiec,C., (1995) *Degradable Polymers: Principles and Applications*, 1st edition, Eds. G.Scott and D.Gilead, Chapman & Hall (Kluwer), Chapter 8.
22. Scott,G. (2001) *Biodegradable Plastics in Agriculture*, ICS-UNIDO, Alexandria, In press.
23 Gilead,D. in *Degradable Polymers: Principles and Applications*, 1st edition, Eds. G.Scott , D.Gilead, Chapman & Hall (Kluwer), 1995, Chapters 10.
24. Fabbri,A. in *Degradable Polymers: Principles and Applications*, 1st edition, Eds. G.Scott and D.Gilead, Chapman and Hall (Kluwer), Chapter 11.
25. Harpaz, R., (2000) Plastor Hazorea, Israel, personal communication.
26. Yang, S.R. (2000), Tainan Agricultural Improvement Station, Taiwan, personal communication.
27. Scott,G, (1999) *Polymers and the Environment*, Royal Society of Chemistry, Chapter 2.
28. Scott, G. (1999) *Wastes Management,* May, 38-39.
29. Lovel, S. (1999) Yorkshire Dales National Park Authority, UK, personal communication.
30. F.Kawai, M.Shibata, S.Yokoyama, S.Maeda, K.Tada and S.Hayashi (1999), *Degradability, Renewability and Recycling, 5th International ScientificWorkshop on biodegradable Plastics and Polymers, Macromolecular Symposia*, Eds. A-C.Albertsson, E.Chiellini, J.Feijen, G.Scott and M.Vert, Wiley-VCH, 73-84
31. *'The Green Report'* (1990) Report of a task force set up by the Attorneys General of USA to investigate 'Green Marketing'
32. Scott,G (2001).in *Environmentally Degradable Plastics: Present Status and Perspectives*, Eds. S.Miertus, E.Chiellini and X.Ren, ICS-UNIDO, Trieste, in press.
33. CEN TC 261 (2000) *EN 13432 Packaging – Requirements for packaging recoverable through composting and biodegradation – Test scheme and evaluation criteria for the final acceptance of packaging*, Commité Européen de Normalisation,
34. *Characterisation of Degradability,* (March 28 2000) **N18**, CEN TC 249 WG 9
35. BSi PKW/4 2002/701329 Draft Standard, *Packaging – Determination of the biodegradability andeco-toxicity of packaging materials based on oxo-biodegradable plastics.*
36. Yang, S.R. (1999) *Degradability*, Renewability and Recycling, 5th International Scientific Workshop on biodegradable Plastics and Polymers, Macromolecular Symposia, eds. A-C.Albertsson, E.Chiellini, J.Feijen, G.Scott and M.Vert, Wiley-VCH, 101-112.

37. Scott, G (1972) *Int. J. Environmental Studies*, **3**, 35.
38. Ottenbrite, R.M., Albertsson, A-C and Scott, G. (1992) in *Biodegradable Polymer and Plastics*, Eds. M.Vert, J.Feijen, A.Albertsson, G.Scott and E.Chiellini, Royal Society of Chemistry, pp.73-92.
39. ISO TC I-94

SUBJECT INDEX

– C –

– D –

– N –

– O –

– X –

– Z –

Lightning Source UK Ltd.
Milton Keynes UK
UKHW021819241019
352249UK00002B/46/P